Probability and Mathematical Statistics (Continued)

ROHATGI · An Introduction to Probability Theory and Mathematical Statistics
SCHEFFE · The Analysis of Variance
SEBER · Linear Regression Analysis
WILKS · Mathematical Statistics
WILLIAMS · Diffusions, Markov Processes, and Martingales, Volume I: Foundations
ZACKS · The Theory of Statistical Inference

Applied Probability and Statistics

BAILEY · The Elements of Stochastic Processes with Applications to the Natural Sciences
BAILEY · Mathematics, Statistics and Systems for Health
BARNETT and LEWIS · Outliers in Statistical Data
BARTHOLOMEW · Stochastic Models for Social Processes, *Second Edition*
BARTHOLOMEW and FORBES · Statistical Techniques in Manpower Planning
BECK and ARNOLD · Parameter Estimation in Engineering and Science
BENNETT and FRANKLIN · Statistical Analysis in Chemistry and the Chemical Industry
BHAT · Elements of Applied Stochastic Processes
BLOOMFIELD · Fourier Analysis of Time Series: An Introduction
BOX · R. A. Fisher, The Life of a Scientist
BOX and DRAPER · Evolutionary Operation: A Statistical Method for Process Improvement
BOX, HUNTER, and HUNTER · Statistics for Experimenters: An Introduction to Design, Data Analysis and Model Building
BROWN and HOLLANDER · Statistics: A Biomedical Introduction
BROWNLEE · Statistical Theory and Methodology in Science and Engineering, *Second Edition*
BURY · Statistical Models in Applied Science
CHAMBERS · Computational Methods for Data Analysis
CHATTERJEE and PRICE · Regression Analysis by Example
CHERNOFF and MOSES · Elementary Decision Theory
CHOW · Analysis and Control of Dynamic Economic Systems
CLELLAND, deCANI, BROWN, BURSK, and MURRAY · Basic Statistics with Business Applications, *Second Edition*
COCHRAN · Sampling Techniques, *Third Edition*
COCHRAN and COX · Experimental Designs, *Second Edition*
COX · Planning of Experiments
COX and MILLER · The Theory of Stochastic Processes, *Second Edition*
DANIEL · Biostatistics: A Foundation for Analysis in the Health Sciences, *Second Edition*
DANIEL · Application of Statistics to Industrial Experimentation
DANIEL and WOOD · Fitting Equations to Data
DAVID · Order Statistics
DEMING · Sample Design in Business Research
DODGE and ROMIG · Sampling Inspection Tables. *Second Edition*
DRAPER and SMITH · Applied Regression Analysis

continued inside back cover

Introduction to
The Theory of
Nonparametric Statistics

Introduction to The Theory of Nonparametric Statistics

RONALD H. RANDLES
University of Iowa
DOUGLAS A. WOLFE
Ohio State University

JOHN WILEY & SONS, New York • Chichester • Brisbane • Toronto

Library of Congress Cataloging in Publication Data

Randles, Ronald H.
Introduction to the theory of nonparametric statistics.

(Wiley series in probability and mathematical statistics)
Includes index.
1. Nonparametric statistics. I. Wolfe, Douglas A., joint author. II. Title.

QA278.8.R36 519.5′3 79-411
ISBN 0-471-04245-5

Printed in the United States of America

10 9 8 7 6 5 4 3 2 1

To our wives

Carolyn Randles and *Marilyn Wolfe*

PREFACE

Although the first appearance of statistics as a science dates to the seventeenth century, it was not until much later that the papers of F. Wilcoxon (1945) and H.B. Mann and D.R. Whitney (1947) provided the initial theoretical foundation for the discipline of nonparametric statistics. In the years that followed, further development took place at a fast pace. However, when a field matures this rapidly, it is often difficult for textbooks to keep pace. Although early texts on the methodology were completed by Kendall (1948) and Siegel (1956), and on the theory by Fraser (1957), it was not until the late 1960s and early 1970s that the newly developed general theory again reached the settled confines of a book. During that period methodological texts appeared by Conover (1971), Hollander and Wolfe (1973), Lehmann (1975), and Gibbons (1976). In addition, nonparametric theory books at the intermediate level were written by Noether (1967), Hájek (1969), and Gibbons (1971), while more advanced texts were written by Hájek and Šidák (1967), and Puri and Sen (1971).

The level of this text is intermediate. The reader is expected to have completed an introductory mathematical statistics sequence at the level of Hogg and Craig (1978), Mood, Graybill, and Boes (1974), or Dudewicz (1976), for example. This, of course, entails that the reader be familiar with the basic concepts of statistical inference and have a good knowledge of advanced calculus. The goal of our approach is to bring the reader to a basic understanding of the concepts and theory that are important to the field of nonparametric statistics. An equally important goal to us, however, is to develop this theory without sacrificing the intuitive flavor that is prevalent in most of the early work in nonparametric statistics and, indeed, remains an important force in current nonparametric usage and research. Thus, for example, we devote the entire Chapter 2 to the introduction of the most important basic approaches that lead to nonparametric distribution-free tests of hypotheses. The emphasis there is on the nature of nonparametric procedures, and the sign, Wilcoxon signed rank, and Mann–Whitney–Wilcoxon rank sum tests are presented as illustrations of the important counting and ranking techniques. In addition, we choose to introduce the reader to the theory behind the nonparametric procedures by way of U-statistics. This again serves to maintain the intuitive nature, and provides a lead-in to the more comprehensive theory of general linear rank statistics included later.

We touch on the most prominent hypothesis testing settings, but detailed coverage is provided only for the one- and two-sample problems. However, the book is not devoted solely to tests. Confidence intervals and point estimation are discussed in detail in Chapters 6 and 7, respectively. We feel their inclusion is vital since they play an increasingly important role in modern nonparametric development.

The text is not intended to be a comprehensive presentation of the existing nonparametric methodology. Such a need is satisfied by the texts by Conover (1971), Hollander and Wolfe (1973), Lehmann (1975), and Gibbons (1976). The purpose of the text is instead to present the important mathematical statistics tools that are fundamental to the development of nonparametric statistics. Thus the text is organized around these tools rather than around methodological topics. We emphasize (1) techniques for making a test distribution-free, (2) U-statistics, (3) asymptotic efficiency, (4) the Hodges–Lehmann technique for creating a confidence interval and a point estimator from a test, and (5) linear rank statistics, to name just a few. We have also included some currently developing areas, such as M-estimation, adaptive testing procedures, rank-like techniques, and partially sequential sampling schemes. We have not attempted to be complete on any of these topics—actually, completeness is impossible in such active arenas—but we have tried to acquaint the reader with the fundamental ideas. Our goal is to present a more comprehensive description of the basic tools of nonparametrics than is available in current intermediate level texts, while remaining well below the mathematical level of Hájek and Šidák (1967), and Puri and Sen (1971).

The material in the book can be used for either a one-semester or a two-quarter course. The underlying core for either of these is found in Chapters 1–10, inclusively. Any remaining time could be devoted to topics from Chapters 11 and 12 most appropriate to an individual instructor. In addition, Sections 3.5, 3.6, 4.3, 6.3, 7.3, and 7.4 could be omitted without affecting the coverage of later material. An instructor who wished to emphasize testing problems could skip Chapters 6 and 7 and include them later as time permitted. This would not destroy the flow of the topics.

The Appendix at the end of this book includes (1) a list of the major distribution types mentioned in the text, (2) a discussion of our representation for integrals, and some theorems for manipulating them, (3) statements of important results in mathematical statistics that are used but not proved in text, and (4) statements of selected mathematical results applied in text.

The reference system within this book lists chapter, section, and quantity within the section, in that order. Thus a citation of equation (11.4.8) refers to the eighth numbered quantity in Section 4 of Chapter 11. Within a given section all quantities assigned reference numbers are numbered sequentially. Thus equations, theorems, lemmas, etc., are included in the same numbering scheme. The only exception is for the exercises, which are numbered separately in a sequential fashion within each section. References to the Appendix use the letter A in place of the chapter number.

We have benefited from the help of numerous people. Michael Fligner read almost all of the manuscript and made many valuable suggestions. Tim Robertson also made a number of important contributions. Kathy Altobelli gave portions of the manuscript a hard line-by-line reading, and her suggestions led to an improved manuscript. James Kepner, Ping Lu, and Tie Hua Ng also contributed suggestions and exercise solutions.

We were able to collaborate on this text during the academic year 1976–1977 while Ronald Randles was a Visiting Associate Professor at Ohio State University. We wish to thank The University of Iowa and The Ohio State University, and particularly Robert V. Hogg and D. Ransom Whitney, for making this arrangement possible.

Preliminary versions of this book were used in five separate offerings of the theoretical nonparametric statistics sequences at The Ohio State University and The University of Iowa. We are appreciative of the contributions of the many students who corrected errors and cleared up ambiguities.

This book was written and typed at The Ohio State University and The University of Iowa. We thank Kathleen Narcross (who did most of the typing), Amy Bernstein, Ada Burns, Rozanne Huff, Barbara Meeder, Carolyn Randles, and Gloria Rudolph for their excellent preparation of various versions of the manuscript.

It was our pleasure to work with Beatrice Shube, editor in Wiley-Interscience, on this project.

RONALD H. RANDLES
DOUGLAS A. WOLFE

Iowa City, Iowa
Columbus, Ohio
January 1979

CONTENTS

1 PRELIMINARIES

1.1. Introduction

The rudiments of nonparametric statistics originate in the latter part of the nineteenth century. Some early contributions are cited in a survey paper by H. Scheffé (1943). However, it was not until the pioneering papers of F. Wilcoxon (1945) and H. B. Mann and D. R. Whitney (1947) that the theoretical foundations of nonparametric statistics began to be assembled. In the ensuing years the advances in this field have been of major proportions, providing in many cases important techniques in mathematical statistics indigenous to nonparametric theory but finding usage in parametric settings as well. The concept of asymptotic relative efficiency proposed by Pitman (1948) and the development of the theory of U-statistics initiated by Hoeffding (1948) are prime examples of such contributions.

The purpose of this text is to introduce the important theoretical foundations of nonparametric statistics, both classical and current. Consequently, the text is organized around them rather than around methodological topics. We are not able to present a complete accounting of these subjects, since many are currently active research arenas, but we have tried to provide the reader with the fundamental concepts and tools.

We begin with a brief description of some of the notation and terms which are used throughout the text. All random variables will assume values on the real line (which is denoted by R). The distribution of a random variable, say X, is most often described in terms of its **cumulative distribution function** (abbreviated **c.d.f.**), say $F(\cdot)$, which is defined to be

$$F(x) = P[X \leqslant x], \qquad -\infty < x < \infty.$$

In this text we emphasize **continuous random variables**, each of which is characterized by a nonnegative **density function**, say $f(x)$, that satisfies

$\int_{-\infty}^{\infty} f(x)\,dx = 1$. The density is related to the c.d.f. in that $F'(x)$ exists and equals $f(x)$ at all but at most a countable number of x-values. We then have

$$P[a<X<b] = P[a \leqslant X \leqslant b] = \int_a^b f(x)\,dx, \text{ for any constants } a < b.$$

The **support** of a continuous random variable is defined to be the closure of the set

$$\{x \mid f(x) > 0\}.$$

On occasion we discuss a **discrete random variable**, say, X. Its distribution is described by the **probability function**

$$p(x) = P[X=x], \qquad -\infty < x < \infty,$$

which is related to its c.d.f. through

$$F(x) = P[X \leqslant x] = \sum_{t \leqslant x} p(t),$$

where the latter sum is over all $t \leqslant x$ such that $p(t) > 0$. The **support** of this discrete random variable is defined as

$$\{t \mid p(t) > 0\}.$$

If $X_1, \ldots, X_n$ denotes a random sample from some underlying population, we say that these random variables are **independent and identically distributed** (abbreviated **i.i.d.**). In such contexts, the **sample mean** refers to

$$\bar{X} = \frac{1}{n} \sum_{i=1}^{n} X_i$$

and the **sample variance** is

$$S^2 = \frac{1}{n-1} \sum_{i=1}^{n} (X_i - \bar{X})^2.$$

A vector of n random variables (not necessarily i.i.d.) is represented by $\mathbf{X} = (X_1, \ldots, X_n)$.

It is also useful to note the definition of an indicator function that occurs repeatedly in this text, namely,

$$\begin{aligned} \Psi(t) &= 0, \qquad \text{if } t \leqslant 0, \\ &= 1, \qquad \text{if } t > 0. \end{aligned}$$

The symbol $\Psi(\cdot)$ is reserved for this purpose. (The only exception occurs in Section 7.4.) One final notation that is widely used in this text is that for the **greatest integer function**. For any real number x we let $[\![x]\!]$ represent the greatest integer less than or equal to x.

1.2. Order Statistics

In many statistical analyses the information from a random sample is utilized through the ordered values of the sample. These ordered sample observations are referred to as **order statistics**, and they play a fundamental role in the development of nonparametric statistics, both for hypothesis tests and for estimators. In this section we develop some of the basic properties of order statistics that are used throughout the rest of the book. For a more detailed accounting, see Sarhan and Greenberg (1962) or David (1970).

The Joint Distribution

Let the continuous random variables $X_1,\ldots,X_n$ denote a random sample from a population with c.d.f. $F(x)$ and density $f(x)$. Let $X_{(i)}$, $i=1,\ldots,n$, be the ith smallest of these sample observations. We refer to $X_{(1)} \leqslant \cdots \leqslant X_{(n)}$ as the **order statistics** for the random sample $X_1,\ldots,X_n$. Unlike the Xs themselves, the order statistics are neither mutually independent nor identically distributed. We obtain their joint distribution by a change of variable.

Theorem 1.2.1. *Let $X_{(1)} \leqslant \cdots \leqslant X_{(n)}$ be the order statistics for a random sample of continuous random variables from a distribution with c.d.f. $F(x)$ and density $f(x)$. The joint density for the order statistics is then*

$$
\begin{aligned}
g(x_{(1)},\ldots,x_{(n)}) &= n!\prod_{i=1}^{n} f(x_{(i)}), \qquad -\infty < x_{(1)} < \cdots < x_{(n)} < \infty \\
&= 0, \qquad \text{elsewhere.}
\end{aligned} \tag{1.2.2}
$$

Proof: We provide the structure of the proof for general n, with a detailed illustration for the specific case $n=2$. Define the sets $A=\{(x_1,\ldots,x_n)| -\infty < x_i < \infty,\ x_i \neq x_j \text{ for } i\neq j\}$ and $B=\{(x_{(1)},\ldots,x_{(n)})| -\infty < x_{(1)} < \cdots < x_{(n)} < \infty\}$. The transformation defining the order statistics maps A onto B, but not in a one-to-one fashion, since each of the $n!$ permutations of the observed values yields the same value for the order statistics. Thus, for example, when $n=2$, $(x_1,x_2)=(1.4,6.9)$ and $(x_1,x_2)=(6.9,1.4)$ both yield $(x_{(1)},x_{(2)})=(1.4,6.9)$. If we partition A into $n!$ subsets, each one corresponding to a particular ordering of the sample observations, we see that the order statistics transformation now maps each of these partitioned sets onto B in a one-to-one fashion. For illustration, when $n=2$, A is partitioned into A_1 and A_2, where $A_1=\{(x_1,x_2)|-\infty<x_1<x_2<\infty\}$ and $A_2=\{(x_1,x_2)|-\infty<x_2<x_1<\infty\}$. On A_1 the order statistics set $x_1=x_{(1)}$ and $x_2=x_{(2)}$. The absolute value of the Jacobian of this one-to-one transformation is

$$|J_1| = \left\|\begin{matrix} 1 & 0 \\ 0 & 1 \end{matrix}\right\| = 1.$$

On A_2 the order statistics set $x_1=x_{(2)}$ and $x_2=x_{(1)}$, yielding the Jacobian with absolute value

$$|J_2| = \left\|\begin{matrix} 0 & 1 \\ 1 & 0 \end{matrix}\right\| = 1.$$

For general n, it follows similarly that the Jacobian of each of the one-to-one transformations from one of the $n!$ partitions of A onto B has an absolute value equal to 1. The joint density of the order statistics is then simply the sum of the contributions from each of the partitions. Specifically, when $n=2$ we have

$$\begin{aligned} g(x_{(1)},x_{(2)}) &= f(x_{(1)})f(x_{(2)})|J_1| + f(x_{(2)})f(x_{(1)})|J_2| \\ &= 2f(x_{(1)})f(x_{(2)}), \qquad -\infty < x_{(1)} < x_{(2)} < \infty. \end{aligned}$$

For general n, the joint density in (1.2.2) follows from the fact that the contribution of each of the $n!$ partitions is $\prod_{i=1}^{n} f(x_{(i)})$. ■

A special case of the joint density of the order statistics that is important to nonparametric statistics corresponds to an underlying uniform distribution on the interval (0, 1). For this case, the joint density (1.2.2) of the order statistics is

$$\begin{aligned} g(x_{(1)},\ldots,x_{(n)}) &= n!, \qquad 0 < x_{(1)} < \cdots < x_{(n)} < 1, \\ &= 0, \qquad \text{elsewhere}. \end{aligned} \tag{1.2.3}$$

Marginal Distributions

Although the marginal distribution of a single order statistic $X_{(j)}$, $1 \leqslant j \leqslant n$, can be obtained directly by the proper integration of (1.2.2) (see Exercise 1.2.1), we will develop these formulas by a more indirect, but instructive, approach based on binomial probabilities. (An expression for the joint density of two order statistics is given in Exercise 1.2.2.)

Theorem 1.2.4. *The marginal density for the jth order statistic* $X_{(j)}$, $1 \leqslant j \leqslant n$, *under the conditions of Theorem* 1.2.1 *is*

$$g_j(t) = \frac{n!}{(j-1)!(n-j)!}[F(t)]^{j-1}[1-F(t)]^{n-j}f(t), \qquad -\infty < t < \infty. \tag{1.2.5}$$

Proof: Let $G_j(t)$ be the distribution function for $X_{(j)}$. Then, for any t,

$$\begin{aligned} G_j(t) &= P(X_{(j)} \leqslant t) = P(\text{at least } j\ X\text{s are} \leqslant t) \\ &= \sum_{i=j}^{n} P(\text{exactly } i\ X\text{s are} \leqslant t) \\ &= \sum_{i=j}^{n} \binom{n}{i}[F(t)]^i[1-F(t)]^{n-i}, \end{aligned} \tag{1.2.6}$$

since whether any particular X is $\leqslant t$ is a Bernoulli event with probability $F(t)$ and the n Bernoulli events in question are mutually independent. From the well-known relation between binomial sums and the incomplete beta function (see Exercise 1.2.8), we can write

$$G_j(t) = \frac{n!}{(j-1)!(n-j)!}\int_0^{F(t)} x^{j-1}(1-x)^{n-j}\,dx. \tag{1.2.7}$$

Letting $H(\cdot)$ be the c.d.f. for a beta distribution with parameters $\alpha = j$ and $\beta = n-j+1$, we have $G_j(t) = H(F(t))$. Differentiating $G_j(t) = H(F(t))$ with respect to t via the chain rule yields $g_j(t) = h(F(t))f(t)$, the desired expression for the marginal density of $X_{(j)}$. ■

When computing cumulative probabilities for $X_{(j)}$, we note that we can use either cumulative binomial sums as in (1.2.6) or tables of the incomplete beta function in (1.2.7). Moreover, although the density form in (1.2.5) depends upon the underlying distribution being continuous, the two cumulative distribution expressions are valid for either continuous or discrete variables.

The Probability Integral Transformation and Uniform Order Statistics

As mentioned previously, the uniform distribution on (0, 1) plays a special role in nonparametric statistics. This is primarily due to a result referred to as the probability integral transformation. For a random variable X with c.d.f. $F(x)$, we define the **inverse distribution function $F^{-1}(\cdot)$** by

$$F^{-1}(y) = \inf\{x | F(x) \geq y\}, \qquad 0 < y < 1. \tag{1.2.8}$$

We note that if $F(x)$ is strictly increasing between 0 and 1, then there is only one x such that $F(x)=y$. In this case the infimum is unnecessary, and $F^{-1}(y)=x$ without ambiguity.

Suppose there is some x such that $F(x)=y$. Since $F(\cdot)$ is continuous from the right, $F(F^{-1}(y))=y$. In particular, this shows that if $F(\cdot)$ is continuous, then $F(F^{-1}(y))=y$ for every y satisfying $0<y<1$. However, if $F(\cdot)$ is the c.d.f. of a discrete distribution, then for a given y there may be no x for which $F(x)=y$. In such cases $F^{-1}(y)$ is the smallest x yielding an $F(\cdot)$ value larger than y, and hence, in general, we have the relationship

$$y \leq F(F^{-1}(y)) \qquad \text{for } 0 < y < 1.$$

Theorem 1.2.9. (Probability Integral Transformation) *Let X be a continuous random variable with distribution function $F(x)$. The random variable $Y=F(X)$ has a uniform distribution on* $(0,1)$.

Proof: As noted above, since $F(\cdot)$ is continuous, $F(F^{-1}(y))=y$ for $0<y<1$. Using the monotonicity of $F(\cdot)$ we see that $\{X \leq F^{-1}(y)\}$ implies $\{F(X) \leq F(F^{-1}(y))=y\}$. Also,

$$\{F(X) \leq y\} = \{X \leq F^{-1}(y)\} \cup \{X > F^{-1}(y) \text{ and } F(X)=y\}.$$

The continuous distribution of X implies that $P(F(X)=y)=0$. Thus

$$P(F(X) \leq y) = P(X \leq F^{-1}(y)).$$

Let $H(y)$ be the distribution function for Y. Since Y assumes values only in $[0,1]$, we know that

$$\begin{aligned} H(y) &= 0, \qquad y < 0 \\ &= 1, \qquad y \geq 1. \end{aligned} \tag{1.2.10}$$

Also,

$$\begin{aligned} H(y) &= P(Y \leq y) = P(F(X) \leq y) = P(X \leq F^{-1}(y)) \\ &= F(F^{-1}(y)) = y, \qquad 0 < y < 1. \end{aligned} \tag{1.2.11}$$

Using (1.2.10), (1.2.11) and the nondecreasing nature of $H(\cdot)$, we see that $H(y)$ is the distribution function for a uniform distribution on $(0,1)$, as desired. ■

We now use this result in connection with the earlier results on order statistics. Let $X_{(1)} \leqslant \cdots \leqslant X_{(n)}$ be the order statistics for a random sample from a continuous distribution with c.d.f. $F(x)$. Then, in view of Theorem 1.2.9, $F(X_{(1)}) \leqslant \cdots \leqslant F(X_{(n)})$ are distributed as the order statistics from a uniform distribution on $(0,1)$. Hence the joint density of $V_i = F(X_{(i)})$, $i=1,\ldots,n$, is given in (1.2.3), and the marginal density for each V_j, $1 \leqslant j \leqslant n$, follows from (1.2.5) and has form

$$
\begin{aligned}
g_j(t) &= \frac{n!}{(j-1)!(n-j)!}\, t^{j-1}(1-t)^{n-j}, \qquad 0<t<1 \\
&= 0, \qquad \text{elsewhere.}
\end{aligned} \tag{1.2.12}
$$

Thus V_j has the beta distribution with parameters $\alpha = j$ and $\beta = n-j+1$.

The moments for this beta distribution are used in later chapters and are easily calculated from (1.2.12). For any positive number r, we have

$$
\begin{aligned}
E(V_j^r) &= \frac{n!}{(j-1)!(n-j)!} \int_0^1 t^{r+j-1}(1-t)^{n-j}\, dt \\
&= \frac{n!\Gamma(r+j)}{(j-1)!\Gamma(n+r+1)} \int_0^1 \frac{\Gamma(n+r+1)}{\Gamma(r+j)(n-j)!}\, t^{r+j-1}(1-t)^{n-j}\, dt \\
&= \frac{n!\Gamma(r+j)}{(j-1)!\Gamma(n+r+1)},
\end{aligned} \tag{1.2.13}
$$

where $\Gamma(k) = (k-1)!$ whenever k is a positive integer. Thus when V_j is the jth order statistic from a uniform distribution

$$
E(V_j) = \frac{j}{n+1}
$$

and (1.2.14)

$$
\operatorname{Var}(V_j) = \frac{j(n-j+1)}{(n+1)^2(n+2)}.
$$

Expected values for order statistics from distributions other than the uniform are also useful in nonparametric testing procedures, but for many important distributions no simple expressions can be given for these expectations. (One notable exception is the exponential distribution—see Exercise 1.2.7.) In particular, such an explicit formula is not available for the normal distribution if n is larger than 3. However, because of the importance of the expectations of order statistics from a standard normal distribution (i.e., mean zero and variance one), these values have been calculated via numerical integration and have been tabulated by several people, including Harter (1961).

Another important result that, in a sense, is a converse to the probability integral transformation, is stated in the following theorem, which, unlike Theorem 1.2.9, is valid for c.d.f.'s $F(\cdot)$ of both discrete and continuous random variables.

Theorem 1.2.15. *Let U be a random variable with a uniform distribution on $(0,1)$. Define $X = F^{-1}(U)$, where $F^{-1}(\cdot)$ denotes the inverse of the distribution function $F(x)$, as given in expression (1.2.8). Then X has c.d.f. $F(x)$.*

Proof: Using the monotonicity of $F(\cdot)$, we see that $\{F^{-1}(U) \leqslant x\}$ implies $\{U \leqslant F(F^{-1}(U)) \leqslant F(x)\}$. Also, the definition of $F^{-1}(\cdot)$ shows that $\{U \leqslant F(x)\}$ implies $\{F^{-1}(U) \leqslant x\}$, and hence

$$P[U \leqslant F(x)] = P[F^{-1}(U) \leqslant x].$$

Let $G(x)$ denote the c.d.f. of X. Then,

$$\begin{aligned} G(x) &= P[X \leqslant x] \\ &= P[F^{-1}(U) \leqslant x] \\ &= P[U \leqslant F(x)] = F(x). \quad \blacksquare \end{aligned}$$

A random number generator on a computer produces values $U_1, \ldots, U_n$ that are approximately independent uniform $(0,1)$ variates. If the c.d.f. $F(\cdot)$ has an inverse function $F^{-1}(\cdot)$, then $F^{-1}(U_1), \ldots, F^{-1}(U_n)$ will be approximately distributed as independent variables with c.d.f. $F(x)$. Note also that $F^{-1}(U_{(1)}), \ldots, F^{-1}(U_{(n)})$ behave like the order statistics for a sample of size n from $F(\cdot)$. Thus to generate order statistics from any specific distribution, we can first generate ordered uniform variates and then transform them using the inverse function of the population c.d.f. In Section 1 of Chapter 4 we illustrate the use of Theorem 1.2.15 in a Monte Carlo study of the relative performances of several test statistics.

Exercises 1.2.

1.2.1. Derive the marginal density expression in (1.2.5) by integrating the joint density in (1.2.2).

1.2.2. Let $X_{(1)} \leqslant \cdots \leqslant X_{(n)}$ be the order statistics for a random sample from a continuous distribution with c.d.f. $F(x)$ and density $f(x)$.

(a) Show that the density for the joint distribution of $X_{(r)}$ and $X_{(s)}$, with $r<s$, has form

$$g_{r,s}(u,v) = \frac{n!}{(r-1)!(s-r-1)!(n-s)!}\left[F(u)\right]^{r-1}\left[F(v)-F(u)\right]^{s-r-1}$$
$$\cdot\left[1-F(v)\right]^{n-s} f(u)f(v), \qquad -\infty < u < v < \infty$$
$$= 0, \qquad \text{elsewhere.}$$

(b) If the distribution function $F(x)$ corresponds to the uniform distribution on (0, 1), obtain an expression for $\operatorname{Cov}(X_{(r)}, X_{(s)})$ and $\operatorname{Corr}(X_{(r)}, X_{(s)})$.

1.2.3. Let $X_{(1)} \leqslant \cdots \leqslant X_{(n)}$ be the order statistics for a random sample from a continuous distribution with c.d.f. $F(x)$ and p.d.f. $f(x)$. Define $W_i = F(X_{(i)})$, $i=1,\ldots,n$. Show that for any $r<s$, W_{s-r} and $W_s - W_r$ have the same distribution.

1.2.4. Let $U_1,\ldots,U_8$ denote 8 i.i.d. random variables, each with a uniform (0, 1) distribution.

(a) Find the probability that the value nearest 1 exceeds .98.

(b) Find the number c such that the probability is .7 that the value nearest zero will exceed c.

1.2.5. Let (X, Y) have a jointly continuous bivariate distribution such that the marginal c.d.f.'s for X and Y are $F(x)$ and $G(y)$, respectively. Prove the following result: The pair $(F(X), G(Y))$ has a uniform distribution over the unit square $(0,1)\times(0,1)$ if and only if X and Y are independent. Comment on this fact in view of Theorem 1.2.9.

1.2.6. Let $X_{(1)} \leqslant \cdots \leqslant X_{(n)}$ be the order statistics for a random sample from a continuous distribution with c.d.f. $F(x)$ and density $f(x)$. Define U_i, $i=1,\ldots,n$, by

$$U_i = \frac{F(X_{(i)})}{F(X_{(i+1)})}, \qquad i = 1,\ldots,n-1, \text{ and } U_n = F(X_{(n)}).$$

(a) Find the joint distribution of $U_1,\ldots,U_n$.

(b) Show that $U_1,\ldots,U_n$ are mutually stochastically independent and find the marginal distribution of U_r, $r=1,\ldots,n$.

(c) Show that $U_1, U_2^2,\ldots,U_n^n$ are i.i.d. uniform $(0,1)$ variables.

1.2.7. Let $X_{(1)} \leqslant \cdots \leqslant X_{(n)}$ be the order statistics for a random sample from the exponential distribution with density

$$f(x) = e^{-x}, \qquad x \geqslant 0,$$
$$= 0, \qquad \text{elsewhere.}$$

(a) Show that $Y_1 = X_{(1)}$ and $Y_i = X_{(i)} - X_{(i-1)}$, $i=2,\ldots,n$, are mutually stochastically independent. In addition, show that Y_i has a gamma distribution with parameters

$$\alpha = 1 \text{ and } \beta = \frac{1}{n+1-i}, \qquad \text{for } i = 1,\ldots,n.$$

(b) Show that

$$E(X_{(r)}) = \sum_{t=1}^{r} \left(\frac{1}{n-t+1}\right)$$

and

$$\operatorname{Var}(X_{(r)}) = \sum_{t=1}^{r} \left(\frac{1}{n-t+1}\right)^2,$$

for $r=1,\ldots,n$.

(c) Verify that

$$\operatorname{Cov}(X_{(r)}, X_{(s)}) = \operatorname{Var}(X_{(r)}), \qquad \text{for any } r < s.$$

Obtain an expression for $\operatorname{Corr}(X_{(r)}, X_{(s)})$, $r<s$.

1.2.8. Establish the equivalence between the cumulative binomial expression in (1.2.6) and the incomplete beta integral in (1.2.7). [Hint: Integration by parts can be used.]

1.2.9. Let $X_{(1)} \leqslant \cdots \leqslant X_{(n)}$ be the order statistics for a random sample from a continuous distribution with c.d.f. $F(x)$ and density $f(x)$. If n is an odd integer, say, $n=2k+1$ for some nonnegative integer k, we call $X_{(k+1)}$

the **sample median.** If n is an even integer, say $n=2k$ for some positive integer k, it is conventional to call $[X_{(k)}+X_{(k+1)}]/2$ the **sample median.**

(a) Find the form of the density of the sample median for a random sample of size $n=11$ from a logistic distribution with density

$$f(x) = \frac{e^{-x}}{(1+e^{-x})^2}, \qquad -\infty < x < \infty.$$

(b) If n is odd and $X_{(k+1)}$ is the sample median for a random sample of size n from a continuous distribution that is symmetric about ξ, show that the distribution of $X_{(k+1)}$ is also symmetric about ξ.

1.2.10. Let $X_{(1)} \leqslant X_{(2)}$ be the order statistics for a random sample of size 2 from a normal distribution with mean μ and variance σ^2. Evaluate $E(X_{(1)})$, $E(X_{(2)})$, $\mathrm{Var}(X_{(1)})$, $\mathrm{Var}(X_{(2)})$ and $\mathrm{Cov}(X_{(1)}, X_{(2)})$.

1.2.11. Let $X_{(1)} \leqslant \cdots \leqslant X_{(n)}$ be the order statistics for a random sample from a continuous distribution that is symmetric about 0. Show that

$$E(X_{(i)}) = -E(X_{(n-i+1)}), \qquad i = 1,\ldots,n,$$

provided the necessary expectations exist. What is the corresponding statement when the underlying distribution is symmetric about μ, not necessarily zero?

1.2.12. Let $X_1,\ldots,X_n$ be independent random variables such that X_i has the geometric distribution with density

$$\begin{aligned} P(X_i=t) &= (1-p_i)^{t-1}p_i, \qquad && t = 1,2,\ldots \\ &= 0, && \text{elsewhere.} \end{aligned}$$

Find the distribution of $X_{(1)} = \text{minimum } \{X_1,\ldots,X_n\}$.

1.2.13. Let $X_{(1)} \leqslant \cdots \leqslant X_{(n)}$ denote the order statistics for a random sample of size n from a continuous distribution that is symmetric about 0. Show that

$$E(X_{(i)}) = \int_0^{\infty} \left[F_{(n+1-i)}(t) - F_{(i)}(t) \right] dt,$$

provided this expectation exists, where $F_{(j)}(t)$ denotes the c.d.f. of $X_{(j)}$, $j=i,\ n+1-i$. [Hint: You may use (without proof) the following fact: If X

has c.d.f. $F(x)$ and $E(X)$ exists, then

$$\lim_{x \to -\infty} xF(x) = 0.]$$

1.2.14. Let $X_{(1)} \leqslant \cdots \leqslant X_{(n)}$ be the order statistics for a random sample from a continuous distribution with c.d.f. $F(x)$. Set $\tau_{(i)} = E(X_{(i)})$, $i = 1, \ldots, n$, where we assume all these expectations exist. Let $Y_{(1)} \leqslant \cdots \leqslant Y_{(n)}$ be the order statistics for a random sample from $F[(x-\mu)/\sigma]$, where μ and $\sigma > 0$ are constants.

(a) Show that $E(Y_{(i)}) = \mu + \sigma\tau_{(i)}$, $i = 1, \ldots, n$.

(b) If $\sigma = 1$ and the underlying density $f(x) = dF(x)/dx$ is uniformly (in x) bounded above by some constant $M > 0$, prove that

$$\left| E(F(Y_{(i)})) - \frac{i}{n+1} \right| \leqslant |\mu| M.$$

[Hint: Consider applying the mean value theorem to $F(x-\mu) - F(x)$.]

(c) Now suppose that we observe the ordered values $Y_{(1)} \leqslant \cdots \leqslant Y_{(n)}$ for a random sample from $F[(x-\mu)/\sigma]$, with both μ and σ unknown, but with $\tau_{(1)}, \ldots, \tau_{(n)}$ known. Find the estimators $\hat{\mu}$ and $\hat{\sigma}$ that minimize

$$\begin{aligned} D(\mu, \sigma) &= \sum_{i=1}^{n} (Y_{(i)} - E(Y_{(i)}))^2 \\ &= \sum_{i=1}^{n} (Y_{(i)} - \mu - \sigma\tau_{(i)})^2. \end{aligned}$$

(d) Show that $\hat{\mu}$ and $\hat{\sigma}$ in **(c)** are unbiased for μ and σ, respectively, with no additional restrictions on $F(x)$.

1.2.15. **(a)** Let $X_{(n)}$ be the largest order statistic for a random sample of size n from a continuous distribution with strictly increasing distribution function $F(x)$ and density $f(x)$. If the expected value μ and the variance σ^2 for this distribution are finite, show that

$$E(X_{(n)}) \leqslant \mu + \sigma(n-1)[2n-1]^{-1/2}.$$

[Hint: Use the change of variable $t = F(y)$ in the appropriate integral and then Schwarz's inequality.]

(b) Under the conditions in **(a)**, deduce a lower bound for $E(X_{(1)})$, where $X_{(1)}$ is the smallest order statistic for the random sample from $F(x)$.

(c) Let $R_n = X_{(n)} - X_{(1)}$, where $X_{(1)}$ and $X_{(n)}$ are as defined in **(a)** and **(b)**. We call R_n the **sample range**. Obtain an upper bound for the expected value of the sample range under the conditions in **(a)**.

1.2.16. Definition 1.2.16. Let $T(X_1,\ldots,X_n)$ be a statistic based on a random sample $X_1,\ldots,X_n$ from a distribution with c.d.f. $F(x)$ and median ξ. We say that T is **median unbiased** for ξ if ξ is a median of the distribution of T.

Let $\tilde{X}$ be the sample median (see Exercise 1.2.9) for a random sample of size m from a continuous distribution with c.d.f. $F(x)$ and median ξ. If m is odd, show that $\tilde{X}$ is median unbiased for ξ.

1.3. The "Equal in Distribution" Technique

An Introduction

There are many problems in statistics where "equal in distribution" arguments provide simple proofs of quite general propositions. The purpose of this section is to acquaint you with this technique which we will use in numerous places throughout the text.

Definition 1.3.1. Two random variables S and T are said to be **equal in distribution** if they have the same c.d.f. (If they are identically dimensioned vectors they must have the same joint c.d.f.) To denote "equal in distribution" we use the notation

$$S \stackrel{d}{=} T,$$

where $\stackrel{d}{=}$ is read "has the same distribution as".

Note that $\stackrel{d}{=}$ is an equivalence relation. That is, it satisfies:

(i) (reflexitivity) $X \stackrel{d}{=} X$
(ii) (symmetry) $X \stackrel{d}{=} Y$ implies $Y \stackrel{d}{=} X$

and

(iii) (transitivity) $X \stackrel{d}{=} Y$ and $Y \stackrel{d}{=} Z$ imply $X \stackrel{d}{=} Z$.

Thus, for example, $X \stackrel{d}{=} Y \stackrel{d}{=} Z$ means that all three random variables have the same c.d.f.

We now consider some common settings where random variables possess this equal in distribution property. This first theorem describes the reasonable notion that for a symmetrically distributed random variable, the random distance (possibly negative) that it is above its point of symmetry has the same distribution as the random distance it is below that value.

Theorem 1.3.2. *A random variable X has a distribution that is symmetric about some number μ if and only if*

$$(X-\mu) \stackrel{d}{=} (\mu - X).$$

Proof: (i) Suppose X is symmetrically distributed about μ. If $F(\cdot)$ denotes the c.d.f. of X, then

$$F(\mu+t) = 1 - F((\mu-t)^-) \tag{1.3.3}$$

for all real t, where $F(x^-)$ denotes $\text{limit}_{\Delta\downarrow 0}F(x-\Delta)$. The c.d.f. of $W = X-\mu$ is

$$G(w) = P[X-\mu \leqslant w] = F(\mu+w)$$

and the c.d.f. of $V = \mu - X$ is

$$H(v) = P[\mu - X \leqslant v] = 1 - F((\mu - v)^-).$$

From (1.3.3) it follows that $G(t) = H(t)$ for all real t. Thus we write $(X-\mu) \stackrel{d}{=} (\mu - X)$.

(ii) If $(X-\mu) \stackrel{d}{=} (\mu-X)$ then

$$\begin{aligned} F(\mu+t) &= P[X \leqslant \mu+t] = P[X-\mu \leqslant t] \\ &= P[\mu - X \leqslant t] = 1 - F((\mu-t)^-) \end{aligned}$$

for all real t. Thus X is symmetrically distributed about μ. ■

Theorem 1.3.4. *Let $X_1,\ldots,X_n$ denote independent random variables such that the distribution of X_i is symmetric about some value μ_i for $i = 1,\ldots,n$. Then*

$$\begin{aligned} (X_1-\mu_1,\ldots,X_n-\mu_n) &\stackrel{d}{=} (\mu_1 - X_1, X_2-\mu_2,\ldots,X_n-\mu_n) \\ &\stackrel{d}{=} \cdots \stackrel{d}{=} (\mu_1 - X_1,\ldots,\mu_n - X_n), \end{aligned}$$

where all 2^n such terms appear in this string of equalities in distribution.

Proof: Exercise 1.3.6. ■

Let $X_1,\ldots,X_n$ be independent and identically distributed random variables, each with c.d.f. $F(\cdot)$. The joint c.d.f. of these n random variables is then

$$G(x_1,\ldots,x_n) = \prod_{i=1}^{n} F(x_i).$$

Now consider a rearrangement of the variables, such as $X_n, X_{n-1},\ldots,X_1$. The joint c.d.f of these variables is

$$H(t_1,\ldots,t_n) = P[X_n \leqslant t_1,\ldots,X_1 \leqslant t_n] = \prod_{i=1}^{n} F(t_i) = G(t_1,\ldots,t_n),$$

which holds for all values of $t_1,\ldots,t_n$. Thus

$$(X_1,\ldots,X_n) \stackrel{d}{=} (X_n, X_{n-1},\ldots,X_1).$$

Theorem 1.3.5. *Let $X_1,\ldots,X_n$ be independent and identically distributed random variables. Let $(\alpha_1,\ldots,\alpha_n)$ denote any permutation of the integers $(1,\ldots,n)$. Then $(X_1,X_2,\ldots,X_n) \stackrel{d}{=} (X_{\alpha_1},X_{\alpha_2},\ldots,X_{\alpha_n})$.*

Proof: See the argument preceding this theorem. ■

This result shows that arranging i.i.d. random variables in any order yields the same joint distribution. That is, for $n=3$,

$$(X_1,X_2,X_3) \stackrel{d}{=} (X_1,X_3,X_2) \stackrel{d}{=} (X_2,X_1,X_3) \stackrel{d}{=} (X_2,X_3,X_1)$$
$$\stackrel{d}{=} (X_3,X_1,X_2) \stackrel{d}{=} (X_3,X_2,X_1).$$

We now generalize the class of variables considered in the previous theorem.

Definition 1.3.6. A collection of random variables $X_1,\ldots,X_n$ is said to be **exchangeable** if for every permutation $(\alpha_1,\ldots,\alpha_n)$ of the integers $(1,\ldots,n)$,

$$(X_1,\ldots;X_n) \stackrel{d}{=} (X_{\alpha_1},\ldots,X_{\alpha_n}).$$

Clearly, a collection of i.i.d. random variables is exchangeable, but the converse is not necessarily true. Consider (X_1,X_2) which assumes the three points $(0,0)$, $(1,0)$, and $(0,1)$ with equal probability. Note that $(X_1,X_2) \stackrel{d}{=} (X_2,X_1)$, but X_1 and X_2 are not independent, since $P((X_1,X_2)=(0,0)) = (1/3) \neq (2/3)\cdot(2/3) = P(X_1=0)\cdot P(X_2=0)$. However, it is true that if a collection of random variables is exchangeable, then each variable in the collection must have the same marginal distribution, as we prove later (see also Exercise 1.3.7). We use this more general idea of exchangeable, but not independent, random variables in Chapter 11 to construct classes of statistics referred to as rank-like statistics.

The following result is important to our understanding of the applications of equal in distribution arguments.

Theorem 1.3.7. *If* $\mathbf{X} \stackrel{d}{=} \mathbf{Y}$ *and* $U(\cdot)$ *is a (measurable) function (possibly vector valued) defined on the common support of these random variables, then*

$$U(\mathbf{X}) \stackrel{d}{=} U(\mathbf{Y}).$$

Proof: A rigorous proof of this theorem requires the use of measure theory. We present only a sketch of the central structure of the proof. Let A denote an arbitrary (measurable) subset of the support of $U(\cdot)$. Then

$$P[U(\mathbf{X}) \in A] = \int \cdot \cdot \int I_{[\mathbf{x} | U(\mathbf{x}) \in A]} dH(\mathbf{x}), \tag{1.3.8}$$

where

$$I_{[\mathbf{x} | U(\mathbf{x}) \in A]} = \begin{cases} 1 & \text{if } U(\mathbf{x}) \in A \\ 0 & \text{otherwise,} \end{cases}$$

and $H(\cdot)$ is the c.d.f. of $\mathbf{X}$. Since $\mathbf{x}$ on the right side of equation (1.3.8) is only a dummy variable of integration, we may as well use the symbol $\mathbf{y}$ in its place. But $\mathbf{X} \stackrel{d}{=} \mathbf{Y}$ implies that $H(\cdot)$ is also the c.d.f. of $\mathbf{Y}$. Thus

$$\begin{aligned} P[U(\mathbf{X}) \in A] &= \int \cdot \cdot \int I_{[\mathbf{y} | U(\mathbf{y}) \in A]} dH(\mathbf{y}) \\ &= P[U(\mathbf{Y}) \in A]. \end{aligned}$$

Since this holds for any such set A,

$$U(\mathbf{X}) \stackrel{d}{=} U(\mathbf{Y}). \quad \blacksquare$$

Thus a function can be applied to both sides of an equal in distribution symbol and the distributional equality is preserved. For example, suppose $(X_1, X_2, \ldots, X_n)$ is a vector of exchangeable random variables. Then the fact that

$$(X_i, X_1, \ldots, X_{i-1}, X_{i+1}, \ldots, X_n) \stackrel{d}{=} (X_j, X_1, \ldots, X_{j-1}, X_{j+1}, \ldots, X_n)$$

begin with a description of some general properties of a statistic $t(X_1,\ldots,X_n)$ based on data $X_1,\ldots,X_n$.

Definition 1.3.13. The statistic $t(\cdot)$ is said to be:

(i) **odd** if

$$t(-x_1,\ldots,-x_n) = -t(x_1,\ldots,x_n)$$

or

(ii) **even** if

$$t(-x_1,\ldots,-x_n) = t(x_1,\ldots,x_n)$$

for every $x_1,\ldots,x_n$.

Definition 1.3.14. The statistic $t(\cdot)$ is said to be

(i) a **translation** statistic if

$$t(x_1+k,\ldots,x_n+k) = t(x_1,\ldots,x_n)+k$$

or

(ii) a **translation-invariant** statistic if

$$t(x_1+k,\ldots,x_n+k) = t(x_1,\ldots,x_n)$$

for every k and $x_1,\ldots,x_n$.

Example 1.3.15. **(a)** The sample mean $\bar{X}$ is an odd translation statistic, since

$$\left(\frac{1}{n}\right)\sum_{i=1}^{n}(-x_i) = -\bar{x}$$

and

$$\left(\frac{1}{n}\right)\sum_{i=1}^{n}(x_i+k) = \bar{x}+k.$$

(b) The sample variance S^2 is an even translation-invariant statistic, since

$$\frac{1}{n-1}\sum_{i=1}^{n}\left(-x_i-\frac{1}{n}\sum_{j=1}^{n}(-x_j)\right)^2 = s^2$$

and

$$\frac{1}{n-1}\sum_{i=1}^{n}\left(x_i+k-\frac{1}{n}\sum_{j=1}^{n}(x_j+k)\right)^2=s^2.$$

(c) The difference $\overline{X}-\text{median}_{1\leqslant i\leqslant n}(X_i)$ is an odd translation-invariant statistic, since

$$\frac{1}{n}\sum_{j=1}^{n}(-x_j)-\underset{1\leqslant i\leqslant n}{\text{median}}(-x_i)=-\left(\bar{x}-\underset{1\leqslant i\leqslant n}{\text{median}}(x_i)\right)$$

and

$$\frac{1}{n}\sum_{j=1}^{n}(x_j+k)-\underset{1\leqslant i\leqslant n}{\text{median}}(x_i+k)=\bar{x}-\underset{1\leqslant i\leqslant n}{\text{median}}(x_i).$$

To investigate conditions under which a statistic will have a symmetric distribution we first prove a general theorem due to Wolfe (1973). Let $\mathcal{D}^*$ be the collection of all transformations from the n-dimensional reals onto the n-dimensional reals.

Theorem 1.3.16. *Let $U(\mathbf{X})=U(X_1,\ldots,X_n)$ be any real-valued statistic. If there exists a transformation $g(\cdot)$ in $\mathcal{D}^*$ and a number μ such that*

$$U(\mathbf{x})-\mu=\mu-U(g(\mathbf{x})) \tag{1.3.17}$$

for every $\mathbf{x}$ *in the support of* $\mathbf{X}$, *and*

$$g(\mathbf{X})\stackrel{d}{=}\mathbf{X}, \tag{1.3.18}$$

then $U(\mathbf{X})$ is symmetrically distributed about μ.

Proof: Expression (1.3.18) and Theorem 1.3.7 imply that

$$U[g(\mathbf{X})]\stackrel{d}{=}U(\mathbf{X}).$$

This together with equation (1.3.17) yields

$$\begin{aligned}U(\mathbf{X})-\mu&=\mu-U(g(\mathbf{X}))\\&\stackrel{d}{=}\mu-U(\mathbf{X}).\end{aligned}$$

Thus, according to Theorem 1.3.2, $U(\mathbf{X})$ is symmetrically distributed about μ. ■

Corollary 1.3.19. Let $X_1,\ldots,X_n$ be i.i.d., each with a distribution symmetric about μ. Then an odd translation statistic $V(X_1,\ldots,X_n)$ is symmetrically distributed about μ.

Proof: From Theorems 1.3.4 and 1.3.7 we see that

$$(X_1-\mu,\ldots,X_n-\mu) \stackrel{d}{=} (\mu-X_1,\ldots,\mu-X_n)$$

and thus

$$(X_1,\ldots,X_n) \stackrel{d}{=} (2\mu-X_1,\ldots,2\mu-X_n). \tag{1.3.20}$$

If we let $g_1(\cdot)$ map $(x_1,\ldots,x_n)$ into $(2\mu-x_1,\ldots,2\mu-x_n)$, the odd translation properties imply that

$$\begin{aligned} V(g_1(\mathbf{x})) &= V(2\mu-x_1,\ldots,2\mu-x_n) \\ &= V(-x_1,\ldots,-x_n)+2\mu \\ &= -V(x_1,\ldots,x_n)+2\mu. \end{aligned}$$

That is,

$$V(\mathbf{x})-\mu=\mu-V(g_1(\mathbf{x})) \tag{1.3.21}$$

for every $x_1,\ldots,x_n$. By Theorem 1.3.16 the statistic $V(X_1,\ldots,X_n)$ is symmetrically distributed about μ. ■

Thus we see that when sampling from a population that is symmetric about μ, a location estimator such as $\overline{X}$ is also symmetrically distributed about μ. So are other odd translation statistics, such as the sample median, mid-range $\{(X_{(n)}+X_{(1)})/2\}$, and mode.

Corollary 1.3.22. Let $X_1,\ldots,X_n$ be i.i.d. according to a distribution that is symmetric about μ. Then an odd translation-invariant statistic $V(X_1,\ldots,X_n)$ is symmetrically distributed about zero.

Proof: It follows in the same fashion as Corollary 1.3.19. ■

Thus when sampling from a symmetric population, $\overline{X}-\text{median}_{1\leqslant i\leqslant n}(X_i)$ is symmetrically distributed about zero. So is the difference between any two odd translation statistics.

We now turn our attention to two-sample statistics, denoting the two samples by $X_1,\ldots,X_m$ and $Y_1,\ldots,Y_n$.

Definition 1.3.23. A two-sample statistic $W(X_1,\ldots,X_m; Y_1,\ldots, Y_n)$ is said to be a **shift** statistic if

$$W(x_1+h,\ldots,x_m+h;y_1+k,\ldots,y_n+k) = (k-h)+ W(x_1,\ldots,x_m;y_1,\ldots,y_n)$$

for every h, k, and $x_1,\ldots,x_m,y_1,\ldots,y_n$.

Note that a statistic satisfying

$$W(x_1+h,\ldots,x_m+h;y_1+k,\ldots,y_n+k) = (h-k)+ W(x_1,\ldots,x_m;y_1,\ldots,y_n)$$

would also be considered a shift statistic with the roles of the two samples reversed.

Definition 1.3.24. When $m=n$, a two-sample statistic $W(X_1,\ldots, X_n; Y_1,\ldots, Y_n)$ is said to be

(i) **interchangeable** if

$$W(y_1,\ldots,y_n;x_1,\ldots,x_n) = W(x_1,\ldots,x_n;y_1,\ldots,y_n)$$

or

(ii) **negative interchangeable** if

$$W(y_1,\ldots,y_n;x_1,\ldots,x_n) = -W(x_1,\ldots,x_n;y_1,\ldots,y_n)$$

for every $x_1,\ldots,x_n,y_1,\ldots,y_n$.

Example 1.3.25. (a) The difference between two odd translation statistics (one computed from each sample) is an odd shift statistic. Examples are: $\overline{Y}-\overline{X}$ and $\text{median}_{1\leqslant i\leqslant n}(Y_i)-[(X_{(n)}+X_{(1)})/2]$.

(b) Statistics such as $\overline{X}+\overline{Y}$ and $S_x^2+S_y^2$ are interchangeable.

Corollary 1.3.26. Let $X_1,\ldots,X_m$ be i.i.d., each with a distribution symmetric about μ_x, and let the independent sample $Y_1,\ldots, Y_n$ be i.i.d., each symmetrically distributed about μ_y. Then if $W(\cdot)$ is an odd shift (two-sample) statistic, $W(X_1,\ldots,X_m; Y_1,\ldots, Y_n)$ is symmetrically distributed about $\mu_y-\mu_x$.

Proof: Theorems 1.3.4 and 1.3.7 imply that

$$(X_1,\ldots,X_m, Y_1,\ldots, Y_n) \stackrel{d}{=} (2\mu_x-X_1,\ldots,2\mu_x-X_m,2\mu_y-Y_1,\ldots,2\mu_y-Y_n).$$

If we let $g_2(\cdot)$ map $(x_1,\ldots,x_m,y_1,\ldots,y_n)$ into $(2\mu_x-x_1,\ldots,2\mu_x-x_m,$

$2\mu_y - y_1,\ldots,2\mu_y - y_n)$, the odd shift properties yield

$$\begin{aligned} W(g_2(x_1,\ldots,x_m,y_1,\ldots,y_n)) &= 2(\mu_y-\mu_x) + W(-x_1,\ldots,-x_m;-y_1,\ldots,-y_n) \\ &= 2(\mu_y-\mu_x) - W(x_1,\ldots,x_m;y_1,\ldots,y_n). \end{aligned}$$

The result then follows from Theorem 1.3.16. ■

Corollary 1.3.27. Let $X_1,\ldots,X_n,Y_1,\ldots,Y_n$ be a random sample of size $2n$ from some population. Then if the two-sample statistic $W(\cdot)$ is negative interchangeable, $W(X_1,\ldots,X_n;Y_1,\ldots,Y_n)$ is symmetrically distributed about zero.

Proof: By Theorem 1.3.5, $(X_1,\ldots,X_n,Y_1,\ldots,Y_n) \stackrel{d}{=} (Y_1,\ldots,Y_n,X_1,\ldots,X_n)$. Define $g_3(x_1,\ldots,x_n,y_1,\ldots,y_n)=(y_1,\ldots,y_n,x_1,\ldots,x_n)$. Thus

$$W(g_3(x_1,\ldots,x_n,y_1,\ldots,y_n)) = -W(x_1,\ldots,x_n;y_1,\ldots,y_n).$$

Applying Theorem 1.3.16 with $\mu=0$ yields that $W(X_1,\ldots,X_n;Y_1,\ldots,Y_n)$ is symmetrically distributed about zero. ■

Note that this corollary requires the Xs and Ys to have the *same* distribution. It is useful in establishing the symmetry of certain test statistics under their null hypotheses. Corollaries 1.3.26 and 1.3.27 establish the symmetry of statistics such as $\bar{Y}-\bar{X}$ and $\text{median}_{1\leqslant i\leqslant n}(Y_i) - \text{median}_{1\leqslant i\leqslant n}(X_i)$ under two different sets of conditions.

We now consider problems concerning the joint behavior of two statistics. The following lemma plays a key role.

Lemma 1.3.28. *If two random variables V and W satisfy*

$$(V-\mu, W) \stackrel{d}{=} (\mu - V, W)$$

for some constant μ, then, if it exists, $\text{Cov}(V,W)$ *is zero.*

Proof: Left as Exercise 1.3.4. ■

Let $\mathbf{X}=(X_1,\ldots,X_n)$ denote a random vector with support set $\mathcal{X}$. Let $\mathcal{D}^*$ again be the collection of transformations from the n-dimensional reals onto the n-dimensional reals.

Theorem 1.3.29. *Let $V(\mathbf{X})$ and $W(\mathbf{X})$ denote two statistics. Suppose there exists a constant μ and some $g(\cdot)$ in $\mathcal{D}^*$ such that*

$$\begin{aligned} V(\mathbf{x}) - \mu &= \mu - V(g(\mathbf{x})) \qquad \textit{for every } \mathbf{x}\in\mathcal{X}, \\ W(g(\mathbf{x})) &= W(\mathbf{x}) \qquad \textit{for every } \mathbf{x}\in\mathcal{X}, \end{aligned} \tag{1.3.30}$$

and

$$g(\mathbf{X}) \stackrel{d}{=} \mathbf{X}. \tag{1.3.31}$$

Then

$$[V(\mathbf{X})-\mu, W(\mathbf{X})] \stackrel{d}{=} [\mu - V(\mathbf{X}), W(\mathbf{X})]. \tag{1.3.32}$$

Proof: Using $g(\cdot)$ we have

$$\begin{aligned}[V(\mathbf{X})-\mu, W(\mathbf{X})] &= [\mu - V(g(\mathbf{X})), W(g(\mathbf{X}))] \\ &\stackrel{d}{=} [\mu - V(\mathbf{X}), W(\mathbf{X})]. \quad \blacksquare\end{aligned}$$

Corollary 1.3.33. Let $X_1,\ldots,X_n$ denote a random sample from a population that is symmetric about some number μ. Let $V(\mathbf{X})$ be an odd translation statistic and $W(\mathbf{X})$ an even translation-invariant statistic. Then $V(\mathbf{X})$ and $W(\mathbf{X})$ satisfy (1.3.32), and, if it exists, $\text{Cov}[V(\mathbf{X}), W(\mathbf{X})]$ is zero.

Proof: Using $g_1(\cdot)$ from the proof of Corollary 1.3.19, we apply Lemma 1.3.28 and Theorem 1.3.29. ■

This corollary shows that, when sampling from a symmetric population, $\overline{X}$ and S (sample standard deviation) are uncorrelated, and so are $\text{median}_{1\leqslant i\leqslant n}(X_i)$ and the mean deviation from the median $\{(1/n)\Sigma_{j=1}^{n}|X_j - \text{median}_{1\leqslant i\leqslant n}(X_i)|\}$, assuming the respective covariances exist.

Definition 1.3.34. A vector of random variables $\mathbf{Z}=(Z_1,\ldots,Z_N)$ is said to be **equally symmetric** about the vector $\boldsymbol{\mu}=(\mu_1,\ldots,\mu_N)$ if $\mathbf{Z}-\boldsymbol{\mu} \stackrel{d}{=} \boldsymbol{\mu}-\mathbf{Z}$.

Remark 1.3.35. Most of the preceding symmetry and zero covariance results can be written in slightly more general forms. For example, Corollaries 1.3.19, 1.3.22, and 1.3.33 are valid if $(X_1,\ldots,X_n)$ is equally symmetric about $(\mu,\ldots,\mu)$ for some number μ. Likewise, Corollary 1.3.26 is valid if $(X_1,\ldots,X_m,Y_1,\ldots,Y_n)$ is equally symmetric about $(\mu_x,\ldots,\mu_x,\mu_y,\ldots,\mu_y)$, and Corollary 1.3.27 is valid as long as $(X_1,\ldots,X_n,Y_1,\ldots,Y_n)$ is a vector of exchangeable variates.

Material in this subsection comes from Hogg (1960), Hodges and Lehmann (1963), and Hollander (1968). The general theorems plus additional references may be found in Wolfe (1973).

Exercises 1.3.

1.3.1. Let $X_1,\ldots,X_n$ denote a random sample from some population. For each statistic D listed below give general conditions under which it has a symmetric distribution and state the associated point of symmetry.

(a) Let $k=[\![n\alpha]\!]$ denote the greatest integer less than or equal to $n\alpha$, where $0<\alpha<1$ and $n\alpha \geqslant 1$. Set

$$D = (1/2)(X_{(k)}+X_{(n+1-k)}).$$

(b) $D=\text{mode}(X_1,\ldots,X_n)$. (If the sample has multiple modes, D denotes the median of these modal values.)

(c) Let $r=[\![(n+1)/2]\!]$ and $D=(X_n-X_1)+(X_{n-1}-X_2)+\cdots+(X_{n+1-r}-X_r)$.

Let $X_1,\ldots,X_m$ and $Y_1,\ldots,Y_n$ denote independent random samples from two populations.

(d) $D=\text{median}_{1\leqslant i\leqslant m}(X_i)-\bar{Y}$.

(e) Let $m=n$ and take

$$D = Y_{(1)} - X_{(1)}.$$

1.3.2. **(a)** Let X and Y be random variables with c.d.f.'s $F(x)$ and $F(x-\Delta)$, respectively, where Δ is some constant. Show that $X \overset{d}{=} Y-\Delta$.

(b) Let X and Y be random variables with c.d.f.'s $F(x)$ and $F(x/\eta)$, respectively, where η is a positive constant. Show that $X \overset{d}{=} Y/\eta$.

1.3.3. Let X_1,X_2,X_3 be i.i.d. continuous random variables, each symmetrically distributed about zero. Use the equal in distribution technique to argue that

$$P[X_1+X_2+X_3>0] = 1/2.$$

1.3.4. Prove Lemma 1.3.28.

1.3.5. Give an example of a pair of dependent random variables (V,W) satisfying

$$(V-\mu,W) \overset{d}{=} (\mu-V,W)$$

for some number μ. Comment on this example in view of Lemma 1.3.28.

1.3.6. Prove Theorem 1.3.4.

1.3.7. Let $(X_1,\ldots,X_n)$ be a vector of exchangeable random variables, and let $(\alpha_1,\ldots,\alpha_m)$ be any subset of the integers $(1,\ldots,n)$, with $m \leqslant n$. Prove that $(X_{\alpha_1},\ldots,X_{\alpha_m})$ is also a vector of exchangeable random variables.

1.3.8. Prove or disprove the following: If X_1 is symmetrically distributed about μ_1 and X_2 is symmetrically distributed about μ_2, then

$$(X_1-\mu_1, X_2-\mu_2) \stackrel{d}{=} (\mu_1-X_1, \mu_2-X_2).$$

1.3.9. Let the statistic $\hat{\theta}(X_1,\ldots,X_n)$ be defined as the value t that minimizes

$$h(t) = \sum_{i=1}^{n} (X_i-t)^4.$$

Without explicitly solving for $\hat{\theta}(X_1,\ldots,X_n)$, show that $\hat{\theta}(X_1,\ldots,X_n)$ is an odd translation statistic.

1.3.10. Let $X_1,\ldots,X_n$ be mutually stochastically independent random variables such that X_i is symmetrically distributed about μ_i, $i=1,\ldots,n$. Use an $\stackrel{d}{=}$ argument to show that $\sum_{i=1}^n X_i$ is symmetrically distributed about $\sum_{i=1}^n \mu_i$.

1.3.11. Let $X_1,\ldots,X_n$ and $Y_1,\ldots,Y_n$ denote independent random samples from two populations.

(a) Find an example of an odd shift statistic that is not negative interchangeable.

(b) Find an example of a negative interchangeable statistic that is not odd shift. (These examples do not have to be well-known statistics.)

1.3.12. Let $t_1(X_1,\ldots,X_n)$ and $t_2(X_1,\ldots,X_n)$ be statistics based on the data $(X_1,\ldots,X_n)$. Prove or give a counterexample to the following.

(a) If $t_1(\cdot)$ and $t_2(\cdot)$ are odd, then so are $t_1(\cdot)+t_2(\cdot)$ and $t_1(\cdot)-t_2(\cdot)$.

(b) If $t_1(\cdot)$ and $t_2(\cdot)$ are translation statistics, then so are $t_1(\cdot)+t_2(\cdot)$ and $t_1(\cdot)-t_2(\cdot)$.

(c) If $t_1(\cdot)$ and $t_2(\cdot)$ are translation-invariant statistics, then so are $t_1(\cdot)+t_2(\cdot)$ and $t_1(\cdot)-t_2(\cdot)$.

(d) If $t_1(\cdot)$ and $t_2(\cdot)$ are even, then so are $t_1(\cdot)+t_2(\cdot)$ and $t_1(\cdot)-t_2(\cdot)$.

(e) Consider each of **(a)**–**(d)** for the statistics $t_1(\cdot)t_2(\cdot)$ and $t_1(\cdot)/t_2(\cdot)$.

1.3.13. Let $t_1(X_1,\ldots,X_n)$ and $t_2(X_1,\ldots,X_n)$ be statistics based on the data $(X_1,\ldots,X_n)$.

(a) If $t_1(\cdot)$ is odd and $t_2(\cdot)$ is even, what, if anything, can be said about the even or odd property of: $t_1(\cdot)+t_2(\cdot)$, $t_1(\cdot)-t_2(\cdot)$, $t_1(\cdot)t_2(\cdot)$, and $t_1(\cdot)/t_2(\cdot)$?

(b) If $t_1(\cdot)$ is a translation statistic and $t_2(\cdot)$ is a translation-invariant statistic what can be said about the translation or translation-invariant property of: $t_1(\cdot)+t_2(\cdot)$, $t_1(\cdot)-t_2(\cdot)$, $t_1(\cdot)t_2(\cdot)$, and $t_1(\cdot)/t_2(\cdot)$?

1.3.14. Consider the two-population problem with independent random samples $X_1,\ldots,X_m$ and $Y_1,\ldots,Y_n$. Suppose that $\mathbf{Z}=(X_1,\ldots,X_m,Y_1,\ldots,Y_n)$ is equally symmetric about $\boldsymbol{\mu}=(\mu,\ldots,\mu)$, an $(m+n)$ dimension vector. Show that if $V_1(\mathbf{Z})$ is an odd translation-invariant statistic and $V_2(\mathbf{Z})$ is an even translation-invariant statistic, then the covariance between V_1 and V_2 is zero, if it exists.

1.3.15. Suppose that $\mathbf{Z}=(X_1,\ldots,X_n,Y_1,\ldots,Y_n)$ is a $2n$-dimension vector of exchangeable random variables. Show that if $V_1(\mathbf{Z})$ is a negative interchangeable statistic and $V_2(\mathbf{Z})$ is interchangeable, then the covariance between V_1 and V_2 is zero, if it exists.

1.3.16. For each pair of statistics (V,W) listed below give general conditions under which their covariance will equal zero when it exists. (Note Exercises 1.3.14 and 1.3.15.) One-sample:

(a) Let $k=[\![n\alpha]\!]$ be the greatest integer less than or equal to $n\alpha$, where α satisfies $0<\alpha<1$ and $n\alpha\geqslant 1$. Set

$$V=\tfrac{1}{2}(X_{(n+1-k)}+X_{(k)}),\qquad W=X_{(n+1-k)}-X_{(k)}.$$

Two-sample:

(b)

$$V=\bar{Y}-\bar{X},\qquad W=\frac{(m-1)S_x^2+(n-1)S_y^2}{(m+n-2)},$$

where $S_x^2[S_y^2]$ is the sample variance of the Xs $[Y\text{s}]$ with an $(m-1)$ $[(n-1)]$ in the denominator.

(c) $m=n$ and

$$V=Y_{(1)}-X_{(1)},\qquad W=\tfrac{1}{2}\{X_{(n)}-X_{(1)}+Y_{(n)}-Y_{(1)}\}.$$

1.3.17. Let $X_{(1)} \leqslant \cdots \leqslant X_{(n)}$ be the order statistics for a random sample from a continuous distribution with c.d.f. $F(x)$. Define

$$U = \sum_{i=1}^{n} a_i X_{(i)}$$

and

$$V = \sum_{i=1}^{n} b_i X_{(i)},$$

where $a_1, \ldots, a_n$ and $b_1, \ldots, b_n$ are constants satisfying

$$\sum_{i=1}^{n} a_i = 1 \quad \text{and} \quad a_i = a_{n+1-i}, \qquad i = 1, \ldots, n,$$

and

$$\sum_{i=1}^{n} b_i = 0 \quad \text{and} \quad b_i = -b_{n+1-i}, \qquad i = 1, \ldots, n.$$

(a) State (and justify) general conditions on $F(x)$ such that U has a distribution that is symmetric about some point μ.

(b) State (and justify) general conditions on $F(x)$ such that $\text{Cov}(U, V) = 0$, provided it exists.

1.3.18. Let $\bar{X}_i, S_i^2$, $i = 1, \ldots, k$, be the sample means and variances, respectively, for independent random samples of sizes $n_1, \ldots, n_k$, respectively, from a single, common distribution with c.d.f. $F(x)$. Set

$$Z = \frac{\sum_{i=1}^{k} \bar{X}_i S_i}{S_1 + \cdots + S_k}.$$

If the underlying $F(\cdot)$ distribution is symmetric about a point μ, show that Z is also symmetrically distributed about μ.

1.3.19. Definition 1.3.36. Two random variables (X_1, X_2) are said to have a **jointly symmetric distribution** if for some numbers μ_1 and μ_2

$$(X_1 - \mu_1, X_2 - \mu_2) \stackrel{d}{=} (X_1 - \mu_1, \mu_2 - X_2) \stackrel{d}{=} (\mu_1 - X_1, X_2 - \mu_2)$$
$$\stackrel{d}{=} (\mu_1 - X_1, \mu_2 - X_2).$$

Theorem 1.3.37. *Let $V(\mathbf{X})$ and $W(\mathbf{X})$ be statistics that satisfy* (1.3.30) *for some $g(\cdot)\in\mathfrak{D}^*$ satisfying* (1.3.31). *If, in addition, they satisfy*

$$V(\mathbf{x}) = V(h(\mathbf{x})) \qquad \text{for all } \mathbf{x}\in\mathfrak{X},$$
$$W(\mathbf{x}) - \mu^* = \mu^* - W(h(\mathbf{x})) \qquad \text{for all } \mathbf{x}\in\mathfrak{X},$$

and

$$\mathbf{X} \stackrel{d}{=} h(\mathbf{X})$$

for some number μ^* and some $h(\cdot)$ in $\mathfrak{D}^*$, then $V(\mathbf{X})$ and $W(\mathbf{X})$ have a jointly symmetric distribution.

(a) Prove Theorem 1.3.37.

(b) Find an example of two statistics that satisfy the conditions of Theorem 1.3.37.

1.3.20. Let $(X_1, Y_1),\ldots,(X_n, Y_n)$ be a random sample from some bivariate distribution, and let r_s be the Pearson product moment correlation coefficient for this sample. Show that r_s is an interchangeable statistic in $(X_1,\ldots,X_n)$ and $(Y_1,\ldots,Y_n)$.

1.3.21. Let $X_1,\ldots,X_n$ and $Y_1,\ldots,Y_n$ be independent random samples from two populations.

(a) If $U(X_1,\ldots,X_n)$ is any statistic, show that $U(Y_1,\ldots,Y_n)-U(X_1,\ldots,X_n)$ is negative interchangeable.

(b) Let $U_1(X_1,\ldots,X_n)$ and $U_2(Y_1,\ldots,Y_n)$ be two odd translation statistics. Is $U_2(Y_1,\ldots,Y_n)-U_1(X_1,\ldots,X_n)$ negative interchangeable? Prove or give a counterexample.

1.3.22. Let $\mathbf{Z}=(X_1,\ldots,X_n,Y_1,\ldots,Y_n)$, where $X_1,\ldots,X_n,Y_1,\ldots,Y_n$ are mutually stochastically independent with each X_i symmetrically distributed about some value μ_1 and each Y_j symmetrically distributed about some value μ_2. (Note that neither the Xs nor the Ys need be identically distributed.) If r_s is Pearson's product moment correlation coefficient on $(X_1,Y_1),\ldots,(X_n,Y_n)$, show that:

(a) r_s is symmetrically distributed about 0.

(b) r_s is uncorrelated (provided the correlation exists) with *any* function of $X_1,\ldots,X_n$ alone and likewise with *any* function of $Y_1,\ldots,Y_n$ alone.

2 DISTRIBUTION-FREE STATISTICS

In the areas of hypothesis testing and confidence intervals, the field of nonparametric statistics relies very heavily on a concept known as the distribution-free property. For example, a distribution-free hypothesis test is one for which the significance level remains constant over a class of underlying distributional assumptions, and a distribution-free confidence interval will have a constant confidence coefficient holding over such a class. In this chapter we introduce you to the concept of a distribution-free statistic and illustrate some of the more important ways by which such a statistic can arise. We also demonstrate the application of this concept to hypothesis testing situations, describing several important distribution-free tests that will be used repeatedly in this text.

2.1. Distribution-Free over a Class

Let $V_1,\ldots,V_n$ be random variables with a joint distribution denoted by D, where D is a member of some collection $\mathcal{Q}$ of possible joint distributions. We use $T(V_1,\ldots,V_n)$ to denote some statistic based on $V_1,\ldots,V_n$.

Definition 2.1.1. The statistic T is said to be **distribution-free over $\mathcal{Q}$** if the distribution of T is the same for **every** joint distribution in $\mathcal{Q}$.

Although the applications of this concept are most useful in nonparametric procedures, the idea of the general definition can perhaps be better understood by first looking at some parametric examples. Let $\mathcal{Q}_1$ denote the collection of joint distributions of n i.i.d. normally distributed variables with known mean μ_0 and some unknown variance $0<\sigma^2<\infty$. Thus joint distributions in $\mathcal{Q}_1$ are indexed by σ^2. Let $\mathcal{Q}_2$ be a second collection of joint distributions of n i.i.d. normals indexed by their common unknown mean $-\infty<\mu<\infty$ and possessing common known variance σ_0^2. Define

$$\bar{V}=(1/n)\sum_{i=1}^{n}V_i \quad \text{and} \quad S^2=\frac{\sum_{i=1}^{n}(V_i-\bar{V})^2}{n-1},$$

the sample mean and variance, respectively, for the Vs, and consider the two statistics

$$U_1=n^{1/2}(\bar{V}-\mu_0)/S \quad \text{and} \quad U_2=\frac{(n-1)S^2}{\sigma_0^2}.$$

Thus U_1 is the well-known t-statistic for testing the null hypothesis H_0: $\mu=\mu_0$ and U_2 is used to test H_0': $\sigma^2=\sigma_0^2$. Since under H_0 or, equivalently, $\mathcal{C}_1$, the statistic U_1 has a t-distribution with $(n-1)$ degrees of freedom (regardless of the value of σ^2), we can say that U_1 is distribution-free over $\mathcal{C}_1$. In a similar manner, U_2 is distribution-free over $\mathcal{C}_2$, since it has a chi-square distribution with $n-1$ degrees of freedom when H_0' (or, equivalently, $\mathcal{C}_2$) is true, regardless of the value of μ. Thus each of these statistics has a distribution-free property over a **parametric** class consisting of i.i.d. normal variables.

If a random variable Y has a distribution function of the form $F(y-\theta)$, then θ is termed a **location parameter**. Exercise 1.3.2 shows that if X is distributed according to $F(x)$ and Y has c.d.f. $F(y-\theta)$, then $X \stackrel{d}{=} Y-\theta$. If a random variable Y has a c.d.f. of the form $F(y/\eta)$, for $0<\eta<\infty$, it is said to have a distribution indexed by the **scale parameter** η. Exercise 1.3.2 also shows that if X is distributed according to $F(x)$ and Y has c.d.f. $F(y/\eta)$, then $X \stackrel{d}{=} Y/\eta$. These facts are useful in arguing that a statistic is distribution-free over a class of distributions indexed by either a location or scale parameter, as illustrated in the following example involving a scale parameter.

Example 2.1.2. Let $X_{(1)}\leqslant\cdots\leqslant X_{(n)}$ denote the order statistics for a random sample of size n from a normal distribution with mean zero and variance $0<\sigma^2<\infty$ (unknown). Consider the statistic

$$Q=\frac{(X_{(n)}+X_{(1)})/2}{X_{(n)}-X_{(1)}},$$

namely, the midrange divided by the range. We note that if $Z_1,\ldots,Z_n$ are i.i.d. standard normal variables, then

$$\left(\frac{X_1}{\sigma},\ldots,\frac{X_n}{\sigma}\right) \stackrel{d}{=} (Z_1,\ldots,Z_n),$$

and thus,

$$\left(\frac{X_{(1)}}{\sigma},\ldots,\frac{X_{(n)}}{\sigma}\right) \stackrel{d}{=} (Z_{(1)},\ldots,Z_{(n)}).$$

We also note that

$$Q=\frac{(X_{(n)}+X_{(1)})/2}{X_{(n)}-X_{(1)}}=\frac{[(X_{(n)}/\sigma)+(X_{(1)}/\sigma)]}{(X_{(n)}/\sigma)-(X_{(1)}/\sigma)} \stackrel{d}{=} \frac{(Z_{(n)}+Z_{(1)})/2}{Z_{(n)}-Z_{(1)}}.$$

Thus for any positive σ the statistic Q has the same distribution as if it were constructed using observations from a standard normal distribution. That is, the statistic Q is distribution-free over the class $\mathcal{Q}_3$ of joint distributions of i.i.d. normal variables with mean 0 and common, but arbitrary, variance $0<\sigma^2<\infty$.

Although the statistics U_1, U_2, and Q in the preceding examples are indeed distribution-free over the classes $\mathcal{Q}_1$, $\mathcal{Q}_2$, and $\mathcal{Q}_3$, respectively, we would not typically consider them to be nonparametric statistics, since their distributions are different if the underlying distribution is something other than a normal one. A statistic, say, T, is said to be **nonparametric distribution-free** if the class, say, $\mathcal{Q}$, of joint distributions for which T has the same distribution includes more distributional forms than one. The remainder of this chapter contains a presentation of several of the more useful nonparametric distribution-free statistics and a discussion of why each has this distribution-free property.

Exercises 2.1.

2.1.1. Let $X_{(1)} \leqslant \cdots \leqslant X_{(n)}$ be the order statistics for a random sample from the distribution with p.d.f.

$$\begin{aligned} f(x;\theta) &= e^{-(x-\theta)}, && \theta < x < \infty, \\ &= 0, && \text{elsewhere,} \end{aligned}$$

where $-\infty<\theta<\infty$. Show that the statistic $Z=\sum_{i=1}^{n}(X_{(i)}-X_{(1)})$ is distribution-free over the class of i.i.d. random variables with common density

$f(x;\theta)$, $-\infty<\theta<\infty$.

2.1.2. Let $X_1,\ldots,X_n$ be a random sample from the normal distribution $n(\theta,\sigma_0^2)$, $-\infty<\theta<\infty$ and σ_0^2 known. Let $Z=\Sigma_{i=1}^n a_i X_i$, with $\Sigma_{i=1}^n a_i=0$. Show that any such Z is distribution-free over the class of i.i.d. normal variables with common variance σ_0^2 and mean θ, $-\infty<\theta<\infty$.

2.1.3. Let $X_{(1)}\leqslant\cdots\leqslant X_{(n)}$ be the order statistics for a random sample from a distribution with p.d.f.

$$f(x;\theta)=\frac{1}{\theta}e^{-(x/\theta)}, \qquad 0<x<\infty,$$
$$=0, \qquad \text{elsewhere},$$

where $0<\theta<\infty$. Show that the statistic $Z=nX_{(1)}/\Sigma_{i=1}^n X_{(i)}$ is distribution-free over the class of i.i.d. variables, each with density $f(x;\theta)$, $0<\theta<\infty$.

2.1.4. Let $X_1,\ldots,X_n$ denote a random sample from a logistic distribution centered at 0 with scale parameter $\eta>0$; that is, the c.d.f. of X_i is

$$F\left(\frac{x}{\eta}\right)=\frac{1}{1+e^{-x/\eta}}, \qquad -\infty<x<\infty.$$

Show that the statistic

$$V=\text{median}\,(X_1,\ldots,X_n)/\sum_{i=1}^{n}|X_i-\text{median}\,(X_1,\ldots,X_n)|$$

is distribution-free over this class of underlying distributions.

2.2. Counting Statistics

Possibly the simplest of all nonparametric distribution-free statistics are those based on counts of whether a particular event happens in each of n independent trials. Let $X_1,\ldots,X_n$ be independent random variables such that X_i has a continuous distribution with c.d.f. $F_i(x)$, $i=1,\ldots,n$. Assume that $F_i(\theta)=p_0$, $0<p_0<1$, for $i=1,\ldots,n$ and some unknown θ (that is, each X_i has the same, but unknown, p_0th quantile). Let θ_0 be a known real number and define the statistics

$$\Psi_i=\Psi(X_i-\theta_0), \qquad i=1,\ldots,n, \tag{2.2.1}$$

where $\Psi(t)=1$, 0 as $t>$, $\leqslant 0$.

Theorem 2.2.2. *Let $\Psi_1,\ldots,\Psi_n$ be defined by (2.2.1), and let $S(\Psi_1,\ldots,\Psi_n)$ be any statistic based on $\Psi_1,\ldots,\Psi_n$ only. Then if $\theta=\theta_0$, the statistics $\Psi_1,\ldots,\Psi_n$ are independent identically distributed Bernoulli random variables with parameter $1-p_0$, and $S(\Psi_1,\ldots,\Psi_n)$ is distribution-free over the nonparametric class $\mathcal{C}_4$ consisting of all joint distributions of independent continuous random variables, each with p_0th quantile equal to θ_0.*

Proof: Since Ψ_i is a function of X_i only and $X_1,\ldots,X_n$ are mutually independent, then $\Psi_1,\ldots,\Psi_n$ are mutually independent variables. Moreover, Ψ_i has a Bernoulli distribution with parameter $1-F_i(\theta_0)$, for $i=1,\ldots,n$. Hence for the class $\mathcal{C}_4$, the Ψ_is are independent and identically distributed Bernoulli variables with common parameter $[1-F_i(\theta_0)]=1-p_0$. By definition, then, any statistic $S(\Psi_1,\ldots,\Psi_n)$, based on $\Psi_1,\ldots,\Psi_n$ only, will be distributed as that function of n i.i.d. Bernoulli variables with parameter $1-p_0$, provided that the joint distribution belongs to $\mathcal{C}_4$. This is simply another way of saying that $S(\Psi_1,\ldots,\Psi_n)$ is distribution-free over $\mathcal{C}_4$. ■

In connection with Theorem 2.2.2, a particular function of the Ψ_is that is useful in the one-sample testing situation is

$$B = B(\Psi_1,\ldots,\Psi_n) = \sum_{i=1}^{n} \Psi_i. \tag{2.2.3}$$

This statistic is distributed as a binomial variable with parameters n and $(1-p_0)$, provided the joint distribution is in $\mathcal{C}_4$. In particular, when $p_0=1/2$ and $\mathcal{C}_4$ corresponds to the null hypothesis H_0: [each X_i, $i=1,\ldots,n$, has median θ_0], then $B=\Sigma_{i=1}^{n}\Psi_i$ is known as the **sign test statistic**. The associated sign test is useful for detecting location alternatives for which each X_i has a median that is greater than θ_0. (We could also consider alternatives where the medians are all less than θ_0 or two-sided alternatives encompassing either possibility holding simultaneously for all the medians.) The sign test statistic is seen to be distribution-free under H_0, with a null distribution that is binomial $(n, 1/2)$. Many tables thus exist to help create exact critical regions for this test, and the asymptotic normality of a binomial variable provides approximate critical regions for large sample sizes.

We note that this null distribution of B is discrete. Thus, for example, a test that rejects H_0 for large values of B has *at most* $n+2$ available α-levels. We refer to these as the **natural α-levels** of the test. (By using randomization we could achieve any α-level of interest, but this would not be a desirable practice.) Throughout this text, whenever a test statistic has a discrete distribution we consider only its natural α-levels.

The structural simplicity of the sign test makes it applicable in a wide variety of settings, including the following example, where we deal with paired data.

Example 2.2.4. Barry (1968) conducted an interesting experiment in parapsychology to see whether individuals are able to use thought processes to retard the growth of fungi. Ten individuals took part in this experiment. For each individual there were five experimental and five control fungi cultures, all kept in separate petri dishes. An individual did not come into physical contact with any of the dishes but was placed in a room with all ten and asked to concentrate on the experimental cultures in order to retard the fungi growths in these dishes. The growth for each culture was later analyzed by an individual who did not know which were control and which were experimental dishes.

As a model for this experiment we let $U_i(V_i)$ be the total growth of the experimental (control) cultures for the ith individual in the experiment. We then form

$$X_i = V_i - U_i,$$

and say that the individual succeeded in retarding growth if $X_i > 0$ and failed if $X_i \leqslant 0$. The null hypothesis is that the thought processes had no effect. This would imply that U_i and V_i are i.i.d. continuous random variables, and hence that X_i is a random variable that is continuous and symmetrically distributed about 0; that is, $\theta_0 = 0$ and $p_0 = 1/2$. The alternative of interest is that the growth was retarded, in which case X_i has a greater tendency to be positive than negative. Under H_0, we note that the statistic

$$B = \sum_{i=1}^{10} \Psi_i$$

has a binomial (10, 1/2) distribution. The results of the experiment show that nine of the ten subjects succeeded in retarding the fungi growths. In only one of the ten did the control culture grow as slow or slower than the experimental one. To assess the statistical significance of this finding we compute

$$\begin{aligned} P[B \geqslant 9 | H_0 \text{ true}] &= \sum_{j=9}^{10} \binom{10}{j} \left(\frac{1}{2}\right)^{10} \\ &= .0107. \end{aligned}$$

Thus, this experiment does seem to indicate that thought processes can influence the growth of living organisms.

Exercises 2.2.

2.2.1. Let $X_1,\ldots,X_n$ be independent random variables such that X_i has continuous distribution function $F_i(x)$, where $F_i(\theta)=1/2$ for $i=1,\ldots,n$, with $-\infty<\theta<\infty$ unknown. To test $H_0:\theta=\theta_0$ against the alternative $H_1:\theta>\theta_0$, where θ_0 is specified, the sign test rejects H_0 in favor of H_1 if and only if $B\geqslant b(\alpha, n)$, where $B=$(number of Xs$>\theta_0$) is the sign statistic of Section 2.2 and $b(\alpha, n)$ is the upper 100αth percentile point for the null (H_0) distribution of B (that is, $P_0(B\geqslant b(\alpha, n))=\alpha$). Find an expression for the probability of rejecting H_0 with this test as a function of $F_i(\theta_0)$, $i=1,\ldots,n$. To what does this simplify when we assume $F_1\equiv\cdots\equiv F_n\equiv F$? For this latter case, evaluate the power function of the $n=10$, $\alpha=.0547$ test for $[1-F(\theta_0)]$ values of 0.55, 0.65, 0.75, 0.85, and 0.95.

2.2.2. Using Exercise 2.2.1 as a guide, discuss how to use the sign statistic B to test hypotheses about percentiles other than the median.

2.2.3. Lamp (1976) studied the age distribution of a common mayfly species, **Stenacron interpunctatum**, among various habitats in Big Darby Creek, Ohio. One of the measurements considered was head width (in micrometer divisions, 1 division$=.0345$ mm) and a subset of the data collected from the mayflies in habitat A follows: 36, 31, 30, 27, 20, 33, 27, 18, 19, 28. Using a sign test procedure with significance level $\alpha=.0107$, test the null hypothesis that the median head width for mayflies from habitat A is 22 micrometer divisions against the alternative that it is greater than 22. Find the smallest significance level at which this data would lead to rejection of the null hypothesis.

2.2.4. For the Lamp data of Exercise 2.2.3, test the null hypothesis that the 70th percentile of the distribution of mayfly head widths in habitat A is equal to 32 micrometer divisions against the alternative that it is less than 32. Conduct the test by finding the smallest level at which the data would lead to rejection of the null hypothesis. With this smallest rejection level as the significance level, what is the power of the test against the alternative that 32 is the 80th percentile?

2.2.5. Let $X_1,\ldots,X_n$ be i.i.d. continuous random variables symmetrically distributed about zero. Suppose we measure scale (spread) by $P(|X|>1)$. Describe a test based on counting statistics for H_0: $P[|X|>1]=.5$ versus H_1: $P[|X|>1]>.5$.

2.2.6. Consider the regression model

$$Y_i = \beta d_i + E_i, \qquad i = 1,\ldots,n,$$

where the Y_is are the observed random variables, β is an unknown parameter, the constants $0<d_1 \leqslant \cdots \leqslant d_n$ are known, and $E_1,\ldots,E_n$ are i.i.d. continuous random variables each having a distribution with median 0. Describe a test based on counting statistics for H_0: $\beta=0$ versus H_1: $\beta>0$.

2.3. Ranking Statistics

In this section we consider another common technique for constructing distribution-free statistics, namely, that of ranking sample observations. We discuss this approach in a general setting and then illustrate in detail the specific application of it to the two-sample location problem.

Let $Z_1,\ldots,Z_N$ be a random sample from a continuous distribution with distribution function $F(z)$, and let $Z_{(1)} \leqslant \cdots \leqslant Z_{(N)}$ be the corresponding order statistics.

Definition 2.3.1. The sample observation Z_i is said to have **rank** R_i^* among $Z_1,\ldots,Z_N$ if $Z_i = Z_{(R_i^*)}$, provided the R_i^*th order statistic is uniquely defined.

Remark 2.3.2. If there are two or more sample observations tied in value over the position of the R_i^*th order statistic, then the corresponding ranks are not well defined. In such cases we say that there are tied ranks, and in specific applications of these ranking ideas we require well defined methods for handling such tied ranks. However, under the continuity assumption on $F(z)$, the probability of obtaining any tied ranks is zero, and hence properties about the ranks of sample observations can be obtained without being concerned with methods for handling tied ranks. For this reason (except where specifically noted otherwise) we henceforth assume that the rank of Z_i is uniquely defined by (2.3.1).

Let $\mathbf{R}^* = (R_1^*,\ldots,R_N^*)$, where R_i^* is the rank of Z_i among the observations $Z_1,\ldots,Z_N$. The following theorem contains the important result that enables us to construct distribution-free rank statistics.

Theorem 2.3.3. *Let $Z_1,\ldots,Z_N$ be a random sample from a continuous distribution, and let $\mathbf{R}^* = (R_1^*,\ldots,R_N^*)$ be the vector of ranks, where R_i^* is the rank of Z_i among $Z_1,\ldots,Z_N$. If $\mathcal{R} = \{\mathbf{r} : \mathbf{r}$ is a permutation of the integers $1,\ldots,N\}$, then $\mathbf{R}^*$ is uniformly distributed over $\mathcal{R}$.*

Proof: We note that there are $N!$ elements in $\mathcal{R}$. Thus the theorem is established if we can show that $\mathbf{R}^*$ assumes each of the permutations of $(1,\ldots,N)$ with probability $1/N!$. Let $\mathbf{r}=(r_1,\ldots,r_N)$ be an arbitrary element of $\mathcal{R}$. Then

$$\begin{aligned} P(\mathbf{R}^*=\mathbf{r}) &= P\big[(Z_1,\ldots,Z_N)=(Z_{(r_1)},\ldots,Z_{(r_N)})\big] \\ &= P(Z_{d_1}<\cdots<Z_{d_N}), \end{aligned}$$

where d_i is the position of the number i in the permutation $\mathbf{r}$, for $i=1,\ldots,N$. From Theorem 1.3.5, we know that $(Z_1,\ldots,Z_N) \stackrel{d}{=} (Z_{d_1},\ldots,Z_{d_N})$, which implies (see Exercise 2.3.1) that $P(Z_{d_1}<\cdots<Z_{d_N})=P(Z_1<\cdots<Z_N)$. Thus we have

$$P(\mathbf{R}^*=\mathbf{r}) = P(Z_1<\cdots<Z_N) = P(\mathbf{R}^*=\mathbf{r}_0),$$

where $\mathbf{r}_0=(1,\ldots,N)$. Since there are $N!$ elements in $\mathcal{R}$, the result follows from the fact that $\mathbf{r}$ is arbitrary. ■

The following are important consequences of Theorem 2.3.3.

Corollary 2.3.4. Let $Z_1,\ldots,Z_N$ denote a random sample from a continuous distribution, and let $\mathbf{R}^*$ be the corresponding vector of ranks, where R_i^* is the rank of Z_i among the N random variables. Then

$$\begin{aligned} P[R_i^*=r] &= \frac{1}{N}, \qquad \text{for} \quad r=1,\ldots,N, \\ &= 0, \qquad \text{elsewhere,} \end{aligned}$$

and if $i\neq j$,

$$\begin{aligned} P[R_i^*=r, R_j^*=s] &= \frac{1}{N(N-1)}, \qquad \text{if } r \text{ and } s \text{ are integers satisfying} \\ & \qquad\qquad r\neq s=1,\ldots,N, \\ &= 0, \qquad \text{elsewhere.} \end{aligned}$$

Proof: Exercise 2.3.3. ■

Corollary 2.3.5. Let $\mathbf{R}^*$ be the vector of ranks corresponding to a random sample of size N from a continuous distribution. Then

$$\begin{aligned} E[R_i^*] &= \frac{N+1}{2}, \qquad i=1,\ldots,N, \\ \operatorname{Var}[R_i^*] &= \frac{(N+1)(N-1)}{12}, \end{aligned}$$

and for $i \neq j$,

$$\operatorname{Cov}\left[R_i^*, R_j^*\right] = \frac{-(N+1)}{12}.$$

Proof: Exercise 2.3.4. ■

Corollary 2.3.6. Let $Z_1, \ldots, Z_N$ be a random sample from a continuous distribution, and let $\mathbf{R}^* = (R_1^*, \ldots, R_N^*)$ be the vector of ranks as defined in Theorem 2.3.3. If $V(\mathbf{R}^*)$ is a statistic based on $(Z_1, \ldots, Z_N)$ only through $\mathbf{R}^*$, then $V(\mathbf{R}^*)$ is distribution-free over the class $\mathcal{Q}_5$ of joint distributions of N i.i.d. continuous univariate random variables.

Proof: The result is immediate from Theorem 2.3.3, since for any $D \in \mathcal{Q}_5$ the statistic $V(\mathbf{R}^*)$ is distributed as that function of an N-dimensional vector distributed uniformly over $\mathfrak{R}$. ■

A statistic such as $V(\mathbf{R}^*)$ that is a function of $Z_1, \ldots, Z_N$ only through the rank vector $\mathbf{R}^*$ is called a **rank statistic**. Much of the field of nonparametric statistics is based on such variables, particularly in the areas of hypothesis testing and confidence intervals. Their importance is amply demonstrated throughout the remainder of the text. Thus for purposes of illustration only it suffices to consider in detail their role in the construction of distribution-free hypothesis tests for the two-sample location problem.

Let $X_1, \ldots, X_m$ and $Y_1, \ldots, Y_n$ be independent random samples from continuous distributions with distribution functions $F(x)$ and $G(x) = F(x - \Delta)$, respectively, where $-\infty < \Delta < \infty$ is an unknown shift parameter. The null hypothesis of interest for this setting is $H_0: \Delta = 0$, and the alternative to H_0 is one of the three alternatives $H_1: \Delta >$, $<$, or $\neq 0$. For ease of illustration we discuss rank tests for this problem only for the alternative $H_1: \Delta > 0$, simply noting important differences for the other alternatives ($\Delta < 0$ or $\Delta \neq 0$), as necessary.

Let Q_i, $i = 1, \ldots, m$ and R_j, $j = 1, \ldots, n$ be the ranks of X_i and Y_j, respectively, among the $N = m + n$ combined X and Y observations. That is, R_j is the rank of Y_j among the m Xs and n Ys combined and treated as a single set of observations, and similarly for Q_i. Of course, this implies that the rank vector $\mathbf{R}^* = (Q_1, \ldots, Q_m, R_1, \ldots, R_n)$ is simply a permutation of $(1, \ldots, N)$ and although random, it must satisfy the constraint

$$\sum_{i=1}^{m} Q_i + \sum_{j=1}^{n} R_j = \sum_{i=1}^{N} i = \frac{N(N+1)}{2}.$$

Now, to construct a hypothesis test of H_0: $\Delta = 0$ (against the alternative

H_1: $\Delta > 0$) that is based on these combined sample rankings, we would naturally be interested in comparisons of those ranks assigned to the Xs, namely the subvector $(Q_1, \ldots, Q_m)$, and those obtained by the Y observations, namely, $(R_1, \ldots, R_n)$. Note that if we know which ranks are assigned to the Ys, then we also know which ones are assigned to the Xs. Hence we may concentrate on one of the subvectors of ranks, and we choose to use the ranks of the Y observations $(R_1, \ldots, R_n)$.

To construct a suitable critical region for testing H_0 against H_1: $\Delta > 0$ that is based on $(R_1, \ldots, R_n)$, we must decide which values of this rank vector are indicative of H_1. Intuitively, we would want to have such a rejection region contain those rank vectors $(R_1, \ldots, R_n)$ that correspond, in some manner, to large ranks for the Ys; that is, to large values for the Ys relative to the Xs. Of course, what we mean by "corresponding" to large Y ranks will depend on the rank statistic we use to create this rejection region. Many two-sample rank statistics for this location problem have been proposed, and we consider a large class of them in detail in Chapter 9. Here, however, we concentrate on a particular member of this class, called the rank sum statistic, first proposed and studied independently by Wilcoxon (1945) and Mann and Whitney (1947) in what were pioneering papers that initiated the era of modern nonparametric statistics.

The test statistic proposed by Wilcoxon is

$$W = \sum_{i=1}^{n} R_i; \tag{2.3.7}$$

that is, W is the sum of the ranks for the Y sample observations when ranked among all $m+n$ observations. The statistic suggested by Mann and Whitney is

$$U = \sum_{i=1}^{m} \sum_{j=1}^{n} \Psi(Y_j - X_i), \tag{2.3.8}$$

where $\Psi(t) = 1, 0$ as $t >, \leqslant 0$. It represents the total number of times a Y observation is larger than an X observation. When there are no ties among X and Y observations, W and U are linearly related by

$$W = U + \frac{n(n+1)}{2}. \tag{2.3.9}$$

(See Exercise 2.3.2.) Therefore, it suffices to examine W in detail.

Remark 2.3.10. The statistic W is clearly a function of the data only through the rank vector $\mathbf{R}^* = (Q_1, \ldots, Q_m, R_1, \ldots, R_n)$. Under H_0: $\Delta = 0$, the observations $X_1,\ldots,X_m$, $Y_1,\ldots,Y_n$ are i.i.d. continuous random variables, and by Corollary 2.3.6 the rank statistic W is nonparametric distribution-free over this broad class.

Theorem 2.3.11. *Let W be the Mann-Whitney-Wilcoxon rank sum statistic for testing H_0: $\Delta=0$ versus H_1: $\Delta>0$, when $X_1,\ldots,X_m$ and $Y_1,\ldots,Y_n$ are independent random samples from $F(x)$ and $F(y-\Delta)$, respectively. Under H_0: $\Delta=0$ the discrete distribution of W is given by*

$$P_0[W=w] = \frac{t_{m,n}(w)}{\binom{N}{n}}, \quad w = \frac{n(n+1)}{2}, \quad \frac{n(n+1)}{2}+1,\ldots,\frac{n(2m+n+1)}{2}$$

$$= 0, \qquad \text{elsewhere,} \tag{2.3.12}$$

where $t_{m,n}(w)$ is the number of unordered subsets of n integers taken (without replacement) from $\{1,\ldots,N\}$ for which the sum is equal to w.

Proof: Exercise 2.3.6. ■

Thus to evaluate $P_0(W=w)$ for given m and n, we need to look at all $\binom{N}{n}$ possible arrangements of the X and Y observations, calculate the value of W for each such arrangement, and tabulate the function $t_{m,n}(w)$. We illustrate this process for the case $m=3$, $n=2$.

Example 2.3.13. For the case $m=3$ and $n=2$, there are $\binom{3+2}{2}=10$ arrangements to be examined in order to obtain the null distribution of W. These arrangements, together with the associated values of W, are:

Arrangement	Value of W
$xxxyy$	$W=9$
$xxyxy$	$W=8$
$xyxxy$	$W=7$
$yxxxy$	$W=6$
$xxyyx$	$W=7$
$xyxyx$	$W=6$
$yxxyx$	$W=5$
$xyyxx$	$W=5$
$yxyxx$	$W=4$
$yyxxx$	$W=3$.

Hence from these facts and expression (2.3.12), the null distribution of W

for $m=3$ and $n=2$ is

$$P_0(W=9) = P_0(W=8) = P_0(W=4) = P_0(W=3) = \frac{1}{10}$$

$$P_0(W=7) = P_0(W=6) = P_0(W=5) = \frac{2}{10}.$$

For any values of m and n, the procedure illustrated in Example 2.3.13 can be used to table the null distribution of W. Fortunately, these values have been computed and tabled for many m and n combinations. For example, the null distribution of W is available in Table A.5 in Hollander and Wolfe (1973) for $(m+n) \leqslant 20$. For large values of m and n there is also a good approximation to the null distribution of W based on the standard normal distribution. We consider such large sample properties for W in more detail in Chapters 3 and 9. In addition, Mann and Whitney developed a recursion formula for the null distribution of their U statistic which, because of the linear relationship between U and W (see Exercise 2.3.2), can be used in a similar capacity for developing the null distribution of W. This recursion relationship is discussed in Exercise 2.3.10.

For testing H_0: $\Delta=0$ against the alternative H_1: $\Delta>0$ we would want to reject H_0 for large values of W; that is, our test would be of the form: reject H_0: $\Delta=0$ in favor of H_1: $\Delta>0$ if and only if $W \geqslant w(\alpha, m, n)$, where $w(\alpha, m, n)$ is the critical value and is dependent on the desired natural significance level α and the sample sizes m and n. In fact, $w(\alpha, m, n)$ is the upper 100 αth percentile for the null distribution of W, a number that does not depend on the specific parametric form of the distribution function $F(x)$ as long as it is the c.d.f. of a continuous random variable. The values $w(\alpha, m, n)$ are tabled for m and n both $\leqslant 50$ and for certain popular α values by Wilcoxon, Katti and Wilcox (1970).

We now illustrate the application of this one-sided rank sum test to some real data.

Example 2.3.14. In developmental psychology researchers are often interested in studying when infants develop or acquire certain behaviors. If a newborn infant is held under his arms and his bare feet are permitted to touch a flat surface, he will perform well-coordinated walking movements similar to those of an adult. If the dorsa of his feet are drawn against the edge of a flat surface, he will perform placing movements, much like those of a kitten. These two reflexes are referred to as the walking and the placing reflex, respectively. Zelzano, Zelzano, and Kolb (1972) reported on a study that investigated, among other things, whether "active-exercise" of these reflexes could, in fact, shorten the time it takes for an infant to learn to walk alone. Twelve white, 1-week-old male infants from middle-class and upper-middle-class families were used in that portion of their study. Six infants were randomly assigned to the experimental active-exercise group and to the no-exercise control group. Infants in the active-exercise group received stimulation of the walking and placing reflexes during four 3-minute sessions each day from the beginning of the second through the end of the eighth week. (The walking and placing reflexes normally disappear by about eight weeks.) Infants in the no-exercise control received no special training. At the end of eight weeks all the parents were informed of the research objectives and asked to report the age at which their infants first walked alone. These results are given in Table 2.3.15.

Calling the no-exercise group the Y sample and the active-exercise group the X sample, we are here interested in testing H_0: $\Delta=0$ against the alternative H_1: $\Delta>0$, where Δ corresponds to the shift parameter discussed previously. Our level α test based on the Wilcoxon rank sum statistic W is to reject H_0 in favor of $\Delta>0$ iff $W \geqslant w(\alpha, 6, 6)$. If $\alpha=.032$, we see from Table A.5 in Hollander and Wolfe (1973) that $w(.032, 6, 6)=51$. Ordering the combined sample of twelve observations yields 9.00, 9.00, 9.50, 9.50, 9.75, 10.00, 11.50, 11.50, 12.00, 13.00, 13.00, 13.25. Since there are some tied observations (because the time of first walking alone was not reported with sufficient accuracy), we have some ambiguity in assigning the ranks. One method of handling this problem is to assign average ranks to a group of tied observations. Using this method on the walking data, we see that the ranks for the Y observations are 1.5, 7.5, 7.5, 9, 10.5, and 12. Hence the value of W for this data is $W=1.5+7.5+7.5+9+10.5+12=48$. Since $48<w(.032, 6, 6)=51$, we would not reject the null hypothesis H_0: $\Delta=0$ with this data. In fact, the smallest level at which the data would have led to rejection of H_0 is $\underline{\alpha}=P(W \geqslant 48)=.090$.

TABLE 2.3.15.
Ages At Which Infants First Walked Alone (in months)

ACTIVE-EXERCISE GROUP	NO-EXERCISE GROUP
9.00	11.50
9.50	12.00
9.75	9.00
10.00	11.50
13.00	13.25
9.50	13.00

Source: P. R. Zelzano, N. A. Zelzano, and S. Kolb (1972).

We now turn our attention to a brief discussion of the corresponding tests of H_0: $\Delta=0$ against the alternatives $\Delta<0$ or $\Delta\neq 0$. To aid in determining the critical regions for these cases, we utilize the following theorem.

Theorem 2.3.16. *Let $W=\Sigma_{i=1}^{n} R_i$ be the Wilcoxon rank sum statistic. When H_0: $\Delta=0$ is true (i.e., $F\equiv G$), the distribution of W is symmetric about the value $\mu=n(m+n+1)/2$.*

Proof: Recall that under H_0 the rank vector $\mathbf{R}^*=(Q_1,\ldots,Q_m, R_1,\ldots,R_n)$ is uniformly distributed over $\mathfrak{R}$, the set of permutations of the integers $(1,\ldots,N)$, where $N=m+n$. Note that under H_0, $-X_1,\ldots,-X_m$, $-Y_1,\ldots,-Y_n$ are also i.i.d. continuous random variables and so their rank vector is also uniform over $\mathfrak{R}$. But multiplying by -1 makes the largest value the smallest and so forth and hence reverses the order of ranking. Thus the rank vector for $-X_1,\ldots,-X_m$, $-Y_1,\ldots,-Y_n$, written in terms of the components of the original $\mathbf{R}^*$, is $(N+1-Q_1,\ldots,N+1-Q_m, N+1-R_1,\ldots,N+1-R_n)$. Therefore,

$$(Q_1,\ldots,Q_m, R_1,\ldots,R_n) \overset{d}{=} (N+1-Q_1,\ldots, N+1-Q_m, N+1-R_1,\ldots,N+1-R_n).$$

Computing the rank sum statistic on each side of this distributional equation and applying Theorem 1.3.7 yields

$$\sum_{i=1}^{n} R_i \overset{d}{=} \sum_{i=1}^{n} (N+1-R_i) = n(N+1) - \sum_{i=1}^{n} R_i.$$

Thus under H_0,

$$W-[n(N+1)/2] \overset{d}{=} [n(N+1)/2]-W. \quad \blacksquare$$

For testing H_0: $\Delta=0$ against H_1: $\Delta<0$ we would want a critical region of the form: reject H_0 in favor of $\Delta<0$ if and only if $W\leqslant w'(\alpha, m, n)$, where α is a natural significance level for the test and $w'(\alpha, m, n)$ is the 100αth percentile for the null distribution of W. From Theorem 2.3.16, we see that $w'(\alpha, m, n)=n(m+n+1)-w(\alpha, m, n)$, and hence our rejection region for the alternative $\Delta<0$ would be $W\leqslant n(m+n+1)-w(\alpha, m, n)$. In a similar fashion, the critical region for an α-level test of H_0: $\Delta=0$ against the two-sided alternative H_1: $\Delta\neq 0$ would have form: reject H_0 in favor of $\Delta\neq 0$ iff $W\geqslant w(\alpha/2, m, n)$ or $W\leqslant n(m+n+1)-w(\alpha/2, m, n)$ or, equivalently,

$$\left|W-\frac{n(m+n+1)}{2}\right| \geqslant \left[w(\alpha/2, m, n)-\frac{n(m+n+1)}{2}\right].$$

Theorem 2.3.16 also provides us with the null mean of W. When H_0: $\Delta=0$ is true, the symmetry of its distribution and the boundedness of W imply that the null (H_0: $\Delta=0$) mean of W is

$$E_0(W) = \frac{n(m+n+1)}{2}. \tag{2.3.17}$$

In addition, it can be shown that the null (H_0: $\Delta=0$) variance of W is

$$\mathrm{Var}_0(W) = \frac{mn(m+n+1)}{12}. \tag{2.3.18}$$

(See Exercise 2.3.8.)

Exercises 2.3.

2.3.1. Let $(X_1,\ldots,X_N) \stackrel{d}{=} (Y_1,\ldots,Y_N)$. Show that $P(X_1<\cdots<X_N)=P(Y_1<\cdots<Y_N)$.

2.3.2. Let W and U be the Wilcoxon and Mann-Whitney statistics, respectively, based on m Xs and n Ys. If there are no ties between the X and Y observations, show that $W=U+(n(n+1)/2)$. (Hence, the two statistics are equivalent for purposes of testing H_0: $\Delta=0$.)

2.3.3. Prove Corollary 2.3.4.

2.3.4. Prove Corollary 2.3.5.

2.3.5. Let $Z_1,\ldots,Z_N$ be i.i.d. continuous random variables with rank vector $\mathbf{R}^*$, where R_i^* denotes the rank of Z_i among $Z_1,\ldots,Z_N$. Assume $N \geqslant 2$. Consider $V=R_1-R_N$. Show that

$$\begin{aligned} P[V=k] &= \frac{N-|k|}{N(N-1)}, && \text{if } |k| = 1,\ldots,N-1, \\ &= 0, && \text{elsewhere.} \end{aligned}$$

2.3.6. Prove Theorem 2.3.11.

2.3.7. Consider the two-sample location problem setting of Theorem 2.3.11. Describe the nonparametric distribution-free property of each of the following statistics under H_0: $\Delta=0$ by verifying that each is a rank statistic. (See Corollary 2.3.6. You need not attempt to obtain the actual H_0 distribution.)

(a) $V_1=$[the number of Ys greater than the largest X].

(b) V_2=[the number of Ys greater than the median of the X observations], when m is an odd integer.

(c) V_3=[the number of Ys greater than the median of the combined sample of all $m+n$ observations].

2.3.8. Use the results of Corollaries 2.3.4 and 2.3.5 to verify the formulas for $E_0(W)$ and $\text{Var}_0(W)$ given in (2.3.17) and (2.3.18), respectively.

2.3.9. Let $X_1,\ldots,X_m$ and $Y_1,\ldots,Y_n$ be independent random samples from continuous distributions with distribution functions $F(x)$ and $G(y)$, respectively, and let $\mathbf{R}=(R_1,\ldots,R_n)$ be the vector of ranks for the Ys in the combined sample ranking of the $N=(m+n)$ X and Y observations. Let $R_{\text{max}}=\text{maximum}\{R_1,\ldots,R_n\}$. Show that R_{max} is distribution-free under the assumption $F\equiv G$. Evaluate this distribution for $m=3$ and $n=3$.

2.3.10. Let $t_{m,n}(w)$ be the counter used in Theorem 2.3.11 to describe the null $H_0:\Delta=0$ distribution of the Mann-Whitney-Wilcoxon rank sum statistic W for sample sizes of m Xs and n Ys.

(a) Show that the recurrence relation

$$t_{m,n}(w) = t_{m,n-1}(w-m-n) + t_{m-1,n}(w)$$

holds for all $w=(n(n+1)/2),\ldots,(n(2m+n+1)/2)$ and all integer valued m and n if the following initial and boundary conditions are defined for all $i=1,\ldots,m$ and $j=1,\ldots,n$:

$$t_{i,j}(w) = 0, \qquad \text{for all } w < \frac{j(j+1)}{2}$$

$$t_{i,0}(0) = 1, \qquad t_{0,j}\left(\frac{j(j+1)}{2}\right) = 1$$

$$t_{i,0}(w) = 0, \qquad \text{for all } w \neq 0$$

$$t_{0,j}(w) = 0, \qquad \text{for all } w \neq \frac{j(j+1)}{2}.$$

(b) Use the result in **(a)** to obtain a corresponding recursion relation for the null ($H_0:\Delta=0$) distribution of W.

(c) Using the formulas from **(b)**, develop the null ($H_0:\Delta=0$) distribution of W for all combinations of $1\leqslant m,\ n\leqslant 3$.

2.3.11. Let $(Z_1,\ldots,Z_N)$ be a vector of jointly continuous random variables such that $P(Z_1<\cdots<Z_N)=P(Z_{d_1}<\cdots<Z_{d_N})$, for every permuta-

tion $(d_1,\ldots,d_N)$ of $(1,\ldots,N)$. (Thus every arrangement of the Z variables is equally likely.)

(a) Show that Theorem 2.3.3 still holds for this less restrictive assumption on the Zs. What about the ensuing results in Corollaries 2.3.4, 2.3.5, and 2.3.6, Remark 2.3.10, Theorem 2.3.11, and Theorem 2.3.16?

(b) Comment on the significance, with regard to applications, of the more general results in **(a)**.

2.3.12. Let $(Z_1,\ldots,Z_N)$ be a vector of exchangeable (see Definition 1.3.6) random variables.

(a) Show that the condition of Exercise 2.3.11 pertains for such Zs (and hence the conclusions of that exercise are applicable).

(b) Let $(Z_1,\ldots,Z_N)$ be a random vector having a multivariate normal distribution with mean vector $\boldsymbol{\mu}=(\mu_1,\ldots,\mu_N)$ and covariance matrix $\boldsymbol{\Sigma}$ with elements $\sigma_{ij}=\sigma^2$ for $i=j=1,\ldots,N$, and $\sigma_{ij}=\rho\sigma^2$, for $i\neq j=1,\ldots,N$. Show that $(Z_1,\ldots,Z_N)$ is exchangeable, provided $\mu_1=\mu_2=\cdots=\mu_N$.

(c) Let $(Z_1,\ldots,Z_m,Z_{m+1},\ldots,Z_{m+n})$ have a multivariate normal distribution with mean vector $\boldsymbol{\mu}=(\mu_1,\ldots,\mu_{m+n})$ and covariance matrix $\boldsymbol{\Sigma}$ having the form given in **(b)**. Suppose

$$\begin{aligned} \mu_i &= \theta_1, & i &= 1,\ldots,m, \\ &= \theta_2, & i &= m+1,\ldots,m+n. \end{aligned}$$

Discuss how the results in **(a)** and **(b)** can be employed to test the null hypothesis $H_0:\theta_1=\theta_2$. Be sure to consider the forms of the appropriate critical regions for various alternatives to H_0.

2.3.13. Let $X_1,\ldots,X_m$ and $Y_1,\ldots,Y_n$ be $(m+n)$ random variables. Consider a random walk starting at the origin $(0,0)$ and proceeding to the point (n,m) by a series of $(m+n)$ steps, where the ith step is one unit up or to the right according to whether the ith order statistic (assuming no tied Xs and/or Ys) from the combined $(m+n)$ observations is an X or a Y, respectively. Let U be the Mann-Whitney-Wilcoxon statistic. What is the relation between the area under the path of the random walk and the value of U?

2.3.14. The data in Table 2.3.19 are a subset of the data obtained by Poland, Smith, Kuntzman, Jacobson, and Conney (1970) in an experiment concerned with the effect of occupational exposure to DDT on human drug and steroid metabolism. The DDT-exposed subjects were employees

TABLE 2.3.19.
6β-Hydroxycortisol Excretion (μg/24 hours)

CONTROL SUBJECTS	DDT FACTORY WORKERS
41	254
208	171
191	345
118	134
355	190
245	447
200	106
76	173
133	449
213	198

Source: A. Poland, D. Smith, R. Kuntzman, M. Jacobson, and A. H. Conney (1970).

of the Montrose Chemical Corporation who had been working in the DDT plant at Torrance, California, for more than five years. All these individuals had received moderate-to-intense occupational exposure to DDT, and all were in good health. The control subjects were volunteer policemen and firemen in Brownsville, Texas, together with officials from the United States Public Health Service in Atlanta, Georgia. One of the measures considered in the study was the 24-hour urinary excretion of 6β-hydroxycortisol. These values are reported in Table 2.3.19 for twenty of the studied subjects.

The intent of the investigation was to examine whether prolonged occupational exposure to DDT stimulated drug and steroid metabolism in the body. Using the rank sum statistic W and data in Table 2.3.19, test the conjecture that DDT exposure increases the hydroxylation of cortisol by the liver, resulting in increased urinary excretion of 6β-hydroxycortisol. Use a significance level of $\alpha = .032$. In addition, find the smallest significance level at which this data would lead to rejection of the appropriate null hypothesis.

2.3.15. Let $X_1, \ldots, X_m$ and $Y_1, \ldots, Y_n$ be independent random samples from continuous distributions with distribution functions $F(x)$ and $G(x) = F(x-\Delta)$, respectively. If Δ_0 is some known number, discuss how you would use the Wilcoxon rank sum statistic to test (in a nonparametric distribution-free manner) the null hypothesis $H_0: \Delta = \Delta_0$ against the alternatives $\Delta <, >$, or $\neq \Delta_0$.

2.4. Statistics Utilizing Counting and Ranking

A third basic technique for constructing distribution-free statistics combines the two techniques previously considered in Sections 2.2 and 2.3, namely, those of counting and ranking. In this section we consider a general setting in which such a combination is useful and then discuss more thoroughly the specific application of it to the one-sample location problem.

Let $Z_1,\ldots,Z_n$ be a random sample from a continuous distribution that is symmetric about zero. (Note: The point of symmetry could be any *known* value. If X is symmetrically distributed about θ_0, $Z=X-\theta_0$ is symmetrically distributed about 0.) Define

$$\Psi_i = \Psi(Z_i), \tag{2.4.1}$$

where $\Psi(t)=1,0$ as $t>, \leqslant 0$. The result that enables us to combine the counting and ranking techniques under the assumption of population symmetry will be established in several steps.

Lemma 2.4.2. *Let Z be a continuous random variable with a distribution that is symmetric about* 0. *Then the random variables $|Z|$ and $\Psi=\Psi(Z)$ are stochastically independent.*

Proof: Let $F(z)$ be the c.d.f. of Z. Then, for $t\geqslant 0$,

$$\begin{aligned}
P(\Psi=1,|Z|\leqslant t) &= P(Z>0,|Z|\leqslant t)\\
&= P(0<Z\leqslant t) = F(t)-F(0), \text{ since } Z \text{ is continuous,}\\
&= \tfrac{1}{2}\left[F(t)-F(-t)\right], \text{ since the distribution of } Z \text{ is}\\
&\qquad \text{symmetric about 0,}\\
&= \tfrac{1}{2}P(-t\leqslant Z\leqslant t) = \tfrac{1}{2}P(|Z|\leqslant t)\\
&= P(\Psi=1)P(|Z|\leqslant t), \text{ since 0 is the median of the } Z\\
&\qquad \text{distribution.}
\end{aligned}$$

Similarly,

$$\begin{aligned}
P(\Psi=0,|Z|\leqslant t) &= P(|Z|\leqslant t)-P(\Psi=1,|Z|\leqslant t)\\
&= P(|Z|\leqslant t)\left[1-P(\Psi=1)\right], \text{ from the previous argument,}\\
&= P(|Z|\leqslant t)P(\Psi=0).
\end{aligned}$$

Hence $|Z|$ and Ψ are stochastically independent, as desired. ∎

Definition 2.4.3. For any random variables $Z_1,\ldots,Z_n$, the **absolute rank** of Z_i, denoted by R_i^+, is the rank of $|Z_i|$ among $|Z_1|,\ldots,|Z_n|$. The **signed rank** of Z_i is then $\Psi_i R_i^+$, where Ψ_i is defined in (2.4.1).

We see that the signed rank of a positive observation is simply its absolute rank, but the signed rank of a nonpositive observation is zero. Many important distribution-free statistics are functions of $\mathbf{\Psi}=(\Psi_1,\ldots,\Psi_n)$ and the absolute rank vector $\mathbf{R}^+=(R_1^+,\ldots,R_n^+)$; often these statistics can be written in terms of the signed ranks $\Psi_1 R_1^+,\ldots,\Psi_n R_n^+$. Such statistics are referred to as **signed rank statistics.** The following theorem establishes an important joint distributional structure for $\mathbf{\Psi}$ and $\mathbf{R}^+$.

Theorem 2.4.4. *Let $Z_1,\ldots,Z_n$ denote a random sample from a continuous distribution that is symmetric about zero. Let $\mathbf{R}^+$ denote the vector of absolute ranks of the Z_is. If Ψ_i is defined by* (2.4.1) *for $i=1,\ldots,n$, then the $n+1$ random variables $\Psi_1,\ldots,\Psi_n$, $\mathbf{R}^+$ are mutually independent. Moreover, each Ψ_i is a Bernoulli random variable with $p=1/2$, and $\mathbf{R}^+$ is uniformly distributed over $\mathcal{R}$, the set of all permutations of the integers* $(1,\ldots,n)$.

Proof: Since $Z_1,\ldots,Z_n$ are independent, it follows from Lemma 2.4.2 that $\Psi_1,|Z_1|,\ldots,\Psi_n,|Z_n|$ are $2n$ mutually independent random variables. Each Ψ_i is a Bernoulli variable with parameter $p_i=P[Z_i>0]=1/2$, since Z_i is continuous and symmetrically distributed about zero. The vector $\mathbf{R}^+$ is independent of $\Psi_1,\ldots,\Psi_n$ because it is a function only of $|Z_1|,\ldots,|Z_n|$. Since it is the rank vector of n i.i.d. continuous random variables, Theorem 2.3.3 shows that $\mathbf{R}^+$ is uniformly distributed over the permutations of the integers $(1,\ldots,n)$. ■

Corollary 2.4.5. Let $S(\mathbf{\Psi},\mathbf{R}^+)$ be a statistic that depends on the observations $Z_1,\ldots,Z_n$ only through $\Psi_1,\ldots,\Psi_n$ and $\mathbf{R}^+$. Then the statistic $S(\cdot)$ is distribution-free over $\mathcal{Q}_6$, the collection of joint distributions of n i.i.d. continuous random variables, each symmetrically distributed about zero.

Proof: It follows from Theorem 2.4.4, since $\mathbf{\Psi}$ and $\mathbf{R}^+$ have the same joint distribution for every joint distribution in $\mathcal{Q}_6$. ■

More often than not, statistics that are functions of $\mathbf{\Psi}$ and $\mathbf{R}^+$ can be written as symmetric functions of the signed ranks $\Psi_1 R_1^+,\ldots,\Psi_n R_n^+$. When this is the case, the following theorem can play a key role in establishing the distribution of such a statistic.

Theorem 2.4.6. *Let $Z_1,\ldots,Z_n$ be a random sample from a continuous distribution that is symmetric about* 0. *Let Q be the number of positive Zs,* *and for $Q=q$, let $S_1 < \cdots < S_q$ denote the ordered absolute ranks of those Zs that are positive (i.e., $S_1 < \cdots < S_q$ are the positive signed ranks in numerical order). Then*

$$P(Q=q, S_1=s_1,\ldots,S_q=s_q) = (1/2)^n, \quad \text{for } q=0,1,\ldots,n \text{ and each of the } q\text{-tuples } (s_1,\ldots,s_q) \text{ such that } s_i \text{ is an integer and } 1 \leqslant s_1 < \cdots < s_q \leqslant n,$$

$$= 0, \quad \text{elsewhere.} \tag{2.4.7}$$

Proof: Let $q \in \{0,1,\ldots,n\}$, and let $(s_1,\ldots,s_q)$ be an arbitrary q-tuple such that s_i is an integer and $1 \leqslant s_1 < \cdots < s_q \leqslant n$. Define $s_{q+1}^* < \cdots < s_n^*$ to be the integers, written in numerical order, in $\{1,\ldots,n\}$ that are not included in $\{s_1,\ldots,s_q\}$. From Theorem 2.4.4 we see that the independence of $\mathbf{\Psi}$ and $\mathbf{R}^+$ and the corresponding distribution of each yields

$$P\big[\Psi_1=1, R_1^+=s_1,\ldots,\Psi_q=1, R_q^+=s_q, \Psi_{q+1}=0, R_{q+1}^+=s_{q+1}^*,\ldots,\Psi_n=0, R_n^+=s_n^*\big] = \left(\frac{1}{2}\right)^n \frac{1}{n!}. \tag{2.4.8}$$

However, this is only one $\mathbf{\Psi}$ and $\mathbf{R}^+$ combination that yields $\{Q=q, S_1=s_1,\ldots,S_q=s_q\}$. There are $\binom{n}{q}q!(n-q)!=n!$ such combinations. Moreover, Theorem 2.4.4 shows that each such combination is equally likely. Hence

$$P\big[Q=q, S_1=s_1,\ldots,S_q=s_q\big] = \left(\frac{1}{2}\right)^n. \quad \blacksquare \tag{2.4.9}$$

Remark 2.4.10. Let $X_1,\ldots,X_n$ be a random sample from a continuous distribution that is symmetric about a known point θ_0. Then the variables $Z_i = X_i - \theta_0$, $i=1,\ldots,n$ form a random sample from a continuous distribution that is symmetric about 0. Hence we again emphasize that all the preceding discussion and theory apply equally well to the $(X_i-\theta_0)$s. Consequently, we talk about distribution-free signed rank statistics over the class of continuous, univariate distributions that are symmetric about θ_0, where θ_0 is any known real number.

Let $X_1,\ldots,X_n$ be a random sample from a continuous distribution that is symmetric about the unknown median θ, $-\infty<\theta<\infty$, and has distribution function $F(x)$. Let θ_0 be known, and consider testing H_0: $\theta=\theta_0$ against one of the three alternatives H_1: $\theta >, <$, or $\neq \theta_0$. (As with the two-sample setting of Section 2.3, we concentrate on the alternative $\theta > \theta_0$, while mentioning the appropriate changes for the other alternatives, as necessary.)

Define $Z_i = X_i - \theta_0$, $i = 1, \ldots, n$, and let $\Psi_1 R_1^+, \ldots, \Psi_n R_n^+$ be the signed ranks of $Z_1, \ldots, Z_n$, respectively. When the null hypothesis H_0: $\theta = \theta_0$ is true, we see from Corollary 2.4.5 and Remark 2.4.10 that any statistic that is a function of $Z_1, \ldots, Z_n$ only through $\Psi_1 R_1^+, \ldots, \Psi_n R_n^+$ is distribution-free over the class of all continuous, univariate distributions that are symmetric about θ_0. In order to base a test of H_0: $\theta = \theta_0$ against the alternative $\theta > \theta_0$ on these signed ranks, we must determine an appropriate rejection region. Intuitively, we would want our rejection region to correspond to "large" and "many" positive signed ranks, since such outcomes are most indicative of $\theta > \theta_0$. The form of the test statistic will determine what is meant by "many" and "large." As with the two-sample location problem, many signed rank statistics have been proposed as test statistics for H_0: $\theta = \theta_0$, and we consider a large class of them in Chapter 10. In this section we consider only a single member of that class, for purposes of illustration. This statistic was also proposed by Wilcoxon (1945) in his fundamental paper.

The Wilcoxon signed rank statistic is

$$W^+ = \sum_{i=1}^{n} \Psi_i R_i^+; \qquad \textbf{(2.4.11)}$$

that is, W^+ is the sum of the signed ranks. For testing H_0: $\theta = \theta_0$, an appropriate critical region containing "large" and "many" positive signed ranks would consist of large values of W^+. Hence an α-level test would have the form: reject H_0: $\theta = \theta_0$ in favor of $\theta > \theta_0$ iff $W^+ \geqslant w^+(\alpha, n)$, where $w^+(\alpha, n)$ is the critical point and is equal to the upper 100αth percentile for the null H_0 distribution of W^+ for a sample of size n. To obtain the necessary values of $w^+(\alpha, n)$, we need to establish a method for obtaining the form of the null distribution of W^+.

Corollary 2.4.12. Let W^+ be the Wilcoxon signed rank statistic for testing H_0: $\theta = \theta_0$. For a sample of size n, the H_0: $\theta = \theta_0$ distribution of W^+ is

$$P_0(W^+ = k) = \frac{c_n(k)}{2^n}, \qquad k = 0, 1, \ldots, \frac{n(n+1)}{2}, \qquad \textbf{(2.4.13)}$$

$$= 0, \qquad \text{elsewhere,}$$

where $c_n(k)$ = number of subsets of integers from $\{1,\ldots,n\}$ for which the sum is equal to k. (Note that these subsets can contain anywhere from 0 to n integers, inclusive.)

Proof: The range of possible values for W^+ is the set of integers from 0 through $(1+\cdots+n)=(n(n+1)/2)$, inclusive. The null distribution (2.4.13) of W^+ then follows at once from Theorem 2.4.6 and the definition of $c_n(k)$. ■

As was the case for the rank sum statistic W, the null distribution of W^+ is not obtained in a closed form. To evaluate $P_0(W^+=k)$ for a particular n, we must calculate the value of W^+ for each of the 2^n possible subsets of $\{1,2,\ldots,n\}$ and tabulate $c_n(k)$. We demonstrate this for the case $n=3$.

Example 2.4.14. With $n=3$, we must consider $2^3=8$ different subsets of $\{1,2,3\}$ to obtain the null distribution of W^+. These subsets, together with the respective values of W^+, are:

Subset of $\{1,2,3\}$	Value of W^+
$\varnothing$(empty set)	0
$\{1\}$	1
$\{2\}$	2
$\{3\}$	3
$\{1,2\}$	3
$\{1,3\}$	4
$\{2,3\}$	5
$\{1,2,3\}$	6

Hence the null distribution of W^+ for $n=3$ is

$$P_0(W^+=0) = P_0(W^+=1) = P_0(W^+=2) = \tfrac{1}{8}$$

$$P_0(W^+=4) = P_0(W^+=5) = P_0(W^+=6) = \tfrac{1}{8}$$

$$P_0(W^+=3) = \tfrac{1}{4}.$$

The null distribution of W^+ has been tabled for various values of n. For example, see Table A.4 in Hollander and Wolfe (1973) for $n\leqslant 15$ and Wilcoxon, Katti, and Wilcox (1970) for $n\leqslant 50$. For sample sizes larger than

15 there is a good approximation to the exact null distribution of W^+. This approximation (based on the standard normal distribution) is discussed, along with other large sample properties, in more detail in Chapters 3 and 10. In addition, there is a recursion formula for developing the exact null distribution of W^+. (See Exercise 2.4.1.)

Before we apply the one-sided Wilcoxon signed rank test to a real situation, we consider the problem of paired replicates data. Let $(X_1, Y_1), \ldots, (X_n, Y_n)$ be a random sample from a continuous bivariate distribution. Define $Z_i = Y_i - X_i$ for $i = 1, \ldots, n$. Then $Z_1, \ldots, Z_n$ is a random sample from a continuous univariate distribution. If θ represents the median of the distribution of Z_i, then statements about θ provide us with information about the relative locations of the individual X and Y variables. Thus, a test of the hypothesis H_0: $\theta = \theta_0$, θ_0 known, would be a test of the hypothesis that a Y observation tends to be θ_0 units larger than its associated X observation. This kind of data is called **paired replicates data** and is very common when we are trying to ascertain whether there has been an effect due to application of a "treatment." For such settings, X is called the **pretreatment variable**, Y is called the **posttreatment variable**, θ is referred to as the **treatment effect**, and the usual null hypothesis is H_0: $\theta = \theta_0 = 0$, corresponding to no treatment effect.

We note that the Wilcoxon signed rank test can be appropriately applied to the paired replicates data, provided the null $(\theta = \theta_0)$ distribution of $Y - X$ is symmetric about θ_0. This assumption is very often inherently satisfied for paired replicates data. (See Exercise 2.4.9.) We also point out that it is not necessary that the X and Y variables be independent. Indeed, for most situations, they will be strongly dependent, often corresponding to pre- and posttreatment observations on the same subject.

Example 2.4.15. In an investigation to determine the effect of aspirin on bleeding time and platelet adhesion, Bick, Adams, and Schmalhorst (1976) studied the reactions of normal subjects to aspirin. A subset of their data is presented in Table 2.4.16, where the X observation for each subject is the bleeding time (in seconds) before ingestion of 600 mg aspirin and the Y observation is the bleeding time (again in seconds) two hours after administration of the aspirin.

Letting $Z_i = Y_i - X_i$, $i = 1, \ldots, 14$, we are interested in testing H_0: $\theta = 0$, corresponding to no aspirin effect on bleeding times, against the alternative H_1: $\theta > 0$, corresponding to an increase in bleeding times as a result of the aspirin. Our Wilcoxon signed rank test, at an $\alpha = .034$ significance level, is then to reject H_0: $\theta = 0$ in favor of H_1: $\theta > 0$ iff $W^+ \geqslant w^+(.034, 14) = 82$ (see, for example, Table A.4 in Hollander and Wolfe (1973)). Ordering the absolute values of the Z differences yields: 25, 30, 45, 55, 75, 80, 85, 105,

TABLE 2.4.16.
Bleeding Times (in seconds)

SUBJECT i	X_i	Y_i
1	270	525
2	150	570
3	270	190
4	420	395
5	202	370
6	255	210
7	165	490
8	220	250
9	305	360
10	210	285
11	240	630
12	300	385
13	300	195
14	70	295

Source: R. L. Bick, T. Adams, and W. R. Schmalhorst (1976).

168, 225, 255, 325, 390, 420. Hence the absolute ranks for the positive Zs written in numerical order are 2, 4, 5, 7, 9, 10, 11, 12, 13, 14 and the corresponding value of W^+ is $2+4+5+7+9+10+11+12+13+14=87$. Since $87>82$, we reject H_0: $\theta=0$ in favor of $\theta>0$ at the $\alpha=.034$ level of significance. The smallest level at which we would have reached this conclusion with the data in Table 2.4.16 is $\underline{\alpha}=P_0(W^+\geqslant 87)=.015$.

To specify the appropriate critical regions for tests of H_0: $\theta=\theta_0$ against the alternatives $\theta<\theta_0$ or $\theta\neq\theta_0$, we establish the symmetry of the null distribution of W^+.

Theorem 2.4.17. *Let W^+ be the Wilcoxon signed rank statistic. When H_0: $\theta=\theta_0$ is true, the distribution of W^+ is symmetric about its mean $\mu=(n(n+1)/4)$.*

Proof: Theorem 2.4.4 shows that $\Psi_1,\ldots,\Psi_n,\mathbf{R}^+$ are mutually independent random variables and that Ψ_i is a Bernoulli variate with $p=1/2$ for $i=1,\ldots,n$. Since $1-\Psi_i$ is also a Bernoulli variate with $p=1/2$, we have

$$(\Psi_1,\ldots,\Psi_n,\mathbf{R}^+)\overset{d}{=}(1-\Psi_1,\ldots,1-\Psi_n,\mathbf{R}^+). \tag{2.4.18}$$

Now computing W^+ on each side of the distributional equation (2.4.18) and applying Theorem 1.3.7 yields

$$\sum_{i=1}^{n} \Psi_i R_i^+ \stackrel{d}{=} \sum_{i=1}^{n} (1-\Psi_i) R_i^+ = \frac{n(n+1)}{2} - \sum_{i=1}^{n} \Psi_i R_i^+$$

or

$$W^+ - \frac{n(n+1)}{4} \stackrel{d}{=} \frac{n(n+1)}{4} - W^+. \quad \blacksquare$$

For testing $H_0:\theta=\theta_0$ against $H_1:\theta<\theta_0$ at the α level of significance we would take a critical region of the form: reject H_0 in favor of $\theta<\theta_0$ iff $W^+ \leqslant w'(\alpha,n)$, where $w'(\alpha,n)$ is the 100αth percentile for the null distribution of W^+. From Theorem 2.4.17 we have $w'(\alpha,n)=(n(n+1)/2)-w^+(\alpha,n)$, and our critical region for the alternative $\theta<\theta_0$ is $W^+ \leqslant [(n(n+1)/2)-w^+(\alpha,n)]$. Similarly, the critical region for an α-level test of H_0: $\theta=\theta_0$ against the two-sided alternative H_1: $\theta\neq\theta_0$ has form: reject H_0: $\theta=\theta_0$ in favor of $\theta\neq\theta_0$ iff $W^+ \geqslant w^+(\alpha/2,n)$ or $W^+ \leqslant [(n(n+1)/2)-w^+(\alpha/2,n)]$, or, equivalently,

$$\left| W^+ - \frac{n(n+1)}{4} \right| \geqslant \left[w^+\left(\frac{\alpha}{2},n\right) - \frac{n(n+1)}{4} \right].$$

Theorem 2.4.17 also yields the null H_0: $\theta=\theta_0$ mean of W^+ to be

$$E_0(W^+) = \frac{n(n+1)}{4}. \tag{2.4.19}$$

You are asked in Exercise 2.4.5 to show that the null variance of W^+ is

$$\mathrm{Var}_0(W^+) = \frac{n(n+1)(2n+1)}{24}. \tag{2.4.20}$$

Exercises 2.4.

2.4.1. Let $c_n(k)$ be the counting function used in Corollary 2.4.12 to describe the null H_0: $\theta=\theta_0$ distribution of the Wilcoxon signed rank statistic W_n^+ for a sample of size n.

(a) Show that the recurrence relation

$$c_n(k) = c_{n-1}(k-n) + c_{n-1}(k)$$

holds for all $k=0,1,\ldots,(n(n+1)/2)$, and all positive integers $n \geqslant 2$, if we take the boundary conditions that $c_1(d)=0$ for $d \neq 0$ or 1 and $c_1(0)=c_1(1)=1$.

(b) Use the result in **(a)** to obtain a corresponding recursion relation for the null H_0: $\theta=\theta_0$ distribution of W_n^+.

(c) Use the formula in **(b)** to develop the null H_0: $\theta=\theta_0$ distribution of W_n^+ for each of the values $n=1,\ldots,4$.

2.4.2. Let W^+ be the Wilcoxon signed rank statistic for testing H_0: $\theta=0$, when $X_1,\ldots,X_n$ are i.i.d. continuous random variables symmetrically distributed about θ. Define W^- to be the sum of absolute ranks for those sample observations that are nonpositive.

(a) Show that $W^+=(n(n+1)/2)-W^-$.

(b) Let $W_0=W^+-W^-$. Show that the null (H_0: $\theta=0$) distribution of W_0 is symmetric about 0.

2.4.3. Let $X_1,\ldots,X_n$ be a random sample from some continuous distribution. Each of the $(n(n+1)/2)$ simple averages $(X_i+X_j)/2$, $i \leqslant j=1,\ldots,n$, is called a **Walsh average**. If W^+ is the Wilcoxon signed rank statistic for testing the null hypothesis that the underlying distribution of each X_i is symmetric about θ_0, show that $W^+=$(number of Walsh averages greater than θ_0), provided there are no tied absolute Xs and no Xs equal to θ_0 (events that have zero probability of occurring).

2.4.4. Let W^+ be the Wilcoxon signed rank statistic for testing H_0: $\theta=\theta_0$. Show that

$$W^+ \stackrel{d}{=} \sum_{j=1}^{n} V_j,$$

when H_0: $\theta=\theta_0$ is true, where $V_1,\ldots,V_n$ are mutually stochastically independent with

$$P(V_j=j) = P(V_j=0) = \tfrac{1}{2}, \qquad j=1,\ldots,n.$$

2.4.5. Use the representation in Exercise 2.4.4 to verify the formulas for $E_0(W^+)$ and $\text{Var}_0(W^+)$ given in (2.4.19) and (2.4.20), respectively.

2.4.6. Obtain an expression for the null hypothesis moment generating function of the Wilcoxon signed rank statistic W^+. Use it to evaluate $E_0(W^+)$ and $\text{Var}_0(W^+)$. [Hint: See Exercise 2.4.4.]

2.4.7. Use the moment generating function in Exercise 2.4.6 to show that the null distribution of W^+ is symmetric about $(n(n+1)/4)$.

2.4.8. Using the Walsh average representation for W^+ (see Exercise 2.4.3), obtain expressions for $E(W^+)$ and $\mathrm{Var}(W^+)$ for an arbitrary value of θ.

2.4.9. Let X and Y be two random variables, and set $Z=Y-X$.

(a) If X and Y are symmetrically distributed and independent, prove that Z is symmetrically distributed. Evaluate the point of symmetry for Z in terms of those for X and Y.

(b) If $(X,Y) \stackrel{d}{=} (Y,X)$, (i.e., X and Y are exchangeable), show that Z is symmetrically distributed about zero. Comment on this result with regard to the paired replicates data setting.

(c) If there exists a number c such that $(X-c, Y-c) \stackrel{d}{=} (c-X, c-Y)$ [i.e., X and Y are equally symmetric about (c,c)], show that Z is symmetrically distributed about zero.

(d) Prove or give a counterexample: The median of Z is equal to the median of Y minus the median of X.

(e) Answer the question in **(d)** under the conditions of **(a)**.

2.4.10. The data in Table 2.4.21 are a subset of the data obtained by Flores and Zohman (1970) in an experiment investigating the effect that the method of bedmaking has on oxygen consumption for patients assigned to complete or modified bedrest. The subjects were inpatients of the Rehabilitation Medicine Service, Montefiore Hospital and Medical Center, Bronx, New York. The measure used was net oxygen consumption for the patients during bedmaking. The two bedmaking procedures studied were the regular procedure, consisting of changing the sheets from side to side while rolling the patient onto his side, and the cardiac "top-to-bottom" procedure, consisting of moving the patient to a sitting position and changing the sheets from the top to the bottom of the bed. The values in Table 2.4.21 are the net oxygen consumption (in cc) for the eight patients in the study for each of the bedmaking procedures. Using the Wilcoxon signed rank procedure, test the hypothesis that cardiac bedmaking actually puts more strain (increases O_2 consumption) on the patient assigned to bedrest then does regular bedmaking. Use a significance level of $\alpha=.039$. In addition, find the smallest significance level at which the data would lead to rejection of the appropriate null hypothesis.

2.4.11. Let $Z_1,\ldots,Z_N$ be a random sample from a continuous distribution that is symmetric about 0. Let m and n (with $N=m+n$) be the

TABLE 2.4.21.
Net O_2 Consumption (in cc)

PATIENT	CARDIAC BEDMAKING	REGULAR BEDMAKING
1	339	175
2	349	105
3	387	411
4	159	170
5	579	295
6	586	240
7	519	319
8	275	125

Source: A. M. Flores and L. R. Zohman (1970).

number of negative and positive Zs, respectively. Also, let $X_1,\ldots,X_m$ and $Y_1,\ldots,Y_n$ denote the absolute values of those Z_is that are negative and positive, respectively, written in order of appearance and not necessarily in numerical order. How does the signed rank statistic W^+ on the Zs relate to the rank sum statistic on the Xs and Ys?

2.4.12. Let X be a random variable. We say that the distribution of X is **weighted-symmetric** about a point c if there exists a number $\lambda > 0$ such that

$$P(X > c + z) = \lambda P(X < c - z) \qquad \textbf{(2.4.22)}$$

for all $z > 0$.

(a) Show that if $\lambda = 1$ in (2.4.22), then the distribution of X is symmetric about c.

(b) Prove that if the distribution of X is weighted-symmetric about c, then $\lambda = (P(X > c)/P(X < c))$, provided $P(X < c) > 0$.

(c) Let X have a distribution (not necessarily continuous) that is weighted-symmetric about c. Show that the variables $|X - c|$ and $\Psi_c = \Psi(X - c)$ are stochastically independent, provided $0 < P(X < c) = P(X \leqslant c) < 1$, where $\Psi(t) = 1,0$ as $t >, \leqslant 0$. [Actually, this condition is necessary and sufficient for the conclusion of Lemma 2.4.2 to hold. Hence we have a characterization of weighted-symmetry about a point c. See Wolfe (1974) for the details.]

2.4.13. Let $X_1,\ldots,X_n$ be a random sample from a continuous distribution that is weighted-symmetric about a point θ. (See Exercise 2.4.12.)

Establish results similar to Theorem 2.4.6 for this more general setting. Obtain the form of the appropriate $H_0: \theta = \theta_0$ distribution of W^+ (as in Corollary 2.4.12) for this setting with $\lambda = .2$.

2.4.14. Let X have density

$$f(x) = \frac{x+1}{2}, \qquad -1 < x < 0,$$
$$= \frac{3(1-x)}{2}, \qquad 0 < x < 1,$$
$$= 0, \qquad \text{elsewhere.}$$

Show that the X distribution is weighted-symmetric (see Exercise 2.4.12) about 0. What is λ?

2.5. Discussion

In this chapter we introduced the concept of a statistic T being distribution-free over a class $\mathcal{Q}$ of joint distributions. When $\mathcal{Q}$ corresponds to conditions in the statement of a null hypothesis H_0 and T is used to test H_0 against appropriate alternatives, we say that the test procedure based on T is distribution-free over $\mathcal{Q}$ (i.e., under H_0). This means, among other things, that the level of significance, say, α, for such a test is constant over $\mathcal{Q}$; that is, the probability of a type I error is α for any underlying joint distribution that belongs to $\mathcal{Q}$. As a result, when using distribution-free procedures we need not specify a model that is any more restrictive than requiring the joint distribution to be in $\mathcal{Q}$ when H_0 is true.

In Sections 2.2, 2.3, and 2.4 we discussed the most common methods for constructing distribution-free statistics. These included statistics based on counters, such as the sign statistic, those computed from rankings, such as the Mann-Whitney-Wilcoxon rank sum statistic, and those, such as the Wilcoxon signed rank statistic, that utilize both the counting and ranking techniques. Although a large percentage of the standard distribution-free test procedures are constructed via one of these basic methods, there are situations in which other, more involved techniques can be employed while still maintaining a distribution-free property. Three such additional methods that have been used are: (1) the permutation approach due to Fisher (1935), (ii) the concept of rank-like statistics first considered by Moses (1963), and (iii) the formulation of adaptive distribution-free procedures. However, some of these additional methods require more preliminary structure than we presently have available, and we defer discussion of them until Chapter 11.

3 | *U*-STATISTICS

3.1. One-Sample *U*-Statistics

In this chapter we study *U*-statistics, a class of unbiased estimators of characteristics (parameters) of a population (or populations). Although the thrust of this effort emphasizes estimation, the *U*-statistics developed are often used as test statistics for hypotheses about these parameters. When used as test statistics, some *U*-statistics are nonparametric distribution-free; others are not. In some instances a statistic is not exactly nonparametric distribution-free, but is approximately so in a limiting (large sample sizes) sense. This is discussed in Section 3.5.

We begin by working toward the definition of a one-sample *U*-statistic.

Definition 3.1.1. A parameter γ is said to be **estimable of degree r for the family of distributions $\mathfrak{F}$** if r is the smallest sample size for which there exists a function $h^*(x_1,\ldots,x_r)$ such that

$$E_F[h^*(X_1,\ldots,X_r)] = \gamma$$

for every distribution $F(\cdot)\in\mathfrak{F}$, where $X_1,\ldots,X_r$ denotes a random sample from $F(x)$ and $h^*(\cdot)$ is a statistic and thus does not depend on $F(\cdot)$.

Thus a parameter is estimable (of some degree) if there is some estimator of it that is unbiased for every $F(\cdot)\in\mathfrak{F}$. The function $h^*(\cdot)$ in Definition 3.1.1 is called a **kernel** of the parameter γ. We note that, without loss of generality, we can assume that a kernel is symmetric in its arguments; that is,

$$h^*(x_1,\ldots,x_r) = h^*(x_{\alpha_1},\ldots,x_{\alpha_r})$$

for every permutation $(\alpha_1,\ldots,\alpha_r)$ of the integers $1,\ldots,r$. For any kernel $h^*(x_1,\ldots,x_r)$ we can always create one that is symmetric in its arguments by using

$$h(x_1,\ldots,x_r) = \frac{1}{r!} \sum_{\alpha \in A} h^*(x_{\alpha_1},\ldots,x_{\alpha_r}), \tag{3.1.2}$$

where the summation is over $A = \{\boldsymbol{\alpha} | \boldsymbol{\alpha}$ is a permutation of the integers $1,\ldots,r\}$. It is easy to see that $h(\cdot)$ is symmetric in its arguments and is an unbiased estimator of γ for each $F(\cdot)$ in $\mathcal{F}$. Hence from now on we emphasize symmetric kernels; that is, ones that are symmetric in their arguments. We denote a symmetric kernel by $h(\cdot)$, dropping the use of an asterisk.

Now, suppose we have a random sample $X_1,\ldots,X_n$, $n \geqslant r$, from a distribution with c.d.f. $F(\cdot) \in \mathcal{F}$. Naturally, we want to use all n observations in constructing an unbiased estimator of γ.

Definition 3.1.3. A ***U*-statistic** for the estimable parameter γ of degree r is created with the symmetric kernel $h(\cdot)$ by forming

$$U(X_1,\ldots,X_n) = \frac{1}{\binom{n}{r}} \sum_{\beta \in B} h(X_{\beta_1},\ldots,X_{\beta_r}),$$

where $B = \{\boldsymbol{\beta} | \boldsymbol{\beta}$ is one of the $\binom{n}{r}$ unordered subsets of r integers chosen without replacement from the set $\{1,\ldots,n\}\}$.

Note that a U-statistic is an unbiased estimator of γ for every $F(\cdot) \in \mathcal{F}$ and is symmetric in its n arguments. In fact, when $\mathcal{F}$ includes all continuous distributions, it can be shown that such a U-statistic is the unique minimum variance unbiased estimator of γ. (See Exercises 3.1.10 and 3.1.11.)

Example 3.1.4. Let $\mathcal{F}$ denote the class of all distributions with finite first moment γ. Then

$$\gamma = E_F[X_1].$$

Thus the mean is an estimable parameter of degree 1 for $\mathcal{F}$. Here $h(x) = x$ is the kernel, and it is (trivially) symmetric in its arguments. The U-statistic estimator of the mean is

$$U_1(X_1,\ldots,X_n) = \frac{1}{\binom{n}{1}} \sum_{i=1}^{n} X_i = \bar{X}.$$

Example 3.1.5. Let $\mathcal{F}$ denote the collection of all distributions with finite variance γ. Note that

$$\gamma = E_F[X_1^2 - X_1 X_2].$$

Thus the variance is an estimable parameter of degree 2. (The fact that the degree is 2 and cannot be 1 is discussed in Exercise 3.1.9.) The associated symmetric kernel is

$$h(x_1,x_2) = \tfrac{1}{2}\left[\left(x_1^2 - x_1x_2\right) + \left(x_2^2 - x_1x_2\right)\right]$$
$$= \tfrac{1}{2}(x_1 - x_2)^2. \tag{3.1.6}$$

The U-statistic estimator of the variance uses the kernel in (3.1.6) to form

$$U_2(X_1,\ldots,X_n) = \frac{1}{\binom{n}{r}} \sum_{\beta\in B} \frac{1}{2}\left[\sum_{i=1}^{2} X_{\beta_i}^2 - 2X_{\beta_1}X_{\beta_2}\right]$$
$$= \frac{1}{n(n-1)}\left[(n-1)\sum_{i=1}^{n} X_i^2 - 2\sum_{i<j}^{n}\sum^{n} X_iX_j\right]$$
$$= \frac{1}{n(n-1)}\left[n\sum_{i=1}^{n} X_i^2 - \left(\sum_{j=1}^{n} X_j\right)^2\right]$$
$$= S^2, \tag{3.1.7}$$

the well-known unbiased estimator of a population variance.

Example 3.1.8. Suppose we desire to estimate the probability that a distribution assigns to positive numbers. We could let

$$\Psi(x) = 1 \qquad \text{if } x > 0$$
$$= 0 \qquad \text{otherwise.} \tag{3.1.9}$$

Then $E_F[\Psi(X_1)] = P_F[X_1 > 0] \equiv p_F$.

Note that this $\Psi(\cdot)$ provides an unbiased estimator of $P_F(X_1 > 0)$ for each $F(\cdot)$ but that the numerical value of the probability p_F is not the same for every $F(\cdot)$. The corresponding U-statistic is

$$U_3(X_1,\ldots,X_n) = \frac{1}{n}\left[\text{the number of } X_i\text{s that are positive}\right]$$
$$= \frac{1}{n}B, \tag{3.1.10}$$

where B is the statistic used in the sign test.

Example 3.1.11. Consider the population parameter

$$\gamma = P[X_1 + X_2 > 0], \tag{3.1.12}$$

where X_1, X_2 are independent observations from $F(x)$. It follows that

$$h(x_1, x_2) = \Psi(x_1 + x_2)$$

provides a symmetric kernel of degree $r=2$, where $\Psi(\cdot)$ is defined in expression (3.1.9). The corresponding U-statistic is

$$\begin{aligned} U_4(X_1,\ldots,X_n) &= \frac{1}{\binom{n}{2}} \sum_{i<j}^{n} \sum^{n} \Psi(X_i + X_j) \\ &= \frac{1}{\binom{n}{2}} \big[\text{the number of pairs } (i,j) \text{ such that} \\ &\qquad\qquad 1 \leqslant i < j \leqslant n \text{ and } X_i + X_j > 0\big]. \end{aligned} \tag{3.1.13}$$

We will see later that this statistic is related to W^+ of the Wilcoxon signed rank test.

The next step is to develop a general expression for the variance of a U-statistic. First, for a symmetric kernel $h(\cdot)$ consider the random variables

$$h(X_1,\ldots,X_c,X_{c+1},\ldots,X_r) \text{ and } h(X_1,\ldots,X_c,X_{r+1},\ldots,X_{2r-c})$$

having exactly c sample observations in common. The covariance of these two variables is given by

$$\begin{aligned} \zeta_c &= \operatorname{Cov}\big[h(X_1,\ldots,X_c,X_{c+1},\ldots,X_r), h(X_1,\ldots,X_c,X_{r+1},\ldots,X_{2r-c})\big] \\ &= E\big[h(X_1,\ldots,X_c,X_{c+1},\ldots,X_r) h(X_1,\ldots,X_c,X_{r+1},\ldots,X_{2r-c})\big] - \gamma^2, \end{aligned} \tag{3.1.14}$$

since both of the variables have expected values equal to γ. Furthermore, since $h(\cdot)$ is symmetric in its arguments and the variables $X_1,\ldots,X_n$ are i.i.d., it follows that

$$\zeta_c = \operatorname{Cov}\big[h(X_{\beta_1},\ldots,X_{\beta_r}), h(X_{\beta'_1},\ldots,X_{\beta'_r})\big] \tag{3.1.15}$$

whenever $\boldsymbol{\beta}=(\beta_1,\ldots,\beta_r)$ and $\boldsymbol{\beta}'=(\beta'_1,\ldots,\beta'_r)$ are subsets of the integers

$\{1,\dots,n\}$ having exactly c integers in common. Note also that if $\boldsymbol{\beta}$ and $\boldsymbol{\beta}'$ have no integers in common, then $h(X_{\beta_1},\dots,X_{\beta_r})$ and $h(X_{\beta_1'},\dots,X_{\beta_r'})$ are independent. Hence, we set

$$\zeta_0 = 0.$$

Exercise 3.1.4 shows that

$$0 \leqslant \zeta_c \leqslant \zeta_r \tag{3.1.16}$$

for $c=1,\dots,r$.

Now, the variance of a U-statistic is

$$\begin{aligned}
\operatorname{Var}[U] &= E\left[\left\{\frac{1}{\binom{n}{r}}\sum_{\boldsymbol{\beta}\in B}\left[h(X_{\beta_1},\dots,X_{\beta_r})-\gamma\right]\right\}^2\right] \\
&= \frac{1}{\left[\binom{n}{r}\right]^2}\sum_{\boldsymbol{\beta}\in B}\sum_{\boldsymbol{\beta}'\in B} E\left[\{h(X_{\beta_1},\dots,X_{\beta_r})-\gamma\}\{h(X_{\beta_1'},\dots,X_{\beta_r'})-\gamma\}\right] \\
&= \frac{1}{\left[\binom{n}{r}\right]^2}\sum_{\boldsymbol{\beta}\in B}\sum_{\boldsymbol{\beta}'\in B} \operatorname{Cov}\left[h(X_{\beta_1},\dots,X_{\beta_r}),h(X_{\beta_1'},\dots,X_{\beta_r'})\right].
\end{aligned} \tag{3.1.17}$$

All terms in (3.1.17) for which $\boldsymbol{\beta}$ and $\boldsymbol{\beta}'$ have exactly c integers in common have the same covariance, namely, ζ_c. The number of such terms is $\binom{n}{r}\binom{r}{c}\binom{n-r}{r-c}$, which is simply: (the number of ways to choose the r integers in $\boldsymbol{\beta}$) times (the number of ways to select c integers among those in $\boldsymbol{\beta}$ to be in common between $\boldsymbol{\beta}$ and $\boldsymbol{\beta}'$) times (the number of ways to choose the remaining $r-c$ integers in $\boldsymbol{\beta}'$ so that they are not in $\boldsymbol{\beta}$). It follows that

$$\begin{aligned}
\operatorname{Var}[U] &= \frac{1}{\left[\binom{n}{r}\right]^2}\sum_{c=0}^{r}\binom{n}{r}\binom{r}{c}\binom{n-r}{r-c}\zeta_c \\
&= \frac{1}{\binom{n}{r}}\sum_{c=1}^{r}\binom{r}{c}\binom{n-r}{r-c}\zeta_c,
\end{aligned} \tag{3.1.18}$$

since $\zeta_0=0$. This provides a general expression for the variance of a U-statistic, where ζ_c is derived via the computational formula in (3.1.14).

Example 3.1.19. Consider the U-statistic estimator $\overline{X}$ of the population mean, as developed in Example 3.1.4. In this case $r=1$, $h(x)=x$, and

$\zeta_1 = \text{Var}[X_1] = \sigma^2$. The general expression for the variance of a U-statistic then reduces correspondingly to

$$\text{Var}[\bar{X}] = \frac{1}{\binom{n}{1}}\binom{1}{1}\binom{n-1}{0}\zeta_1$$
$$= \frac{\sigma^2}{n},$$

as is well known.

Example 3.1.20. Let us examine the U-statistic $U_4(\cdot)$ described in (3.1.13). Here $r=2$ and $h(x_1,x_2)=\Psi(x_1+x_2)$. From the computational formula (3.1.14) for the covariances we see that

$$\begin{aligned}\zeta_1 &= E[\Psi(X_1+X_2)\Psi(X_1+X_3)] - \gamma^2\\ &= P[X_1+X_2>0, X_1+X_3>0] - \{P[X_1+X_2>0]\}^2\end{aligned}$$

and

$$\begin{aligned}\zeta_2 &= E[\Psi(X_1+X_2)\Psi(X_1+X_2)] - \gamma^2\\ &= \gamma(1-\gamma).\end{aligned}$$

The variance of the U-statistic $U_4(\cdot)$ is then

$$\text{Var}[U_4(X_1,\ldots,X_n)]$$
$$= \frac{4(n-2)\{P[X_1+X_2>0, X_1+X_3>0]-\gamma^2\}+2\gamma(1-\gamma)}{n(n-1)}. \tag{3.1.21}$$

In Section 3.3 we study a general theory for the asymptotic distribution of U-statistics. The following theorem will be useful in that development.

Theorem 3.1.22. *Let $U(X_1,\ldots,X_n)$ be the U-statistic for a symmetric kernel $h(x_1,\ldots,x_r)$. If $E[h^2(X_1,\ldots,X_r)]<\infty$, then*

$$\operatorname*{limit}_{n\to\infty} n\cdot \text{Var}[U(X_1,\ldots,X_n)] = r^2\zeta_1.$$

Proof: The hypothesis implies that $\zeta_r = \text{Var}[h(X_1,\ldots,X_r)]$ exists. Thus from expressions (3.1.16) and (3.1.18) we see that $\text{Var}(U(X_1,\ldots,X_n))$ exists.

Define

$$K_c = \frac{[r!]^2}{c![(r-c)!]^2}, \qquad c = 1,\ldots,r.$$

Multiplying (3.1.18) by n, we see that a general term in the resulting sum is

$$\frac{n\binom{r}{c}\binom{n-r}{r-c}}{\binom{n}{r}}\zeta_c = K_c \cdot n \cdot \frac{(n-r)(n-r-1)\cdots(n-2r+c+1)}{n(n-1)\cdots(n-r+1)}\zeta_c.$$

We note that there are $r+1-c$ factors in the numerator involving n, and r such factors in the denominator. Thus if $c=1$, this term goes to $K_1\zeta_1$ as $n\to\infty$. For $c>1$, the term goes to zero. Since $K_1 = r^2$, the proof is complete. ■

Exercises 3.1.

3.1.1. Assume that $X_1,\ldots,X_n$ is a random sample from some population. Each of the following defines a parameter γ for the underlying F. (i) Determine the degree r of γ. (It is not necessary to show that it is a minimum.) (ii) Construct a symmetric kernel, $h(\cdot)$, for γ. (iii) Describe the corresponding U-statistic and (iv) the family of appropriate distributions, $\mathcal{F}$.

(a) $P[|X|>1]$

(b) $P[X_1+X_2+X_3>0]$

(c) $E[(X-\mu)^3]$, where μ is the mean associated with F.

(d) $E[(X_1-X_2)^4]$.

3.1.2. Let U_1 and U_2 be one-sample U-statistics for parameters γ_1 and γ_2, each for the same $\mathcal{F}$.

(a) Let r_i be the degree of γ_i, $i=1,2$. Show that the degree of the parameter $\gamma_1+\gamma_2$ is less than or equal to $\max(r_1,r_2)$. By means of an example, show that it can indeed be less than $\max(r_1,r_2)$.

(b) If the degree of $\gamma_1+\gamma_2$ is equal to $\max(r_1,r_2)$, show that $V=U_1+U_2$ is a U-statistic estimator of $\gamma_1+\gamma_2$. [Comment: This result can be generalized to a sum of k U-statistics.]

3.1.3. Consider the parameter $\gamma = P(X_1 + X_2 > 0)$ of Example 3.1.11, where X_1 and X_2 are i.i.d. continuous random variables with c.d.f. $F(\cdot)$. Define

$$h_1(x) = 1 - F(-x)$$

for $F(\cdot) \in \mathcal{F}$, and show that $E[h_1(X_1)] = \gamma$. Is $h_1(x_1)$ a symmetric kernel for γ? Justify your answer.

3.1.4. **(a)** Note that Jensen's inequality (Theorem A.3.3 in the Appendix) implies

$$E\left[Z_1^2\right] \geqslant E\left[Z_1\right] \cdot E\left[Z_2\right]$$

for any i.i.d. random variables Z_1 and Z_2 with finite second moment. Use this fact to show that if $E[h^2(X_1, \ldots, X_r)] < \infty$ then

$$\zeta_c \leqslant \zeta_r$$

for each $c = 1, \ldots, r$, where ζ_c is defined by (3.1.14). [Hint: First fix the values $x_1, \ldots, x_c$ and apply Jensen's inequality to the i.i.d. random variables

$$h(x_1, \ldots, x_c, X_{c+1}, \ldots, X_r) \text{ and } h(x_1, \ldots, x_c, X_{r+1}, \ldots, X_{2r-c}).]$$

(b) Define

$$h_c(x_1, \ldots, x_c) = E\left[h(x_1, \ldots, x_c, X_{c+1}, \ldots, X_r)\right].$$

Show that $\operatorname{Var}[h_c(X_1, \ldots, X_c)] = \zeta_c$ and hence that

$$\zeta_c \geqslant 0$$

for $c = 1, \ldots, r$.

3.1.5. Consider U_4, the U-statistic of Examples 3.1.11 and 3.1.20. Show that

$$1 > P(X_1 > 0) > 0$$

is a necessary and sufficient condition for $\operatorname{Var}(U_4)$ to be positive.

3.1.6. Consider the regression model in which the observations are $(c_1, X_1), \ldots, (c_n, X_n)$, with

$$X_i = \alpha + \beta c_i + E_i,$$

where $E_1,\ldots,E_n$ are independent and identically distributed random variables with mean zero. Assume that $c_1 < \cdots < c_n$ are known constants and α and β are unknown parameters. Note that this setting is not that of a typical U-statistic, since the X_is are not identically distributed. However, use the principles involved in constructing U-statistics to create unbiased estimators of both α and β. That is, find the smallest number of (c_i, X_i)'s necessary to yield an unbiased estimator of each parameter and use that "kernel" to obtain an unbiased estimator that treats all n observations symmetrically.

3.1.7. For each γ defined below, determine $\zeta_1, \zeta_2, \ldots, \zeta_r$, $\mathrm{Var}(U)$, and $\lim_{n\to\infty} n\,\mathrm{Var}(U)$ of the corresponding U-statistic. What assumptions on the underlying distribution are necessary for $\mathrm{Var}(U)$ to be finite?

(a) $P[|X|>1]$ (Exercise 3.1.1, part **a**).

(b) $P[X_1+X_2+X_3>0]$ (Exercise 3.1.1, part **b**).

(c) $\mathrm{Var}(X)$ (Example 3.1.5).

(d) μ^3, where μ is the mean of the underlying population and $\mathcal{F}$ consists of all populations with a finite first moment.

3.1.8. Consider Gini's mean difference statistic

$$D = \frac{1}{\binom{n}{2}} \sum_{i<j}^{n} \sum^{n} |X_i - X_j|.$$

(a) For what parameter γ is this the appropriate U-statistic?

(b) Find expressions for ζ_1, ζ_2 and $\mathrm{Var}(D)$. What moments must exist for $\mathrm{Var}(D)$ to be finite?

3.1.9. In the context of Example 3.1.5 we wish to show that the variance σ^2 is not an estimable parameter of degree 1 when $\mathcal{F}$ is the collection of all distributions with finite variance. Suppose $h_1(X)$ is an unbiased estimator of σ^2. Then

$$h(X) = X^2 - h_1(X)$$

is an unbiased estimator of μ^2 for all $F(\cdot) \in \mathcal{F}$. Show via contradictory examples that there is no such $h(\cdot)$. [Hint: You might use uniform distributions.]

3.1.10. Let U be a U-statistic for a parameter γ. If $\mathcal{F}$ includes all continuous distributions, show that U is the minimum variance unbiased estimator of γ. [Hint: First note that in this setting the order statistics are sufficient and complete for $F \in \mathcal{F}$. (See pages 48 and 133 of Lehmann (1959).) Then apply the Lehmann–Scheffé Theorem. (See, for example, Zacks (1971), page 104.)]

3.1.11. Suppose that U_1 and U_2 are two U-statistics for the same parameter γ. If $\mathcal{F}$ includes all continuous distributions, show that $P(U_1 = U_2) = 1$ (i.e., $U_1 = U_2$ with probability one). Thus, in this setting, a U-statistic for a parameter γ is essentially unique. What does this say about the uniqueness of a symmetric kernel for γ? [Hint: See Exercise 3.1.10.]

3.2. Some Convergence Results

In this section we develop many of the results needed to assess the asymptotic properties of certain statistics in this and later chapters. We begin with standard definitions and results.

Definition 3.2.1. The sequence of random variables $\{W_n\}$ is said to **converge in probability** to the constant c, if for every $\epsilon > 0$,

$$\operatorname*{limit}_{n\to\infty} P[\,|W_n - c| < \epsilon\,] = 1.$$

When $\{W_n\}$ is a sequence of estimators of a parameter γ, then W_n is termed a **consistent** estimator of γ if W_n converges to γ in probability for each value of γ.

Definition 3.2.2. The sequence $\{W_n\}$ is said to **converge in quadratic mean** to c if

$$\operatorname*{limit}_{n\to\infty} E\left[(W_n - c)^2\right] = 0.$$

Theorem 3.2.3. *Convergence in quadratic mean implies convergence in probability.*

Proof: By Chebyshev's inequality (see Theorem A.3.1 of the Appendix),

$$P[\,|W_n - c| \geqslant \epsilon\,] \leqslant \frac{E\left[(W_n - c)^2\right]}{\epsilon^2},$$

and the result follows. ∎

Example 3.2.4. Let $X_1,\ldots,X_n$ be i.i.d. from a distribution with mean μ and finite variance σ^2. Then, for the sample mean $\overline{X}$ we have

$$E\left[\left(\overline{X}-\mu\right)^2\right]=\frac{\sigma^2}{n}.$$

Thus $\overline{X}$ converges in quadratic mean (and hence in probability) to μ.

This result is a special case of the following important corollary.

Corollary 3.2.5. Let $X_1,\ldots,X_n$ denote a random sample from a population with c.d.f. $F(x)$ and let γ be an estimable parameter of degree r. For $n \geqslant r$, let $U(X_1,\ldots,X_n)$ denote the U-statistic estimator of γ. Then $U(X_1,\ldots,X_n)$ converges in quadratic mean to γ, provided $E[h^2(X_1,\ldots,X_r)]$ is finite.

Proof: Using Theorem 3.1.22, $E[\{U(X_1,\ldots,X_n)-\gamma\}^2]=\text{Var}(U(X_1,\ldots,X_n))\to 0$ as $n\to\infty$. ■

This shows, for example, that the U-statistic U_4 of Example 3.1.20 converges in probability to $\gamma=P[X_1+X_2>0]$ as $n\to\infty$.

The following result plays a prominent role in applications of the convergence in probability concept.

Theorem 3.2.6. *If $\{W_n\}$ converges in probability to c and if $k(t)$ is a function that is continuous at $t=c$, then $k(W_n)$ converges in probability to $k(c)$.*

Proof: Exercise 3.2.2. ■

Definition 3.2.7. A sequence of random variables $\{W_n\}$ is said to have a **limiting distribution** $F(w)$ [or to be asymptotically distributed according to $F(w)$] if

$$\operatorname*{limit}_{n\to\infty} P\left[W_n \leqslant w\right]=F(w)$$

for all w values at which the c.d.f. $F(w)$ is continuous.

We note that W_n converges in probability to c if and only if W_n has a limiting distribution that is degenerate at c, that is, puts probability 1 on the value c. The proof of this fact is left as Exercise 3.2.4.

The most famous limit theorem in statistics is probably the Central Limit Theorem, a generalization of which is given in Theorem A.3.7 in the Appendix. The following theorem is also very useful.

Theorem 3.2.8 (Slutsky's Theorem). *Let $\{W_n\}$ be a sequence of random variables with limiting distribution $F(w)$. Let $\{X_n\}$ denote a sequence of random variables that converges in probability to the constant c. Then the first and second members of each pair listed below have the same limiting distribution:*

(a) $(W_n + X_n)$ *and* $(W_n + c)$
(b) $X_n W_n$ *and* cW_n
(c) W_n / X_n *and* W_n / c, *if* $c \neq 0$.

Proof: See Theorem A.3.13 of the Appendix. ■

Note that in Slutsky's theorem, $\{W_n\}$ and $\{X_n\}$ are not necessarily independent.

Example 3.2.9. Suppose that $X_1, \ldots, X_n$ is a random sample of size n from a distribution with mean μ, variance $\sigma^2 > 0$, and finite fourth moments. Note that S/σ converges in probability to one. (See Exercise 3.2.3.) Hence by Slutsky's theorem 3.2.8 the test statistic for the one-sample t-test of H_0: $\mu = \mu_0$, namely,

$$\frac{\sqrt{n}\,(\bar{X} - \mu_0)}{S} = \frac{\sqrt{n}\,(\bar{X} - \mu_0)}{\sigma(S/\sigma)},$$

has the same asymptotic distribution as

$$\frac{\sqrt{n}\,(\bar{X} - \mu_0)}{\sigma},$$

which, according to the Central Limit Theorem, is asymptotically standard normal when H_0 is true.

Example 3.2.10. Consider Example 3.1.8, where $U_3(X_1, \ldots, X_n)$ is an unbiased estimator of $p_F = P[X > 0]$. The variance of $U_3(X_1, \ldots, X_n)$ is $p_F(1 - p_F)/n$, which goes to zero as $n \to \infty$. Thus $U_3(X_1, \ldots, X_n)$ converges to p_F in probability. If $0 < p_F < 1$, then it follows from Theorem 3.2.6 and Slutsky's theorem 3.2.8 that

$$\frac{(U_3(X_1, \ldots, X_n) - p_F)}{\sqrt{\dfrac{U_3(X_1, \ldots, X_n)[1 - U_3(X_1, \ldots, X_n)]}{n}}} \tag{3.2.11}$$

has the same asymptotic distribution as

$$\frac{\sqrt{n}\left[U_3(X_1,\ldots,X_n)-p_F\right]}{\sqrt{p_F(1-p_F)}},$$

which by the Central Limit Theorem is asymptotically standard normal. This limiting distribution for (3.2.11) is often used to construct approximate confidence intervals for p_F.

Theorem 3.2.12. *If the sequence of random variables* $\{V_n\}$ *has an asymptotic distribution with c.d.f.* $F(v)$ *and if* $\{W_n\}$ *is a sequence of random variables such that* $\{W_n - V_n\}$ *converges in probability to* 0, *then the limiting distribution of* $\{W_n\}$ *is also given by the c.d.f.* $F(w)$.

Proof: Note that $W_n = V_n + (W_n - V_n)$. Since $W_n - V_n$ converges in probability to 0, Theorem 3.2.8 implies that W_n has the same limiting distribution as V_n. ■

Often the conditions of this theorem are verified by showing that $W_n - V_n$ converges in quadratic mean to zero; that is,

$$\operatorname*{limit}_{n\to\infty} E\left[(W_n - V_n)^2\right] = 0.$$

Example 3.2.13. Let $X_1,\ldots,X_n$ denote a random sample from a population with mean μ, variance $\sigma^2>0$, and $\tau^2 = E[(X_i-\mu)^4]-\sigma^4$ satisfying $0<\tau^2<\infty$. We are interested in the limiting distribution of

$$W_n = \sqrt{n}\left[\frac{\sum_{i=1}^{n}(X_i-\bar{X})^2}{n} - \sigma^2\right].$$

Setting

$$V_n = \sqrt{n}\left[\frac{\sum_{i=1}^{n}(X_i-\mu)^2}{n} - \sigma^2\right],$$

we see that

$$W_n - V_n = -\sqrt{n}\,(\bar{X}-\mu)^2 = -\left[n^{1/4}(\bar{X}-\mu)\right]^2.$$

According to Theorem 3.2.6, $W_n - V_n$ will converge to zero in probability provided that $n^{1/4}(\bar{X}-\mu)$ converges to zero in probability. This fact follows from Theorem 3.2.3, since

$$E\left[\left\{n^{1/4}(\bar{X}-\mu)\right\}^2\right] = \sqrt{n}\, E\left[(\bar{X}-\mu)^2\right] = \frac{\sigma^2}{\sqrt{n}}$$

goes to zero as $n \to \infty$. Thus Theorem 3.2.12 implies that W_n and V_n have the same limiting distribution. Finally, the Central Limit Theorem shows that V_n has a limiting normal distribution with mean zero and variance τ^2.

The following results are due to Hájek. [See, for example, Hájek and Šidák (1967).]

Lemma 3.2.14. *For any random variables S and T,*

$$\left|[\mathrm{Var}(S)]^{1/2} - [\mathrm{Var}(T)]^{1/2}\right| \leqslant [\mathrm{Var}(S-T)]^{1/2}.$$

Proof: Note that

$$\frac{\mathrm{Cov}(S, T)}{\sqrt{\mathrm{Var}(S)\mathrm{Var}(T)}} \leqslant 1$$

implies

$$\mathrm{Var}(S) + \mathrm{Var}(T) - 2\sqrt{\mathrm{Var}(S)\mathrm{Var}(T)} \leqslant \mathrm{Var}(S) + \mathrm{Var}(T) - 2\mathrm{Cov}(S, T).$$

It follows that

$$\left\{[\mathrm{Var}(S)]^{1/2} - [\mathrm{Var}(T)]^{1/2}\right\}^2 \leqslant \mathrm{Var}(S-T),$$

and the desired result is obtained by taking square roots. ■

Theorem 3.2.15. *Let $\{V_n\}$ be a sequence of random variables with finite, nonzero variances such that $V_n/\sqrt{\mathrm{Var}(V_n)}$ has a limiting distribution with c.d.f. $F(v)$. If*

$$\frac{E\left[\{W_n - V_n\}^2\right]}{\mathrm{Var}(V_n)} \to 0 \qquad \textbf{(3.2.16)}$$

as $n \to \infty$, then $W_n/\sqrt{\mathrm{Var}(W_n)}$ has a limiting distribution with c.d.f. $F(w)$.

Proof: Note that

$$\frac{W_n}{\sqrt{\mathrm{Var}(V_n)}} = \frac{V_n}{\sqrt{\mathrm{Var}(V_n)}} + \frac{W_n - V_n}{\sqrt{\mathrm{Var}(V_n)}}. \tag{3.2.17}$$

By Theorem 3.2.3, the second term on the right-hand side of (3.2.17) converges to zero in probability. Therefore, by Theorem 3.2.12 $W_n/\sqrt{Var(V_n)}$ and $V_n/\sqrt{Var(V_n)}$ have the same limiting distribution. Note that

$$\frac{\mathrm{Var}(W_n - V_n)}{\mathrm{Var}(V_n)} = \frac{E\{[W_n - V_n]^2\} - \{E[W_n - V_n]\}^2}{\mathrm{Var}(V_n)}$$

$$\leqslant \frac{E\{[W_n - V_n]^2\}}{\mathrm{Var}(V_n)} \to 0$$

as $n \to \infty$. From Lemma 3.2.14 it follows that

$$\left| \sqrt{\frac{\mathrm{Var}(W_n)}{\mathrm{Var}(V_n)}} - 1 \right| \leqslant \sqrt{\frac{\mathrm{Var}(W_n - V_n)}{\mathrm{Var}(V_n)}} \to 0$$

as $n \to \infty$. Thus, by Theorem 3.2.8, $W_n/\sqrt{\mathrm{Var}(W_n)}$ has the same limiting distribution as $W_n/\sqrt{\mathrm{Var}(V_n)}$ with c.d.f. $F(w)$. ■

Exercises 3.2.

3.2.1. Assume that the sequence of random variables $\{T_n\}$ converges in probability to $c \neq 0$. Apply the definition and prove directly that T_n/c converges in probability to 1.

3.2.2. Prove Theorem 3.2.6.

3.2.3. In the setting of Example 3.2.9, show that S/σ converges in probability to 1. [Hint: See Corollary 3.2.5, Theorem 3.2.6, and Exercise 3.2.1.]

3.2.4. Show that $\{W_n\}$ converges in probability to c iff W_n has a limiting distribution that puts probability one at the number c.

3.2.5. Let $\{X_n\}$ and $\{Y_n\}$ denote sequences of random variables that converge in probability to c and k, respectively. Use Theorem 3.2.8 to show that

(i) $X_n + Y_n$ converges in probability to $c+k$,

(ii) $X_n Y_n$ converges in probability to ck.

3.2.6. In the context of Example 3.2.13, show that

$$\sqrt{n}\,[S^2 - \sigma^2]$$

has a limiting normal distribution with mean 0 and variance τ^2, where S^2 is the unbiased estimator of the population variance given in (3.1.7).

3.2.7. Give an example of a sequence of random variables $\{W_n\}$ that converges in probability to a number c but does not converge in quadratic mean to c.

3.2.8. Let $\{\mu_n\}, \{\mu_n^*\}, \{\sigma_n\}$, and $\{\sigma_n^*\}$ be sequences of constants such that

$$\operatorname*{limit}_{n\to\infty}\left(\frac{\sigma_n}{\sigma_n^*}\right) = 1 \text{ and } \operatorname*{limit}_{n\to\infty}\frac{\mu_n - \mu_n^*}{\sigma_n} = 0.$$

Then if $\{T_n\}$ is a sequence of random variables such that

$$\frac{T_n - \mu_n}{\sigma_n}$$

has a limiting standard normal distribution, show that

$$\frac{T_n - \mu_n^*}{\sigma_n^*}$$

also has a limiting standard normal distribution.

3.2.9. Let $X_1,\ldots,X_n$ be a random sample from a population with c.d.f. $F(x)$ and consider the sample skewness

$$T_1 = \frac{1}{n}\sum_{i=1}^{n}\left(X_i - \bar{X}\right)^3 / S^3,$$

where S^2 is the sample variance given in (3.1.7), and the related random variable

$$T_2 = \frac{1}{n}\sum_{i=1}^{n}\left(\frac{X_i - \mu}{\sigma}\right)^3,$$

where μ and σ are the mean and standard deviation, respectively, of the population. Assuming the existence of appropriate moments, show that $T_1 - T_2$ converges in probability to zero.

3.2.10. Let $F_n(x)$ denote the empirical c.d.f. for a random sample $X_1,\ldots,X_n$ from $F(x)$. That is, for every x, $F_n(x) =$ (the number of X_is $\leqslant x)/n$. Show that, for any x, $F_n(x)$ converges in quadratic mean to $F(x)$.

3.2.11. Let $\{X_n^{(1)}\},\ldots,\{X_n^{(k)}\}, k \geqslant 2$, be k sequences of random variables such that $X_n^{(i)}$ converges in probability to $c^{(i)}$. Prove that $\sup_{1\leqslant i\leqslant k} X_n^{(i)}$ converges in probability to $\sup_{1\leqslant i\leqslant k} c^{(i)}$. [Hint: First argue that it suffices to prove it for $k=2$. (Comment: A similar statement is valid with inf replacing sup.)]

3.2.12. Let $X_1,\ldots,X_n$ denote a random sample from a population with mean 0, variance 1, and a finite fourth moment. Find the limiting distribution of

$$\frac{\sum_{i=1}^{n} X_i}{\sqrt{\sum_{i=1}^{n} X_i^2}}.$$

3.3. The Projection Principle and the One-Sample U-Statistic Theorem

The purpose of this section is to show that, under certain conditions, standardized one-sample U-statistics have limiting normal distributions. For each U-statistic the method entails obtaining a related random variable that (i) has an easily established limiting normal distribution and (ii) is asymptotically equivalent to the U-statistic of interest in the sense that their difference converges to zero in quadratic mean (see Theorem 3.2.12). We begin the development with a discussion of the projection principle, a technique that yields the related random variable.

Let $X_1,\ldots,X_n$ denote i.i.d. random variables, each with c.d.f. $F(\cdot)$. Suppose we desire to show that a standardized version of a statistic $W = W(X_1,\ldots,X_n)$ has a limiting normal distribution, where $W(\cdot)$ is such that it treats the n random variables $X_1,\ldots,X_n$ symmetrically. Since standardization will be necessary, we actually work with

$$W^* = W - E(W)$$

throughout this development; hence, $E(W^*) = 0$.

Consider a class of random variables, each member of which is a sum of i.i.d. random variables. Specifically, let

$$\mathcal{V} = \left\{ V \mid V = \sum_{i=1}^{n} k(X_i), \text{ where } k(\cdot) \text{ is some real-valued function} \right\}. \tag{3.3.1}$$

The **projection** of W^* onto $\mathcal{V}$, this class of sums of independent and identically distributed random variables, is given by

$$V^* = \sum_{i=1}^{n} k^*(X_i), \tag{3.3.2}$$

where

$$k^*(x) = E[W^* | X_i = x].$$

As we shall see shortly, the projection V^* is the member of $\mathcal{V}$ that is the closest in some sense to W^*. Note that the projection V^* need not be a statistic, since it may depend on parameters or properties of the underlying distribution $F(\cdot)$.

Example 3.3.3. Consider Example 3.1.5, where $\gamma = \mathrm{Var}_F(X)$ and

$$U_2(X_1,\ldots,X_n) = \frac{1}{n(n-1)}\left[(n-1)\sum_{j=1}^{n} X_j^2 - 2\sum_{j<k}^{n}\sum^{n} X_j X_k\right].$$

We now fix $X_i = x$ and take the expected value of $U_2 - \gamma$ with respect to the other $(n-1)$ Xs, yielding

$$\begin{aligned} E[U_2(X_1,\ldots,X_n) - \gamma | X_i = x] &= \frac{1}{n(n-1)}\left[(n-1)\{(n-1)(\gamma+\mu^2)+x^2\}\right. \\ &\qquad \left. -2\binom{n-1}{2}\mu^2 - 2(n-1)\mu x\right] - \gamma \\ &= \frac{1}{n}[(x-\mu)^2 - \gamma], \end{aligned}$$

where μ is the mean of the distribution $F(\cdot)$. Hence the projection of $U_2(X_1,\ldots,X_n) - \gamma$ onto the space (3.3.1) is

$$V_2^* = \frac{1}{n}\sum_{i=1}^{n}(X_i - \mu)^2 - \gamma.$$

Note that like $U_2(X_1,\ldots,X_n)-\gamma$, V_2^* has expected value 0. However, the asymptotic distribution of V_2^* is easily determined from the usual Central Limit Theorem, since its random term is a constant times a sum of independent and identically distributed random variables.

We now construct a general expression for the projection of a one-sample U-statistic. If $h(x_1,\ldots,x_r)$ denotes a symmetric kernel, let

$$h_1(x) = E[h(x,X_2,\ldots,X_r)]. \tag{3.3.4}$$

Since $h(\cdot)$ is symmetric in its r arguments and the random variables $X_1,\ldots,X_n$ are i.i.d., it follows that

$$h_1(x) = E[h(X_1,x,X_3,\ldots,X_r)] = \cdots = E[h(X_1,\ldots,X_{r-1},x)]; \tag{3.3.5}$$

that is, the same function results no matter which of the n X variables is fixed. Note also that

$$\begin{aligned} E[h_1(X_1)] &= E_{X_1}[E[h(X_1,\ldots,X_r)|X_1]] \\ &= E[h(X_1,\ldots,X_r)] = \gamma \end{aligned} \tag{3.3.6}$$

and

$$\begin{aligned} \operatorname{Var}[h_1(X_1)] &= E[h_1^2(X_1)] - \gamma^2 \\ &= E_{X_1}[E[h(X_1,X_2,\ldots,X_r)h(X_1,X_{r+1},\ldots,X_{2r-1})|X_1]] - \gamma^2 \\ &= E[h(X_1,\ldots,X_r)h(X_1,X_{r+1},\ldots,X_{2r-1})] - \gamma^2 \\ &= \zeta_1 \end{aligned} \tag{3.3.7}$$

by (3.1.14).

The following lemma yields a general expression, in terms of the function $h_1(\cdot)$, for the projection of a one-sample U-statistic onto the class $\mathcal{V}$.

Lemma 3.3.8. *If $U(\cdot)$ is a one-sample U-statistic for the parameter γ with symmetric kernel $h(\cdot)$, the projection of $U(\cdot)-\gamma$ onto the class $\mathcal{V}$ is given by*

$$V^* = \frac{r}{n}\sum_{i=1}^{n}\{h_1(X_i)-\gamma\}, \tag{3.3.9}$$

where

$$h_1(x) = E[h(x,X_2,\ldots,X_r)].$$

Proof:

$$E[U(X_1,\ldots,X_n)-\gamma|X_i=x] = E\left[\frac{1}{\binom{n}{r}}\sum_{\beta\in B} h(X_{\beta_1},\ldots,X_{\beta_r})-\gamma|X_i=x\right]$$

$$= \frac{1}{\binom{n}{r}}\left[\binom{n-1}{r}\gamma + \sum_{\beta'\in B'} E\{h(X_{\beta'_1},\ldots,X_{\beta'_r})|X_i=x\}\right]-\gamma$$

$$= \frac{r}{n}[h_1(x)-\gamma],$$

where B' is the collection of all subsets of r of the integers $1,\ldots,n$ that include the integer i. The form of the projection then follows from (3.3.2). ■

Example 3.3.10. Consider Example 3.1.11, where the U-statistic $U_4(X_1,\ldots,X_n)$ was constructed for the parameter $\gamma=P[X_1+X_2>0]$. Here the symmetric kernel is $h(x_1,x_2)=\Psi(x_1+x_2)$, where $\Psi(t)=1,0$ as $t>, \leqslant 0$. The associated $h_1(\cdot)$ function is thus

$$\begin{aligned} h_1(x) &= E[\Psi(x+X_2)] \\ &= P[x+X_2>0] \\ &= 1-F(-x). \end{aligned}$$

Lemma 3.3.8 then shows that the projection of $U_4-\gamma$ onto the class $\mathcal{V}$ of sums of i.i.d. random variables is given by

$$V_4^* = \frac{2}{n}\sum_{i=1}^{n}\{1-F(-X_i)-\gamma\}.$$

The following lemma shows that the projection principle produces a random variable in the class $\mathcal{V}$ that is as close as possible to the original statistic when closeness is measured by the expected squared difference.

Lemma 3.3.11. *Suppose $W^*=W^*(X_1,\ldots,X_n)$ treats the n i.i.d. random variables $X_1,\ldots,X_n$ symmetrically and $E[W^*]=0$. Let V^* denote the projection of W^* onto $\mathcal{V}$. Specifically,*

$$V^* = \sum_{i=1}^{n} k^*(X_i),$$

where $k^*(x) = E[W^*|X_i = x]$. *Then, for any* V *in* $\mathcal{V}$,

$$E[(W^* - V^*)^2] \leqslant E[(W^* - V)^2].$$

Proof: Consider

$$\begin{aligned} E[(W^* - V)^2] &= E[\{(W^* - V^*) + (V^* - V)\}^2] \\ &= E[(W^* - V^*)^2] + E[(V^* - V)^2] \\ &\quad + 2E[(W^* - V^*)(V^* - V)]. \end{aligned} \tag{3.3.12}$$

Using $V = \sum_{i=1}^{n} k(X_i)$ and $V^* = \sum_{i=1}^{n} k^*(X_i)$, we write

$$E[(W^* - V^*)(V^* - V)] = \sum_{i=1}^{n} E[(W^* - V^*)(k^*(X_i) - k(X_i))].$$

Now,

$$\begin{aligned} &E[(W^* - V^*)\{k^*(X_i) - k(X_i)\}] \\ &\quad = E_{X_i}[E[(W^* - V^*)\{k^*(X_i) - k(X_i)\}|X_i]] \\ &\quad = E_{X_i}[\{k^*(X_i) - k(X_i)\}E[(W^* - V^*)|X_i]]. \end{aligned}$$

However,

$$\begin{aligned} E[(W^* - V^*)|X_i = x] &= E[W^*|X_i = x] - k^*(x) - (n-1)E[k^*(X_j)] \\ &= k^*(x) - k^*(x) = 0, \end{aligned}$$

since $E[k^*(X_j)] = E[W^*] = 0$ for any j satisfying $1 \leqslant j \leqslant n$ and $j \neq i$. Thus (3.3.12) becomes

$$E[(W^* - V)^2] = E[(W^* - V^*)^2] + E[(V^* - V)^2],$$

and the lemma follows from the nonnegativity of $E[(V^* - V)^2]$. ■

We are now in a position to prove the important theorem due to Hoeffding (1948) that establishes the asymptotic normality of standardized one-sample U-statistics. It provides an example of a Central Limit Theorem for dependent variables.

Theorem 3.3.13 (One-Sample U-Statistic Theorem). *Let $X_1,\ldots,X_n$ denote a random sample from some population. Let γ be an estimable parameter of degree r with symmetric kernel $h(x_1,\ldots,x_r)$. If $E[h^2(X_1,\ldots,X_r)]<\infty$ and if*

$$U(X_1,\ldots,X_n)=\frac{1}{\binom{n}{r}}\sum_{\beta\in B}h(X_{\beta_1},\ldots,X_{\beta_r}),$$

where B consists of the subsets of r integers chosen without replacement from $\{1,\ldots,n\}$, then

$$\sqrt{n}\,[\,U(X_1,\ldots,X_n)-\gamma]$$

has a limiting normal distribution with mean 0 and variance $r^2\zeta_1$, provided

$$\zeta_1=E[\,h(X_1,X_2,\ldots,X_r)h(X_1,X_{r+1},\ldots,X_{2r-1})\,]-\gamma^2$$

is positive.

Proof: Denoting $U(X_1,\ldots,X_n)$ by U_n, we write

$$nE[\,(U_n-\gamma-V_n^*)^2\,]=nE[\,(U_n-\gamma)^2\,]+nE[\,(V_n^*)^2\,]-2nE[\,(U_n-\gamma)V_n^*\,],\tag{3.3.14}$$

where V_n^* is the projection of $U_n-\gamma$ given by (3.3.9). Consider

$$\begin{aligned}nE[\,(U_n-\gamma)V_n^*\,]&=nE\left[\left\{\frac{1}{\binom{n}{r}}\sum_{\beta\in B}[\,h(X_{\beta_1},\ldots,X_{\beta_r})-\gamma\,]\right\}\right.\\&\qquad\left.\times\left\{\frac{r}{n}\sum_{i=1}^{n}[\,h_1(X_i)-\gamma\,]\right\}\right]\\&=\frac{r}{\binom{n}{r}}\sum_{i=1}^{n}\sum_{\beta\in B}E[\,\{h(X_{\beta_1},\ldots,X_{\beta_r})-\gamma\}\{h_1(X_i)-\gamma\}\,].\end{aligned}$$

Note that if $i\notin\{\beta_1,\ldots,\beta_r\}$, then

$$E[\,\{h(X_{\beta_1},\ldots,X_{\beta_r})-\gamma\}\{h_1(X_i)-\gamma\}\,]=0,$$

and if $i \in \{\beta_1, \ldots, \beta_r\}$, the term equals

$$E_{X_i}\left[E\left[\{h(X_{\beta_1}, \ldots, X_{\beta_r}) - \gamma\}\{h_1(X_i) - \gamma\} | X_i\right]\right] = E_{X_i}\left[\{h_1(X_i) - \gamma\}^2\right] = \zeta_1.$$

Since each $i = 1, \ldots, n$ occurs in $\binom{n-1}{r-1}$ of the possible sets $\{\beta_1, \ldots, \beta_r\}$, we have

$$nE\left[(U_n - \gamma)V_n^*\right] = \frac{rn\binom{n-1}{r-1}}{\binom{n}{r}}\zeta_1 = r^2\zeta_1. \qquad \textbf{(3.3.15)}$$

Using (3.3.6) and (3.3.7), we see that $E[V_n^*] = 0$ and

$$nE\left[(V_n^*)^2\right] = n\operatorname{Var}(V_n^*) = n^2\operatorname{Var}\left(\frac{r}{n}(h_1(X_1) - \gamma)\right) = r^2\zeta_1. \qquad \textbf{(3.3.16)}$$

Therefore (3.3.14) becomes

$$nE\left[(U_n - \gamma - V_n^*)^2\right] = nE\left[(U_n - \gamma)^2\right] - r^2\zeta_1,$$

which, according to Theorem 3.1.22, goes to zero as $n \to \infty$. Thus Theorems 3.2.3 and 3.2.12 show that $\sqrt{n}\,[U_n - \gamma]$ and $\sqrt{n}\,V_n^*$ have the same limiting distribution. The Central Limit Theorem and (3.3.16) show that $\sqrt{n}\,V_n^*$ has a limiting normal distribution with mean zero and variance $r^2\zeta_1$, provided $\zeta_1 > 0$; hence, the theorem follows. ■

Example 3.3.17 (Signed Rank Statistic). Let $W^+(X_1, \ldots, X_n)$ denote the Wilcoxon signed rank test statistic as described in (2.4.11). Note that for $i < j$, $\Psi(X_{(i)} + X_{(j)}) = 1$ iff $X_{(j)} > 0$ and $|X_{(i)}| < |X_{(j)}|$. Also, $\Psi(X_{(j)} + X_{(j)}) = 1$ iff $X_{(j)} > 0$. Thus

$$\sum_{i=1}^{j} \Psi(X_{(i)} + X_{(j)})$$

is just the signed rank of $X_{(j)}$. It follows that

$$\begin{aligned}
W^+ &= \sum\sum_{1 \leqslant i \leqslant j \leqslant n} \Psi(X_{(i)} + X_{(j)}) \\
&= \sum\sum_{1 \leqslant i \leqslant j \leqslant n} \Psi(X_i + X_j) \\
&= \sum_{j=1}^{n} \Psi(2X_j) + \sum\sum_{1 \leqslant i < j \leqslant n} \Psi(X_i + X_j) \\
&= nU_3(X_1, \ldots, X_n) + \binom{n}{2} U_4(X_1, \ldots, X_n),
\end{aligned}$$

where $U_3(\cdot)$ is the U-statistic estimator of $P[X_1>0]$ as given in Example 3.1.8 and $U_4(\cdot)$ is the U-statistic estimator of $P[X_1+X_2>0]$ as described in Example 3.1.11. Therefore,

$$\frac{\sqrt{n}}{\binom{n}{2}}[W^+ - E(W^+)] = \frac{n^{3/2}}{\binom{n}{2}}[U_3 - E(U_3)] + \sqrt{n}\,[U_4 - E(U_4)].$$

Now, Corollary 3.2.5 and Theorem 3.2.3 show that $U_3 - E(U_3)$ converges in probability to zero. Since $n^{3/2}/\binom{n}{2}$ goes to zero as $n\to\infty$, it follows from Slutsky's Theorem 3.2.8 that

$$\frac{\sqrt{n}}{\binom{n}{2}}[W^+ - E(W^+)] \text{ and } \sqrt{n}\,[U_4 - E(U_4)]$$

have the same limiting distribution. Since the kernel for U_4 is $h(x_1,x_2)=\Psi(x_1+x_2)$, we note that

$$E[\Psi^2(X_1+X_2)] = E[\Psi(X_1+X_2)] < \infty.$$

Thus, the one-sample U-statistic theorem (3.3.13) is applicable to $\sqrt{n}\,[U_4 - E(U_4)]$ and, since ζ_1 (see Example 3.1.20) is positive, it shows that the limiting distribution of

$$\frac{\sqrt{n}\,[W^+ - E(W^+)]}{\binom{n}{2}}$$

is normal with mean 0 and variance

$$4P[X_1+X_2>0, X_1+X_3>0] - 4\{P[X_1+X_2>0]\}^2.$$

This fact can be used to determine large sample critical values for a test based on W^+. Under H_0, the underlying distribution is continuous and symmetric about zero; hence, $F(x)=1-F(-x)$ for all x. Therefore, under H_0,

$$\begin{aligned} P[X_1+X_2>0] &= \int_{-\infty}^{\infty}[1-F(-x)]\,dF(x) \\ &= \frac{1}{2}F^2(x)\Big]_{-\infty}^{\infty} = \frac{1}{2}, \\ P[X_1+X_2>0, X_1+X_3>0] &= \int_{-\infty}^{\infty}[1-F(-x)]^2\,dF(x) \\ &= \frac{1}{3}F^3(x)\Big]_{-\infty}^{\infty} = \frac{1}{3}, \end{aligned}$$

and consequently $r^2\zeta_1 = \frac{1}{3}$. Thus

$$\frac{W^+ - E_0(W^+)}{\binom{n}{2}\sqrt{\frac{1}{3n}}}$$

has a limiting standard normal distribution under H_0. The null hypothesis standard deviation of W^+ is $\sigma_n = [n(n+1)(2n+1)/24]^{1/2}$ and

$$\left[\frac{n(n+1)(2n+1)}{24}\right]^{1/2} \Big/ \left[\binom{n}{2}\sqrt{\frac{1}{3n}}\right] \to 1$$

as $n \to \infty$. Hence, from Slutsky's theorem (3.2.8) (see Exercise 3.2.8), the standardized Wilcoxon signed rank statistic, namely,

$$\frac{W^+ - E_0(W^+)}{\sqrt{\mathrm{Var}_0(W^+)}} = \frac{W^+ - \left[\frac{n(n+1)}{4}\right]}{\left[\frac{n(n+1)(2n+1)}{24}\right]^{1/2}},$$

is asymptotically standard normal under H_0.

Example 3.3.18. Let the nonnegative continuous random variable T with c.d.f. $F(t)$ represent the lifetime of a new unit (electrical equipment, biological organism, etc.). The function

$$\bar{F}(t) = P[T > t] = 1 - F(t)$$

is called the **survival function**, since it represents the probability that the unit survives past age t. If the unit has already lived past age s, then its conditional survival function is

$$P[T > s + t \mid T > s] = \frac{\bar{F}(s+t)}{\bar{F}(s)},$$

representing the probability of surviving at least t additional time units. It is of practical importance to determine whether

$$\frac{\bar{F}(s+t)}{\bar{F}(s)} \leqslant \bar{F}(t), \qquad \text{for all } s > 0 \text{ and } t > 0, \tag{3.3.19}$$

with strict inequality for some s_0 and t_0. This condition is termed **new better than used** because the survival function of a new unit is then better than that of a used unit of any age s. The boundary of this condition, namely

$$\bar{F}(s+t) = \bar{F}(s)\bar{F}(t) \qquad \text{for all } s>0 \text{ and } t>0, \qquad \textbf{(3.3.20)}$$

characterizes (see Exercise 3.3.3) the class of exponential lifetime distributions, for which

$$\bar{F}(t) = \exp(-t/\beta), \qquad t>0 \text{ and } \beta>0. \qquad \textbf{(3.3.21)}$$

To develop a test of the null hypothesis H_0 that $\bar{F}(t)$ is given by (3.3.21) against "new better than used" alternatives, Hollander and Proschan (1972) consider the measure

$$\begin{aligned}\tau &= \int_0^\infty \int_0^\infty \{\bar{F}(s)\bar{F}(t) - \bar{F}(s+t)\}\, dF(s)\, dF(t) \\ &= \frac{1}{4} - \int_0^\infty \int_0^\infty \bar{F}(s+t)\, dF(s)\, dF(t).\end{aligned}$$

The "new better than used" alternatives correspond to positive values of τ, or equivalently, to small values of

$$\begin{aligned}\gamma &= \int_0^\infty \int_0^\infty \bar{F}(s+t)\, dF(s)\, dF(t) \\ &= \int_0^\infty \int_0^\infty P[T_1 > s+t]\, dF(s)\, dF(t) \\ &= P[T_1 > T_2 + T_3],\end{aligned}$$

where T_1, T_2, T_3 are continuous, independent and identically distributed according to $F(\cdot)$.

If $T_1, \ldots, T_n$ denotes a random sample of lifetimes distributed according to $F(t)$, the U-statistic estimator of γ is

$$\begin{aligned}U_5 &= \frac{6}{n(n-1)(n-2)} \sum_{i<j<k}^{n}\sum^{n}\sum^{n} \frac{1}{3}\big[\Psi(T_i - T_j - T_k) \\ &\qquad + \Psi(T_j - T_i - T_k) + \Psi(T_k - T_i - T_j)\big] \\ &= \frac{2}{n(n-1)(n-2)} \sum_{i<j<k}^{n}\sum^{n}\sum^{n} \Psi(T_{(k)} - T_{(j)} - T_{(i)}).\end{aligned}$$

Now, if each T_i is exponential with parameter β, then $T_1/\beta,\ldots,T_n/\beta$ are independent exponentials with parameter 1. Since

$$U_5 = U_5(T_1,\ldots,T_n) = U_5\left(\frac{T_1}{\beta},\ldots,\frac{T_n}{\beta}\right),$$

it follows that U_5 is distribution-free over the exponential family. Under H_0,

$$\begin{aligned}\gamma &= P[T_1 > T_2 + T_3]\\ &= \int_0^{\infty}\int_0^{\infty} e^{-(s+t)}e^{-s}e^{-t}\,ds\,dt\\ &= \frac{1}{4}.\end{aligned}$$

Hollander and Proschan (1972) proposed using U_5 to test the null hypothesis that the lifetime distribution is a member of the exponential family against the alternative that it is "new better than used." Their paper includes tables of critical values for the test statistic when n is small. Since the symmetric kernel

$$h(t_1,t_2,t_3) = \tfrac{1}{3}\left[\Psi(t_1-t_2-t_3)+\Psi(t_2-t_1-t_3)+\Psi(t_3-t_1-t_2)\right]$$

assumes only the values $\frac{1}{3}$ and 0, we see that $E[h^2(T_1,T_2,T_3)] < \infty$. Thus since $\zeta_1 > 0$, the one-sample U-statistic theorem (3.3.13) shows that, under H_0, $\sqrt{n}\,(U_5 - \frac{1}{4})$ has a limiting normal distribution with mean 0 and variance $9\zeta_1$. It can be shown (see Exercise 3.3.2) that under H_0

$$9\zeta_1 = \frac{5}{432}.$$

These facts yield approximate critical values for the test based on U_5 when n is large.

Exercises 3.3.

3.3.1. If U denotes a U-statistic satisfying the conditions of Theorem 3.3.13, show that

$$\frac{[U-\gamma]}{\sqrt{\operatorname{Var}(U)}}$$

is asymptotically standard normal, provided $\zeta_1 > 0$.

3.3.2. In Example 3.3.18 show that under H_0

$$9\zeta_1 = \frac{5}{432}.$$

3.3.3. Let us show that (3.3.20) characterizes the exponential distribution.

(a) If T is exponential, show that (3.3.20) is satisfied.

(b) Assume that T has a continuous distribution over the nonnegative portion of the real line; thus $F(0)=0$. Assume that (3.3.20) is satisfied and define

$$k(t) = \ln\left[\bar{F}(t)\right] - t\ln\left[\bar{F}(1)\right].$$

(i) Show that

$$k(s+t) = k(s) + k(t), \qquad \text{for all } s > 0 \text{ and } t > 0.$$

(ii) Use (i) to show that $k(t)$ has the same behavior on the interval $[J, J+1]$ as it does on $[0,1]$, where J is any nonnegative integer.

(iii) Show that $k(t)$ is bounded on $[0,1]$.

(iv) Assume there exists a $t_0>0$ such that $k(t_0)\neq 0$. Use $t_n = nt_0$ to conclude that $k(t)=0$ for all $t\geqslant 0$ and thus that T has an exponential distribution.

3.3.4. Let $X_1,\ldots,X_n$ be i.i.d. random variables with c.d.f. $F(x)$. Let $p=P[X_1>0]$ and set $\gamma=p(1-p)$.

(a) Find a U-statistic estimator of γ, say, U^*.

(b) Find the exact (small-sample) variance of this U-statistic in terms of p.

(c) Apply the one-sample U-statistic theorem (3.3.13) to U^* and determine the form of the limiting variance of $\sqrt{n}\,(U^* - \gamma)$. [Note that the one-sample U-statistic theorem does not apply when $p=1/2$.]

3.3.5. Let $X_1,\ldots,X_n$ denote a random sample from a continuous distribution that is symmetric about some number θ. Suppose that we wish to measure whether or not $\theta=\theta_0$ (a known value). Consider the parameter

$$\gamma = P\left[X_1+X_2+X_3>3\theta_0\right].$$

(a) Describe a U-statistic estimator of γ.

(b) If H_0: $\theta=\theta_0$ obtains, show that $\gamma=1/2$.

(c) Find the projection of $[U-\gamma]$ onto the class of random variables $\mathcal{V}$.

(d) Apply the one-sample U-statistic theorem to this estimator from **(a)** and show that the asymptotic variance of $\sqrt{n}\,[U-\gamma]$ is given by

$$9\{P[X_1+X_2+X_3>3\theta_0, X_1+X_4+X_5>3\theta_0]-\gamma^2\}.$$

3.3.6. Let $X_1,\ldots,X_n$ be a random sample from a continuous distribution. Suppose we measure the scale of the underlying distribution with

$$\gamma = P[|X_1-X_2|>.954].$$

[The value .954 is the median (to three decimal places) of the distribution of $|X_1-X_2|$ when the X_is have a standard normal distribution.]

(a) Describe a U-statistic estimator of γ.

(b) Find the projection of $[U-\gamma]$ onto the class of random variables $\mathcal{V}$.

(c) Apply the one-sample U-statistic theorem to this estimator and show that the asymptotic variance of $\sqrt{n}\,[U-\gamma]$ is equal to:

$$4\{P[|X_1-X_2|>.954, |X_1-X_3|>.954]-\gamma^2\}.$$

3.3.7. Let U_1 and U_2 be one-sample U-statistic estimators of the parameters γ_1 and γ_2 of degrees r_1 and r_2, respectively. Assume $\gamma_1+\gamma_2$ is of degree $r=\max(r_1,r_2)$ and hence that $V=U_1+U_2$ is the U-statistic estimator of $\gamma_1+\gamma_2$. (See Exercise 3.1.2.) Describe conditions under which $\sqrt{n}\,(V-\gamma_1-\gamma_2)$ has a limiting normal distribution with mean 0. Determine the form of the variance of this limiting distribution.

3.4. Two-Sample U-Statistics

The technique for constructing unbiased estimators based on a single sample, as described in Section 3.1, extends directly to many other settings. For the sake of simplicity we confine detailed consideration in this section to estimators based on random samples from two populations and leave brief discussions of other classes of U-statistics to Section 3.6. Let $X_1,\ldots,X_m$ and $Y_1,\ldots,Y_n$ be independent random samples from distributions with c.d.f.'s $F(x)$ and $G(y)$, respectively. A parameter γ is said to be **estimable of degree (r,s)**, for distributions (F,G) in a family $\mathcal{F}$ if r and s are the smallest sample sizes for which there exists an estimator of γ that is unbiased for every $(F,G)\in\mathcal{F}$. That is, there is a function $h^*(\cdot)$ such that

$$E_{(F,G)}\left[h^*(X_1,\dots,X_r;Y_1,\dots,Y_s)\right]=\gamma,$$

for every $(F,G)\in\mathcal{F}$. Without loss of generality the two-sample **kernel** $h^*(\cdot)$ can be assumed to be symmetric in its X_i components and separately symmetric in its Y_j components (Exercise 3.4.1). Letting $h(\cdot)$ denote such a symmetric two-sample kernel, we have the following direct extension of the concept of a *U*-statistic to this two-sample setting.

Definition 3.4.1. For an estimable parameter γ of degree (r,s) and with symmetric kernel $h(\cdot)$, a **two-sample *U*-statistic** has, for $m\geqslant r$ and $n\geqslant s$, the form

$$U(X_1,\dots,X_m;Y_1,\dots,Y_n)=\frac{1}{\binom{m}{r}\binom{n}{s}}\sum_{\alpha\in A}\sum_{\beta\in B}h(X_{\alpha_1},\dots,X_{\alpha_r};Y_{\beta_1},\dots,Y_{\beta_s}),$$

where $A[B]$ is the collection of all subsets of $r[s]$ integers chosen without replacement from the integers $\{1,\dots,m\}[\{1,\dots,n\}]$.

Example 3.4.2. Let $\mathcal{F}$ denote the collection of all pairs of distributions (F,G) such that each has a finite first moment. The difference in the means μ_Y and μ_X is an estimable parameter of degree $(1,1)$, since

$$E[Y_1-X_1]=\mu_Y-\mu_X$$

for all $(F,G)\in\mathcal{F}$. The corresponding two-sample *U*-statistic is

$$U_6(X_1,\dots,X_m;Y_1,\dots,Y_n)=\bar{Y}-\bar{X}.$$

Example 3.4.3. Suppose that we seek to estimate

$$\gamma=P[X<Y].$$

The appropriate kernel is

$$\Psi(y-x),$$

where $\Psi(\cdot)$ is defined in (3.1.9). The corresponding two-sample *U*-statistic is

$$\begin{aligned}U_7(X_1,\dots,X_m;Y_1,\dots,Y_n)&=\frac{1}{mn}\sum_{j=1}^{n}\sum_{i=1}^{m}\Psi(Y_j-X_i)\\&=\frac{1}{mn}\sum_{j=1}^{n}\left[\text{the number of } X_i\text{s}<Y_j\right].\end{aligned}\qquad\textbf{(3.4.4)}$$

This you will recognize as $(1/mn)$ times the Mann–Whitney form (2.3.8) of the test statistic for the Mann–Whitney–Wilcoxon test.

The variance expressions for two-sample U-statistics are analogous to those for one-sample U-statistics, but they are often considerably more complex. For integers c and d such that $0 \leqslant c \leqslant r$ and $0 \leqslant d \leqslant s$, let $\zeta_{c,d}$ denote the covariance between two kernel random variables with exactly c X_is and d Y_js in common. That is, define

$$\begin{aligned}\zeta_{c,d} &= \operatorname{Cov}\big[h(X_1,\ldots,X_c,X_{c+1},\ldots,X_r;Y_1,\ldots,Y_d,Y_{d+1},\ldots,Y_s),\\ &\qquad h(X_1,\ldots,X_c,X_{r+1},\ldots,X_{2r-c};Y_1,\ldots,Y_d,Y_{s+1},\ldots,Y_{2s-d})\big]\\ &= E\big[h(X_1,\ldots,X_r;Y_1,\ldots,Y_s)\\ &\qquad \times h(X_1,\ldots,X_c,X_{r+1},\ldots,X_{2r-c};Y_1,\ldots,Y_d,Y_{s+1},\ldots,Y_{2s-d})\big]-\gamma^2,\end{aligned} \tag{3.4.5}$$

and set

$$\zeta_{0,0}=0. \tag{3.4.6}$$

It follows in analogy to the one-sample case that

$$\begin{aligned}&\operatorname{Var}\big[U(X_1,\ldots,X_m;Y_1,\ldots,Y_n)\big]\\ &\qquad=\frac{1}{\binom{m}{r}\binom{n}{s}}\sum_{c=0}^{r}\sum_{d=0}^{s}\binom{r}{c}\binom{m-r}{r-c}\binom{s}{d}\binom{n-s}{s-d}\zeta_{c,d}.\end{aligned} \tag{3.4.7}$$

The derivation of this expression is left as Exercise 3.4.2.

To study the asymptotic behavior of a two-sample U-statistic, let $N=m+n$, and index the sample sizes by N. The sequences $\{m_N\}$ and $\{n_N\}$ are then assumed to satisfy $\operatorname{limit}_{N\to\infty}(m_N/N)=\lambda$ and $\operatorname{limit}_{N\to\infty}(n_N/N)=1-\lambda$, where $0<\lambda<1$. However, for notational simplicity, we still denote the sample sizes by m and n, with the dependence on N being implicit.

Theorem 3.4.8. *If* $E[h^2(X_1,\ldots,X_r;Y_1,\ldots,Y_s)]<\infty$, *then*

$$\operatorname*{limit}_{N\to\infty} N\operatorname{Var}\big[U(X_1,\ldots,X_m;Y_1,\ldots,Y_n)\big]=\frac{r^2\zeta_{1,0}}{\lambda}+\frac{s^2\zeta_{0,1}}{(1-\lambda)}.$$

Proof: Exercise 3.4.4. ■

Corollary 3.4.9. If $E[h^2(X_1,\ldots,X_r;Y_1,\ldots,Y_s)]<\infty$, then $U(X_1,\ldots,X_m;Y_1,\ldots,Y_n)$ converges in probability to γ.

Proof: This follows directly from Theorem 3.4.8, as in the proof of Corollary 3.2.5. ■

To establish the asymptotic normality of a general two-sample U-statistic, we need to find a random variable which has an easily established limiting normal distribution and which is asymptotically equivalent to $\sqrt{N}\,[U-\gamma]$ in the sense of Theorem 3.2.12. For this purpose we consider random variables of the type

$$\sum_{i=1}^{m} k_1(X_i) + \sum_{j=1}^{n} k_2(Y_j). \tag{3.4.10}$$

The random variable within this class that is closest to $[U-\gamma]$, as measured by expected squared difference, is

$$\sum_{i=1}^{m} k_1^*(X_i) + \sum_{j=1}^{n} k_2^*(Y_j), \tag{3.4.11}$$

where

$$k_1^*(x) = E[\,U-\gamma|X_1=x\,]$$

and

$$k_2^*(y) = E[\,U-\gamma|Y_1=y\,]. \tag{3.4.12}$$

This projection of $U-\gamma$ onto the class described in (3.4.10) is used in the proof of the following important extension, due to Lehmann (1951), of Hoeffding's (1948) U-statistic theorem to two-sample statistics.

Theorem 3.4.13. (Two-Sample *U*-Statistic Theorem). *Let $X_1,\ldots,X_m$ and $Y_1,\ldots,Y_n$ denote independent random samples from populations with c.d.f.'s $F(x)$ and $G(y)$, respectively. Let $h(\cdot)$ be a symmetric kernel for an estimable parameter, γ, of degree (r,s). If $E[h^2(X_1,\ldots,X_r;\,Y_1,\ldots,Y_s)]<\infty$, then*

$$\sqrt{N}\,\left[\,U(X_1,\ldots,X_m;\,Y_1,\ldots,Y_n)-\gamma\,\right]$$

has a limiting normal distribution with mean 0 and variance $[r^2\zeta_{1,0}/\lambda]+[s^2\zeta_{0,1}/(1-\lambda)]$, provided this variance is positive, where $0<\lambda=\lim_{N\to\infty}(m/N)<1$, and $\zeta_{1,0}$ and $\zeta_{0,1}$ are defined by equation (3.4.5).

Proof: Exercise 3.4.6. ■

Example 3.4.14. Continuing Example 3.4.2, in which the kernel is $h(x,y) = y - x$, we find

$$\zeta_{1,0} = E[(Y_1 - X_1)(Y_2 - X_1)] - (\mu_Y - \mu_X)^2$$
$$= \mu_Y^2 - 2\mu_X\mu_Y + \sigma_X^2 + \mu_X^2 - (\mu_Y - \mu_X)^2 = \sigma_X^2$$

and similarly

$$\zeta_{0,1} = E[(Y_1 - X_1)(Y_1 - X_2)] - (\mu_Y - \mu_X)^2 = \sigma_Y^2,$$

the variances of the X_is and Y_js, respectively. It follows from the two-sample *U*-statistic theorem that

$$\sqrt{N}\left(\bar{Y} - \bar{X} - (\mu_Y - \mu_X)\right)$$

is asymptotically normal with mean zero and variance $[\sigma_X^2/\lambda] + [\sigma_Y^2/(1-\lambda)]$, provided that $E[(Y-X)^2] < \infty$ and $\max(\sigma_X^2, \sigma_Y^2) > 0$.

Example 3.4.15. Consider Example 3.4.3 in which $\gamma = P[X < Y]$, the symmetric kernel is $\Psi(y-x)$, and the *U*-statistic is $(1/mn)$ times the Mann–Whitney form of the Mann–Whitney–Wilcoxon test statistic. For this setting we have

$$\zeta_{1,0} = E[\Psi(Y_1 - X_1)\Psi(Y_2 - X_1)] - \gamma^2$$
$$= P[X_1 < Y_1, X_1 < Y_2] - \{P[X_1 < Y_1]\}^2, \qquad \textbf{(3.4.16)}$$

and similarly

$$\zeta_{0,1} = P[X_1 < Y_1, X_2 < Y_1] - \{P[X_1 < Y_1]\}^2.$$

It follows from the two-sample *U*-statistic theorem that $\sqrt{N}[U_7 - \gamma]$ has a limiting normal distribution with mean zero and variance $(\zeta_{1,0}/\lambda) + (\zeta_{0,1}/(1-\lambda))$. The rank sum form of the Mann–Whitney–Wilcoxon test statistic is

$$W = \sum_{j=1}^{n} R_j = mnU_7 + \frac{n(n+1)}{2},$$

where R_j is the rank of Y_j among all $m+n$ observations. [See equation

(2.3.9).] Therefore,

$$\frac{\sqrt{N}}{mn}[W-E(W)]$$

has a limiting normal distribution with mean zero and variance $(\zeta_{1,0}/\lambda)+(\zeta_{0,1}/(1-\lambda))$, where $\zeta_{1,0}$ and $\zeta_{0,1}$ are given in (3.4.16). Under the usual null hypothesis H_0 for the Mann–Whitney–Wilcoxon test, the X_is and the Y_js are all i.i.d. continuous random variables. Hence, under H_0,

$$P[X_1<Y_1,X_1<Y_2] = P[X_1<X_2,X_1<X_3] = \frac{1}{3},$$

$$P[X_1<Y_1,X_2<Y_1] = P[Y_2<Y_1,Y_3<Y_1] = \frac{1}{3},$$

and

$$P[X_1<Y_1] = \frac{1}{2}.$$

Therefore, under $H_0, \zeta_{1,0}=\zeta_{0,1}=1/12$, and the null variance of W, namely, $\sigma_W^2=mn(m+n+1)/12$, satisfies

$$\lim_{N\to\infty}\left[\frac{mn(m+n+1)}{12}\right]\Big/\left[\frac{m^2n^2}{N}\left(\frac{1}{12\lambda}+\frac{1}{12(1-\lambda)}\right)\right] = 1.$$

Thus, by Slutsky's theorem (see Exercise 3.2.8), the standardized versions of the Mann–Whitney–Wilcoxon test statistic, namely,

$$\frac{U_7-\frac{1}{2}}{\sqrt{\frac{m+n}{12mn}}} \quad \text{and} \quad \frac{W-n(m+n+1)/2}{\sqrt{\frac{mn(m+n+1)}{12}}} \tag{3.4.17}$$

both have a limiting standard normal distribution under H_0. This fact is often used to provide large sample critical values for the associated test.

Example 3.4.18. Consider the two-sample scale problem, where $X_1,\ldots,X_m$ and $Y_1,\ldots,Y_n$ are independent random samples from continuous distributions with c.d.f.'s

$$F\left(\frac{x-\theta_1}{\eta_1}\right) \text{ and } F\left(\frac{x-\theta_2}{\eta_2}\right),$$

respectively, and where the objective is to detect whether the Y_js are more spread out than the X_is. That is, we are interested in testing the null hypothesis H_0: $\eta_1 = \eta_2$ against the alternative H_1: $\eta_1 < \eta_2$, with the parameters θ_1 and θ_2 both unknown. For this problem, Lehmann (1951) proposed using a test based on the statistic

$$U_8(X_1,\ldots,X_m; Y_1,\ldots,Y_n) = \frac{1}{\binom{m}{2}\binom{n}{2}} \sum_{i<j}^{m}\sum^{m} \sum_{h<k}^{n}\sum^{n} \Psi(|Y_h - Y_k| - |X_i - X_j|),$$

which measures $\gamma = P[|Y_1 - Y_2| > |X_1 - X_2|]$. (Note that U_8 can be viewed as a scale analogue to the Mann–Whitney–Wilcoxon statistic.) Under H_0, $Y_1 - Y_2$ and $X_1 - X_2$ are i.i.d. and thus $\gamma = 1/2$. Theorem 3.4.13 shows that $\sqrt{N}\,[U_8 - \gamma]$ is asymptotically normal with mean zero and variance $\sigma^2 = 4[(\zeta_{1,0}/\lambda) + (\zeta_{0,1}/(1-\lambda))]$, where

$$\zeta_{1,0} = P\big[|Y_1 - Y_2| > |X_1 - X_2|, |Y_3 - Y_4| > |X_1 - X_3|\big] - \gamma^2$$

and **(3.4.19)**

$$\zeta_{0,1} = P\big[|Y_1 - Y_2| > |X_1 - X_2|, |Y_1 - Y_3| > |X_3 - X_4|\big] - \gamma^2.$$

Sukhatme (1957) showed that this asymptotic variance depends on the form of the underlying F. Consequently, the statistic U_8 is not nonparametric distribution-free in either a small sample or asymptotic sense, and we cannot base a nonparametric distribution-free test of H_0 versus H_1 on U_8. However, in Section 3.5 we discuss how to modify this procedure in order to get a test that is asymptotically nonparametric distribution-free.

Exercises 3.4.

3.4.1. Show that from any two-sample kernel $h^*(\cdot)$ we can construct a corresponding kernel $h(\cdot)$ that is symmetric in its X_i components and separately in its Y_j components. [Note that both $h^*(\cdot)$ and $h(\cdot)$ are kernels for the same parameter, say, γ.]

3.4.2. Derive the variance formula in (3.4.7).

3.4.3. For the usual two-sample formulation, find a U-statistic estimator, say, U, of

$$\gamma = \text{Var}(X) + \text{Var}(Y),$$

and derive the limiting variance of $\sqrt{N}\,[U - \gamma]$.

3.4.4. Using (3.4.7), prove Theorem 3.4.8.

3.4.5. Let $U(\cdot)$ be a two-sample U-statistic for the parameter γ with symmetric kernel $h(\cdot)$. Show that the projection of $U-\gamma$ onto the class given by (3.4.10) is

$$V_n^* = \frac{r}{m}\sum_{i=1}^{m}\{h_{1,0}(X_i)-\gamma\} + \frac{s}{n}\sum_{j=1}^{n}\{h_{0,1}(Y_j)-\gamma\},$$

where

$$h_{1,0}(x) = E[h(x,X_2,\ldots,X_r;Y_1,\ldots,Y_s)]$$

and

$$h_{0,1}(y) = E[h(X_1,\ldots,X_r;y,Y_2,\ldots,Y_s)].$$

3.4.6. Use Theorem 3.4.8 and the result of Exercise 3.4.5 to prove Theorem 3.4.13. [Hint: The asymptotic normality of the standardized V_n^* of Exercise (3.4.5) uses the Central Limit Theorem and Theorem A.3.15 of the Appendix.]

3.4.7. If U denotes a two-sample U-statistic satisfying the conditions of Theorem 3.4.13, show that

$$\frac{[U-\gamma]}{\sqrt{\mathrm{Var}(U)}}$$

is asymptotically standard normal, provided

$$\left[\frac{r^2\zeta_{1,0}}{\lambda}+\frac{s^2\zeta_{0,1}}{(1-\lambda)}\right] > 0.$$

3.4.8. Let $X_1,\ldots,X_m$ and $Y_1,\ldots,Y_n$ be independent random samples from continuous distributions $F(x)$ and $G(y)$, respectively. As a measure of the difference in locations, Hollander (1967) considered

$$\gamma = P[X_1+X_2<Y_1+Y_2].$$

(a) Construct a U-statistic estimator, say, U, for γ.

(b) Under H_0: $[F(x)=G(x)$ for all $x]$, show that $\gamma=1/2$.

(c) Apply the two-sample U-statistic theorem to $\sqrt{N}\,[U-\gamma]$ and find an expression for its limiting variance.

(d) Show that, under H_0, this limiting variance of $\sqrt{N}\,[U-1/2]$ is

$$\sigma_0^2 = 4\left(\frac{\zeta_{1,0}}{\lambda} + \frac{\zeta_{0,1}}{(1-\lambda)}\right),$$

where

$$\zeta_{1,0} = P(X_1 < X_2 + X_3 - X_4, X_1 < X_5 + X_6 - X_7) - 1/4$$

and

$$\zeta_{0,1} = P(X_1 > X_2 + X_3 - X_4, X_1 > X_5 + X_6 - X_7) - 1/4.$$

(e) Prove that, under $H_0, \zeta_{1,0} = \zeta_{0,1}$ and hence that $\sigma_0^2 = 4\zeta_{0,1}/\lambda(1-\lambda)$. Lehmann (1964) showed that the quantity $\zeta_{0,1}$ depends on the underlying F even under H_0. (Hence the test based on this U-statistic estimator of γ is not nonparametric distribution-free either for small sample sizes or in an asymptotic sense.)

3.4.9. Suppose $X_1,\ldots,X_{2m}$ and $Y_1,\ldots,Y_{2n}$ are independent random samples from continuous distributions $F(x-\theta)$ and $F((x-\theta)/\eta)$, respectively, where the location parameter θ is unknown. Randomly pair the X_is, forming $X_1^* = X_i - X_j, \ldots, X_m^* = X_{i'} - X_{j'}$. (Each X_i belongs to one and only one such pair.) Do likewise with the Y_js, forming the differences $Y_1^*,\ldots,Y_n^*$, where $Y_j^* = Y_{i'} - Y_{j'}$. Use a U-statistic to construct a nonparametric distribution-free test of H_0: $\eta = 1$ versus H_1: $\eta > 1$ based on $X_1^*,\ldots,X_m^*, Y_1^*,\ldots,Y_n^*$. Be sure to justify the nonparametric distribution-free property of your test.

3.4.10. Show that the projection given in (3.4.11) is the closest to $[U-\gamma]$ among random variables in the class (3.4.10) in the same sense as was used in Lemma 3.3.11.

3.4.11. Consider the scale problem in which $X_1,\ldots,X_m$ and $Y_1,\ldots,Y_n$ are independent random samples from continuous distributions $F(x-\theta_x)$ and $F((y-\theta_y)/\eta)$, respectively, where both θ_x and θ_y are known and $F(\cdot)$ satisfies $F(x) = 1 - F(-x)$ for all x. We wish to test H_0: $\eta = 1$ versus H_1: $\eta > 1$. Let $X_i^* = |X_i - \theta_x|$ for $i = 1,\ldots,m$ and $Y_j^* = |Y_j - \theta_y|$ for $j = 1,\ldots,n$. Suppose we base a test of H_0 versus H_1 on the U-statistic estimator of $\gamma = P[X_i^* < Y_j^*]$. Show that this test is nonparametric distribution-free and find general expressions, in terms of $F(\cdot)$ and η, for the mean of U and the limiting variance of $\sqrt{N}\,[U-\gamma]$. Simplify these expressions for the case when H_0 is true.

3.4.12. Let U_1 and U_2 be two-sample U-statistics for the parameters γ_1 and γ_2 of degrees (r_1,s_1) and (r_2,s_2), respectively, for the same family $\mathfrak{F}$. If the degree of $\gamma_1+\gamma_2$ is $(\max[r_1,r_2], \max[s_1,s_2])$, show that $V = U_1 + U_2$ is a U-statistic estimator of $\gamma_1+\gamma_2$.

3.5. Asymptotically Nonparametric Distribution-Free Test Statistics

As seen in Chapter 2, a test statistic that is nonparametric distribution-free has the special advantage of exact α-levels over the class of distributions for which the property obtains. Some test statistics are not nonparametric distribution-free but do have this property in a limiting sense.

Definition 3.5.1. The test statistic W_n is said to be **asymptotically distribution-free over the class of distributions** $\mathcal{C}$ if the limiting distribution of W_n is the same for all distributions in $\mathcal{C}$. If the class $\mathcal{C}$ is a nonparametric one (i.e., it contains more than one parametric form), then W_n is said to be **asymptotically nonparametric distribution-free.**

Example 3.5.2. Consider the one-sample t-statistic for testing H_0: $\mu = \mu_0$, namely,

$$T^+ = \frac{\sqrt{n}\,(\bar{X} - \mu_0)}{S}.$$

For any finite n, it is distribution-free over the parametric class $\mathcal{C}$ consisting of normal distributions with mean μ_0 and any finite, positive variance, but it is not nonparametric distribution-free. However, as was shown in Example 3.2.9, the statistic T^+ is asymptotically standard normal whenever, for each n, the X_is are independent and identically distributed according to a distribution with mean μ_0, positive variance, and finite fourth moment. Since this class is a nonparametric one, T^+ is asymptotically nonparametric distribution-free.

Example 3.5.3. Lehmann's U-statistic measure of the difference in scales of two populations was described in Example 3.4.18. That U-statistic, U_8, is an unbiased estimator of $\gamma = P[|X_1 - X_2| < |Y_1 - Y_2|]$. As observed in that previous example, the null hypothesis $\eta_1 = \eta_2$ implies $\gamma = 1/2$, but if the scales differ we have $\gamma \neq 1/2$. Thus it is natural to want to test H_0: $\eta_1 = \eta_2$ against H_1: $\eta_1 \neq \eta_2$ by using U_8. In spite of the fact that U_8 is a scale difference analogue to the Mann–Whitney–Wilcoxon test statistic, it is not nonparametric distribution-free under H_0 for any m and n. Moreover, under H_0 the variance of the limiting normal distribution of $\sqrt{N}\,(U_8 - 1/2)$, namely,

$$\sigma^2 = \frac{4\zeta_{1,0}}{\lambda} + \frac{4\zeta_{0,1}}{(1-\lambda)}$$

also depends on $F(\cdot)$. [The quantities $\zeta_{1,0}$ and $\zeta_{0,1}$ are described in (3.4.19).] However, if $\hat{\sigma}^2$ is any consistent, under H_0, estimator of σ^2, then Slutsky's theorem and the two-sample U-statistic theorem show that $\sqrt{N}(U_8 - 1/2)/\hat{\sigma}$ has a limiting standard normal distribution under H_0. This approach provides an asymptotically nonparametric distribution-free test of H_0: $\eta_1 = \eta_2$ versus H_1: $\eta_1 \neq \eta_2$ based on U_8.

For a consistent estimator of σ^2 under H_0 we see from Exercise 3.2.5 that we can use

$$\hat{\sigma}^2 = \frac{4\hat{\zeta}_{1,0}}{(m/N)} + \frac{4\hat{\zeta}_{0,1}}{(n/N)},$$

provided $\hat{\zeta}_{1,0}$ and $\hat{\zeta}_{0,1}$ are consistent estimators of $\zeta_{1,0}$ and $\zeta_{0,1}$, respectively, under H_0. To define a consistent estimator of $\zeta_{1,0}$, we consider the kernel

$$h_9(x_1,x_2,x_3;y_1,y_2,y_3,y_4) = \Psi(|y_1-y_2|-|x_1-x_2|)\Psi(|y_3-y_4|-|x_1-x_3|).$$

The corresponding U-statistic, say, U_9, is a consistent estimator of

$$P[|Y_1-Y_2|>|X_1-X_2|, |Y_3-Y_4|>|X_1-X_3|] = \zeta_{1,0}+\gamma^2.$$

A consistent estimator of $\zeta_{1,0}$ under H_0 is then

$$\hat{\zeta}_{1,0} = U_9 - \left(\frac{1}{2}\right)^2.$$

An alternative estimator of $\zeta_{1,0}$ is

$$\hat{\hat{\zeta}}_{1,0} = U_9 - (U_8)^2,$$

which will consistently estimate $\zeta_{1,0}$ under both H_0 and H_1. Consistent estimators of $\zeta_{0,1}$ can be constructed in a similar manner.

Example 3.5.4. Let $X_1,\ldots,X_n$ denote a random sample from a continuous population with c.d.f. $F(x-\theta)$, where the parameter θ is the unknown population median. Suppose we wish to test the null hypothesis that the underlying population is symmetric around θ against the alternative that it is asymmetric. One instance where this issue is of importance occurs when trying to decide whether to use the signed rank procedure or the sign procedure to test the null hypothesis $\theta = \theta_0$, where θ_0 is known.

Our test statistic for this problem will examine the data for evidence of asymmetry. To "see" signs of asymmetry requires at least three observations. In particular, such a triple of observations would appear to exhibit

x x x	Right triple
x x x	Left triple

FIGURE 3.5.5

skewness to the right if the middle observation is closer to the smallest than to the largest. With this in mind, if we let $X_{(1)}<X_{(2)}<X_{(3)}$ denote the ordered values of three arbitrary observations, we say that they form a **right triple** if $(X_{(3)}-X_{(2)})>(X_{(2)}-X_{(1)})$ and a **left triple** if $(X_{(3)}-X_{(2)})<(X_{(2)}-X_{(1)})$ (see Figure 3.5.5). A preponderance of right triples over left ones among all possible triples in the data would be indicative that the underlying distribution is skewed to the right, but if the left triples dominate in number, a left-skewed distribution would be indicated.

These considerations lead to using

$$\gamma = P[X_1+X_2>2X_3]-P[X_1+X_2<2X_3]$$

as a measure of asymmetry. Its relationship to the right and left triples is seen by forming the U-statistic estimator of γ. The associated symmetric kernel is

$$h(x_1,x_2,x_3) = \tfrac{1}{3}[\text{sign}(x_1+x_2-2x_3)$$

$$+\text{sign}(x_1+x_3-2x_2)+\text{sign}(x_2+x_3-2x_1)],$$

where $\text{sign}(t)=1,0,-1$ as $t>,=,<0$. Notice that $h(x_1,x_2,x_3)$ is $1/3$ if (x_1,x_2,x_3) forms a right triple, $-1/3$ if it forms a left triple, and zero otherwise. Hence the U-statistic estimator of γ is

$$U_{10} = \frac{1}{\binom{n}{3}} \sum\sum\sum_{1\leqslant i<j<k\leqslant n} h(x_i,x_j,x_k)$$

$$= \frac{2}{n(n-1)(n-2)}[(\text{number of right triples})-(\text{number of left triples})].$$

Clearly, $E[U_{10}]=\gamma$ and the one-sample U-statistic theorem shows that

$$\frac{\sqrt{n}\,(U_{10}-\gamma)}{\tau} \qquad \textbf{(3.5.6)}$$

has a limiting standard normal distribution, where

$$\tau^2 = 9\zeta_1$$
$$= 9\,\mathrm{Cov}\left[h(X_1,X_2,X_3),h(X_1,X_4,X_5)\right].$$

Moreover, we note that $V = X_1 + X_2 - 2X_3$ is an odd translation-invariant statistic. Thus if the underlying distribution is symmetric about θ, Corollary 1.3.22 shows that V is symmetrically distributed about zero, and hence that $\gamma = 0$.

To test H_0: [the underlying distribution is symmetric about the unknown median θ] against H_1: [the underlying distribution is asymmetric with $\gamma \neq 0$], Randles, Fligner, Policello, and Wolfe (1979) proposed basing a test on

$$\frac{\sqrt{n}\,U_{10}}{\hat{\tau}}, \tag{3.5.7}$$

where $\hat{\tau}^2$ is any consistent estimator of τ^2 under H_0. This **triples test**, as it is called, is asymptotically nonparametric distribution-free. There are numerous possible choices for $\hat{\tau}^2$. One such consistent estimator would be

$$\hat{\tau}_*^2 = 9\left[\frac{1}{n}\sum_{i=1}^{n} h_*^2(X_i) - (U_{10})^2\right], \tag{3.5.8}$$

where

$$h_*(X_i) = \frac{1}{\binom{n-1}{2}} \sum\sum_{\substack{1 \leq j < k \leq n \\ j \neq i \neq k}} h(X_i, X_j, X_k).$$

(See Exercise 3.5.6.) Other possible choices may be found in Randles, Fligner, Policello, and Wolfe (1979).

In a number of important testing problems, knowledge of a nuisance parameter (one that is not directly of interest) would enable construction of a nonparametric distribution-free test. If it is unknown, one can often substitute a consistent estimator for the parameter, but the resulting test will not generally be nonparametric distribution-free. For example, Exercise 3.4.11 deals with a distribution-free test for the two-sample scale problem when the location parameters θ_X and θ_Y are assumed known. However, if θ_X and θ_Y are unknown, Sukhatme (1958) shows that replacing them with consistent estimators yields a procedure that is not nonparametric distribution-free for any sample sizes m and n but is at least asymptotically nonparametric distribution-free under certain regularity conditions. Other references in which nuisance parameters are estimated to obtain asymptotically nonparametric distribution-free tests include Crouse (1964), Raghavachari (1965), Gross (1966), Gupta (1967), and Fligner and Hettmansperger (1979).

Exercises 3.5.

3.5.1. Consider the two-sample difference in locations problem as described in Exercise 3.4.8. Show how to construct an asymptotically nonparametric distribution-free test of H_0: $[F(x)=G(x)$ for all $x]$ versus an alternative in which $\gamma>1/2$ based on the Hollander U-statistic of that exercise. Include a description of an appropriate variance estimator.

3.5.2. Let $X_1,\ldots,X_n$ denote a random sample from a population with finite fourth moment and positive variance. As noted in Example 3.1.5, the unbiased sample variance, S^2, is a U-statistic estimator of the population variance, σ^2. Under our assumptions, the one-sample U-statistic theorem shows that $\sqrt{n}\,(S^2-\sigma^2)$ has a limiting normal distribution with mean zero and variance $\tau^2=\sigma^4\{K-1\}$, where K denotes the population kurtosis, $E[(X-\mu)^4/\sigma^4]$. Use these facts to construct an asymptotically nonparametric distribution-free test of H_0: $\sigma^2=1$ versus H_1: $\sigma^2>1$. Specify an appropriate estimator of τ^2 and argue that it is consistent, being sure to state any additional assumptions on the underlying distribution which are required.

3.5.3. Consider the location problem of Exercise 3.3.5. Construct an asymptotically nonparametric distribution-free test of H_0: $\theta=\theta_0$ versus H_1: $\theta\neq\theta_0$ based on the U-statistic discussed in that exercise. Describe an appropriate asymptotic variance estimator.

3.5.4. Exercise 3.3.6 describes a problem in which scale is the quantity of interest. Show how to use the results of that exercise to construct an asymptotically nonparametric distribution-free test of H_0: $\gamma=1/2$ versus H_1: $\gamma>1/2$. Specify an appropriate asymptotic variance estimator.

3.5.5. Let $X_1,\ldots,X_m$ and $Y_1,\ldots,Y_n$ denote independent random samples from continuous populations with respective c.d.f.'s $F(x+\theta)$ and $F(y-\theta)$. In addition, assume that these distributions have a support which is an interval on the real line and that they are symmetric about $-\theta$ and θ, respectively. We wish to test H_0: $\theta=0$ versus H_1: $\theta>0$. For this purpose we consider the parameter

$$\gamma = P[X<0<Y].$$

(a) Find a U-statistic estimator, U, of this parameter. Be sure to display the kernel and the degree and to express U in as simple a form as possible.

(b) Determine the asymptotic variance of $\sqrt{N}\,[U-\gamma]$, where $N=m+n$. Evaluate γ and this asymptotic variance expression under H_0.

(c) Is the test of H_0 versus H_1 based on U from part **(a)** nonparametric distribution-free, or can it be made asymptotically nonparametric distribution-free? Justify your answer.

3.5.6. Show that the estimator $\hat{\tau}_*^2$ in (3.5.8) is a consistent estimator of τ^2, the limiting variance of $\sqrt{n}\,[U_{10}]$ in Example 3.5.4. [Hint: Express the first term in $\hat{\tau}_*^2$ as a linear combination of U-statistics.]

3.5.7. Express and simplify the expression for the exact small-sample variance of the U-statistic U_{10} in Example 3.5.4. Indicate how to construct a consistent estimator for each ζ_i term in the variance. (The estimators should be consistent under both H_0 and H_1.)

3.5.8. Let $X_1,\ldots,X_m$ and $Y_1,\ldots,Y_n$ be independent random samples from continuous populations with c.d.f.'s $F(x)$ and $F(x/\eta)$, respectively. Both distributions are assumed to be symmetric about zero. We are interested in testing H_0: $\eta=1$ versus H_1: $\eta>1$. Suppose we measure the difference in the scales (about zero) with the parameter

$$\gamma = P[|X_1|+|X_2|<|Y_1|+|Y_2|].$$

(a) Find a U-statistic estimator, say, U, of this parameter. Be sure to display the kernel and the degree.

(b) What is the value of γ under H_0?

(c) Determine the form of the limiting variance of $\sqrt{N}\,[U-\gamma]$, with $N=m+n$.

(d) Show how to construct an asymptotically distribution-free test of H_0 versus H_1 based on the U-statistic in part **(a)**. Specify an appropriate asymptotic variance estimator.

3.5.9. Let $X_1,\ldots,X_m$ and $Y_1,\ldots,Y_n$ denote independent random samples from continuous populations with c.d.f.'s $F(x-\theta)$ and $F((x-\theta)/\eta)$, respectively, where the unknown location parameter θ is the point of symmetry for both distributions. Suppose that we wish to test H_0: $\eta=1$ versus H_1:

$\eta>1$; that is, that the Ys are more spread out than the Xs. For this purpose we consider the parameter

$$\gamma = P[Y_1<X_1<Y_2].$$

(a) Describe a U-statistic estimator, say, U, of γ.

(b) What is the value of γ under H_0?

(c) Determine the form of the limiting variance of $\sqrt{N}\,[\mathrm{U}-\gamma]$, with $N=m+n$.

(d) Is the test of H_0 versus H_1 based on U from part **(a)** nonparametric distribution-free or can it be made asymptotically nonparametric distribution-free? Justify your answer.

3.6. Some Additional *U*-Statistics Theorems

In this section we consider briefly some further extensions of the one- and two-sample U-statistics results of Sections 3.3 and 3.4, respectively. The first of these theorems deals with a general k-sample ($k \geqslant 1$) setting. Let X_{ij}, $j=1,\ldots,n_i$; $i=1,\ldots,k$ be k independent random samples, with the ith sample having c.d.f. $F_i(x)$. For this setting, a parameter γ is said to be **estimable of degree $(r_1,\ldots,r_k)$** for distributions $(F_1,\ldots,F_k)$ in some family $\mathcal{F}$, if $(r_1,\ldots,r_k)$ are the smallest sample sizes for which there exists an estimator of γ which is unbiased for every $(F_1,\ldots,F_k)\in\mathcal{F}$. In direct analogy to the one- and two-sample cases, let $h(x_{11},\ldots,x_{1r_1}; x_{21},\ldots,x_{2r_2};\ldots;x_{k1},\ldots,x_{kr_k})$ be a function that is separately symmetric in the arguments $(x_{i1},\ldots,x_{ir_i})$, for each $i=1,\ldots,k$. If

$$E_{(F_1,\ldots,F_k)}\left[h(X_{11},\ldots,X_{1r_1};\ldots;X_{k1},\ldots,X_{kr_k})\right] = \gamma$$

for all $(F_1,\ldots,F_k)\in\mathcal{F}$, then $h(\cdot)$ is called a **k-sample symmetric kernel** for γ.

Definition 3.6.1. For an estimable parameter γ of degree $(r_1,\ldots,r_k)$ and symmetric kernel $h(\cdot)$, the **k-sample U-statistic** has, for $n_i \geqslant r_i$, the form

$$U(X_{11},\ldots,X_{1n_1};\ldots;X_{k1},\ldots,X_{kn_k})$$

$$= \frac{1}{\binom{n_1}{r_1}\cdots\binom{n_k}{r_k}} \sum_{\alpha_1\in A_1}\cdots\sum_{\alpha_k\in A_k} h\left(X_{1\alpha_{11}},\ldots,X_{1\alpha_{1r_1}};\ldots;X_{k\alpha_{k1}},\ldots,X_{k\alpha_{kr_k}}\right),$$

where $\boldsymbol{\alpha}_i=(\alpha_{i1},\ldots,\alpha_{ir_i})$ and A_i is the collection of all subsets of r_i integers chosen without replacement from the integers $\{1,\ldots,n_i\}$, for $i=1,\ldots,k$.

Example 3.6.2. Let $\mathcal{F}$ be the collection of all k-tuples of distributions $(F_1,\ldots,F_k)$, where each distribution is continuous. Set

$$\gamma = \sum_{u<v}^{k}\sum^{k} P(X_u<X_v),$$

where X_u and X_v are independently distributed as $F_u(\cdot)$ and $F_v(\cdot)$, respectively. Then γ is an estimable parameter of degree $(1,1,\ldots,1)$, since

$$h(x_{11};x_{21};\ldots;x_{k1}) = \sum_{u<v}^{k}\sum^{k} \Psi(x_{v1}-x_{u1})$$

is an appropriate k-sample symmetric kernel, where $\Psi(\cdot)$ is defined in (3.1.9). The corresponding k-sample U-statistic is

$$U_{11}(X_{11},\ldots,X_{1n_1};\ldots;X_{k1},\ldots,X_{kn_k})$$
$$= \frac{1}{n_1\cdots n_k}\sum_{\beta_1=1}^{n_1}\cdots\sum_{\beta_k=1}^{n_k} h(X_{1\beta_1};\ldots;X_{k\beta_k}) = \sum_{u<v}^{k}\sum^{k}\frac{V_{uv}}{n_u n_v}, \tag{3.6.3}$$

where V_{uv} is the Mann–Whitney form of the Mann–Whitney–Wilcoxon statistic for the uth and vth samples. The statistic U_{11} is similar to the statistic $J=\sum\sum_{1\leqslant u<v\leqslant k}V_{uv}$ proposed by Terpstra (1952) and Jonckheere (1954) for testing the null hypothesis H_0: $[F_1(x)=\cdots=F_k(x)$ for all $x]$ against ordered alternatives of the form H_1: $[F_1(x)\geqslant\cdots\geqslant F_k(x)$ for all x, with at least one strict inequality for at least one $x]$. (See Exercise 3.6.1.) We note that both U_{11} and J are nonparametric distribution-free under H_0, for any $n_1,\ldots,n_k$. This follows from the facts that each V_{uv} is a function of the observations only through their ranks in the combined ranking of all $N=n_1+\cdots+n_k$ observations and the combined rank vector $\mathbf{R}^*=(R_{11}^*,\ldots,R_{1n_1}^*,\ldots,R_{k1}^*,\ldots,R_{kn_k}^*)$, where the respective ranks correspond to $(X_{11},\ldots,X_{1n_1},\ldots,X_{k1},\ldots,X_{kn_k})$, is uniformly distributed over the permutations of the integers $1,\ldots,N$ under H_0. (See Theorem 2.3.3.)

To obtain the asymptotic normality of a properly standardized k-sample U-statistic, we proceed as in the two-sample setting. Let i be a fixed integer satisfying $1\leqslant i\leqslant k$ and define

$$H_{i1} = h\left(X_{1\alpha_{11}},\ldots,X_{1\alpha_{1r_1}};\ldots;X_{k\alpha_{k1}},\ldots,X_{k\alpha_{kr_k}}\right)$$

and

$$H_{i2} = h\left(X_{1\beta_{11}},\ldots,X_{1\beta_{1r_1}};\ldots;X_{k\beta_{k1}},\ldots,X_{k\beta_{kr_k}}\right), \tag{3.6.4}$$

where the two sets $(\alpha_{j1},\ldots,\alpha_{jr_j})$ and $(\beta_{j1},\ldots,\beta_{jr_j})$ have no integers in common whenever $j \neq i$ and exactly one integer in common when $j = i$. Now, set

$$\begin{aligned}\zeta_{0,\ldots,0,1,0,\ldots,0} &= \operatorname{Cov}[H_{i1}, H_{i2}] \\ &= E[H_{i1}H_{i2}] - \gamma^2,\end{aligned} \tag{3.6.5}$$

where the only 1 in the subscript of $\zeta_{0,\ldots,0,1,0,\ldots,0}$ is in the ith position. Taking $N = \sum_{i=1}^{k} n_i$, we have the following k-sample U-statistic theorem due to Lehmann (1963a).

Theorem 3.6.6. *Let $U(X_{11},\ldots,X_{1n_1};\ldots;X_{k1},\ldots,X_{kn_k})$ be a k-sample U-statistic for the parameter γ of degree $(r_1,\ldots,r_k)$. If $\lim_{n\to\infty}(n_i/N) = \lambda_i$, $0 < \lambda_i < 1$, for $i = 1,\ldots,k$, and if $E[h^2(X_{11},\ldots,X_{1r_1};\ldots;X_{k1},\ldots,X_{kr_k})] < \infty$, then $\sqrt{N}(U-\gamma)$ has a limiting normal distribution with mean 0 and variance*

$$\sigma^2 = \sum_{i=1}^{k} \frac{r_i^2}{\lambda_i}\zeta_{0,\ldots,0,1,0,\ldots,0},$$

provided $\sigma^2 > 0$.

Proof: Exercise 3.6.3. ■

Example 3.6.7. For the statistic U_{11} of Example 3.6.2, Theorem 3.6.6 shows that $\sqrt{N}(U_{11}-\gamma)$ is asymptotically normal with mean 0 and variance

$$\sigma^2 = \sum_{i=1}^{k}\left[\frac{\zeta_{0,\ldots,0,1,0,\ldots,0}}{\lambda_i}\right].$$

As we have seen in previous examples, this asymptotic normality can be used to determine large sample cutoffs for a test of H_0: $[F_1(x) = \cdots = F_k(x)$ for all $x]$ based on U_{11}. Under H_0, the asymptotic null variance of $\sqrt{N}(U_{11}-\gamma)$ is

$$\sigma_0^2 = \sum_{i=1}^{k}\left[\frac{(2i-1-k)^2}{12\lambda_i}\right]$$

(see Exercise 3.6.2), and thus

$$\frac{\sqrt{N}\,(U_{11}-E_0(U_{11}))}{\sigma_0}=\frac{\sqrt{N}\left[U_{11}-\dfrac{k(k-1)}{4}\right]}{\sqrt{\displaystyle\sum_{i=1}^{k}\frac{(2i-1-k)^2}{12\lambda_i}}} \tag{3.6.8}$$

is asymptotically standard normal.

As a second extension of the one- and two-sample U-statistics results, we consider the joint limiting distribution of several U-statistics. The following theorem, stated without proof, is also presented in Lehmann (1963a).

Theorem 3.6.9. *Let $U_1,\ldots,U_t$ be k-sample ($k\geqslant 1$) U-statistics, with U_a corresponding to a parameter γ_a of degree $(r_1^{(a)},\ldots,r_k^{(a)})$ and symmetric kernel $h^{(a)}(\cdot)$, for $a=1,\ldots,t$. The joint limiting distribution of $\sqrt{N}\,(U_1-\gamma_1),\ldots,\sqrt{N}\,(U_t-\gamma_t)$, with $N=n_1+\cdots+n_k$, is t-variate normal with zero mean vector and covariance matrix $\boldsymbol{\Sigma}=((\sigma^{(a,b)}))$, where*

$$\sigma^{(a,b)}=\sum_{i=1}^{k}\frac{r_i^{(a)}r_i^{(b)}}{\lambda_i}\zeta_i^{(a,b)}, \tag{3.6.10}$$

for $\lambda_i=\lim_{N\to\infty}(n_i/N)$. The quantities $\zeta_i^{(a,b)}$ are given by

$$\begin{aligned}\zeta_i^{(a,b)}&=\operatorname{Cov}\left[H_{i1}^{(a)},H_{i2}^{(b)}\right]\\&=E\left[H_{i1}^{(a)}H_{i2}^{(b)}\right]-\gamma_a\gamma_b,\end{aligned}$$

where

$$H_{i1}^{(a)}=h^{(a)}\left(X_{1\alpha_{11}},\ldots,X_{1\alpha_{1r_1^{(a)}}};\ldots;X_{k\alpha_{k1}},\ldots,X_{k\alpha_{kr_k^{(a)}}}\right)$$

and

$$H_{i2}^{(b)}=h^{(b)}\left(X_{1\beta_{11}},\ldots,X_{1\beta_{1r_1^{(b)}}};\ldots;X_{k\beta_{k1}},\ldots,X_{k\beta_{kr_k^{(b)}}}\right),$$

and the two sets $(\alpha_{j1},\ldots,\alpha_{jr_j^{(a)}})$ and $(\beta_{j1},\ldots,\beta_{jr_j^{(b)}})$ have no integers in common if $j\neq i$ and exactly one integer in common when $j=i$. Here a and b represent integers between 1 and t, inclusive.

Proof: See Lehmann (1963a). ■

Example 3.6.11. Consider the $t=(k(k-1)/2)$ k-sample U-statistics V_{uv}, $u<v=2,\ldots,k$, of Example 3.6.2. Then, the parameter $\gamma_{uv}=P(X_u<X_v)$ corresponds to V_{uv} and has degree $(0,\ldots,0,1,0,\ldots,0,1,0,\ldots,0)$, where the 1's are in the uth and vth positions. The kernel associated with γ_{uv} (and V_{uv}) is $h^{(u,v)}(x_{u1};x_{v1})=\Psi(x_{v1}-x_{u1})$. Theorem 3.6.9 then shows that the joint limiting distribution of $\sqrt{N}\,(V_{uv}-\gamma_{uv})$, $u<v$, is $(k(k-1)/2)$-variate normal with zero mean vector and covariance matrix $\boldsymbol{\Sigma}=((\sigma^{(a,b)}))$ given by (3.6.10). The evaluation of this limiting covariance matrix is left as Exercise 3.6.6. Theorem 3.6.9 has often been used in the literature. For example, Hollander (1968) utilized this result to establish asymptotic independence between pairs of suitably normed U-statistics.

For our final extension of the one- and two-sample U-statistics results, we simply wish to point out that the arguments used in establishing Theorems 3.3.13 and 3.4.13 and, more generally, 3.6.6 and 3.6.9, do not require that the involved samples be univariate in nature. Hence we can apply these theorems to vector-valued random variables as well. We illustrate such an application in the following example.

Example 3.6.12. Let $\mathbf{X}_1,\ldots,\mathbf{X}_n$, with $\mathbf{X}_i=(X_{1i},X_{2i})$, be a random sample from a continuous bivariate distribution with c.d.f. $F(x_1,x_2)$. Kendall (1938) considers the parameter

$$\tau = 2P((X_{22}-X_{21})(X_{12}-X_{11})>0)-1$$

as a measure of the correlation between the variables X_{1i} and X_{2i}, where (X_{1i},X_{2i}) has c.d.f. $F(x_1,x_2)$. Setting $\gamma=(\tau+1)/2$, we see that γ is estimable of degree 2, with symmetric kernel

$$h(\mathbf{x}_1,\mathbf{x}_2) = \Psi((x_{22}-x_{21})\cdot(x_{12}-x_{11})). \qquad \textbf{(3.6.13)}$$

The corresponding one-sample U-statistic is

$$U_{12}(\mathbf{X}_1,\ldots,\mathbf{X}_n) = \frac{1}{\binom{n}{2}}\sum_{i<j}^{n}\sum^{n} h(\mathbf{X}_i,\mathbf{X}_j)$$

$$= \frac{1}{\binom{n}{2}}\Big[\text{number of pairs } (i,j) \text{ such that } 1\leqslant i<j\leqslant n$$

$$\text{and } (X_{2j}-X_{2i})(X_{1j}-X_{1i})>0\Big]. \qquad \textbf{(3.6.14)}$$

Using the designation that a pair $(\mathbf{X}_i,\mathbf{X}_j)$ is **concordant** if and only if

$(X_{2j}-X_{2i})(X_{1j}-X_{1i})>0$, we see that the U-statistic U_{12} can be written as

$$U_{12}=\frac{1}{\binom{n}{2}}\left[\text{number of concordant pairs }(\mathbf{X}_i,\mathbf{X}_j),\ i<j\right].$$

(The corresponding estimator of τ, namely, $K=2U_{12}-1$, is referred to as **Kendall's sample correlation coefficient**.) The extension of Theorem 3.3.13 to vector-valued random variables implies that $\sqrt{n}\,(U_{12}-\gamma)$ has a limiting normal distribution with mean 0 and variance $4\zeta_1$, where ζ_1 is defined in (3.1.14). The specific evaluation of ζ_1 for this correlation example is left as Exercise 3.6.9. The U-statistic, U_{12}, is often used to test H_0: [independence between X_{1i} and X_{2i} for all i] against H_1: $\gamma\neq 1/2$. The test based on U_{12} is nonparametric distribution-free. (See Exercise 3.6.10.)

These three extensions of the one- and two-sample U-statistics theorems provide ample illustration of the variety of settings where these techniques find application. However, many other general U-statistics results can be found in the statistical literature. For example, Hoeffding (1948) shows that, under certain conditions, Theorem 3.3.13 remains true even if the independent one-sample observations $X_1,\ldots,X_n$ are allowed to have individually different distributions.

Exercises 3.6.

3.6.1. Let $F_i(x)=F(x-\Delta_i)$, $i=1,\ldots,k$, where $F(\cdot)$ is a distribution function and $\Delta_1,\ldots,\Delta_k$ are shift parameters such that $\Delta_1\leqslant\cdots\leqslant\Delta_k$. Show that these distribution functions belong to the ordered alternatives H_1 considered in Example 3.6.2.

3.6.2. Consider the U-statistic U_{11} of Example 3.6.2. Verify that the asymptotic null variance of

$$\sqrt{N}\left(U_{11}-\frac{k(k-1)}{4}\right)$$

is

$$\sigma_0^2=\sum_{i=1}^{k}\left[\frac{(2i-1-k)^2}{12\lambda_i}\right].$$

If $\lambda_1=\cdots=\lambda_k=1/k$, simplify this formula for σ_0^2.

3.6.3. Prove Theorem 3.6.6. [Hint: Consider the projection Q^* of $Q=\sqrt{N}\,(U-\gamma)$ onto the class of random variables of the type

$$\sum_{\alpha_1=1}^{n_1} g_1(X_{1\alpha_1})+\cdots+\sum_{\alpha_k=1}^{n_k} g_k(X_{k\alpha_k}),$$

and show that $E[(Q-Q^*)^2]\to 0$ as $N\to\infty$. Then proceed as in the one- and two-sample settings.]

3.6.4. Let $(X_1,\ldots,X_{n_1})$, $(Y_1,\ldots,Y_{n_2})$, and $(Z_1,\ldots,Z_{n_3})$ be three independent random samples from continuous distributions $F_1(x)$, $F_2(y)$, and $F_3(z)$, respectively. Consider the parameter

$$\gamma = P(X_1<Y_1<Z_1).$$

(a) Construct a *U*-statistic estimator, say, U, for γ.

(b) Under H_0: $[F_1(x)=F_2(x)=F_3(x)$ for all $x]$, show that $\gamma=1/6$.

(c) Find an expression for the limiting variance of $\sqrt{N}\,(U-\gamma)$, with $N=n_1+n_2+n_3$. Simplify this expression when H_0 is true.

(d) Is the test of H_0 based on U nonparametric distribution-free or can it be made asymptotically nonparametric distribution-free? Justify your answer.

3.6.5. Let X_{ij}, $j=1,\ldots,n_i$; $i=1,\ldots,k$ be k independent random samples from continuous distributions, with the ith sample having c.d.f. $F_i(x)$. Let l be some integer between 1 and k, inclusive, and consider the parameter

$$\gamma = \sum\sum_{1\leqslant u<v\leqslant l} P(X_u<X_v)+\sum\sum_{l\leqslant u<v\leqslant k} P(X_v<X_u),$$

where X_u and X_v are independently distributed as $F_u(\cdot)$ and $F_v(\cdot)$, respectively.

(a) Construct a *U*-statistic estimator, say, U^*, for γ.

(b) Under H_0: $[F_1(x)=\cdots=F_k(x)$ for all $x]$, evaluate γ.

(c) Argue that U^* in **(a)** is nonparametric distribution-free under H_0.

(d) Find an expression for the limiting variance of $\sqrt{N}\,(U^*-\gamma)$ under H_0, where $N=\sum_{i=1}^k n_i$.

[Comment: Suppose $F_i(x)=F(x-\Delta_i)$, $i=1,\ldots,k$, where $\Delta_1,\ldots,\Delta_k$ are shift parameters such that $\Delta_1\leqslant\cdots\leqslant\Delta_l\geqslant\Delta_{l+1}\geqslant\cdots\geqslant\Delta_k$. (Archambault, Mack and Wolfe (1977) referred to these as "umbrella" alternatives.) A test based on U^* is designed to test H_0 against such alternatives (with at least one strict inequality among the Δ's) when l is known.]

3.6.6. Using (3.6.10), evaluate the limiting covariance matrix in Example 3.6.11.

3.6.7. Let U_3 and U_4 be the one-sample U-statistics of Examples 3.1.8 and 3.1.11, respectively. Use Theorem 3.6.9 to investigate the joint limiting distribution of $\sqrt{n}\,(U_3-\gamma_3)$ and $\sqrt{n}\,(U_4-\gamma_4)$, where $\gamma_3=P(X_1>0)$ and $\gamma_4=P(X_1+X_2>0)$. How does this limiting distribution simplify when H_0: $[F(x)+F(-x)=1$ for all $x]$ is true, where $F(\cdot)$ is the c.d.f. of a continuous distribution?

3.6.8. Let U_7 and U_8 be the two-sample U-statistics of Examples 3.4.3 and 3.4.18, respectively. Use Theorem 3.6.9 to investigate the joint limiting distribution of $\sqrt{N}\,(U_7-\gamma_7)$ and $\sqrt{N}\,(U_8-\gamma_8)$, where $\gamma_7=P(X<Y)$ and $\gamma_8=P(|X_1-X_2|<|Y_1-Y_2|)$. [Hollander (1968) discusses the implications of this result regarding the asymptotic independence of U_7 and U_8.]

3.6.9. Evaluate the limiting variance of $\sqrt{n}\,(U_{12}-\gamma)$ in Example 3.6.12. Using the general variance expression (3.1.18), find the exact (small-sample) variance of U_{12}.

3.6.10. The statistic U_{12} of Example 3.6.12 can be used to test H_0: $[X_{1i}$ and X_{2i} are independent] against the alternative H_1: $[X_{1i}$ and X_{2i} are dependent], where (X_{1i},X_{2i}) are jointly continuous random variables. What is the value of γ (and thus of τ) under H_0? Show that U_{12} can be written as

$$U_{12}=\frac{1}{\binom{n}{2}}\sum_{i<j}^{n}\sum^{n}\Psi((R_j-R_i)(Q_j-Q_i)), \tag{3.6.15}$$

where R_i is the rank of X_{1i} among $X_{11},\ldots,X_{1n}$ and Q_i is the rank of X_{2i} among $X_{21},\ldots,X_{2n}$. Argue that U_{12} is nonparametric distribution-free under H_0.

3.6.11. Consider Kendall's correlation coefficient K of Example 3.6.12. Show that the distribution of K is symmetric about 0 when H_0: $[X_{1i}$ and X_{2i} are independent] is true. [Hint: Use the equal in distribution arguments of Section 1.3.]

3.6.12. Let $\mathbf{X}_1,\ldots,\mathbf{X}_n$, with $\mathbf{X}_i=(X_{1i},X_{2i})$, be a random sample from a continuous bivariate distribution with c.d.f. $F(x,y)$. Consider the parameter

$$\gamma = P(|X_{22}-X_{21}|<|X_{12}-X_{11}|),$$

which measures the difference in scales for the X_{1i} components and the X_{2i} components.

(a) Describe a U-statistic estimator of γ, say, U.

(b) Find the asymptotic variance of $\sqrt{n}\,(U-\gamma)$.

3.6.13. Let $X_1,\ldots,X_n$ denote a random sample from a population with c.d.f. $F(\cdot)$. Let U_i denote a U-statistic estimator for the parameter γ_i of degree r_i based on the symmetric kernel $h^{(i)}(x_1,\ldots,x_{r_i})$, $i=1,2$. Show that the covariance between U_1 and U_2 is given by

$$\operatorname{Cov}(U_1,U_2)=\frac{1}{\binom{n}{r^*}}\sum_{c=1}^{r_*}\binom{r_*}{c}\binom{n-r_*}{r^*-c}\zeta_c^{(1,2)},$$

provided that

$$\zeta_c^{(1,2)}=\operatorname{Cov}\Big[h^{(1)}(X_1,\ldots,X_c,X_{c+1},\ldots,X_{r_1}),\\ h^{(2)}(X_1,\ldots,X_c,X_{r_1+1},\ldots,X_{r_1+r_2-c})\Big]$$

exists for $c=1,\ldots,r_*$, where $r_*=\min(r_1,r_2)$, $r^*=\max(r_1,r_2)$, and $n\geqslant r^*$.

4 POWER FUNCTIONS AND THEIR PROPERTIES

4.1. Power Functions

In Chapter 2 we discussed several important methods for constructing a test statistic with a nonparametric distribution-free property. Such a property provides assurance that the testing procedure maintains the designated α-level over a wide variety of distributional assumptions. A second and equally important consideration for testing procedures is their effectiveness in detecting alternative hypotheses. Chapters 4 and 5 are devoted to this topic. The power function and its properties are developed in this chapter, and in Chapter 5 we discuss large sample comparisons of the effectiveness of various testing procedures.

Definition 4.1.1. Suppose that a model is indexed by parameter(s) ξ. The **power function** of a test of hypothesis relevant to this model is given by

$$\mathcal{P}(\xi) = P_{\xi}[\text{the null hypothesis is rejected}].$$

Consider the one-sample problem in which $X_1,\ldots,X_n$ is a random sample from a continuous distribution that is symmetric about θ and has c.d.f. $F(x-\theta)$. This model is indexed by two parameters, θ and the underlying distribution $F(\cdot)$. The appropriate null hypothesis is that the point of symmetry is a particular value, which, without loss of generality, is taken to be 0. Thus we test H_0: $\theta=0$ versus, for example, H_1: $\theta>0$.

Example 4.1.2. The sign test, with test statistic $B=$[number of X_is>0], is appropriate for this problem. Under H_0, B is a binomial random variable with probability of success equal to $\frac{1}{2}$. More generally, for any θ, B is a binomial random variable with probability of success equal to

$1-F(-\theta)=F(\theta)$. So if the rule rejects whenever $B \geqslant b_0$, the power function is

$$\mathcal{P}(\theta,F) = \sum_{b=b_0}^{n} \binom{n}{b}\{F(\theta)\}^b\{1-F(\theta)\}^{n-b}. \tag{4.1.3}$$

Suppose that we let $F(x)$ correspond to a normal distribution with mean 0 and variance σ^2. For $n=18$ and rejection region $B \geqslant 13$, we have a test with level $\alpha=.048$. A picture of the power curve (4.1.3) as a function of θ/σ for this particular situation is shown as a solid line in Figure 4.1.4.

Example 4.1.5. Alternatively, we could use a one-sample Student t-test for this location problem, rejecting if

$$T^+ = \frac{\bar{X}}{S/\sqrt{n}}$$

is greater than $t(\alpha,n-1)$, the upper 100αth percentile of a Student t-distribution with $n-1$ degrees of freedom. If the underlying distribution is normal with mean θ and variance σ^2, then the power function of the test based on T^+ becomes

$$\mathcal{P}(\theta/\sigma,\text{normal}) = 1-\Phi^*\big(t(\alpha,n-1)|n-1,\sqrt{n}\,\theta/\sigma\big), \tag{4.1.6}$$

where $\Phi^*(x|r,c)$ is the c.d.f. of a noncentral t-distribution with r degrees of

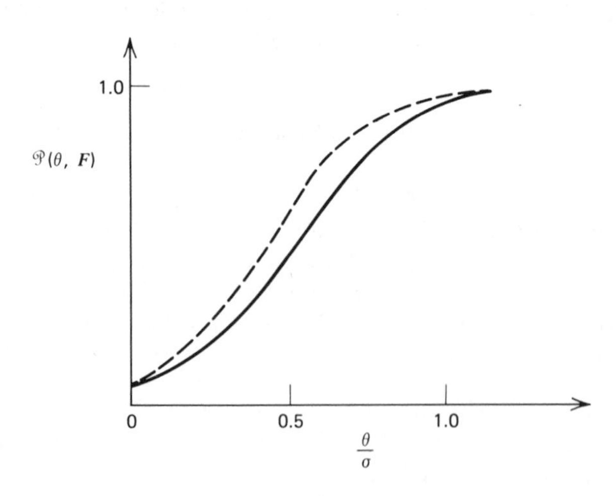

FIGURE 4.1.4

freedom and noncentrality parameter c. The power function (4.1.6) is shown as a hashed line in Figure 4.1.4 for the case in which $n=18$ and $\alpha=.05$. It is not surprising that the power function for the t-test is better than that of the sign test in this situation, since we are trying to detect a shift in a normal distribution. Unfortunately, the power function of the one-sample t-test is very difficult to express when the underlying distribution is other than normal. Thus analogous power comparisons between the one-sample t-test and the sign test for nonnormal distributions are not easily made.

For the Wilcoxon signed rank test the form of the power function is even more complex. A representation of each rank configuration under alternatives is discussed in Lemma 4.3.18, but the power function simplifies in only a few special cases. Klotz (1963) has computed some exact powers of the Wilcoxon signed rank test for shifts in a normal distribution. He used samples of sizes 5 through 10. So did Arnold (1965) when he studied its power against shifts in t-distributions (including the Cauchy). However, the limited number of distributions for which exact powers are available makes it difficult to compare the Wilcoxon signed rank test to the t-test and the sign test with regard to exact powers. In order to examine and relate their power properties, we resort to a Monte Carlo study. That is, we simulate the selection of random samples on a computer and observe the relative frequency with which a particular test rejects the null hypothesis (i.e., we compute the empirical power of the test). The results of one such Monte Carlo study involving 5000 replications of the random sampling process are shown in Table 4.1.7. It displays the empirical powers (times 1000) of the three one-sample tests: Student t, Wilcoxon signed rank (W^+), and sign test (B), for shifts in five different types of underlying distributions (uniform, normal, logistic, double exponential, and Cauchy) and three sample sizes ($n=10, 15$, and 20). Randomization was used for W^+ and B so that all three tests have a nominal $\alpha=.05$ level. The studied shift amounts, θ, were $0\sigma, .2\sigma, .4\sigma, .6\sigma$, and $.8\sigma$, where σ denotes the standard deviation of the distribution. (When the underlying distribution is a Cauchy centered at 0, σ denotes the value such that the probability between $-\sigma$ and σ is the same as the probability between -1 and 1 for a standard normal distribution.) Generation of observations from these distributions is discussed in Exercises 4.1.4 and 4.1.5. Additional details on these distributions may be found in Section A.1 of the Appendix.

Certain comparisons among the tests can be made using Table 4.1.7. When the underlying distribution is uniform, the t-test is somewhat better than the Wilcoxon signed rank test, and the sign test does rather poorly in comparison. For normal distributions, the t-test is only slightly better than the signed rank test; although the sign test does better in comparison, it is

TABLE 4.1.7.
Empirical Power Times 1000 of One-Sample Tests with $\alpha = .05$

DISTRIBUTION:		UNIFORM					NORMAL					LOGISTIC					DOUBLE EXPONENTIAL					CAUCHY				
n	TEST \ θ/σ	0	.2	.4	.6	.8	0	.2	.4	.6	.8	0	.2	.4	.6	.8	0	.2	.4	.6	.8	0	.2	.4	.6	.8
	T^+	051	136	294	512	746	049	146	330	543	758	049	156	345	571	770	047	172	374	602	781	028	095	197	309	414
10	W^+	049	133	277	474	681	050	144	315	527	741	049	157	348	564	771	048	190	412	633	804	049	166	332	493	623
	B	049	101	188	303	453	054	132	263	440	633	048	141	300	502	699	048	197	407	617	782	047	183	390	579	720
	T^+	051	169	408	703	914	048	181	424	716	906	051	189	452	727	905	049	202	473	739	898	025	094	210	321	418
15	W^+	051	163	383	642	852	047	178	418	693	893	050	188	456	739	912	050	226	532	786	926	050	196	423	622	750
	B	053	124	230	390	590	048	149	331	564	777	051	173	387	648	846	051	239	528	775	907	052	228	498	733	861
	T^+	049	214	522	829	971	048	225	546	833	967	046	228	553	831	962	044	238	571	835	955	026	099	214	329	433
20	W^+	050	205	479	768	935	049	218	531	813	962	049	236	565	844	969	049	284	652	885	975	049	234	514	730	849
	B	055	133	278	487	703	056	186	417	677	873	054	198	484	755	921	052	297	644	869	962	055	274	608	831	930

One-sample tests: T^+ = Student t-test, W^+ = Wilcoxon signed rank test, and B = sign test

Values near 050 or 950 have standard deviation 3.1, near 200 or 800 it is 5.7, near 350 or 650 it is 6.7, and for 500 it is 7.1.

Maximum estimated standard deviation of value differences between tests for $\theta/\sigma = 0$ is 3.4, for $\theta/\sigma = .2$ is 6.4, and for other θ/σ values it is 7.5. (Estimated standard deviation of $\hat{p}_1 - \hat{p}_2$ is $[\widehat{\mathrm{Var}}(\hat{p}_1) + \widehat{\mathrm{Var}}(\hat{p}_2) - 2\widehat{\mathrm{Cov}}(\hat{p}_1, \hat{p}_2)]^{1/2}$.)

still a poor third place. The signed rank test and the t-test are quite close when the underlying distribution is logistic, with the signed rank procedure doing better for larger sample sizes. The sign test remains in third place but is more competitive. When the distribution is double exponential, the sign test is best for small amounts of shift, but the signed rank test is better for larger shift amounts. The t-test is in third place, but remains competitive. With a Cauchy distribution, the t-test becomes very conservative (the true level is much lower than the prescribed level), and its power drops off dramatically. The sign test is best for this case, with the signed rank test a reasonable second place.

Now consider the two-sample location problem in which $X_1,\ldots,X_m$ and $Y_1,\ldots,Y_n$ are independent random samples from $F(x)$ and $F(x-\Delta)$, respectively, and we wish to test H_0: $\Delta=0$ versus H_1: $\Delta>0$.

Example 4.1.8. The test of H_0 versus H_1 can be performed using the two-sample t-test which rejects if

$$T=\frac{\bar{Y}-\bar{X}}{\sqrt{\left(\dfrac{1}{n}+\dfrac{1}{m}\right)\dfrac{(m-1)S_X^2+(n-1)S_Y^2}{m+n-2}}} \tag{4.1.9}$$

is greater than $t(\alpha,m+n-2)$, the upper 100αth percentile of a Student t-distribution with $m+n-2$ degrees of freedom. If the underlying $F(\cdot)$ is the c.d.f. of a normal distribution, then the power function for this test is

$$\mathcal{P}(\Delta/\sigma,\text{ normal})=1-\Phi^*\big(t(\alpha,m+n-2)|m+n-2,\Delta\sqrt{mn/(m+n)}\;/\sigma\big),$$

where $\Phi^*(x|r,c)$ is the c.d.f. of a noncentral t-distribution with r degrees of freedom and noncentrality parameter c.

A nonparametric distribution-free test for shift in the two-sample problem can be based on the Mann–Whitney–Wilcoxon test statistic

$$W=\sum_{i=1}^{n}R_i,$$

where R_i denotes the rank of Y_i among all $m+n$ observations. Like the one-sample signed rank test, the power function of this rank procedure is not easily expressed. Hoeffding's (1951) representation, which is discussed in Lemma 4.3.12, is of some help here, but only in limited cases. Milton (1970) has computed tables of the exact power of the test based on W for shifts in a normal distribution with sample sizes m and n between 1 and 7.

TABLE 4.1.10.

Empirical Power Times 1000 of Two-Sample Tests with $\alpha=.05$

DISTRIBUTION:		UNIFORM					NORMAL					LOGISTIC				
m, n	TEST \ Δ/σ	0	.3	.6	.9	1.2	0	.3	.6	.9	1.2	0	.3	.6	.9	1.2
$m=5$, $n=5$	T	051	105	202	330	520	044	111	213	356	523	047	119	226	391	559
	W	049	111	203	332	492	046	108	208	346	503	048	117	229	384	545
$m=15$, $n=15$	T	047	196	478	763	940	052	206	497	785	947	046	208	494	779	933
	W	046	193	450	712	895	054	205	479	766	933	046	214	515	787	943
$m=5$, $n=15$	T	048	128	270	487	710	047	144	303	511	724	046	131	295	518	722
	W	047	123	253	447	646	048	141	287	492	694	047	134	296	520	720
$m=15$, $n=5$	T	048	131	287	493	712	053	149	313	518	729	049	136	305	521	727
	W	047	129	270	448	647	050	140	296	499	703	046	140	313	528	724

Two-sample tests: T=Student t-test, W=Mann–Whitney–Wilcoxon test

Values near 050 or 950 have standard deviation 3.1, near 200 or 800 it is 5.7, near 350 or 650 it is 6.7, near 500 it is 7.1.

Maximum estimated standard deviation of value differences between tests for $\Delta/\sigma=0$ is 3.5, for $\Delta/\sigma=.2$ is 5.4 and for other Δ/σ values it is 7.1.

Bell, Moser, and Thompson (1966) consider shifts in uniform distributions with $m+n$ assuming values 9 through 16 and 20 and $m \doteq n$. Hayman and Govindarajulu (1966) study shifts in an exponential distribution with $m+n=11$, 15, 21. However, the lack of available exact powers for various comparable distributions and sample sizes for both the t-test and W led us to perform a second Monte Carlo study. Table 4.1.10 displays this study of the powers of the Mann–Whitney–Wilcoxon test and the two-sample Student t-test for the two-sample problem of a shift in location. Randomization was used for the test based on W to ensure that it achieves a nominal $\alpha=.05$ level. The details of this study are similar to those of the

TABLE 4.1.10. (Continued)

DOUBLE EXPONENTIAL					CAUCHY					WEIBULL ($\alpha=2$)					EXPONENTIAL				
0	.3	.6	.9	1.2	0	.3	.6	.9	1.2	0	.3	.6	.9	1.2	0	.3	.6	.9	1.2
045	127	255	419	588	024	066	132	207	288	049	113	221	374	545	044	133	288	465	629
049	132	269	426	589	051	118	218	323	408	049	117	219	369	537	049	163	332	515	650
046	218	507	785	928	030	079	153	243	333	046	202	488	768	935	045	226	524	783	927
046	252	594	845	962	046	210	484	700	839	046	204	488	765	927	046	339	700	903	977
046	136	304	536	733	056	087	137	205	282	041	129	289	520	723	026	137	354	587	768
047	153	351	581	768	046	133	284	441	576	049	133	290	497	688	047	229	454	646	784
050	142	315	537	732	061	097	146	209	279	053	137	293	505	714	064	159	323	537	734
047	165	366	589	764	046	140	297	457	590	046	126	283	491	711	046	179	432	698	871

one-sample Monte Carlo described earlier in this section. Two additional distributions, both skewed to the right, were included for completeness. The shift amounts, Δ, were 0σ, $.3\sigma$, $.6\sigma$, $.9\sigma$ and 1.2σ, where σ denotes the same quantity (usually the standard deviation) that it did in the previous study. Details concerning the distributions used are found in Exercises 4.1.4 and 4.1.5 and in Section A.1 of the Appendix.

When the underlying distributions are uniforms, we notice that the test based on T is a bit better than the one based on W. The test based on T is only slightly better than W when the distributions are normals. For logistic distributions there is no clear winner with respect to power, but W may have a very slight edge. With double exponential or Cauchy distributions the W test is better than T and in the Cauchy case the test based on T is very poor. The performance of the two tests is very similar for this particular Weibull distribution, but when the underlying distributions are exponential W is superior to T.

Exercises 4.1.

4.1.1. Let X_1, X_2, X_3, X_4 be i.i.d. Poisson (λ) (i.e., $P[X=x]=(\lambda^x e^{-\lambda})/x!$ for $x=0,1,\ldots$). Suppose we test H_0: $\lambda=\frac{1}{4}$ versus H_1: $\lambda>\frac{1}{4}$ with a test that rejects H_0 if $\Sigma_{i=1}^{4} X_i \geqslant 3$. What is the level of this test? Evaluate its power function at $\lambda=2/5$, $1/2$, $3/5$, $3/4$. Sketch its power function.

4.1.2. Consider the class of distributions $\mathfrak{F}$ that are symmetric about zero. One possible measure of spread for such distributions is $p=P[|X|>1]$. Suppose that we wish to test H_0: $p=1/2$ versus H_1: $p>1/2$ using the statistic $V=$[number of X_is for which $|X_i|>1$], where $X_1,\ldots,X_n$ are i.i.d. $F\in\mathfrak{F}$. Assume $n=10$ and a critical region of $V\geqslant 8$. What is the α-level of this test? Sketch the power function of this test for the case in which the underlying distribution F is normal by computing the power of the test against $\sigma=1.75, 2.0, 2.5$, and 3.5.

4.1.3. Let X_1, X_2 be independent Gamma $(1,\theta)$ random variables; that is, exponentials with scale parameter θ. Suppose we test H_0: $\theta=1$ versus H_1: $\theta>1$ with a test that rejects H_0 if $2(X_1+X_2)\geqslant 9.49$. What is the size of this test? Sketch the power function of this test by evaluating the power at $\theta=1.25, 1.5, 1.75$, and 2.0.

4.1.4. Random number generators produce independent uniform $(0,1)$ random variables. Recall from Theorem 1.2.15 that if U is such a uniform variate, then $X=F^{-1}(U)$ has c.d.f. $F(x)$. Find an appropriate $F^{-1}(\cdot)$ to generate each of the following random variables:

(a) X having a uniform distribution with mean zero and variance 1.

(b) X having a logistic distribution with mean 0 and variance 1.

(c) X having a double exponential distribution with mean 0 and variance 1.

(d) X having a Cauchy distribution centered at 0 with scale such that the probability between -1 and 1 is the same as the probability placed in that region by a standard normal.

(e) X having an exponential distribution shifted to the left so that it has mean 0 and variance 1.

(f) X having a Weibull distribution with $\alpha=2$ that has been shifted to the left so that it has mean 0 and variance 1.

4.1.5. Box and Muller (1958) discovered that if U_1, U_2 are independent uniform (0, 1) random variables, then X_1, X_2 defined by

$$X_1 = [-2\ln(U_1)]^{1/2}\sin(2\pi U_2)$$
$$X_2 = [-2\ln(U_1)]^{1/2}\cos(2\pi U_2),$$

are independent standard normal variables. This provides a useful method of generating normally distributed random variables. Prove their result by deriving the joint density of X_1, X_2. [Hint: First show that the inverses are

$$U_1 = \exp[-(X_1^2 + X_2^2)/2]$$
$$U_2 = \frac{1}{2\pi}\operatorname{Arctan}\left(\frac{X_1}{X_2}\right) \qquad \text{if } X_1 > 0 \text{ and } X_2 > 0$$
$$= \frac{1}{2} + \frac{1}{2\pi}\operatorname{Arctan}\left(\frac{X_1}{X_2}\right) \qquad \text{if } X_2 < 0$$
$$= 1 + \frac{1}{2\pi}\operatorname{Arctan}\left(\frac{X_1}{X_2}\right) \qquad \text{if } X_1 < 0 \text{ and } X_2 > 0.]$$

4.1.6. Tukey (1960) suggested a class of symmetric distributions defined in terms of inverse c.d.f.'s. Subsequently this class has been developed and generalized to include asymmetric distributions by Ramberg and Schmeiser (1972, 1974). The **Ramberg-Schmeiser-Tukey** (abbreviated RST) **lambda distributions** are of the form

$$F^{-1}(u) = \lambda_1 + \lambda_2^{-1}[u^{\lambda_3} - (1-u)^{\lambda_4}],$$

for $0 < u < 1$. Since it is defined in terms of the inverse of a c.d.f., this family is particularly convenient for Monte Carlo studies. The quantities λ_1 and λ_2 are location and scale parameters, respectively, while λ_3 and λ_4 are shape parameters. If $\lambda_3 = \lambda_4$ the distribution is symmetric. Although other $\boldsymbol{\lambda}$ configurations define legitimate inverse c.d.f.'s [see Table I of Ramberg and Schmeiser (1974)], we assume for simplicity that λ_2, λ_3, and λ_4 are all positive. Derive the following formulae for the mean and variance of an RST lambda distribution:

$$\mu = \lambda_1 + \frac{\lambda_4 - \lambda_3}{\lambda_2(\lambda_3+1)(\lambda_4+1)}$$

and

$$\sigma^2 = \left[\frac{1}{2\lambda_3+1} + \frac{1}{2\lambda_4+1} - \frac{2\Gamma(\lambda_3+1)\Gamma(\lambda_4+1)}{\Gamma(\lambda_3+\lambda_4+2)} - \frac{(\lambda_4-\lambda_3)^2}{(\lambda_3+1)^2(\lambda_4+1)^2} \right] \Big/ \lambda_2^2.$$

4.1.7. (O'Meara (1976)) Show that the expected value of the kth order statistic for a random sample of size n from an RST lambda distribution (see Exercise 4.1.6) is

$$\lambda_1 + \lambda_2^{-1} n! \left[\frac{\Gamma(k+\lambda_3)}{(k-1)!\Gamma(n+\lambda_3+1)} - \frac{\Gamma(n+\lambda_4+1-k)}{(n-k)!\Gamma(n+\lambda_4+1)} \right].$$

(Assume for simplicity that λ_2, λ_3 and λ_4 are positive.)

4.2. Properties of a Power Function

Many important properties of a test procedure are described in terms of its power function. Let $\omega \subset \Omega$ and suppose we are testing H_0: $\xi \in \omega$ versus H_1: $\xi \in \Omega - \omega$.

Definition 4.2.1. The **size** of a test with power function $\mathcal{P}(\xi)$ is $\sup_{\xi \in \omega} \mathcal{P}(\xi)$. The test is said to be of **level** α if its size is less than or equal to α.

The tests introduced in Chapter 2 are classic examples of nonparametric distribution-free tests of hypotheses. They have the common property that the null hypothesis distributions of their test statistics are discrete. Hence the size of a test based on such a statistic is restricted to a finite collection of possible values (unless extraneous randomization is used). This finite collection is referred to as the **natural levels** of the test. In this book we are concerned with broadly stated hypotheses that are often indexed by a parameter, say, θ, with a specific role such as location, and an underlying distribution, say, $F(\cdot)$. Thus the parameters for such a problem are $\xi = (\theta, F)$, and the corresponding null hypothesis is composite, as, for example, $\{\xi | \xi = (\theta_0, F)$, where θ_0 is fixed and F is in some class $\mathcal{F}\}$. When such a null class includes many underlying distribution types, a nonparametric distribution-free test has the special advantage that its natural level is achieved for every $F(\cdot)$ in $\mathcal{F}$; that is, the probability of a type I error is α for every point in the composite null hypothesis. A test with this property is referred to as an **exact** test of size α.

Definition 4.2.2. The α-level test with power function $\mathcal{P}(\xi)$ is said to be **unbiased at level α** if

$$\mathcal{P}(\xi) \geqslant \alpha \text{ for all } \xi \in \Omega - \omega.$$

In one sense the property of unbiasedness is weak in that it requires the power function only to be greater than or equal to the level at all points in the alternative. Yet this property does place a strong emphasis on a test's ability to achieve the prescribed level α at all boundary points of the null hypothesis. For example, consider the one-sample t-test for a change in location as described in Example 4.1.5. It can be shown that this test is unbiased when the underlying distribution is known to be normal. See, for example, Figure 4.1.4. However, if the underlying distribution is symmetric and continuous, but not necessarily normal, the test may be biased. For detecting a location change in a Cauchy distribution, for example, the one-sample t-test is very conservative; that is, its size is much smaller than the desired α-level. This fact was illustrated in Table 4.1.7. Since this power function is a continuous function of θ (the amount of change in location), it follows that for very small $\theta > 0$ the power of the test will be less than α when the underlying distribution is Cauchy. Hence for this broader problem the one-sample t-test is not unbiased.

An important topic that is related to the property of bias is monotonicity of a power function. Consider the one-sample problem where

$$X_1, \ldots, X_n \text{ are i.i.d. } F(x-\theta), \tag{4.2.3}$$

and the location parameter θ is the median of the underlying continuous distribution. We wish to test

$$H_0\colon \theta = \theta_0 \text{ versus } H_1\colon \theta > \theta_0. \tag{4.2.4}$$

Theorem 4.2.5. *Suppose that for testing H_0 versus H_1 we reject H_0 for large (small) values of a test statistic $S(X_1, \ldots, X_n)$ that satisfies*

$$S(x_1+k, \ldots, x_n+k) \geqslant (\leqslant) S(x_1, \ldots, x_n) \tag{4.2.6}$$

for every $k \geqslant 0$ and $(x_1, \ldots, x_n)$. Then the test has a monotone power function in θ for the one-sample location problem; that is,

$$\mathcal{P}_S(\theta, F) \leqslant \mathcal{P}_S(\theta', F) \text{ for } \theta \leqslant \theta',$$

and any continuous distribution with c.d.f. $F(\cdot)$.

Proof: If $X_1,\ldots,X_n$ are i.i.d. $F(x-\theta)$, then $X_1+(\theta'-\theta),\ldots,X_n+(\theta'-\theta)$ are i.i.d. $F(x-\theta')$. Consider the case where we reject for large values of $S(\cdot)$ and let c be the corresponding critical value of the test. Then for any $\theta'\geqslant\theta$,

$$\begin{aligned}
\mathcal{P}_S(\theta,F) &= P[S(X_1,\ldots,X_n)\geqslant c|\text{each } X_i\sim F(x-\theta)]\\
&\leqslant P[S(X_1+(\theta'-\theta),\ldots,X_n+(\theta'-\theta))\geqslant c|\text{each } X_i\sim F(x-\theta)]\\
&= P[S(X_1^*,\ldots,X_n^*)\geqslant c|\text{each } X_i^*\sim F(x-\theta')]\\
&= \mathcal{P}_S(\theta',F).
\end{aligned}$$

The case where we reject for small values of $S(\cdot)$ follows similarly. ■

Example 4.2.7. Consider the one-sample Student t-test which rejects H_0 for large values of

$$T^+(X_1,\ldots,X_n) = \frac{\bar{X}-\theta_0}{S/\sqrt{n}},$$

where S denotes the sample standard deviation. It follows that

$$\begin{aligned}
T^+(x_1+k,\ldots,x_n+k) &= \frac{\bar{x}-\theta_0+k}{s/\sqrt{n}}\\
&= T^+(x_1,\ldots,x_n)+\frac{k}{s/\sqrt{n}}.
\end{aligned}$$

Thus property (4.2.6) is satisfied, and the one-sample t-test has a monotone power function in θ.

Example 4.2.8. The sign test rejects for large values of

$$B(X_1,\ldots,X_n) = [\text{number of } X_i\text{s} > \theta_0].$$

Thus for $k\geqslant 0$,

$$\begin{aligned}
B(x_1+k,\ldots,x_n+k) &= [\text{number of } (x_i+k)\text{s} > \theta_0]\\
&= [\text{number of } x_i\text{s} > (\theta_0-k)] \geqslant B(x_1,\ldots,x_n),
\end{aligned}$$

and the test has a monotone power function in θ. Note also that the sign test achieves its natural levels for every distribution described in (4.2.3).

Therefore, it is an unbiased test of H_0: $\theta=\theta_0$ versus H_1: $\theta>\theta_0$ at each natural level α.

Example 4.2.9. The Wilcoxon signed rank test rejects H_0 in (4.2.4) for large values of

$$W^+(X_1,\ldots,X_n) = \sum\sum_{1\leqslant i\leqslant j\leqslant n} \Psi(X_i+X_j-2\theta_0), \tag{4.2.10}$$

where $\Psi(x)=1,0$ as $x>, \leqslant 0$. (See Exercise 2.4.3.) In this form it is easy to see that $W^+(\cdot)$ satisfies (4.2.6) and hence has a monotone power function in θ. If consideration is restricted to continuous distributions that are symmetric about θ, then W^+ provides an exact test at each of its natural levels and is thus an unbiased test of (4.2.4) at these levels.

The development of monotonicity of a power function for the two-sample problem is analogous to that for the one-sample setting. Suppose we have two independent samples with

$$X_1,\ldots,X_m \text{ i.i.d. } F(x) \text{ and } Y_1,\ldots,Y_n \text{ i.i.d. } F(x-\Delta), \tag{4.2.11}$$

where $F(\cdot)$ is the c.d.f. of a continuous distribution. We wish to test

$$H_0\text{: } \Delta=0 \text{ versus } H_1\text{: } \Delta>0. \tag{4.2.12}$$

Theorem 4.2.13. *Suppose that we reject H_0 in favor of H_1 for large (small) values of the test statistic $S(X_1,\ldots,X_m; Y_1,\ldots,Y_n)$ satisfying*

$$S(x_1,\ldots,x_m;y_1+k,\ldots,y_n+k) \geqslant (\leqslant) S(x_1,\ldots,x_m;y_1,\ldots,y_n) \tag{4.2.14}$$

for every $k\geqslant 0$ and every $(x_1,\ldots,x_m;y_1,\ldots,y_n)$. Then for the problem described in (4.2.11) *and* (4.2.12), *the power function of the test based on $S(\cdot)$ is monotone in Δ; that is,*

$$\mathcal{P}_S(\Delta,F) \leqslant \mathcal{P}_S(\Delta',F)$$

for all $\Delta\leqslant\Delta'$.

Proof: Exercise 4.2.2. ∎

Example 4.2.15. The two-sample t-test uses the statistic

$$T(X_1,\ldots,X_m;Y_1,\ldots,Y_n) = \frac{\overline{Y}-\overline{X}}{\sqrt{\left(\dfrac{1}{m}+\dfrac{1}{n}\right)\dfrac{\left[(m-1)S_X^2+(n-1)S_Y^2\right]}{m+n-2}}},$$

which satisfies (4.2.14). Thus the two-sample t-test has a monotone power function in Δ.

Example 4.2.16. The Mann–Whitney–Wilcoxon test can be based on the statistic

$$U_7(X_1,\ldots,X_m;Y_1,\ldots,Y_n)=\sum_{i=1}^{m}\sum_{j=1}^{n}\Psi(Y_j-X_i),$$

where $\Psi(t)=1,0$ as $t>,\leqslant 0$. In this form it is easy to see that (4.2.14) is satisfied and hence that the test has a power function that is monotone in Δ. In addition, since it is nonparametric distribution-free and is thus an exact test for the problem described in (4.2.11) and (4.2.12), it is also unbiased for its natural α-levels.

Another important property of a test of hypothesis concerns the asymptotic behavior of its power function. Let T_N denote a test statistic based on a sample of size N. (In two-sample problems, $N=m+n$.) The sequence of tests $\{T_N\}$ yields a power function $\mathcal{P}_{T_N}(\xi)$, for each N.

Definition 4.2.17. If $\{T_N\}$ denotes a sequence of α-level tests of H_0: $\xi\in\omega$ versus H_1: $\xi\in\Omega-\omega$, then the sequence is said to be **consistent against the class $\Omega-\omega$** if

$$\lim_{N\to\infty}\mathcal{P}_{T_N}(\xi)=1$$

for every $\xi\in\Omega-\omega$.

The idea is that with enough sample information (i.e., $N\to\infty$) we should be able to detect any point in the alternative $\Omega-\omega$.

Theorem 4.2.18. [*Lehmann* (1951)] *Let $\{T_N\}$ denote a sequence of test statistics for an α-level test of H_0: $\xi\in\omega$ versus H_1: $\xi\in\Omega-\omega$, such that the test based on T_N rejects H_0 if $T_N\geqslant c_N$. Suppose there exists a function $k(\xi)$ such that T_N converges in probability to $k(\xi)$ for every $\xi\in\Omega$. If, in addition,*

$$k(\xi)=k_0 \qquad \text{for all } \xi\in\omega,$$
$$k(\xi)>k_0 \qquad \text{for all } \xi\in\Omega-\omega,$$

and

$$\lim_{N\to\infty}c_N\leqslant k_0,$$

then $\{T_N\}$ is a consistent sequence of tests for all alternatives in H_1: $\xi\in\Omega-\omega$.

Proof: Let ξ^* denote an arbitrary point in the alternative $\Omega-\omega$. Define $\epsilon=[k(\xi^*)-k_0]/2>0$. For N sufficiently large, we have $c_N \leqslant k_0+\epsilon=k(\xi^*)-\epsilon$. Therefore, the limit of the power function at ξ^* is

$$\lim_{N\to\infty} \mathcal{P}_{T_N}(\xi^*) = \lim_{N\to\infty} P_{\xi^*}[T_N \geqslant c_N] \geqslant \lim_{N\to\infty} P_{\xi^*}[T_N \geqslant k(\xi^*)-\epsilon]$$
$$\geqslant \lim_{N\to\infty} P_{\xi^*}[|T_N - k(\xi^*)| \leqslant \epsilon] = 1,$$

since T_N converges in probability to $k(\xi^*)$. ■

Note that the convergence in probability of T_N to $k(\xi)$ is often determined by showing convergence in quadratic mean for each $\xi\in\Omega$. (See Theorem 3.2.3.)

Example 4.2.19. Consider the problem in which we have two independent random samples from continuous distributions, namely, $X_1,\ldots,X_m$ i.i.d. $F(x)$ and $Y_1,\ldots,Y_n$ i.i.d. $G(y)$. Here the parameter of interest is the pair of distributions $[F(\cdot),G(\cdot)]$, and the associated null hypothesis is that the distributions are identical. It is, in fact, a composite null hypothesis, indexed by the common continuous distribution, say, $F(\cdot)$. In the more specific location problem we might be interested in alternatives for which the Ys tend to be larger than the Xs. For such settings, we could use the Mann–Whitney–Wilcoxon test which rejects for large values of the U-statistic, U_7, displayed in (3.4.4). We know that $E[U_7]=P[X<Y]$ and, by Corollary 3.4.9, that U_7 converges in probability to $P[X<Y]=k(F(\cdot),G(\cdot))$ as $N=m+n\to\infty$ in such a way that $(m/N)\to\lambda$, where $0<\lambda<1$. Under H_0, the two continuous distributions are identical, and $P_0[X<Y]=k(F(\cdot),F(\cdot))=\frac{1}{2}$. In addition, the asymptotic normality of a properly standardized U_7 under H_0 was shown in Example 3.4.15. This result [specifically note (3.4.17)] shows that, for large N, we reject H_0 if

$$U_7 = T_N \geqslant c_N \doteq \frac{1}{2} + z_\alpha\sqrt{\frac{m+n+1}{12mn}}\,,$$

where z_α is the upper 100αth percentile of a standard normal distribution. As $N\to\infty$, $c_N\to\frac{1}{2}$. Thus Theorem 4.2.18 shows that this one-sided Mann–Whitney–Wilcoxon test is consistent against all alternatives $[F(\cdot),G(\cdot)]$ for which $k(F(\cdot),G(\cdot))=P[X<Y]>\frac{1}{2}$.

We note that Theorem 4.2.18 can be extended to handle one-sided tests that reject for small T_N values and two-sided tests as well. See Exercise 4.2.9 for a statement of the appropriate version for two-sided tests.

Exercises 4.2.

4.2.1. Let $X_1,\ldots,X_n$ be i.i.d. $F(x/\sigma)$, where $F(x/\sigma)$ is the c.d.f. of a continuous distribution that is symmetric about 0 with standard deviation $\sigma>0$. Suppose that we test

$$H_0\colon \sigma = 1 \text{ versus } H_1\colon \sigma > 1$$

by rejecting for large (small) values of a test statistic $S(\cdot)$ satisfying

$$S(dx_1,\ldots,dx_n) \geqslant (\leqslant) S(x_1,\ldots,x_n)$$

for every $(x_1,\ldots,x_n)$ and $d \geqslant 1$. Prove that such a test will have a monotone power function in σ for any fixed distribution type $F(\cdot)$.

4.2.2. Prove Theorem 4.2.13.

4.2.3. Consider the two-sample shift in location problem as defined in (4.2.11) and (4.2.12). Suppose that we test this hypothesis by rejecting H_0: $\Delta=0$ for large values of a rank statistic of the form

$$S = \sum_{i=1}^{n} a(R_i),$$

where $a(1)\leqslant \cdots \leqslant a(m+n)$ are known values (not all the same) and R_i is the rank of Y_i among all $m+n$ observations.

(a) Show that the test based on S will have a monotone power function in Δ for this problem.

(b) Using **(a)** argue that the test based on S will be unbiased when the significance level is one of its natural levels.

4.2.4. Consider a regression problem in which

$$Y_i = \beta c_i + E_i, \qquad i = 1,\ldots,n,$$

where $E_1,\ldots,E_n$ are i.i.d. continuous random variables with mean zero, and $c_1,\ldots,c_n$ are known constants (not all the same). Suppose that we test

$$H_0\colon \beta = 0 \text{ versus } H_1\colon \beta > 0$$

by rejecting H_0 for large (small) values of a test statistic $S(c_1,y_1,\ldots,c_n,y_n)$ satisfying

$$S(c_1,y_1+\delta c_1,\ldots,c_n,y_n+\delta c_n) \geqslant (\leqslant) S(c_1,y_1,\ldots,c_n,y_n)$$

for every $(c_1,y_1,\ldots,c_n,y_n)$ and $\delta \geqslant 0$. Prove that the test based on $S(\cdot)$ will have a power function that is monotone in the parameter β for any fixed distribution of the errors E_i.

4.2.5. Consider the use of the sign test for the one-sample change of location problem described in (4.2.3) and (4.2.4). Suppose that the continuous distribution $F(\cdot)$ has support over the whole real line. With this condition on $F(\cdot)$, show that the sign test is consistent against all alternatives $\theta > \theta_0$.

4.2.6. Let $X_1,\ldots,X_n$ denote a random sample from a continuous distribution that is symmetric about a location parameter θ and has c.d.f. $F(x-\theta)$. To test H_0: $\theta=0$ versus H_1: $\theta>0$, we could use the one-sided Wilcoxon signed rank test. Show that this test is consistent against all alternatives for which $P[X_1+X_2>0]>\frac{1}{2}$. [Hint: See Example 3.3.17.]

4.2.7. In Example 4.2.19 we argued that the one-sided Mann–Whitney–Wilcoxon test is consistent against any alternative for which $P[X<Y]>\frac{1}{2}$. Let $N=m+n\to\infty$ in such a way that $(m/N)\to\lambda$, where $0<\lambda<1$, and assume that $\lim_{N\to\infty}\text{Var}(\sqrt{N}\,U_7)>0$. Under these conditions, show that $P[X<Y]>\frac{1}{2}$ is also a necessary condition for consistency.

4.2.8. Let X and Y be independent, continuous random variables with c.d.f.'s $F(x)$ and $G(y)$ and medians θ_1 and θ_2, respectively.

(a) Suppose that the distributions of X and Y are symmetric about θ_1 and θ_2, respectively, and that there exists an interval (a,b), $a<b$, such that $f(x)=dF(x)/dx>0$ and $g(x)=dG(x)/dx>0$ for all $x\in(a,b)$. Show that, under these conditions,

$$P(X<Y) = \tfrac{1}{2} \text{ if and only if } \theta_1 = \theta_2.$$

Demonstrate, by means of an example, that the existence of such an interval (a,b) is necessary for this if and only if result.

(b) Find two asymmetric distributions for which $P(X<Y)=\frac{1}{2}$, but $\theta_1\neq\theta_2$.

(c) Find two distributions $F(\cdot)$ and $G(\cdot)$ for which $P(X<Y)=\frac{1}{2}$ and $\theta_1=\theta_2$, but $F(\cdot)\not\equiv G(\cdot)$.

(d) Find two distributions for which $\theta_1=\theta_2$, but $P(X<Y)>\frac{1}{2}$.

(e) What do the first four parts of this exercise say about the consistency class of the one-sided Mann–Whitney–Wilcoxon test? (See Exercise 4.2.7.)

4.2.9. Theorem 4.2.20. *Let* $\{T_N\}$ *denote a sequence of test statistics for an* α*-level test of* H_0: $\xi \in \omega$ *versus* H_1: $\xi \in \Omega - \omega$ *that rejects* H_0 *if* $T_N \geqslant c_N$ *or if* $T_N \leqslant c'_N$. *Suppose there exists a function* $k(\xi)$ *such that* T_N *converges in probability to* $k(\xi)$ *for every* $\xi \in \Omega$. *If, in addition,*

$$k(\xi) = k_0 \qquad \text{for all } \xi \in \omega,$$

$$k(\xi) \neq k_0 \qquad \text{for all } \xi \in \Omega - \omega,$$

$$\lim_{N \to \infty} c_N \leqslant k_0$$

and

$$\lim_{N \to \infty} c'_N \geqslant k_0,$$

then $\{T_N\}$ *is a consistent sequence of tests for all alternatives in* H_1: $\xi \in \Omega - \omega$.

Prove this theorem.

4.2.10. **(a)** Let $U(X_1, \ldots, X_n)$ be a one-sample U-statistic for the parameter γ. Consider testing H_0: $\gamma = \gamma_0$ versus H_1: $\gamma > \gamma_0$ by rejecting H_0 for large values of U and assume that U is distribution-free under H_0. Under what conditions will we be able to apply Theorem 4.2.18 and obtain the consistency of the test for alternatives $\gamma > \gamma_0$? Be sure to justify your statements. [Hint: Consider restrictions on the family $\mathcal{F}$ associated with U.]

(b) Discuss a result similar to that in **(a)**, but for a two-sample U-statistic.

4.3. Nonparametric Alternatives

In this section we consider two structures for alternative hypotheses that are nonparametric in nature. Both types have played important roles in the investigation of properties of rank procedures. The simplest and most natural description of these alternatives occurs in the two-sample location problem, so we will emphasize their use in this context. We begin with stochastically ordered random variables.

Definition 4.3.1. The random variable X with c.d.f. $F(x)$ is said to be **stochastically smaller** than the random variable Y with c.d.f. $G(x)$ if

$$F(x) \geqslant G(x)$$

for every x, with strict inequality for at least one x-value. If X is stochastically smaller than Y, then Y is said to be **stochastically larger** than X and the two distributions are said to be **stochastically ordered** with X less than Y. To simplify matters, we often say $F(\cdot)$ is stochastically smaller [larger] than $G(\cdot)$ instead of saying X with c.d.f. $F(\cdot)$ is stochastically smaller [larger] than Y with c.d.f. $G(\cdot)$.

Consider the two-sample setting for which $X_1,\ldots,X_m$ and $Y_1,\ldots,Y_n$ are independent random samples from continuous distributions with c.d.f.'s $F(x)$ and $G(y)$, respectively, and where the problem of interest is to test

$$H_0\text{: } F(x) = G(x) \text{ for every } x$$

versus **(4.3.2)**

$$H_1\text{: } F(\cdot) \text{ is stochastically smaller than } G(\cdot).$$

This alternative H_1 is similar to the shift alternatives studied earlier, in that when it pertains the Ys tend to be larger than the Xs. However, the stochastically ordered alternative is less rigid in its specification of the relationship between $F(\cdot)$ and $G(\cdot)$ and includes shift alternatives as a special case. It is desirable that two-sample tests for location have good power (or at least be unbiased) against such stochastically ordered alternatives.

The following result, due to Lehmann (1959), gives conditions that yield a monotone power function for stochastically ordered alternatives.

Theorem 4.3.3. *Consider a test of* (4.3.2) *that rejects for large (small) values of the statistic* $S(\cdot)$ *satisfying*

$$S(x_1,\ldots,x_m;y_1',\ldots,y_n') \geqslant (\leqslant) S(x_1,\ldots,x_m;y_1,\ldots,y_n) \qquad \textbf{(4.3.4)}$$

for every $x_1,\ldots,x_m$ *and* $y_1 \leqslant y_1',\ldots,y_n \leqslant y_n'$. *Then the power function of the test based on* S *is monotone, in the sense that if* $G(\cdot)$ *is stochastically smaller than* $H(\cdot)$ *and both are c.d.f.'s of continuous distributions, then*

$$\mathcal{P}_S(F,G) \leqslant \mathcal{P}_S(F,H).$$

Proof: We establish the result for the rejection region consisting of large values of S. (The proof for rejection with small values of S then follows immediately.) First we note that $G(x) \geqslant H(x)$ for all x implies $\{x|H(x) \geqslant u\} \subset \{x|G(x) \geqslant u\}$. Therefore, $H^{-1}(u) = \inf\{x|H(x) \geqslant u\} \geqslant \inf\{x|G(x) \geqslant u\} = G^{-1}(u)$ for every $0<u<1$. Let c denote the critical value of the test based on $S(\cdot)$, and let $U_1,\ldots,U_n$ denote i.i.d. uniform $(0,1)$

random variables, each with c.d.f. $J(x)$. Then using the above relationship on the inverse c.d.f.'s and Theorem 1.2.15, we see that

$$\begin{aligned}
\mathcal{P}_S(F,G) &= P\big[\, S(X_1,\dots,X_m;Y_1,\dots,Y_n) \geqslant c \,|\, \text{each } X_i \text{ has c.d.f.}\\
&\qquad F(\cdot) \text{ and each } Y_j \text{ has c.d.f. } G(\cdot)\big]\\
&= P\big[\, S(X_1,\dots,X_m;G^{-1}(U_1),\dots,G^{-1}(U_n)) \geqslant c \,|\, \text{each } X_i\\
&\qquad \text{has c.d.f. } F(\cdot) \text{ and each } U_j \text{ has c.d.f. } J(\cdot)\big]\\
&\leqslant P\big[\, S(X_1,\dots,X_m;H^{-1}(U_1),\dots,H^{-1}(U_n)) \geqslant c \,|\, \text{each } X_i\\
&\qquad \text{has c.d.f. } F(\cdot) \text{ and each } U_j \text{ has c.d.f. } J(\cdot)\big]\\
&= P\big[\, S(X_1,\dots,X_m;Y_1^*,\dots,Y_n^*) \geqslant c \,|\, \text{each } X_i \text{ has}\\
&\qquad \text{c.d.f. } F(\cdot) \text{ and each } Y_j^* \text{ has c.d.f. } H(\cdot)\big]\\
&= \mathcal{P}_S(F,H). \quad \blacksquare
\end{aligned}$$

Corollary 4.3.5. Let $S(\cdot)$ be a test statistic that is nonparametric distribution-free under H_0: $F(x) \equiv G(x)$ over the class of all continuous distributions. If, in addition, the test based on $S(\cdot)$ satisfies the conditions of Theorem 4.3.3, then it is an unbiased test of (4.3.2).

Proof: It follows at once from Theorem 4.3.3. ■

Thus, for example, the one-sided Mann–Whitney–Wilcoxon test is unbiased against the stochastically ordered alternatives in (4.3.2). It is also consistent against these stochastically ordered alternatives, provided that both distributions are of the continuous type. To see this we need only note the result in Example 4.2.19 and the fact that for such alternatives

$$P[X<Y] = \int_{-\infty}^{\infty} F(y)\,dG(y) > \int_{-\infty}^{\infty} G(y)\,dG(y) = \frac{1}{2}.$$

A second type of nonparametric alternative was introduced by Lehmann (1953). Let $G_1(u)$ and $G_2(u)$ be c.d.f.'s of continuous random variables that assume values only on the interval $(0,1)$. Let $X_1,\dots,X_m$ and $Y_1,\dots,Y_n$ be independent random samples from populations with c.d.f.'s $G_1(F(x))$ and $G_2(F(x))$, respectively, where $F(\cdot)$ is the c.d.f. for some continuous distribution. We wish to test

$$H_0\text{: } G_1(u) = G_2(u) = u \text{ for all } u \in (0,1)$$

versus **(4.3.6)**

$$H_1\text{: } G_1(\cdot) \text{ is stochastically smaller than } G_2(\cdot).$$

Generally, there is a parameter defining the relationship between $G_1(\cdot)$ and $G_2(\cdot)$, as is illustrated in the following examples.

Example 4.3.7. Take $G_1(u)=1-(1-u)^{1+\theta}$ and $G_2(u)=u^{1+\theta}$ and test

$$H_0: \theta = 0 \text{ versus } H_1: \theta > 0.$$

Example 4.3.8. Set $G_1(u)=u$ and $G_2(u)=(1-\theta)u+\theta u^2$, where $0 \leqslant \theta \leqslant 1$. We wish to test

$$H_0: \theta = 0 \text{ versus } H_1: 0 < \theta \leqslant 1.$$

Example 4.3.9. Let $G_1(u)=u$ and $G_2(u)=u^k$, where k is a positive integer, and test

$$H_0: k = 1 \text{ versus } H_1: k > 1.$$

In this last example, X has c.d.f. $F(\cdot)$ and Y has c.d.f. $F^k(\cdot)$, which is the c.d.f. of the maximum of k independent observations from $F(\cdot)$.

The importance of these so-called **Lehmann alternatives** derives from the fact that the distribution of the rank vector $\mathbf{R}^* = (Q_1,\ldots,Q_m,R_1,\ldots,R_n)$, where $Q_i(R_j)$ denotes the rank of $X_i(Y_j)$ among all $m+n$ observations, depends only on $G_1(\cdot)$ and $G_2(\cdot)$ and does not depend on the distribution $F(\cdot)$ so long as it is of the continuous type. Thus every rank statistic will have a nonparametric distribution-free property under a Lehmann alternative, similar to the nonparametric distribution-free property that pertains under H_0, corresponding to $X_1,\ldots,X_m, Y_1,\ldots,Y_n$ being i.i.d. $F(x)$.

Theorem 4.3.10. *Let $X_1,\ldots,X_m$ and $Y_1,\ldots,Y_n$ be independent random samples from continuous distributions with respective c.d.f.'s $G_1(F(x))$ and $G_2(F(y))$ and densities $g_1(F(x))f(x)$ and $g_2(F(y))f(y)$. Then, for any $\mathbf{r}^* = (q_1,\ldots,q_m,r_1,\ldots,r_n)$, a permutation of the integers $(1,\ldots,m+n)$, the probability $P[\mathbf{R}^*=\mathbf{r}^*]$ depends only on $G_1(\cdot)$ and $G_2(\cdot)$ and not on $F(\cdot)$.*

Proof: Let

$$A = \{(\mathbf{x},\mathbf{y}) | x_i \text{ has rank } q_i \text{ and } y_j \text{ has rank } r_j \text{ among all } m+n \text{ observations}, i=1,\ldots,m \text{ and } j=1,\ldots,n\}. \quad \textbf{(4.3.11)}$$

Then

$$P[\mathbf{R}^*=\mathbf{r}^*] = \int \cdots_A \int \left\{ \prod_{i=1}^{m} g_1(F(x_i)) \right\} \left\{ \prod_{j=1}^{n} g_2(F(y_j)) \right\} \times \prod_{i=1}^{m} f(x_i) \prod_{j=1}^{n} f(y_j)\, dx_1 \cdots dx_m\, dy_1 \cdots dy_n.$$

Letting $t_k = F(x_k)$ for $k = 1,\dots,m$ and $t_k = F(y_{k-m})$, for $k = m+1,\dots,m+n$, we obtain

$$P[\mathbf{R}^* = \mathbf{r}^*] = \int \underset{A^*}{\cdots} \int \left\{ \prod_{i=1}^{m} g_1(t_i) \right\} \left\{ \prod_{j=1}^{n} g_2(t_{m+j}) \right\} dt_1 \cdots dt_{m+n},$$

where A^* is the same as A only defined in terms of the t_k's instead of $\mathbf{x}$ and $\mathbf{y}$. The theorem follows from the fact that the resulting expression does not depend on $F(\cdot)$. ■

Since the distribution of the rank vector does not depend on $F(\cdot)$, we might as well let $F(\cdot)$ correspond to a uniform distribution on $(0,1)$; that is, take $F(x) = x$. We can then evaluate the probability of each rank configuration, $\mathbf{r}^*$, using the following important lemma due to Hoeffding (1951). Note that this lemma is more general than our current context, requiring only that $G_1(\cdot)$ and $G_2(\cdot)$ be c.d.f.'s of continuous random variables, while not restricting their support to the interval $(0,1)$. Hence, the result can be used to evaluate the powers of two-sample rank tests for a variety of alternatives and will, in particular, be used in Chapter 9 to derive locally most powerful rank tests.

Lemma 4.3.12. *Let $X_1,\dots,X_m$ and $Y_1,\dots,Y_n$ be independent random samples from continuous distributions with c.d.f.'s $G_1(x)$ and $G_2(y)$, respectively, and corresponding densities $g_1(x)$ and $g_2(y)$. Let $H(z)$ and $h(z)$ be the c.d.f. and p.d.f., respectively, of a continuous distribution with support that contains the support of both $G_1(\cdot)$ and $G_2(\cdot)$. Then, if $N = m+n$ and $\mathbf{r}^* = (q_1,\dots,q_m,r_1,\dots,r_n)$ is any permutation of the integers $\{1,\dots,N\}$, we have*

$$P[\mathbf{R}^* = \mathbf{r}^*] = \frac{1}{N!} E\left[\frac{\left\{ \prod_{i=1}^{m} g_1(V_{(q_i)}) \right\} \left\{ \prod_{j=1}^{n} g_2(V_{(r_j)}) \right\}}{\left\{ \prod_{i=1}^{m} h(V_{(q_i)}) \right\} \left\{ \prod_{j=1}^{n} h(V_{(r_j)}) \right\}} \right],$$

where $V_{(1)} < \cdots < V_{(N)}$ are the order statistics for a sample of size N from $H(\cdot)$.

Proof: Let A be as defined in (4.3.11). Then

$$P[\mathbf{R}^* = \mathbf{r}^*] = \int \underset{A}{\cdots} \int \left\{ \prod_{i=1}^{m} g_1(x_i) \right\} \left\{ \prod_{j=1}^{n} g_2(y_j) \right\} dx_1 \cdots dx_m dy_1 \cdots dy_n.$$

If we rename the variables by letting $v_{(q_i)}=x_i$, for $i=1,\ldots,m$, and $v_{(r_j)}=y_j$, for $j=1,\ldots,n$, we obtain

$$P[\mathbf{R}^*=\mathbf{r}^*] = \int_{v_{(1)}<\cdots<v_{(N)}}\cdots\int \left\{\prod_{i=1}^{m} g_1(v_{(q_i)})\right\}\left\{\prod_{j=1}^{n} g_2(v_{(r_j)})\right\} dv_{(1)}\cdots dv_{(N)}$$

$$= \frac{1}{N!}\int_{v_{(1)}<\cdots<v_{(N)}}\cdots\int \left[\frac{\left\{\prod_{i=1}^{m} g_1(v_{(q_i)})\right\}\left\{\prod_{j=1}^{n} g_2(v_{(r_j)})\right\}}{\left\{\prod_{i=1}^{m} h(v_{(q_i)})\right\}\left\{\prod_{j=1}^{n} h(v_{(r_j)})\right\}}\right]$$

$$\times N!\prod_{i=1}^{m} h(v_{(q_i)})\prod_{j=1}^{n} h(v_{(r_j)})\,dv_{(1)}\cdots dv_{(N)}.$$

The result now follows, since the far-right-hand portion of the integrand is the density of the order statistics for a sample of size N from $H(\cdot)$. ■

Example 4.3.13. Let $G_1(u)=u$ and $G_2(u)=u^k$, for $0\leqslant u\leqslant 1$, where k is a positive integer. Then, taking $H(u)=G_1(u)$ in Lemma 4.3.12 yields

$$P[\mathbf{R}^*=\mathbf{r}^*] = \frac{k^n}{N!}E\left[\prod_{j=1}^{n}\{V_{(r_j)}\}^{k-1}\right],$$

where $V_{(1)}<\cdots<V_{(N)}$ are the order statistics for a sample of size N from a uniform $(0,1)$ distribution. Let $s_1<\cdots<s_n$ be the ordered values of $r_1,\ldots,r_n$. Then, the joint density of $V_{(s_1)},\ldots,V_{(s_n)}$ is

$$h(t_1,\ldots,t_n) = \frac{N!}{\prod_{i=0}^{n}(s_{i+1}-s_i-1)!}\prod_{j=0}^{n}(t_{j+1}-t_j)^{s_{j+1}-s_j-1},$$

$0\leqslant t_1<\cdots<t_n\leqslant 1$, where $t_0\equiv 0$, $t_{n+1}\equiv 1$, $s_0\equiv 0$ and $s_{n+1}\equiv N+1$. If we set $Z_i=V_{(s_i)}/V_{(s_{i+1})}$, for $i=1,\ldots,n-1$, and $Z_n=V_{(s_n)}$, it follows that the joint density of $Z_1,\ldots,Z_n$ is

$$k(z_1,\ldots,z_n) = \frac{N!}{\prod_{i=0}^{n}(s_{i+1}-s_i-1)!}\prod_{j=1}^{n}\left\{z_j^{s_j-1}(1-z_j)^{s_{j+1}-s_j-1}\right\}$$

over $0\leqslant z_j\leqslant 1$ for $j=1,\ldots,n$. That is, the Z_js are independent random variables with Z_j having a Beta $(s_j, s_{j+1} - s_j)$ distribution. Since

$$\prod_{j=1}^{n} V_{(r_j)} = Z_1 Z_2^2 \cdots Z_n^n,$$

we see that

$$P[\mathbf{R}^* = \mathbf{r}^*] = \frac{k^n}{N!} \prod_{j=1}^{n} E\left[Z_j^{j(k-1)}\right]$$

$$= \frac{k^n}{\Gamma(s_1)} \prod_{j=1}^{n} \left[\frac{\Gamma(s_j + jk - j)}{\Gamma(s_{j+1} + jk - j)} \right]. \tag{4.3.14}$$

Thus, at least in this special case, the probability of each rank configuration can be expressed relatively conveniently in terms of the ordered ranks assigned to the Y_js.

For completeness, we briefly indicate the analogous results for nonparametric alternatives in the one-sample change of location problem. Let $X_1, \ldots, X_n$ denote a random sample from a continuous distribution with c.d.f. $F(x)$. The null hypothesis is that the underlying distribution is located at a known value θ_0. Here we interpret "is located at" to mean that its median is θ_0. Without loss of generality we assume θ_0 is zero, since otherwise we could describe the entire problem in terms of $X_1^*, \ldots, X_n^*$, where $X_i^* = X_i - \theta_0$, for $i = 1, \ldots, n$. Thus we seek to test

$$H_0\colon F(0) = \tfrac{1}{2} \tag{4.3.15}$$

against alternatives for which the underlying distribution has a median greater than zero.

Theorem 4.3.16. *Suppose that we reject* H_0 *in* (4.3.15) *for large (small) values of a test statistic* $S(\cdot)$ *satisfying*

$$S(x_1', \ldots, x_n') \geqslant (\leqslant) S(x_1, \ldots, x_n) \tag{4.3.17}$$

for every $\mathbf{x}$ *and* $\mathbf{x}'$ *such that* $x_i \leqslant x_i'$, *for* $i = 1, \ldots, n$. *Then the test based on* $S(\cdot)$ *has a monotone power function in the sense that*

$$\mathcal{P}_S(F) \leqslant \mathcal{P}_S(G)$$

whenever $F(\cdot)$ *is stochastically smaller than* $G(\cdot)$ *and both are c.d.f.'s of continuous distributions.*

Proof: Exercise 4.3.8. ■

The analogues of Lehmann alternatives for the one-sample location problem have not played as prominent a role in the development of nonparametric statistics as have the two-sample versions previously described. They are considered briefly in Exercise 4.3.10. For completeness, however, we state the extension of Hoeffding's lemma (4.3.12) in the form given by Fraser (1957b). Let $F(x|1)=P[X\leqslant x|X>0]$ and $F(x|0)=P[-X\leqslant x|X<0]$, where X has a continuous distribution with c.d.f. $F(x)$. They represent the c.d.f.'s of $|X|$ given that X is positive or negative, respectively. Let $F^*(\cdot)$ denote the c.d.f. of a continuous distribution over the positive real numbers with support that contains the support of both $F(\cdot|1)$ and $F(\cdot|0)$. Further, let $f(\cdot)$ and $f^*(\cdot)$ be the densities corresponding to $F(\cdot)$ and $F^*(\cdot)$, respectively.

Lemma 4.3.18. *Assume that $X_1,\dots,X_n$ are i.i.d. continuous random variables each with c.d.f. $F(x)$. Let R_i^+ denote the rank of $|X_i|$ among $|X_1|,\dots,|X_n|$ and set $\Psi_i=\Psi(X_i)$, where $\Psi(x)=1,0$ as $x>,\leqslant 0$. Then for any $\mathbf{d}=(d_1,\dots,d_n)$, a vector of 0's and 1's, and any $\mathbf{r}^+=(r_1^+,\dots,r_n^+)$, a permutation of the integers $\{1,2,\dots,n\}$, we have*

$$P[\Psi=\mathbf{d},\mathbf{R}^+=\mathbf{r}^+]=[F(0)]^n[(1-F(0))/F(0)]^{\sum_{i=1}^n d_i}$$

$$\times\frac{1}{n!}E\left[\frac{\prod_{i=1}^n f(V_{(r_i^+)}|d_i)}{\prod_{j=1}^n f^*(V_{(r_j^+)})}\right],$$

where $V_{(1)}<\cdots<V_{(n)}$ denote the order statistics for a random sample of size n from $F^*(\cdot)$, and $f(\cdot|d_i)$ is the density corresponding to $F(\cdot|d_i)$.

Proof: Exercise 4.3.9. ■

Exercises 4.3.

4.3.1. Consider the family of RST lambda distributions defined in Exercise 4.1.6. For fixed values of $\lambda_1,\lambda_2>0$, and $\lambda_4>0$, show that the RST lambda distribution is stochastically ordered by the parameter $\lambda_3>0$. [Comment: Stochastic ordering can also be described in terms of λ_1 or λ_4 for fixed values of the remaining three parameters.]

4.3.2. Let $F(x)$ and $G(x)$ be c.d.f.'s of continuous distributions that are stochastically ordered. Specifically, assume $F(x) \geqslant G(x)$ for every x with strict inequality at some x_0. Let $k(x)$ denote a nondecreasing function. Assuming both integrals are finite, prove that

$$\int_{-\infty}^{\infty} k(x)\,dF(x) \leqslant \int_{-\infty}^{\infty} k(x)\,dG(x),$$

with strict inequality if $k(\cdot)$ is strictly increasing. What does this result say about the means of the distributions $F(\cdot)$ and $G(\cdot)$? [Hint: Note that

$$\int_{-\infty}^{\infty} k(x)\,dF(x) = \int_{0}^{1} k(F^{-1}(u))\,du.]$$

4.3.3. For each of the following, draw graphs showing the densities $g_1(F(x))f(x)$ and $g_2(F(x))f(x)$ for $F(\cdot)$ corresponding to: (i) a uniform (0, 1) distribution, (ii) an exponential distribution with $F(x) = 1 - e^{-x}$, for $x > 0$.

(a) Use $G_1(\cdot)$ and $G_2(\cdot)$ as defined in Example 4.3.7 with $\theta = 1$.

(b) Use $G_1(\cdot)$ and $G_2(\cdot)$ as defined in Example 4.3.8 with $\theta = 1/2$.

(c) Use $G_1(\cdot)$ and $G_2(\cdot)$ as defined in Example 4.3.9 with $k = 3$.

4.3.4. Consider the two-sample problem described by (4.3.6), where $F(\cdot)$ is the c.d.f. of some continuous distribution and $G_1(\cdot)$ and $G_2(\cdot)$ are defined as in Example 4.3.9. Consider two rank tests for this problem. The first, the Mann–Whitney–Wilcoxon, rejects for large values of

$$W = \sum_{j=1}^{n} R_j,$$

and the second rejects for large values of

$$V = \underset{1 \leqslant j \leqslant n}{\text{median}}\, R_j,$$

where R_j is the rank of Y_j among all $m+n$ observations.

(a) Find the critical region for each test when $m = n = 3$ and $\alpha = .20$.

(b) For the critical regions of part **(a)**, use expression (4.3.14) to find the power of each test against the alternatives corresponding to $k = 2$ and $k = 3$.

4.3.5. Consider the testing problem described in Example 4.3.8. Find an expression (not actual numerical values) for $P[\mathbf{R}^* = \mathbf{r}^*]$ for the case in which $m = n = 3$. [Hint: Example 4.3.13 may be of some help.]

4.3.6. Let $Z_1,\ldots,Z_N$ be independent, continuous random variables with Z_i having c.d.f. $F^{\Delta_i}(x)$. Show that

$$P[Z_1 < \cdots < Z_N] = \frac{\prod_{i=1}^{N} \Delta_i}{\prod_{j=1}^{N} \left(\sum_{k=1}^{j} \Delta_k \right)}.$$

4.3.7. Let $X_1,\ldots,X_m$ and $Y_1,\ldots,Y_n$ be independent random samples from continuous distributions with c.d.f.'s $F^{\Delta_1}(x)$ and $F^{\Delta_2}(x)$, respectively. Use the result of Exercise 4.3.6 to show that

$$P[\mathbf{R}^* = \mathbf{r}^*] = \frac{\delta^n}{\prod_{i=1}^{N} (c_i + d_i\delta)},$$

where $\delta = \Delta_2/\Delta_1$, $c_i = \{$the number of Xs with a combined sample rank less than or equal to $i\}$, $d_i = i - c_i = \{$the number of Ys with a combined sample rank less than or equal to $i\}$, and $\mathbf{R}^*$ is the usual vector of ranks.

4.3.8. Prove Theorem 4.3.16.

4.3.9. Prove Lemma 4.3.18.

4.3.10. Let $X_1,\ldots,X_n$ be i.i.d. $F(x)$, where $F(\cdot)$ is the c.d.f. for some continuous distribution. Let $F(\cdot|1)$ and $F(\cdot|0)$ be the c.d.f.'s of $|X|$ given that X is positive or negative, respectively, where X has c.d.f. $F(x)$. The **Lehmann alternatives** for the one-sample location problem correspond to taking $F(x|1) = G_1(H(x))$ and $F(x|0) = G_2(H(x))$, where $G_1(\cdot)$ and $G_2(\cdot)$ are c.d.f.'s of random variables assuming values only on the interval $(0, 1)$, and $H(\cdot)$ is a c.d.f. for a non-negative, continuous random variable.

(a) Find the forms for $F(\cdot|1)$ and $F(\cdot|0)$ in terms of the underlying distribution $F(\cdot)$.

(b) For the Lehmann alternatives $F(x|1) = G_1(H(x))$ and $F(x|0) = G_2(H(x))$, show that $P(\mathbf{\Psi} = \mathbf{d}, \mathbf{R}^+ = \mathbf{r}^+)$ depends only on $G_1(\cdot)$ and $G_2(\cdot)$ and not on $H(\cdot)$, where $\mathbf{\Psi}$, $\mathbf{d}$, $\mathbf{R}^+$, and $\mathbf{r}^+$ are as given in Lemma 4.3.18.

(c) Using the result in **(b)**, obtain an expression (not actual numerical values) for $P(\mathbf{\Psi} = \mathbf{d}, \mathbf{R}^+ = \mathbf{r}^+)$ for the case in which $G_1(\cdot)$ and $G_2(\cdot)$ are given by Example 4.3.9 with $k = 2$.

4.3.11. Let $X_1,\ldots,X_n$ be a random sample from a continuous distribution that is weighted-symmetric about 0, with weighting factor λ. (See Exercise 2.4.12.)

(a) If $F(\cdot)$ denotes the underlying distribution function for the Xs, show that $F(x|1)=F(x|0)$ for every x, where $F(\cdot|1)$ and $F(\cdot|0)$ are the c.d.f.'s of $|X|$, given that X is positive and that it is negative, respectively.

(b) Use the results in **(a)** and Lemma 4.3.18 to obtain an expression for $P(\boldsymbol{\Psi}=\mathbf{d},\mathbf{R}^+=\mathbf{r}^+)$ under these weighted-symmetry assumptions, where $\boldsymbol{\Psi}$, $\mathbf{d}$, $\mathbf{R}^+$ and $\mathbf{r}^+$ are as given in Lemma 4.3.18. What can you conclude about the joint distribution of $\boldsymbol{\Psi}$ and $\mathbf{R}^+$?

(c) Let W^+ be the Wilcoxon signed rank statistic. Discuss how to obtain the distribution of W^+ for these weighted-symmetric assumptions. From this, deduce that W^+ is nonparametric distribution-free over the class of continuous distributions that are weighted-symmetric about 0 with a fixed weight factor λ.

5 ASYMPTOTIC RELATIVE EFFICIENCY OF TESTS

Let $X_1,\ldots,X_n$ be a random sample from a continuous distribution with c.d.f. $F(x-\theta)$, where $F(x)+F(-x)=1$ for all x. (Thus the underlying distribution for the Xs is symmetric about θ.) In Sections 2.2 and 2.4 we considered two nonparametric distribution-free procedures for testing the null hypothesis H_0: $\theta=0$ against the alternative hypothesis H_1: $\theta>0$. These tests are based on the sign statistic B and the Wilcoxon signed rank statistic W^+, respectively. A question that naturally arises when we have competing tests for the same hypotheses is how to compare the relative effectiveness of the two procedures.

As we saw in Section 4.1, tests are generally rated according to their relative abilities to detect an alternative hypothesis. Thus if both test I and test II have the same level of significance and test II has a power curve that is always at least as great as that of test I for parameter values in the alternative, then we would prefer test II. However, in nonparametric statistics the underlying distributional structure is not rigid enough to provide a theory analogous to that used in parametric settings (e.g., the Neyman-Pearson lemma) to generate uniformly most powerful tests. In fact, it is seldom the case that one nonparametric distribution-free test procedure is uniformly more powerful than another competitor.

In the absence of such optimality theory, it would be natural to consider simply obtaining expressions for the power functions of two competing test procedures and using them to describe the relative properties of the two tests. However, even this basic idea fails for most nonparametric distribution-free tests. In Chapter 4 we discussed some of the aspects of the power curves for the sign and Wilcoxon signed rank tests. We noted that although a nice closed form expression could be obtained for the power of the sign test, such was not the case for the signed rank procedure. Of course, we could use Lemma 4.3.18 to tabulate the power of the W^+ test and then compare the values with those for the sign test. However, this comparison would depend on (i) the sample size n, (ii) the value of the alternative θ, (iii) the form of the distribution $F(x)$, and (iv) the specified significance level α. Clearly, such a tabulation would be voluminous (since all four quantities must be varied), and it would be very difficult—if not impossible—to reach any general conclusions about the relative strengths and weaknesses of the B and W^+ procedures from such data, even if we had the stamina and computer time to create them. The Monte Carlo study shown in Table 4.1.7 provides one comparison of their powers, but caution must be exercised not to state definitive conclusions based on such studies in which random elements could affect the results.

As an alternative, in this chapter we consider the comparison of two test procedures on the basis of their asymptotic distributional properties. With asymptotics the sample size dependence is no longer a problem, and the relevant limiting distributions are usually continuous in nature, thereby making comparisons easier. However, this large-sample approach immediately creates another problem. In the case of the sign and signed rank tests, for example, we would like to compare the relative asymptotic ($n \to \infty$) power of the two procedures for values of $\theta > 0$. However, each of these tests has been shown (Chapter 4) to be consistent for any $\theta > 0$, which implies that the asymptotic power of both tests is 1 for any such θ in the alternative. Thus it is impossible to use asymptotic power to provide a meaningful comparison between B and W^+ for fixed α and $\theta > 0$. Since consistency against an alternative is such a highly desirable property for a test, this same problem with asymptotic power comparisons is common to all test procedures. (If a test is not consistent against a particular alternative we know immediately that we would not use it.) Pitman (1948) suggested a solution to this problem of test comparisons based on asymptotics. To overcome the consistency dilemma, he proposed considering asymptotic properties of test procedures for sequences of alternatives that are themselves converging to the null hypothesis value. (Thus the resulting comparisons are local in nature, corresponding to alternatives close to the null hypothesis value.) For appropriately chosen sequences of this type the behavior of the tests will not be degenerate (as is the case for a fixed alternative). Moreover, it happens that this method of comparison does not depend on the level α, and we are able to asymptotically choose between two tests in a way that depends only on the form of the underlying distributions [$F(x)$ in the case of B and W^+]. This results in considerable savings in tabulation and a clear verdict (depending on the underlying distributional form) as to the relative merits of the two tests for alternatives near the null hypothesis value.

5.1. Pitman Asymptotic Relative Efficiency

Consider the problem of testing the null hypothesis H_0: $\theta \in \omega$ against the alternative H_1: $\theta \in \Omega - \omega$, where θ is a parameter known to belong to Ω. Let $\{S_n\}$ and $\{T_{n'}\}$ be sequences of statistics for testing H_0 against H_1, where the indexing corresponds to the number of sample observations n and n', respectively, used for the tests. Let θ^* be an arbitrary element in the alternative $\Omega - \omega$ and denote by C_n and $D_{n'}$ the critical regions of size α, $0 < \alpha < 1$, for the S_n and $T_{n'}$ tests, respectively; that is,

$$P_\theta(S_n \in C_n) \leqslant \alpha, \qquad \text{for all } \theta \in \omega \text{ and all } n$$

and (5.1.1)

$$P_\theta(T_{n'} \in D_{n'}) \leqslant \alpha, \qquad \text{for all } \theta \in \omega \text{ and all } n'.$$

Let β, $\alpha < \beta < 1$, be arbitrary, but fixed, and define N and N' to be the smallest values of n and n', respectively, for which

$$P_{\theta^*}(S_N \in C_N) \geqslant \beta \text{ and } P_{\theta^*}(T_{N'} \in D_{N'}) \geqslant \beta. \tag{5.1.2}$$

Thus N and N' are the respective minimum numbers of sample observations for which the α-level tests based on S_n and $T_{n'}$ achieve a power of at least β against the alternative θ^*. Actually, $N = N(\alpha, \beta, \theta^*$, underlying distributions) is a function of α, β, θ^* and the form(s) of the underlying distributions, and similarly for $N' = N'(\alpha, \beta, \theta^*$, underlying distributions).

Definition 5.1.3. Let $\{S_n\}$, $\{T_{n'}\}$, α, β and θ^* be as previously defined. Then the **finite sample size relative efficiency** of $\{S_n\}$ with respect to (or relative to) $\{T_{n'}\}$ at the alternative θ^* is taken to be

$$e(S,T|\alpha,\beta,\theta^*, \text{underlying distributions})$$

$$= \frac{N'(\alpha,\beta,\theta^*, \text{underlying distributions})}{N(\alpha,\beta,\theta^*, \text{underlying distributions})}. \tag{5.1.4}$$

We refer to $e(S,T|\alpha,\beta,\theta^*$, underlying distributions) as simply the efficiency of S_n relative to $T_{n'}$, and if $e(S,T|\alpha,\beta,\theta^*$, underlying distributions) > 1, we say that a test based on S_n is more efficient than one based on $T_{n'}$. If it is less than 1, the opposite is true.

We note that the finite sample size efficiency of one test relative to another is local in nature, in the sense that the value of the efficiency usually depends upon α, β, θ^*, and the form(s) of the underlying distributions. This makes the actual comparison (and resulting judgment about which test is better) very difficult. To eliminate the dependence of our evaluation criterion on such things as α, β, and θ^*, we turn to consideration of an asymptotic comparison of two tests. However, to circumvent the previously mentioned problem resulting from our desire that any candidate test be consistent against the alternatives in H_1, we also allow the alternatives at which we are comparing the two tests to converge, as the sample sizes get large, to the null hypothesis value of the parameter. With this in mind we now define what we mean by the asymptotic relative efficiency of one test relative to another.

Definition 5.1.5. Let $\{S_{n_i}\}$ and $\{T_{n_i'}\}$ be two sequences of test statistics for testing H_0: $\theta = \theta_0$, where θ_0 is some specified parameter value, against a class of alternatives H_1 at size α, and let $\{\theta_i\}$ be a sequence of alternatives to H_0 such that $\lim_{i\to\infty} \theta_i = \theta_0$. In addition, let $\beta_{S_{n_i}}(\theta_i)$ and $\beta_{T_{n_i'}}(\theta_i)$ be the powers of the tests based on S_{n_i} and $T_{n_i'}$, respectively, at the alternative θ_i. Let $\{n_i\}$ and $\{n_i'\}$ be increasing sequences of positive integers such that the two sequences of tests have the same limiting significance level α and

$$\alpha < \lim_{i\to\infty} \beta_{S_{n_i}}(\theta_i) = \lim_{i\to\infty} \beta_{T_{n_i'}}(\theta_i) < 1. \qquad \textbf{(5.1.6)}$$

Then the **asymptotic relative efficiency (ARE) of $\{S_{n_i}\}$ relative to $\{T_{n_i'}\}$** (or simply of S relative to T) is

$$\text{ARE}(S, T) = \lim_{i\to\infty} \frac{n_i'}{n_i}, \qquad \textbf{(5.1.7)}$$

provided that this limit is the same for all such sequences $\{n_i\}$ and $\{n_i'\}$, and independent of the $\{\theta_i\}$ sequence.

We know that n_i and n_i' correspond to the numbers of sample observations used by the S and T tests, respectively. Thus ARE(S, T) is the limiting ratio of sample sizes required to achieve the same limiting power against the same sequence of alternatives (converging to the null) when the limiting significance levels of the two tests are equal. This concept is based on the work of Pitman (1948), and ARE(S, T) is often called the **Pitman efficiency** of S relative to T. (Other concepts of efficiency have been considered in the literature, and two of these are discussed briefly in Section 5.5.)

We note that a rough interpretation of the statement ARE(S,T)=1.2, for example, is that we would need approximately 1.2 times as many observations for the T test as for the S test to achieve the same asymptotic power for the sequence of alternatives $\{\theta_i\}$. We also note again that such ARE comparisons are local in nature, since they correspond to evaluating the tests' relative performances as $\theta_i \to \theta_0$.

5.2. Methods for Evaluating ARE(S, T)—Noether's Theorem

Although the definition (5.1.7) of ARE(S,T) is easily interpreted, it is often useful to turn to equivalent expressions in order to evaluate the Pitman efficiency for a given pair of test procedures in a particular setting.

Let $\{S_{n_i}\}$ and $\{T_{n_i'}\}$ be as given in Definition 5.1.5 for testing H_0: $\theta=\theta_0$ against some collection of alternatives H_1, and assume that each procedure rejects for large values of its test statistic. (Similar arguments will hold for two-sided rejection regions and are left as Exercise 5.2.2.) For arbitrary θ, let $\{\mu_{S_{n_i}}(\theta)\}$, $\{\mu_{T_{n_i'}}(\theta)\}$, $\{\sigma^2_{S_{n_i}}(\theta)\}$, and $\{\sigma^2_{T_{n_i'}}(\theta)\}$ be sequences of numbers associated with $\{S_{n_i}\}$ and $\{T_{n_i'}\}$. (We note that in many applications these sequences will be the means and variances of $\{S_{n_i}\}$ and $\{T_{n_i'}\}$.)

Let $\{\theta_i\}$, $\{n_i\}$, and $\{n_i'\}$ also be as given in Definition 5.1.5. Assume that the following four quantities all have the same continuous limiting ($i\to\infty$) distribution with c.d.f. $H(w)$ and interval support: when the true parameter value is θ_i,

$$\frac{S_{n_i}-\mu_{S_{n_i}}(\theta_i)}{\sigma_{S_{n_i}}(\theta_i)} \quad \text{and} \quad \frac{T_{n_i'}-\mu_{T_{n_i'}}(\theta_i)}{\sigma_{T_{n_i'}}(\theta_i)}$$

and when the true parameter value is θ_0,

$$\frac{S_{n_i}-\mu_{S_{n_i}}(\theta_0)}{\sigma_{S_{n_i}}(\theta_0)} \quad \text{and} \quad \frac{T_{n_i'}-\mu_{T_{n_i'}}(\theta_0)}{\sigma_{T_{n_i'}}(\theta_0)}. \tag{5.2.1}$$

[We note that in most applications $H(w)$ will be the standard normal distribution function, $\Phi(w)$.] Let α, $0<\alpha<1$, be fixed but arbitrary, and let $\{c_{n_i}\}$ and $\{d_{n_i'}\}$ be sequences of critical values for the respective tests such that

$$\lim_{i\to\infty} P_{\theta_0}\left(S_{n_i} \geqslant c_{n_i}\right) = \lim_{i\to\infty} P_{\theta_0}\left(T_{n_i'} \geqslant d_{n_i'}\right) = \alpha, \tag{5.2.2}$$

where P_{θ_0} indicates that the probabilities are evaluated under the null H_0: $\theta = \theta_0$. Thus the two tests are required to have the same limiting significance level.

Now, if $\beta_{S_{n_i}}(\theta_i)$ and $\beta_{T_{n'_i}}(\theta_i)$ are the respective power functions for S_{n_i} and $T_{n'_i}$, we have

$$\beta_{S_{n_i}}(\theta_i) = P_{\theta_i}\left[\frac{S_{n_i} - \mu_{S_{n_i}}(\theta_i)}{\sigma_{S_{n_i}}(\theta_i)} \geqslant \frac{c_{n_i} - \mu_{S_{n_i}}(\theta_i)}{\sigma_{S_{n_i}}(\theta_i)}\right]$$

and **(5.2.3)**

$$\beta_{T_{n'_i}}(\theta_i) = P_{\theta_i}\left[\frac{T_{n'_i} - \mu_{T_{n'_i}}(\theta_i)}{\sigma_{T_{n'_i}}(\theta_i)} \geqslant \frac{d_{n'_i} - \mu_{T_{n'_i}}(\theta_i)}{\sigma_{T_{n'_i}}(\theta_i)}\right],$$

where P_{θ_i} indicates probabilities evaluated for the alternative θ_i. In view of equation (5.2.3) and the common form for the limiting distribution, as assumed in (5.2.1), we see that in order for the tests to have the same limiting power under $\{\theta_i\}$, as required in Definition 5.1.5 (i.e., to have

$$\lim_{i\to\infty} \beta_{S_{n_i}}(\theta_i) = \lim_{i\to\infty} \beta_{T_{n'_i}}(\theta_i)), \tag{5.2.4}$$

we must have (see Lemma A.3.12 in the Appendix)

$$\lim_{i\to\infty}\left[\frac{c_{n_i} - \mu_{S_{n_i}}(\theta_i)}{\sigma_{S_{n_i}}(\theta_i)}\right] = \lim_{i\to\infty}\left[\frac{d_{n'_i} - \mu_{T_{n'_i}}(\theta_i)}{\sigma_{T_{n'_i}}(\theta_i)}\right]. \tag{5.2.5}$$

For the tests to have the same limiting significance level, as in (5.2.2), we must also have

$$\begin{aligned}\alpha &= \lim_{i\to\infty} P_{\theta_0}\left[\frac{S_{n_i} - \mu_{S_{n_i}}(\theta_0)}{\sigma_{S_{n_i}}(\theta_0)} \geqslant \frac{c_{n_i} - \mu_{S_{n_i}}(\theta_0)}{\sigma_{S_{n_i}}(\theta_0)}\right] \\ &= \lim_{i\to\infty} P_{\theta_0}\left[\frac{T_{n'_i} - \mu_{T_{n'_i}}(\theta_0)}{\sigma_{T_{n'_i}}(\theta_0)} \geqslant \frac{d_{n'_i} - \mu_{T_{n'_i}}(\theta_0)}{\sigma_{T_{n'_i}}(\theta_0)}\right].\end{aligned}$$

This fact, along with condition (5.2.1) and Lemma A.3.12, requires that

$$\lim_{i\to\infty} \frac{c_{n_i} - \mu_{S_{n_i}}(\theta_0)}{\sigma_{S_{n_i}}(\theta_0)} = \lim_{i\to\infty} \frac{d_{n'_i} - \mu_{T_{n'_i}}(\theta_0)}{\sigma_{T_{n'_i}}(\theta_0)} = h_\alpha, \tag{5.2.6}$$

where $H(h_\alpha)=1-\alpha$ [i.e., h_α is the upper 100αth percentile for the common limiting distribution $H(\cdot)$].

In the next theorem, due to Noether (1955), we use the relationships in equations (5.2.5) and (5.2.6) to establish a computationally useful asymptotic relative efficiency expression for test procedures when certain conditions are satisfied.

Theorem 5.2.7. *Let $\{S_{n_i}\}$ and $\{T_{n_i'}\}$ be two sequences of tests, with associated sequences of numbers $\{\mu_{S_{n_i}}(\theta)\}$, $\{\mu_{T_{n_i'}}(\theta)\}$, $\{\sigma^2_{S_{n_i}}(\theta)\}$ and $\{\sigma^2_{T_{n_i'}}(\theta)\}$, and satisfying the following Assumptions* A1–A6:

A1.
$$\frac{S_{n_i}-\mu_{S_{n_i}}(\theta_i)}{\sigma_{S_{n_i}}(\theta_i)} \quad \text{and} \quad \frac{T_{n_i'}-\mu_{T_{n_i'}}(\theta_i)}{\sigma_{T_{n_i'}}(\theta_i)}$$

have the same continuous limiting $(i\to\infty)$ distribution with c.d.f. $H(\cdot)$ and interval support when θ_i is the true value of θ.

A2. *Same assumption as in* A1 *but with θ_i replaced by θ_0 throughout.*

A3.
$$\lim_{i\to\infty}\frac{\sigma_{S_{n_i}}(\theta_i)}{\sigma_{S_{n_i}}(\theta_0)}=\lim_{i\to\infty}\frac{\sigma_{T_{n_i'}}(\theta_i)}{\sigma_{T_{n_i'}}(\theta_0)}=1.$$

A4.
$$\frac{d}{d\theta}\left[\mu_{S_n}(\theta)\right]=\mu'_{S_n}(\theta) \quad \text{and} \quad \frac{d}{d\theta}\left[\mu_{T_{n'}}(\theta)\right]=\mu'_{T_{n'}}(\theta)$$

are assumed to exist and be continuous in some closed interval about $\theta=\theta_0$, with $\mu'_{S_n}(\theta_0)$ and $\mu'_{T_{n'}}(\theta_0)$ both nonzero.

A5.
$$\lim_{i\to\infty}\frac{\mu'_{S_{n_i}}(\theta_i)}{\mu'_{S_{n_i}}(\theta_0)}=\lim_{i\to\infty}\frac{\mu'_{T_{n_i'}}(\theta_i)}{\mu'_{T_{n_i'}}(\theta_0)}=1.$$

A6.
$$\lim_{n\to\infty}\frac{\mu'_{S_n}(\theta_0)}{\sqrt{n\sigma^2_{S_n}(\theta_0)}}=K_S \quad \text{and} \quad \lim_{n'\to\infty}\frac{\mu'_{T_{n'}}(\theta_0)}{\sqrt{n'\sigma^2_{T_{n'}}(\theta_0)}}=K_T,$$

where K_S and K_T are positive constants.
Then

$$\text{ARE}(S,T)=\frac{K_S^2}{K_T^2}. \tag{5.2.8}$$

Proof: Under Assumption A4 we perform a Taylor series expansion of $\mu_{S_{n_i}}(\theta)$ around $\theta=\theta_0$, yielding

$$\mu_{S_{n_i}}(\theta_i) = \mu_{S_{n_i}}(\theta_0) + (\theta_i - \theta_0)\mu'_{S_{n_i}}(\theta_i^*), \qquad \theta_0 < \theta_i^* < \theta_i. \tag{5.2.9}$$

Similarly for $\mu_{T_{n_i'}}(\theta_i)$, we obtain

$$\mu_{T_{n_i'}}(\theta_i) = \mu_{T_{n_i'}}(\theta_0) + (\theta_i - \theta_0)\mu'_{T_{n_i'}}(\theta_i^{**}), \qquad \theta_0 < \theta_i^{**} < \theta_i. \tag{5.2.10}$$

Now, we note that

$$\lim_{i\to\infty}\left[\frac{c_{n_i} - \mu_{S_{n_i}}(\theta_i)}{\sigma_{S_{n_i}}(\theta_i)}\right] = \lim_{i\to\infty}\left[\left\{\frac{c_{n_i} - \mu_{S_{n_i}}(\theta_0) + \mu_{S_{n_i}}(\theta_0) - \mu_{S_{n_i}}(\theta_i)}{\sigma_{S_{n_i}}(\theta_0)}\right\}\left\{\frac{\sigma_{S_{n_i}}(\theta_0)}{\sigma_{S_{n_i}}(\theta_i)}\right\}\right].$$

Using Assumption A3 and the condition in equation (5.2.6), which follows from Assumption A2, we obtain

$$\lim_{i\to\infty}\left[\frac{c_{n_i} - \mu_{S_{n_i}}(\theta_i)}{\sigma_{S_{n_i}}(\theta_i)}\right] = h_\alpha + \lim_{i\to\infty}\left[\frac{\mu_{S_{n_i}}(\theta_0) - \mu_{S_{n_i}}(\theta_i)}{\sigma_{S_{n_i}}(\theta_0)}\right]. \tag{5.2.11}$$

Proceeding in exactly the same manner for the $\{T_{n_i'}\}$ sequence, we also have

$$\lim_{i\to\infty}\left[\frac{d_{n_i'} - \mu_{T_{n_i'}}(\theta_i)}{\sigma_{T_{n_i'}}(\theta_i)}\right] = h_\alpha + \lim_{i\to\infty}\left[\frac{\mu_{T_{n_i'}}(\theta_0) - \mu_{T_{n_i'}}(\theta_i)}{\sigma_{T_{n_i'}}(\theta_0)}\right]. \tag{5.2.12}$$

Combining equations (5.2.11) and (5.2.12) with equation (5.2.5), the validity of which is a consequence of Assumption A1, we see that for the condition in (5.2.4) to hold (as is required in Definition 5.1.5) we must have

$$\lim_{i\to\infty}\left[\left\{\frac{\mu_{S_{n_i}}(\theta_i) - \mu_{S_{n_i}}(\theta_0)}{\sigma_{S_{n_i}}(\theta_0)}\right\} \div \left\{\frac{\mu_{T_{n_i'}}(\theta_i) - \mu_{T_{n_i'}}(\theta_0)}{\sigma_{T_{n_i'}}(\theta_0)}\right\}\right] = 1. \tag{5.2.13}$$

If we apply the Taylor series expansions in (5.2.9) and (5.2.10), this limit condition becomes

$$1 = \lim_{i\to\infty}\left[\left\{\frac{(\theta_i-\theta_0)\mu'_{S_{n_i}}(\theta_i^*)}{\sigma_{S_{n_i}}(\theta_0)}\right\}\div\left\{\frac{(\theta_i-\theta_0)\mu_{T_{n'_i}}(\theta_i^{**})}{\sigma_{T_{n'_i}}(\theta_0)}\right\}\right]$$

$$= \lim_{i\to\infty}\left[\left\{\frac{\mu'_{S_{n_i}}(\theta_i^*)}{\sigma_{S_{n_i}}(\theta_0)}\right\}\div\left\{\frac{\mu'_{T_{n'_i}}(\theta_i^{**})}{\sigma_{T_{n'_i}}(\theta_0)}\right\}\right],$$

which in turn yields

$$1 = \lim_{i\to\infty}\left[\left\{\frac{\sqrt{n_i}}{\sqrt{n'_i}}\right\}\left\{\frac{\mu'_{S_{n_i}}(\theta_i^*)}{\sqrt{n_i}\,\sigma_{S_{n_i}}(\theta_0)}\right\}\div\left\{\frac{\mu'_{T_{n'_i}}(\theta_i^{**})}{\sqrt{n'_i}\,\sigma_{T_{n'_i}}(\theta_0)}\right\}\right].$$

Finally, using Assumption A6 and the continuity portion of Assumption A4, we have, since $n_i\to\infty$ and $n'_i\to\infty$ as $i\to\infty$, that

$$\text{ARE}(S,T) = \lim_{i\to\infty}\left[\frac{n'_i}{n_i}\right] = \frac{K_S^2}{K_T^2},$$

as desired. ■

Thus to evaluate ARE(S,T) under Assumptions A1–A6 we need only calculate the quantities K_S and K_T.

Definition 5.2.14. The quantity

$$K_S = \lim_{n\to\infty}\frac{\mu'_{S_n}(\theta_0)}{\sqrt{n\sigma^2_{S_n}(\theta_0)}}$$

is called the **efficacy of the test based on S_n** and is denoted by **eff(S).**

Using similar notation and terminology for $T_{n'}$, we have the expression

$$\text{ARE}(S,T) = \left[\frac{\text{eff}(S)}{\text{eff}(T)}\right]^2. \tag{5.2.15}$$

In Section 5.4 we consider applications of these asymptotic relative efficiency results to specific statistics. However, we need to point out two important general properties of the Pitman approach to asymptotic relative efficiency, as utilized in Noether's theorem (5.2.7). First, with regard to the efficacy expression eff(S) of a test statistic S_n, we note that the derivative $\mu'_{S_n}(\theta_0)$ provides a measure of the rate of change in $\mu_{S_n}(\theta)$ for θ values near θ_0; that is, it measures how fast the distribution of S_n is changing its location in response to θ values in the alternative that are close to θ_0. Of course, to properly interpret this rate of change, we must compare it with a measure of the variation in the S_n distribution, and thus we divide $\mu'_{S_n}(\theta_0)$ by $\sigma_{S_n}(\theta_0)$ in the expression for eff(S). Thus ARE(S,T) is simply the square of a limiting ratio of the standardized rates of change for the locations of the S_n and $T_{n'}$ distributions for alternatives to θ_0.

A second important property of the Pitman–Noether approach relates to the sequence of alternatives $\{\theta_i\}$. In first discussing this sequence (see Definition 5.1.5) we required that $\lim_{i\to\infty}\theta_i=\theta_0$; that is, θ_i must converge to the null θ_0 as $i\to\infty$. However, at that time we did not describe the rate of this convergence. Indeed, Assumptions A1–A6 create an important implicit relationship between θ_i and the sample sizes. Using arguments similar to those used in establishing (5.2.6) and (5.2.11), but now applied to the single sequence of test statistics $\{S_{n_i}\}$, we see that

$$\lim_{i\to\infty}\left\{\frac{\mu_{S_{n_i}}(\theta_i)-\mu_{S_{n_i}}(\theta_0)}{\sigma_{S_{n_i}}(\theta_0)}\right\}=h_\alpha-h_\beta, \tag{5.2.16}$$

where h_a is the upper $100a$th percentile for the limiting distribution $H(w)$. Employing the Taylor series expansion for $\mu_{S_{n_i}}(\theta_i)$, as in the proof of Theorem 5.2.7, we see, since $\alpha<\beta$, that

$$\lim_{i\to\infty}\left[n_i^{1/2}(\theta_i-\theta_0)\left\{\frac{\mu'_{S_{n_i}}(\theta_i^*)}{n_i^{1/2}\sigma_{S_{n_i}}(\theta_0)}\right\}\right]=h_\alpha-h_\beta>0, \tag{5.2.17}$$

where $\theta_0<\theta_i^*<\theta_i$. Combining this with Assumptions A5, A6, and the continuity portion of Assumption A4, we have

$$\lim_{i\to\infty}\left[n_i^{1/2}(\theta_i-\theta_0)\right]=\frac{h_\alpha-h_\beta}{K_S}. \tag{5.2.18}$$

Thus although Assumptions A1–A6 display no explicit statement concerning the rate of convergence of the sequence $\{\theta_i\}$, equation (5.2.18) shows that θ_i must go to θ_0 as a constant over the square root of the sample size

n_i; that is, we must have

$$\theta_i = \theta_0 + \frac{C_S}{\sqrt{n_i}} + g_S(n_i), \tag{5.2.19}$$

where C_S is a constant and $g_S(n_i)$ is a function of n_i such that $\lim_{i\to\infty} n_i^{1/2} g_S(n_i) = 0$. These alternatives are often called **Pitman translation alternatives.** As we have seen, although we need not check an explicit condition regarding their status for a particular statistical setting, they are implicit in the Pitman–Noether efficiency structure. We note that although our derivation of ARE made n_i a function of θ_i, equation (5.2.19) views θ_i as a function of n_i. It is particularly useful when computing AREs to take this viewpoint. Hence, in future sections the alternatives are described as functions of the sample size(s).

Two additional comments regarding the Pitman efficiency structure are important. Under Assumptions A1–A6 we have seen that θ_i converges to θ_0 at the rate of the square root of the inverse of the sample size. However, under different sets of assumptions, asymptotic relative efficiency results similar to Theorem 5.2.7 can be obtained for situations where the proper rate of convergence for θ_i is something other than the square root of the inverse of the sample size. These and other related results can be found in Noether (1955).

Finally, for statistical settings where more than a single parameter is involved in both the null and alternative hypotheses (e.g., in analysis of variance problems) the approach of Theorem 5.2.7 cannot be employed. However, general arguments similar to those used in developing equations (5.2.5) and (5.2.6) can be used to derive asymptotic relative efficiency expressions in such cases. See, for example, Andrews (1954), Puri (1964), and Skillings and Wolfe (1977, 1978).

Exercises 5.2.

5.2.1. Establish an equation similar to (5.2.19), but with a different constant, say, C_T, for the sequence of test statistics $\{T_{n_i'}\}$. Use these two representations for θ_i to find an expression for ARE(S,T) in terms of C_S and C_T.

5.2.2. Consider the asymptotic relative efficiency derivation for two-sided tests. Suppose we reject H_0: $\theta = \theta_0$ in favor of the alternative H_1: $\theta \neq \theta_0$ if $S_n(T_n)$ is greater than or equal to $c_n^U(d_n^U)$ or if it is less than or equal to $c_n^L(d_n^L)$, where $c_n^U, d_n^U, c_n^L, d_n^L$ are constants such that

$$\lim_{i\to\infty} P_{\theta_0}\left(S_{n_i} \leqslant c_{n_i}^L\right) = \lim_{i\to\infty} P_{\theta_0}\left(T_{n_i'} \leqslant d_{n_i'}^L\right) = \alpha_1$$

and

$$\lim_{i\to\infty} P_{\theta_0}\left(S_{n_i} \geqslant c_{n_i}^U\right) = \lim_{i\to\infty} P_{\theta_0}\left(T_{n_i'} \geqslant c_{n_i'}^U\right) = \alpha_2,$$

where $\alpha = \alpha_1 + \alpha_2$. [This is similar to the assumption in (5.2.2) for the one-sided tests and corresponds to the two tests having the same limiting size α.] Making assumptions similar to those employed in Section 5.2, show that the asymptotic relative efficiency expression for such two-sided tests is also given by (5.2.8).

5.2.3. Let S, T, and V be three statistics for testing H_0 against H_1. Show that:

(a) $0 \leqslant \text{ARE}(S, T) \leqslant \infty$,

and

(b) $\text{ARE}(S, V) = \text{ARE}(S, T)\ \text{ARE}(T, V) = \dfrac{1}{\text{ARE}(V, S)}$, where we take $(1/\infty) = 0$ and $(1/0) = \infty$.

5.2.4. Let $\{S_n\}$ and $\{T_n\}$ denote two sequences of statistics for testing H_0: $\theta = \theta_0$ against an alternative H_1. Suppose there exist sequences of standardizing constants $\{\mu_n(\theta)\}$ and $\{\sigma_n(\theta)\}$ such that

(i) $(S_n - T_n)/\sigma_n(\theta)$ converges to zero in probability uniformly in θ

and

(ii) $(S_n - \mu_n(\theta))/\sigma_n(\theta)$ satisfies both Assumptions A1 and A2 (in Theorem 5.2.7) for the sequence of alternatives $\{\theta_i\}$ and the null sequence $\{\theta_0\}$. If, in addition, $\{\mu_n(\theta)\}$ and $\{\sigma_n(\theta)\}$ satisfy Assumptions A3–A6 (in Theorem 5.2.7), show that the tests based on S_n and T_n have the same efficacy. Use this result to show that a test based on the Wilcoxon signed rank statistic W^+ (2.4.11) has the same efficacy as one based on the U-statistic U_4 given in Example 3.1.11.

5.2.5. Let $\{S_n\}$ be a sequence of statistics for testing H_0: $\theta = \theta_0$ against an alternative H_1. Suppose there exist sequences of constants $\{\mu_n(\theta)\}$ and $\{\sigma_n(\theta)\}$ such that

$$\frac{S_n - \mu_n(\theta)}{\sigma_n(\theta)}$$

satisfies both Assumptions A1 and A2 in Theorem 5.2.7 for the sequence of alternatives $\{\theta_i\}$ and the null sequence $\{\theta_0\}$. Suppose further that $\{\mu_n(\theta)\}$ and $\{\sigma_n(\theta)\}$ satisfy Assumptions A3–A6 in Theorem 5.2.7. Set

$$T_n = \frac{S_n - a_n}{b_n},$$

where $\{a_n\}$ and $\{b_n\}$ are sequences of constants ($b_n \neq 0$ for every n) that do not depend on θ. Show that the tests based on S_n and T_n have the same efficacy.

5.3. Extended U-Statistic Theorems

We turn our attention now to the problem of using the results of Sections 5.1 and 5.2 to obtain asymptotic relative efficiency expressions for some of the one- and two-sample tests that we have previously discussed. To do this, we need to know something about the asymptotic distributions, under both null and alternative hypotheses, of the various test statistics. Thus far the primary results we have considered for this purpose are the U-statistics Theorems 3.3.13 for one-sample settings and 3.4.13 for two-sample settings. These theorems are adequate for the asymptotic properties of our test statistics under the null hypothesis. However, to obtain ARE values we need to know asymptotic properties of test statistics under a sequence of alternatives converging to the null, and thus depending on the sample size(s). Since both of the U-statistics Theorems 3.3.13 and 3.4.13 are for fixed alternatives [not depending on the sample size(s)], we must first consider extensions of these results before proceeding to obtain any ARE expressions. We provide the details for one-sample U-statistics and leave the corresponding two-sample extensions as exercises.

Let $X_{1:n},\ldots,X_{n:n}$ be a random sample of size n from a distribution with distribution function $F_n(x)$, where here the underlying distribution can depend upon the sample size n. Thus, for example, the distributions for sample sizes 3 and 7, namely, $F_3(x)$ and $F_7(x)$, need not be the same. Let

$$U(X_{1:n},\ldots,X_{n:n}) = \frac{1}{\binom{n}{r}} \sum_{\beta \in B} h(X_{\beta_1:n},\ldots,X_{\beta_r:n}) \tag{5.3.1}$$

be the U-statistic for the symmetric kernel $h(x_1,\ldots,x_r)$, where B is the collection of all unordered subsets of r integers chosen without replacement from $\{1,\ldots,n\}$. Also let $\gamma_n = E(U(X_{1:n},\ldots,X_{n:n}))$. [Note that in this setting the parameter estimated by $U(X_{1:n},\ldots,X_{n:n})$, namely, γ_n, may depend on the sample size n.]

Let

$$h_{1:n}(x) = E\left[h(x, X_{2:n}, \ldots, X_{r:n})\right] \tag{5.3.2}$$

and set

$$\zeta_{1:n} = \mathrm{Var}(h_{1:n}(X_{1:n})). \tag{5.3.3}$$

Note that $h_{1:n}(x)$ and $\zeta_{1:n}$ are analogous to the quantities in (3.3.4) and (3.3.7), respectively, except that $h_{1:n}(x)$ and $\zeta_{1:n}$ may now be functions of n, since the distribution of every $X_{i:n}$ is allowed to change as n changes. For a fixed value of n, the following extension of Lemma 3.3.8 is immediate.

Lemma 5.3.4. *For $U(X_{1:n}, \ldots, X_{n:n}) - \gamma_n$, with $U(\cdot)$ given by (5.3.1), the projection V_n^*, defined in (3.3.9), is given by*

$$V_n^* = \frac{r}{n} \sum_{i=1}^{n} \{h_{1:n}(X_{i:n}) - \gamma_n\}, \tag{5.3.5}$$

where $h_{1:n}(x)$ is defined in (5.3.2).

Proof: For fixed n, the proof is exactly analogous to that of Lemma 3.3.8 with $h_{1:n}(x)$ and γ_n playing the parts of $h_1(x)$ and γ, respectively. ■

To establish the required extension of Theorem 3.3.13, we first consider two important lemmas.

Lemma 5.3.6. *Let $X_{1:n}, \ldots, X_{n:n}$ denote a random sample from a distribution with c.d.f. $F_n(x)$, possibly changing as n changes. Let $U(X_{1:n}, \ldots, X_{n:n})$ and $\zeta_{1:n}$ be as defined in (5.3.1) and (5.3.3), respectively. If*

(i) $\nu_n = E[h^2(X_{1:n}, \ldots, X_{r:n})] < \infty$ *for every $n \geqslant r$*

and

(ii) $\lim_{n\to\infty}[n\,\mathrm{Var}\{U(X_{1:n}, \ldots, X_{n:n})\} - r^2\zeta_{1:n}] = 0$,

then $\sqrt{n}\,\{U(X_{1:n}, \ldots, X_{n:n}) - \gamma_n\}$ and $\sqrt{n}\, V_n^*$ have the same limiting distribution.

Proof: Denoting $U(X_{1:n}, \ldots, X_{n:n})$ by U_n, it follows in direct analogy to the proof of Theorem 3.3.13 that for fixed n we have

$$nE\left[(U_n - \gamma_n - V_n^*)^2\right] = n\,\mathrm{Var}(U_n) - r^2\zeta_{1:n}, \tag{5.3.7}$$

where V_n^* is the projection given in (5.3.5). [Here, γ_n, $h_{1:n}(x)$, and $\zeta_{1:n}$ assume the roles of γ, $h_1(x)$ and ζ_1, respectively, in the proof of Theorem 3.3.13.] Now, assumption (i) implies that $\text{Var}(U_n) < \infty$ for every $n \geqslant r$, and assumption (ii) and (5.3.7) yield

$$\lim_{n\to\infty} nE\left[(U_n - \gamma_n - V_n^*)^2\right] = 0.$$

Hence, Theorems 3.2.3 and 3.2.12 show that $\sqrt{n}\,(U_n - \gamma_n)$ and $\sqrt{n}\, V_n^*$ have the same limiting distribution. ■

Lemma 5.3.8. *Let $X_{1:n}, \ldots, X_{n:n}$ denote a random sample from a distribution with c.d.f. $F_n(x)$, possibly changing as n changes. Let V_n^* be given by* (5.3.5), *where $U(X_{1:n}, \ldots, X_{n:n})$ and $h_{1:n}(\cdot)$ are as defined in* (5.3.1) *and* (5.3.2), *respectively. If*

(i) $E[|h_{1:n}(X_{1:n}) - \gamma_n|^3] < M$, *for every* $n \geqslant r$,

and

(ii) $\lim_{n\to\infty} r^2\zeta_{1:n} = \eta$

for some constants $M > 0$ and $\eta > 0$, then $\sqrt{n}\, V_n^*/\sqrt{r^2\zeta_{1:n}}$ has a limiting standard normal distribution.

Proof: Note that $E[h_{1:n}(X_{1:n}) - \gamma_n] = 0$ and $\text{Var}\{h_{1:n}(X_{1:n}) - \gamma_n\} = \zeta_{1:n}$. Conditions (i) and (ii) show that

$$\lim_{n\to\infty} \frac{E\left[|h_{1:n}(X_{1:n}) - \gamma_n|^3\right]}{\sqrt{n}\,\zeta_{1:n}^{3/2}} = 0.$$

By the Berry–Esséen theorem (A.3.14) in the Appendix, we have

$$|G_n(x) - \Phi(x)| \leqslant A_0 \frac{E\left[|h_{1:n}(X_{1:n}) - \gamma_n|^3\right]}{\sqrt{n}\,\zeta_{1:n}^{3/2}} \tag{5.3.9}$$

$$\text{for all } -\infty < x < \infty \text{ and } n \geqslant r,$$

where $G_n(x)$ is the c.d.f. for $\sqrt{n}\, V_n^*/\sqrt{r^2\zeta_{1:n}}$ and $\Phi(x)$ is the c.d.f. for the standard normal distribution. It follows that

$$\lim_{n\to\infty} G_n(x) = \Phi(x), \qquad -\infty < x < \infty,$$

and the lemma is established. ■

We are now in a position to prove the desired extension of the one-sample U-statistic theorem (3.3.13).

Theorem 5.3.10. *Let $X_{1:n},\ldots,X_{n:n}$ denote a random sample from a distribution with c.d.f. $F_n(x)$, possibly changing as n changes. Let $U(X_{1:n},\ldots,X_{n:n})$, $h_{1:n}(\cdot)$, $\zeta_{1:n}$, and V_n^* be as given in* (5.3.1), (5.3.2), (5.3.3), *and* (5.3.5), *respectively. If*

(i) $\nu_n = E[h^2(X_{1:n},\ldots,X_{r:n})] < \infty$ *for every* $n \geqslant r$,

(ii) $\lim_{n\to\infty}[n\,\mathrm{Var}\{U(X_{1:n},\ldots,X_{n:n})\}] = \lim_{n\to\infty} r^2\zeta_{1:n} = \eta$

and

(iii) $E[|h_{1:n}(X_{1:n}) - \gamma_n|^3] < M$, *for every* $n \geqslant r$,

where $M > 0$ and $\eta > 0$ are constants, then

$$\sqrt{n}\,(U(X_{1:n},\ldots,X_{n:n}) - \gamma_n)$$

has a limiting normal distribution with mean zero and variance η.

Proof: From Lemma 5.3.8 and Slutsky's theorem (3.2.8), we see that $\sqrt{n}\,V_n^*$ has a limiting normal distribution with mean 0 and variance η. The theorem then follows at once from Lemma 5.3.6. ■

Before proceeding to some applications of Theorem 5.3.10, we establish two lemmas which simplify matters for many nonparametric U-statistics.

Lemma 5.3.11. *If $\nu_n = E[h^2(X_{1:n},\ldots,X_{r:n})] \leqslant M$ for every $n \geqslant r$ and some constant $M > 0$, then*

$$\lim_{n\to\infty}\{n\,\mathrm{Var}[U(X_{1:n},\ldots,X_{n:n})] - r^2\zeta_{1:n}\} = 0. \qquad \textbf{(5.3.12)}$$

Proof: Let $\zeta_{c:n}$ be ζ_c in (3.1.14) for these sample-size-dependent alternatives $F_n(x)$. From Exercise 5.3.8 and the boundedness of ν_n, we see that $\zeta_{i:n} \leqslant \zeta_{r:n} = (\nu_n - \gamma_n^2) \leqslant M$ for $i = 1,\ldots,r$ and every $n \geqslant r$. Proceeding as in the proof of Theorem 3.1.22, for given n we have

$$n\,\mathrm{Var}[U(X_{1:n},\ldots,X_{r:n})] - r^2\zeta_{1:n} = \frac{n}{\binom{n}{r}} \sum_{c=1}^{r} \binom{r}{c}\binom{n-r}{r-c}\zeta_{c:n} - r^2\zeta_{1:n}.$$

If we use the fact that $\zeta_{i:n} \leqslant M$ for $i = 1,\ldots,r$ and every $n \geqslant r$, the proof is completed as in Theorem 3.1.22. ■

Lemma 5.3.13. *Let $h_{1:n}(x)$, $F_n(x)$ and $\zeta_{1:n}$ be as defined in Theorem 5.3.10. Suppose there exists a real-valued function $k(x)$ and a c.d.f. $F(x)$ for a continuous distribution such that*

(i) $\lim_{n\to\infty} h_{1:n}(x) = k(x)$ *for every* x

and

(ii) $\lim_{n\to\infty} F_n(x) = F(x)$ *for every* x.

If, in addition,

(iii) *there exists an* $M^* > 0$ *such that* $|h_{1:n}(x)| < M^*$ *for every* x *and every* $n \geqslant r$

and

(iv) $E[k^2(X)] < \infty$,

where X *is a random variable with c.d.f.* $F(\cdot)$, *then*

$$\lim_{n\to\infty} r^2\zeta_{1:n} = \eta = r^2 \operatorname{Var}\left[k(X)\right] < \infty.$$

Proof: The lemma follows at once from two applications of Theorem A.3.6 in the Appendix, once with $t_n(x) = h_{1:n}(x)$ and the second time with $t_n(x) = h_{1:n}^2(x)$. ■

This result is applied in dealing with a particular type of one-sample alternative. For arbitrary n_i, let $X_{1:n_i}, \ldots, X_{n_i:n_i}$ be independent and identically distributed continuous random variables with c.d.f.

$$F(x-\theta_i) = F\left(x-\theta_0-\frac{c}{\sqrt{n_i}}\right), \qquad i = 1, 2, \ldots,$$

where $\lim_{i\to\infty} n_i = \infty$ and c is a nonnegative constant. [Although we could include a function $g(n_i)$ such that $\lim_{i\to\infty} n_i^{1/2} g(n_i) = 0$ in our sequence of Pitman alternatives $\{\theta_i\}$, it would add nothing instructional to the examples and is commonly omitted in applications of Pitman's concept of ARE.] This sequence of location parameters $\{\theta_i = \theta_0 + c/\sqrt{n_i}\ \}$ is referred to as **one-sample Pitman translation alternatives.** Note that if $c = 0$, this "alternative" is really just the null hypothesis value θ_0.

Example 5.3.14. Without loss of generality take $\theta_0 = 0$. Let $U_4(X_{1:n_i}, \ldots, X_{n_i:n_i})$ be the U-statistic estimator of $\gamma_n = P[X_{1:n} + X_{2:n} > 0]$

given in Example 3.1.11 under these Pitman translation alternatives. Conditions (i) and (iii) of Theorem 5.3.10 are immediate since the kernel $h(x_1, x_2) = \Psi(x_1 + x_2)$ for U_4 is bounded between 0 and 1. Moreover, from Example 3.3.10 we see that

$$|h_{1:n_i}(x)| = \left|1 - F\left(-x - \frac{c}{\sqrt{n_i}}\right)\right| \leqslant 1$$

for every $n_i \geqslant 2$ and every x. In addition,

$$\lim_{i\to\infty} h_{1:n_i}(x) = \lim_{i\to\infty}\left[1 - F\left(-x - \frac{c}{\sqrt{n_i}}\right)\right] = 1 - F(-x),$$

for every x, since $F(x)$ is continuous. Thus the conditions of Lemmas 5.3.11 and 5.3.13 are satisfied. This in turn implies that the final condition (ii) of Theorem 5.3.10 is satisfied with $\eta = 4\,\mathrm{Var}[1 - F(-X)] = 4\,\mathrm{Var}[F(-X)]$, where X has c.d.f. $F(\cdot)$. Since the distribution of X is symmetric about the value 0, we know that $X \stackrel{d}{=} -X$, which, from Remark 1.3.11, implies that $\mathrm{Var}(F(-X)) = \mathrm{Var}(F(X))$. Thus $\eta = 4(\frac{1}{12}) = \frac{1}{3}$, since $F(X)$ is uniformly distributed over the interval $(0, 1)$. Consequently, Theorem 5.3.10 shows that

$$\sqrt{n_i}\left[U_4(X_{1:n_i}, \ldots, X_{n_i:n_i}) - P(X_{1:n_i} + X_{2:n_i} > 0)\right]$$

has a limiting $(i \to \infty)$ normal distribution with mean zero and variance 1/3.

We note that condition (iii) of Theorem 5.3.10 is frequently not a serious restriction for the type of U-statistics used in nonparametric problems, since they often correspond to parameters that are probability statements, and hence the associated kernels are generally bounded between 0 and 1. However, for U-statistic estimators of parameters that are not probabilities, condition (iii) can be more restrictive than is necessary. For the sample mean U-statistic $U_1(X_{1:n_i}, \ldots, X_{n_i:n_i}) = \overline{X}$ of Example 3.1.4, for example, condition (iii) would require that

$$E\left[\left|X_{1:n_i} - \theta_0 - \frac{c}{\sqrt{n_i}}\right|^3\right]$$

be bounded uniformly in $n_i \geqslant 1$. The next example shows that this condition is not necessary to reach the conclusion of Theorem 5.3.10 for U_1.

Example 5.3.15. Let $X_{1:n_i}, \ldots, X_{n_i:n_i}$ be independent and identically distributed continuous random variables with c.d.f.

$$F\left(x-\theta_0-\frac{c}{\sqrt{n_i}}\right).$$

Let W represent a random variable with c.d.f. $F(x)$, and assume $E(W)=0$ and $0<\mathrm{Var}(W)=\sigma^2<\infty$. Then it follows at once that

$$E\left[X_{1:n_i}\right]=\theta_0+\frac{c}{\sqrt{n_i}}$$

and $\mathrm{Var}(X_{1:n_i})=\sigma^2$. Moreover, letting

$$V_{n_i}=\frac{X_{1:n_i}-\theta_0-\dfrac{c}{\sqrt{n_i}}}{\sigma},$$

we see that

$$G_{n_i}(v)=P(V_{n_i}\leqslant v)=F(\sigma v), \qquad \text{for every } v.$$

Hence, $\lim_{i\to\infty} G_{n_i}(v)=F(\sigma v)$ for every v, and V_{n_i} has a limiting ($i\to\infty$) distribution with mean 0 and variance 1. Thus from Theorem A.3.7 in the Appendix it follows that

$$\sqrt{n_i}\left[U_1(X_{1:n_i},\ldots,X_{n_i:n_i})-\left(\theta_0+\frac{c}{\sqrt{n_i}}\right)\right]=\sqrt{n_i}\left[\bar{X}-\left(\theta_0+\frac{c}{\sqrt{n_i}}\right)\right]$$

has a limiting ($i\to\infty$) normal distribution with mean zero and variance σ^2, provided only that $0<\sigma^2<\infty$.

We now consider the corresponding two-sample U-statistics properties for Pitman alternatives. Let $X_{1:m},\ldots,X_{m:m}$ and $Y_{1:n},\ldots,Y_{n:n}$ be independent random samples from distributions with c.d.f.'s $F_m(x)$ and $G_n(y)$, respectively, possibly depending on the associated sample sizes. Let $U(X_{1:m},\ldots,X_{m:m};Y_{1:n},\ldots,Y_{n:n})$ be a two-sample U-statistic for the symmetric kernel $h(x_1,\ldots,x_r;y_1,\ldots,y_s)$, and let $\gamma_{m,n}=E[U(X_{1:m},\ldots,X_{m:m};Y_{1:n},\ldots,Y_{n:n})]$. As in the one-sample setting, here $\gamma_{m,n}$ may depend on the sample sizes m and n. For given m and n, let

$$h_{1,0:m,n}(x) = E\left[h(x, X_{2:m}, \ldots, X_{r:m}; Y_{1:n}, \ldots, Y_{s:n})\right] \tag{5.3.16}$$

and

$$h_{0,1:m,n}(y) = E\left[h(X_{1:m}, \ldots, X_{r:m}; y, Y_{2:n}, \ldots, Y_{s:n})\right]. \tag{5.3.17}$$

Finally, set

$$\zeta_{1,0:m,n} = \operatorname{Var}(h_{1,0:m,n}(X_{1:m})) \tag{5.3.18}$$

and

$$\zeta_{0,1:m,n} = \operatorname{Var}(h_{0,1:m,n}(Y_{1:n})). \tag{5.3.19}$$

As in the one-sample setting these quantities depend on the sample sizes m and n because the distribution of every $X_{i:m}(Y_{j:n})$ depends on $m(n)$. Letting $N = m+n$ and making the usual assumption that $\lim_{N\to\infty}(m/N) = \lambda$ and $\lim_{N\to\infty}(n/N) = (1-\lambda)$, with $0<\lambda<1$, we state the two-sample U-statistics theorem for Pitman alternatives, leaving the proof as an exercise.

Theorem 5.3.20. *Let $X_{1:m}, \ldots, X_{m:m}$ and $Y_{1:n}, \ldots, Y_{n:n}$ denote independent random samples from distributions with c.d.f.'s $F_m(x)$ and $G_n(y)$, respectively. Let $U(X_{1:m}, \ldots, X_{m:m}; Y_{1:n}, \ldots, Y_{n:n})$ be a two-sample U-statistic with symmetric kernel $h(x_1, \ldots, x_r; y_1, \ldots, y_s)$, and let $\gamma_{m,n}$, $\zeta_{1,0:m,n}$ and $\zeta_{0,1:m,n}$ be as defined for the sample-size-dependent distributions $F_m(x)$ and $G_n(y)$. If*

(i) $$\nu_{m,n} = E\left[h^2(X_{1:m}, \ldots, X_{m:m}; Y_{1:n}, \ldots, Y_{s:n})\right] < \infty$$
for every $m \geqslant r$ and $n \geqslant s$,

(ii) $$\lim_{N\to\infty} N \operatorname{Var}\left[U(X_{1:m}, \ldots, X_{m:m}; Y_{1:n}, \ldots, Y_{n:n})\right]$$
$$= \lim_{N\to\infty}\left[\frac{r^2\zeta_{1,0:m,n}}{(m/N)} + \frac{s^2\zeta_{0,1:m,n}}{(n/N)}\right] = \eta,$$

and

(iii) $E\left[|h_{1,0:m,n}(X_{1:m}) - \gamma_{m,n}|^3\right] < M_1,$ *for every $m \geqslant r$,*

and

$$E\left[|h_{0,1:m,n}(Y_{1:n})-\gamma_{m,n}|^3\right] < M_2, \qquad \text{for every } n \geqslant s,$$

where $\eta>0$, $M_1>0$, *and* $M_2>0$ *are constants, then* $\sqrt{N}\,[U(X_{1:m},\ldots, X_{m:m}; Y_{1:n},\ldots,Y_{n:n})-\gamma_{m,n}]$ *has a limiting normal distribution with mean* 0 *and variance* η.

Proof: Exercise 5.3.2. ■

In analogy with the one-sample setting, applications of Theorem 5.3.20 to many two-sample U-statistics for nonparametric problems are enhanced by the following result.

Lemma 5.3.21. *If* $\nu_{m,n}=E[h^2(X_{1:m},\ldots,X_{r:m}; Y_{1:n},\ldots,Y_{s:n})]\leqslant M$ *for every* $m\geqslant r$, $n\geqslant s$, *and some constant* $M>0$, *then*

$$\lim_{N\to\infty}\left\{N\operatorname{Var}\left[U(X_{1:m},\ldots,X_{m:m}; Y_{1:n},\ldots,Y_{n:n})\right]\right.$$

$$\left.-\left[\frac{r^2\zeta_{1,0:m,n}}{(m/N)}+\frac{s^2\zeta_{0,1:m,n}}{(n/N)}\right]\right\}=0.$$

Proof: Exercise 5.3.3. ■

To illustrate the application of these results to two-sample U-statistics for the Pitman alternatives of Section 5.2, we consider the following model. For arbitrary m_i, n_i, and N_i, let $X_{1:m_i},\ldots,X_{m_i:m_i}$ and $Y_{1:n_i},\ldots,Y_{n_i:n_i}$ be independent random samples from distributions with c.d.f.'s $F(x)$ and $F(x-(c/\sqrt{N_i}\,))$, respectively, where $\lim_{i\to\infty}N_i=\infty$, $\lim_{i\to\infty}(m_i/N_i)=\lambda$ and $\lim_{i\to\infty}(n_i/N_i)=(1-\lambda)$, for some $0<\lambda<1$. This sequence of shift parameters $\{\Delta_i=c/\sqrt{N_i}\,\}$ is referred to as **two-sample Pitman shift alternatives**. Note again that if $c=0$, this "alternative" is merely the null hypothesis value $\Delta=0$.

Example 5.3.22. Let $U_7(X_{1:m_i},\ldots,X_{m_i:m_i}; Y_{1:n_i},\ldots,Y_{n_i:n_i})$ be the two-sample U-statistic associated with the Mann–Whitney–Wilcoxon test (see Example 3.4.3) under these two-sample Pitman shift alternatives. Also, let $P_{\Delta_i}(X_{1:m_i}<Y_{1:n_i})$ denote the probability associated with the independent random variables $X_{1:m_i}, Y_{1:n_i}$ with respective c.d.f.'s $F(x)$ and $F(x-\Delta_i)$. Then, Lemma 5.3.13 (twice), Lemma 5.3.21 and Theorem 5.3.20 can be

used to show that $\sqrt{N_i}\,[U_7(X_{1:m_i},\ldots,X_{m_i:m_i};Y_{1:n_i},\ldots,Y_{n_i:n_i})-P_{\Delta_i}(X_{1:m_i}<Y_{1:n_i})]$ has a limiting normal distribution with mean 0 and variance $\eta=1/12\lambda(1-\lambda)>0$. (See Exercise 5.3.6.)

As was the case in the one-sample setting, Theorem 5.3.20 does not pose serious restrictions for many of the two-sample U-statistics used in nonparametric problems, since their kernels are generally bounded. On the other hand, for U-statistics with kernels that are not bounded, condition (iii) of Theorem 5.3.20 can again be more stringent than necessary. For example, for the two-sample U-statistic $U_6(X_{1:m_i},\ldots,X_{m_i:m_i};Y_{1:n_i},\ldots,Y_{n_i:n_i})=\bar{Y}-\bar{X}$ based on the sample means and considered in Example 3.4.2, the conclusion of Theorem 5.3.20 is valid under the milder [than (iii)] condition that both population variances are finite (see Exercise 5.3.7).

Exercises 5.3.

5.3.1. In Example 5.3.14, show that we can indeed assume $\theta_0=0$ without loss of generality.

5.3.2. Prove Theorem 5.3.20.

5.3.3. Prove Lemma 5.3.21.

5.3.4. Let $U_3(X_{1:n_i},\ldots,X_{n_i:n_i})$ be the U-statistic associated with the sign test (see Example 3.1.8). Find the form of the limiting ($i\to\infty$) distribution of $\sqrt{n_i}\,(U_3(X_{1:n_i},\ldots,X_{n_i:n_i})-\gamma_{n_i})$ under the sequence of one-sample Pitman alternatives $\{\theta_i\}$.

5.3.5. Let $W^+(X_{1:n_i},\ldots,X_{n_i:n_i})$ be the Wilcoxon signed rank statistic given in (2.4.11). Find the form of the limiting ($i\to\infty$) distribution of

$$\frac{\sqrt{n_i}}{\binom{n_i}{2}}\left[W^+(X_{1:n_i},\ldots,X_{n_i:n_i})-\binom{n_i}{2}P(X_{1:n_i}+X_{2:n_i}>2\theta_0)\right]$$

under the sequence of one-sample Pitman alternatives $F(x-\theta_0-(c/\sqrt{n_i}\,))$, where $F(\theta_0+x)+F(\theta_0-x)=1$ for all x. [Hint: Examples 3.3.17 and 5.3.14 are useful.]

5.3.6. Prove that the conditions of Lemma 5.3.21 and Theorem 5.3.20 are satisfied for the setting in Example 5.3.22, when $F(\cdot)$ is continuous. Verify the form of the limiting variance $\eta=1/12\lambda(1-\lambda)$ and express $P_{\Delta_i}(X_{1:m_i}<Y_{1:n_i})$ in terms of Δ_i and $F(\cdot)$. [Hint: Consider two applications of Lemma 5.3.13.]

5.3.7. Let $U_6(X_{1:m_i},\ldots,X_{m_i:m_i}; Y_{1:n_i},\ldots,Y_{n_i:n_i}) = \overline{Y} - \overline{X}$ be the two-sample U-statistic considered in Example 3.4.2. Show that the conclusion of Theorem 5.3.20 remains valid for U_6 under Pitman alternatives and the milder [than (iii) of that theorem] condition that the common population variance is finite. [Hint: Use the result of Example 5.3.15 and Theorem A.3.15 of the Appendix.]

5.3.8. For the setting of Lemma 5.3.11, show that $\zeta_{i:n} \leqslant \zeta_{r:n}$, for $i = 1,\ldots,r$. [Hint: Exercise 3.1.4 is helpful.]

5.4. Examples of Pitman's ARE for Translation Alternatives

We turn now to some specific examples of Pitman's ARE for translation alternatives, and consider first the one-sample location setting. Let $X_1,\ldots,X_n$ be a random sample from a continuous distribution that is symmetric about θ with c.d.f. $F(x-\theta)$, where $F(x)+F(-x)=1$ for all x. For θ_0 a fixed number, we are interested in tests of H_0: $\theta=\theta_0$ against, for example, the alternatives H_1: $\theta>\theta_0$. For ARE illustrations we study the following three competitive test procedures under the one-sample Pitman translation alternatives $F(x-\theta_i)=F(x-\theta_0-(c/\sqrt{n_i}))$, $i=1,2,\ldots,$ where we freely interchange the sample size designation as n_i or n, depending on whether Pitman alternatives are directly involved in the discussion.

Test 1: Reject H_0 in favor of H_1: $\theta>\theta_0$ iff

$$T^+ = \frac{\sqrt{n}\,(\overline{X}-\theta_0)}{S} \geqslant t_{\alpha,n-1},$$

where T^+ is the one-sample Student t-statistic and $t_{\alpha,n-1}$ is the upper 100αth percentile for the t-distribution with $(n-1)$ degrees of freedom,

Test 2: Reject H_0 in favor of H_1: $\theta>\theta_0$ iff

$$W^+ \geqslant w^+(\alpha,n),$$

where W^+ is the Wilcoxon signed rank statistic and $w^+(\alpha,n)$ is the upper 100αth percentile for the null distribution of W^+,

and

Test 3: Reject H_0 in favor of H_1: $\theta > \theta_0$ iff

$$B \geqslant b\left(\alpha, n, \tfrac{1}{2}\right),$$

where B is the sign statistic and $b(\alpha, n, \frac{1}{2})$ is the upper 100αth percentile for the binomial distribution with parameters n and $p = 1/2$.

We use Theorem 5.2.7 [expression (5.2.8) in particular] to evaluate the AREs for these three tests. Hence we first find the three efficacy expressions. In the case of the parametric t-test (Test 1), we assume $0 < \sigma^2 = \text{Var}(X_1) < \infty$, and we must check Assumptions A1–A6 of Theorem 5.2.7 for the t-statistic. With $T_n^+ = \sqrt{n}\,(\bar{X} - \theta_0)/S$, let $\mu_n(\theta) = \sqrt{n}\,((\theta - \theta_0)/\sigma)$ and $\sigma_n(\theta) \equiv 1$. (Note that in this section the standardizing constants for a test statistic like T_n^+ are denoted by $\mu_n(\theta)$ and $\sigma_n(\theta)$ instead of $\mu_{T_n^+}(\theta)$ and $\sigma_{T_n^+}(\theta)$, as was done in Section 5.2. This will simplify the notation and should not cause confusion, since we deal with only one test statistic at a time in this section.) Define $V_{n_i} = \sqrt{n_i}\,(\bar{X} - \theta_i)/S$, where $\theta_i = \theta_0 + c/\sqrt{n_i}$. Note that V_{n_i} is asymptotically standard normal when $X_1, \ldots, X_{n_i}$ are i.i.d. $F(x - \theta_i)$ for $i = 1, 2, \ldots$. (See Example 3.2.9 for the proof in the null case when $c = 0$. The proof for $c \neq 0$ follows similarly from Example 5.3.15 and is left as an exercise.) Notice also that

$$\begin{aligned} \frac{T_{n_i}^+ - \mu_{n_i}(\theta_i)}{\sigma_{n_i}(\theta_i)} - V_{n_i} &= \sqrt{n_i}\,(\theta_i - \theta_0)\left(\frac{1}{S} - \frac{1}{\sigma}\right) \\ &= c\left(\frac{1}{S} - \frac{1}{\sigma}\right). \end{aligned} \qquad \textbf{(5.4.1)}$$

From Khintchin's weak law of large numbers [see, for example, p. 463 of Bickel and Doksum (1977)], it follows that $1/S$ converges in probability to $1/\sigma$ under both the null hypothesis and under Pitman alternatives, provided $0 < \sigma^2 < \infty$. (In Exercise 5.4.1 you are asked to prove this result under the stronger assumption of finite fourth moments.) Hence, using Theorem 3.2.12 we see that, like V_{n_i}, $(T_{n_i}^+ - \mu_{n_i}(\theta_i))/\sigma_{n_i}(\theta_i)$ has a limiting standard normal distribution under both the null ($c = 0$) and a sequence of Pitman alternatives. Thus Assumptions A1 and A2 are satisfied. In addition, with our $\sigma_{n_i}(\theta_i)$ and $\mu_{n_i}(\theta_i)$, we see that Assumption A3 is immediate for the t-statistic and that

$$\frac{d\mu_n(\theta)}{d\theta} = \frac{\sqrt{n}}{\sigma}$$

clearly satisfies A4. Assumption A5 is also immediate since $\mu'_{n_i}(\theta_i) \equiv \mu'_{n_i}(\theta_0)$. Finally, we have

$$\lim_{n\to\infty} \frac{\mu'_n(\theta_0)}{\sqrt{n\sigma_n^2(\theta_0)}} = \frac{\frac{\sqrt{n}}{\sigma}}{\sqrt{n}} = \frac{1}{\sigma} = K_{T^+} \tag{5.4.2}$$

in the final Assumption A6. Hence the conditions for Theorem 5.2.7 are satisfied with standardizing sequences $\mu_{n_i}(\theta_i)$ and $\sigma_{n_i}(\theta_i)$—which, we note, are not the mean and variance, respectively, of $T_{n_i}^+$—and the efficacy expression for the one-sample t-test is

$$\text{eff}(T^+) = \frac{1}{\sigma}. \tag{5.4.3}$$

For the Wilcoxon signed rank statistic W_n^+ in Test 2, we assume (without loss of generality) that $\theta_0 = 0$. We also assume that $f(x) = dF(x)/dx$ is bounded for all x-values and continuous at all but at most a countable number of x-values, including continuity at $x=0$. Let $V_n^+ = \left[W_n^+/\binom{n}{2}\right] =$ (sum of the signed ranks)$/\binom{n}{2}$. Clearly, a test based on V_n^+ is equivalent to one based on W_n^+. (See Exercise 5.2.5.) Thus by finding the efficacy of the test which uses V_n^+ we simultaneously find it for one using W_n^+. Setting $\mu_n(\theta) = E_\theta[V_n^+] = [2/(n-1)]P_\theta(X_1 > 0) + P_\theta(X_1 > -X_2)$ and $\sigma_n(\theta) = 1/\sqrt{3n}$, where X_1, X_2 are i.i.d. with c.d.f. $F(x-\theta)$, we see from Example 5.3.14 and Theorems 3.2.8 and 3.2.12 (used as in Example 3.3.17 but here applied to these Pitman alternatives) that

$$\frac{V_{n_i}^+ - \mu_{n_i}(\theta_i)}{\sigma_{n_i}(\theta_i)}$$

has a limiting distribution that is standard normal under both the null hypothesis ($c=0$) and Pitman alternatives, and hence Assumptions A1 and A2 are satisfied. In addition, Assumption A3 is obviously satisfied for the sequence $\sigma_n(\theta) = 1/\sqrt{3n}$. For Assumptions A4–A6 we have

$$\mu_n(\theta) = \frac{2}{(n-1)}[1 - F(-\theta)] + \int_{-\infty}^{\infty} [1 - F(-x-\theta)]\, dF(x-\theta),$$

which by setting $y = x - \theta$ in the integral becomes

$$\mu_n(\theta) = \frac{2}{(n-1)}[1 - F(-\theta)] + \int_{-\infty}^{\infty} [1 - F(-y-2\theta)]\, dF(y).$$

Since $f(x)=dF(x)/dx$ is bounded by some positive number M, we see that

$$\frac{d\mu_n(\theta)}{d\theta}=\frac{2}{(n-1)}f(-\theta)+2\int_{-\infty}^{\infty}f(-y-2\theta)\,dF(y), \tag{5.4.4}$$

where we have moved the derivative inside the integral using Theorem A.2.4 of the Appendix with $K^*(y)\equiv M$. Since $f(-y)=f(y)$, we see that

$$\left.\frac{d\mu_n(\theta)}{d\theta}\right|_{\theta=0}=2\left[\frac{f(0)}{(n-1)}+\int_{-\infty}^{\infty}f^2(y)\,dy\right]>0.$$

Also, the continuity of $d\mu_n(\theta)/d\theta$ in some closed interval about $\theta=0$ is a consequence of Theorem A.2.3, and we have that Assumption A4 is satisfied. Condition A5 follows from (5.4.4) by using Theorem A.2.3 of the Appendix and the fact that $f(\cdot)$ is bounded and is continuous at 0 and at all but at most a countable number of other x-values. Finally,

$$\lim_{n\to\infty}\frac{\mu_n'(0)}{\sqrt{n\sigma_n^2(0)}}=\lim_{n\to\infty}\frac{2\left[\dfrac{f(0)}{(n-1)}+\displaystyle\int_{-\infty}^{\infty}f^2(y)\,dy\right]}{\sqrt{n}\,(3n)^{-1/2}}$$

$$=2\sqrt{3}\int_{-\infty}^{\infty}f^2(y)\,dy=K_{W^+},$$

as needed in A6. [This expression has been derived under less stringent conditions on $f(\cdot)$, namely, only that $\int_{-\infty}^{\infty}f^2(y)\,dy<\infty$, by Olshen (1967).] Thus all the assumptions of Theorem 5.2.7 are satisfied for the signed rank statistic with the designated sequences $\mu_n(\theta)$ and $\sigma_n(\theta)$, and the resulting efficacy expression for W^+ is

$$\text{eff}(W^+)=2\sqrt{3}\int_{-\infty}^{\infty}f^2(y)\,dy. \tag{5.4.5}$$

Using the formulas for $\text{eff}(T^+)$ in (5.4.3) and $\text{eff}(W^+)$ in (5.4.5), we are now in a position to apply Theorem 5.2.7 to obtain the ARE (W^+,T^+) expression

$$\text{ARE}(W^+,T^+)=\left[\frac{\text{eff}(W^+)}{\text{eff}(T^+)}\right]^2=12\sigma^2\left[\int_{-\infty}^{\infty}f^2(x)\,dx\right]^2. \tag{5.4.6}$$

Values of $\text{ARE}(W^+,T^+)$ for some specific symmetric densities are given in Table 5.4.7. We illustrate the calculations involved in obtaining the

TABLE 5.4.7.
ARE(W^+, T^+) Values for Selected Distributions

DISTRIBUTION	ARE(W^+, T^+)
Uniform	1
Normal	$3/\pi$
Logistic	$\pi^2/9$
Double exponential	1.5

ARE of $3/\pi$ for the normal distribution. Let

$$f(x) = \frac{1}{\sqrt{2\pi\sigma^2}} e^{-x^2/2\sigma^2}.$$

Then,

$$\begin{aligned} \text{ARE}(W^+, T^+) &= 12\sigma\left[\int_{-\infty}^{\infty} \frac{1}{2\pi\sigma^2} e^{-x^2/\sigma^2}\,dx\right]^2 \\ &= \frac{12\sigma^2}{4\pi\sigma^2}\left[\int_{-\infty}^{\infty} \frac{1}{\sqrt{2\pi\left(\frac{\sigma^2}{2}\right)}} e^{-(x^2/2(\sigma^2/2))}\,dx\right]^2 \\ &= \frac{3}{\pi}. \end{aligned}$$

On first examining the ARE(W^+, T^+) values in Table 5.4.7, we might be surprised at the consistent merit demonstrated by the signed rank test. When the underlying distribution is normal, little is lost by using the W^+ test as opposed to the optimal t-test. Roughly speaking, for every 95 observations used in a t-test, we would require 100 observations to do as well (asymptotically) with W^+—a loss of approximately 5% in effectiveness. For all the other distributions listed, the Wilcoxon test scores as well or better than the t-test. A natural question to ask is whether this phenomenon will carry over to other distributions. Hodges and Lehmann (1956) answered this question by showing that ARE(W^+, T^+) $\geqslant .864$ for every continuous density $f(x)$ satisfying the conditions leading to expression (5.4.6). [For more detail, see Lehmann (1975).] This fact establishes the Wilcoxon signed rank test as a solid competitor to the t-test even when a large sample is available. Of course, the advantage of exact α-levels (for *any* continuous distribution) for small samples still belongs to W^+.

By analogous arguments (left as an exercise), we can establish the efficacy expression for the sign statistic in Test 3 to be

$$\text{eff}(B) = 2f(0), \tag{5.4.8}$$

where we assume (without loss of generality) that the point of symmetry is $\theta_0 = 0$ and, in addition, that the density $f(x)$ is positive and continuous at 0. The corresponding asymptotic relative efficiency expressions for B relative to T^+ and W^+, respectively, are

$$\text{ARE}(B, T^+) = 4\sigma^2 f^2(0) \tag{5.4.9}$$

and

$$\text{ARE}(B, W^+) = \frac{f^2(0)}{3\left[\int_{-\infty}^{\infty} f^2(x)\,dx\right]^2}. \tag{5.4.10}$$

Table 5.4.11 provides the efficiency values $\text{ARE}(B, T^+)$ for several distributions.

It is obvious from Tables 5.4.7 and 5.4.11 that, in general, a test based on the sign statistic is not as efficient as a test based on the signed rank statistic. This is not surprising, since less sample information is used in the sign test than in the signed rank test. However, for heavier-tailed distributions (e.g., the double exponential), the quality of this lesser information can lead to a better performance by the sign test than by the signed rank, so that we cannot discard the sign test as being uniformly inferior. For a lower bound on the sign test efficiencies, Hodges and Lehmann (1956) demonstrated that $\text{ARE}(B, T^+) \geqslant 1/3$ for all continuous, unimodal sym-

TABLE 5.4.11.
$\text{ARE}(B, T^+)$ Values for Selected Distributions

DISTRIBUTION	$\text{ARE}(B, T^+)$
Uniform	$1/3$
Normal	$2/\pi$
Logistic	$\pi^2/12$
Double exponential	2

metric distributions. The value in Table 5.4.11 corresponding to the uniform distribution demonstrates that this lower bound is actually achieved.

We now turn our attention to obtaining similar ARE results for the two-sample location tests based on the Mann–Whitney–Wilcoxon and two-sample t statistics. Let $X_1,\ldots,X_m$ and $Y_1,\ldots,Y_n$ be independent random samples from continuous distributions with distribution functions $F(x)$ and $F(x-\Delta)$, respectively. For this setting, we consider tests of the null H_0: $\Delta=0$ against, for example, the alternative H_1: $\Delta>0$. The Pitman alternatives in this case are given by $F(x)$ and $F(x-\Delta_i)=F(x-(c/\sqrt{N_i}))$, $i=1,\ldots,$ where $c\geqslant 0$ is a constant and $\lim_{i\to\infty} N_i=\infty$ in such a way that $\lim_{i\to\infty}(m_i/N_i)=\lambda$ and $\lim_{i\to\infty}(n_i/N_i)=(1-\lambda)$, $0<\lambda<1$. [Again we freely vary the use of (m_i,n_i,N_i) and (m,n,N) for denoting the sample sizes, depending on whether Pitman alternatives are directly involved in the discussion.] The Mann–Whitney–Wilcoxon and two-sample t tests have forms:

Test 4: Reject H_0 in favor of H_1: $\Delta>0$ iff

$$W \geqslant w(\alpha,m,n),$$

where W is the Mann–Whitney–Wilcoxon rank sum statistic and $w(\alpha,m,n)$ is the upper 100αth percentile for the null distribution of W.

Test 5: Reject H_0 in favor of H_1: $\Delta>0$ iff

$$T=\frac{(\bar{Y}-\bar{X})}{S\sqrt{\dfrac{m+n}{mn}}}\geqslant t_{\alpha,m+n-2},$$

where T is the two-sample Student t-statistic,

$$S^2=\frac{\sum_{i=1}^{m}\left(X_i-\bar{X}\right)^2+\sum_{j=1}^{n}\left(Y_j-\bar{Y}\right)^2}{m+n-2}$$

is the pooled sample variance, and $t_{\alpha,m+n-2}$ is the upper 100αth percentile for the t-distribution with $(m+n-2)$ degrees of freedom. For this test, we make the additional assumption that the common variance of the Xs and Ys, say σ^2, is positive and finite.

As in the one-sample setting, we assume that the density $f(x)=dF(x)/dx$ is bounded for all x-values and continuous at all but a countable number of x-values. For the Mann–Whitney–Wilcoxon Test 4, recall that $W_N=$(sum of the joint ranks for the Y observations)$=mnU_7+(n(n+1)/2)$, where U_7 is the U-statistic estimator of $P[X<Y]$ first introduced in Example 3.4.3. Defining $\mu_N(\Delta)=mnP_\Delta(X_1<Y_1)+(n(n+1)/2)$ and $\sigma_N(\Delta)=(mnN/12)^{1/2}$, where X_1, Y_1 are independent with c.d.f.'s $F(x)$ and $F(x-\Delta)$, respectively, we see from Example 5.3.22 and Slutsky's theorem 3.2.8 that $[W_{N_i}-\mu_{N_i}(\Delta_i)]/\sigma_{N_i}(\Delta_i)$ has a limiting distribution that is standard normal under both Pitman alternatives and the null hypothesis (corresponding to $c=0$). Now, setting $x=t-\Delta$ in the integral expression of $\mu_N(\Delta)$ we change

$$\mu_N(\Delta) = mn\int_{-\infty}^{\infty} F(t)\,dF(t-\Delta) + \frac{n(n+1)}{2}$$

into

$$\mu_N(\Delta) = mn\int_{-\infty}^{\infty} F(x+\Delta)\,dF(x) + \frac{n(n+1)}{2}.$$

Thus since the density $f(x)$ is continuous at all but at most a countable number of x-values and is bounded by some positive constant M for all x-values, we have

$$\frac{d\mu_N(\Delta)}{d\Delta} = mn\int_{-\infty}^{\infty} f(x+\Delta)f(x)\,dx, \tag{5.4.12}$$

by applying Theorem A.2.4 of the Appendix, as was done in creating (5.4.4). When the null H_0: $\Delta=0$ is true we obtain

$$\left.\frac{d\mu_N(\Delta)}{d\Delta}\right|_{\Delta=0} = mn\int_{-\infty}^{\infty} f^2(x)\,dx > 0.$$

Condition A5 of Theorem 5.2.7 follows from (5.4.12) by using Theorem A.2.3 in the Appendix and the fact that $f(x)$ is bounded and continuous at all but at most a countable number of x-values. In addition, Assumption A6 follows, since

$$\lim_{N\to\infty}\frac{\mu_N'(0)}{\sqrt{N\sigma_N^2(0)}} = \lim_{N\to\infty}\left[mn\int_{-\infty}^{\infty} f^2(x)\,dx\Big/\{mnN^2/12\}^{1/2}\right]$$

$$=\sqrt{12\lambda(1-\lambda)}\int_{-\infty}^{\infty} f^2(x)\,dx = K_W. \tag{5.4.13}$$

Hence all the conditions for Theorem 5.2.7 are satisfied, and the associated efficacy expression for W is

$$\mathrm{eff}(W) = \sqrt{12\lambda(1-\lambda)} \int_{-\infty}^{\infty} f^2(x)\,dx. \qquad \textbf{(5.4.14)}$$

Using arguments similar to those employed for the one-sample t-statistic, it can be shown (details are left as an exercise) that the conditions for Theorem 5.2.7 are also satisfied for the two-sample t-statistic T in Test 5, with appropriately chosen $\mu_N(\Delta)$ and $\sigma_N(\Delta)$. The corresponding efficacy expression is

$$\mathrm{eff}(T) = K_T = \{\lambda(1-\lambda)\}^{1/2}/\sigma. \qquad \textbf{(5.4.15)}$$

Hence, from Theorem 5.2.7, we have

$$\mathrm{ARE}(W,T) = \frac{K_W^2}{K_T^2} = 12\sigma^2\left[\int_{-\infty}^{\infty} f^2(x)\,dx\right]^2. \qquad \textbf{(5.4.16)}$$

We note that this ARE(W,T) expression is identical with that between the Wilcoxon signed rank statistic and the one-sample t-statistic [see equation (5.4.6)]. Hence the efficiency values in Table 5.4.7 apply as well to this two-sample setting, as does the lower bound ARE$(W,T) \geqslant .864$ obtained by Hodges and Lehmann (1956). However, there is one major difference between the two settings. Although the one-sample efficiency expression for ARE(W^+,T^+) requires $f(x)$ to be symmetric about 0, such is not the case for the two-sample analogue ARE(W,T). Thus in the two-sample problem we can evaluate expression (5.4.16) for an asymmetric distribution like the exponential, but this evaluation does not pertain to the one-sample problem. The general form

$$12\sigma^2\left[\int_{-\infty}^{\infty} f^2(x)\,dx\right]^2$$

occurs repeatedly in nonparametric efficiency evaluations for situations ranging from these simple one- and two-sample location problems to more complicated analysis of variance settings.

Exercises 5.4.

5.4.1. Let $S_{n_i}^2$ be the sample variance for a random sample of size n_i from a distribution with c.d.f. $F(x-\theta_0-(c/\sqrt{n_i}\,))$, $i=1,\ldots$. If X has c.d.f. $F(x)$

such that $0<\sigma^2=\mathrm{Var}(X)$ and $E(X^4)<\infty$, show that $1/S_{n_i}$ converges $(i\to\infty)$ in probability to $1/\sigma$. [Hint: See Example 3.2.9 and Exercise 3.2.3.]

5.4.2. As in the discussion preceding (5.4.1), let $V_{n_i}=\sqrt{n_i}\,(\bar{X}-\theta_i)/S$ and define $V^*_{n_i}=\sqrt{n_i}\,(\bar{X}-\theta_i)/\sigma$. Show that V_{n_i} and $V^*_{n_i}$ have the same limiting $(i\to\infty)$ distribution under the sequence of Pitman alternatives $F(x-\theta_0-(c/\sqrt{n_i}\,))$. Under the conditions of Exercise 5.4.1, use this to establish the limiting distribution of V_{n_i} for such a sequence of alternatives.

5.4.3. Show that the conditions of Theorem 5.2.7 are satisfied for the two-sample t-statistic T in Test 5. Confirm the expressions for eff(T) given in (5.4.15). [Hint: Exercises 5.3.7 and 5.4.1 might be useful.]

5.4.4. Evaluate ARE(W,T) in (5.4.16) for the exponential distribution with density

$$\begin{aligned} f(x;\theta,\delta) &= \frac{1}{\delta}e^{-\frac{(x-\theta)}{\delta}}, \qquad x>\theta, \\ &= 0, \qquad \text{elsewhere.} \end{aligned}$$

5.4.5. Evaluate ARE(W,T) in (5.4.16) for the Weibull distribution with density

$$\begin{aligned} f(x;\alpha,\delta) &= \frac{\alpha x^{\alpha-1}}{\delta^\alpha}e^{-(x/\delta)^\alpha}, \qquad x>0 \\ &= 0, \qquad \text{elsewhere,} \end{aligned}$$

where $\alpha>1/2$ and $\delta>0$.

5.4.6. Show that the expression for ARE(W,T) given in (5.4.16) is invariant under location and scale changes; that is, the value of ARE(W,T) is not affected by replacing the c.d.f. $F(x)$ by $F((x-b)/a)$, with $a>0$.

5.4.7. Show that the expression for ARE(B,T^+) in (5.4.9) is invariant under location and scale changes. (See Exercise 5.4.6.)

5.4.8. Verify the values of ARE(W^+,T^+) in Table 5.4.7.

5.4.9. Verify the values of ARE(B,T^+) in Table 5.4.11.

5.4.10. Construct a table of values (similar to Tables 5.4.7 and 5.4.11) for ARE(B,W^+). [Hint: Exercise 5.2.3 is useful.]

5.4.11. Verify the necessary conditions leading to the expressions in (5.4.8), (5.4.9), and (5.4.10) for the sign test.

5.4.12. Show that ARE$(W^+, T^+) = .864$ for the density

$$f(x) = \frac{3}{20\sqrt{5}}(5 - x^2), \qquad -\sqrt{5} < x < \sqrt{5}$$
$$= 0, \qquad \text{elsewhere.}$$

(This demonstrates that the lower bound given by Hodges and Lehmann (1956) is achievable.)

5.4.13. Let $X_1, \ldots, X_n$ and $Y_1, \ldots, Y_n$ be independent random samples from continuous distributions with c.d.f.'s $F(x)$ and $F(x - \Delta)$, respectively. Define the statistic

$$V = \sum_{i=1}^{n} \Psi(Y_i - X_i),$$

where $\Psi(t) = 1, 0$ as $t >, \leqslant 0$. Consider the test of H_0: $\Delta = 0$ against H_1: $\Delta > 0$ that rejects H_0 for large values of V. Obtain expressions for eff(V) and ARE(V, W), where W represents the test of H_0: $\Delta = 0$ against H_1: $\Delta > 0$ based on the Mann–Whitney–Wilcoxon rank sum statistic. Evaluate ARE(V, W) for $F(\cdot)$ corresponding to the exponential distribution in Exercise 5.4.4.

5.5. Discussion—Deficiency and Bahadur Efficiency

We wish to emphasize again that the standardizing constants $\mu_n(\theta)$ and $\sigma_n^2(\theta)$ used in obtaining efficiency expressions for Pitman alternatives are not required to be the mean and variance, respectively, for the test statistic being studied. For the nonparametric examples considered in Section 5.4, the various exact mean expressions under Pitman alternatives were used, but this was not the case for the parametric one-sample t-statistic, where an expression asymptotically equivalent to the exact mean was used. In none of the examples did we take $\sigma_n^2(\theta)$ to be the actual variances. In fact, in all of those examples we took expressions for $\sigma_n^2(\theta)$ that were asymptotically equivalent to the *null* variances of the test statistics. In general, this approach often makes the verification of the conditions for Theorem 5.2.7 much simpler. Of course, the choice of the particular forms for the test statistics themselves can also have a lot to do with the difficulty one encounters when computing the efficiency values. This fact was evidenced in the calculations for the two t-statistics in Section 5.4.

Before we leave the topic of efficiency, we briefly comment about two additional efficiency concepts. The first concept, **deficiency**, was introduced by Hodges and Lehmann (1970) to handle cases where $\mathrm{ARE}(S,T) \equiv 1$. Essentially, their approach is to consider expansions similar to those in the Pitman efficiency calculations, but now these expansions are carried out to higher-order terms in the sample size(s). Thus, instead of remaindering a series at terms of order $n^{-1/2}$, for example, terms of order n^{-1} would also be included. Using these ideas, Hodges and Lehmann investigate for example, the loss one incurs by using the one-sample t-statistic as opposed to $\sqrt{n}\,\overline{X}/\sigma$, when σ is known. They show that although there is a slight deficiency from using the t-statistic, the magnitude of the difference is so small as to be overshadowed by the protection afforded by the t-statistic against possible errors in the presumed value of σ. Hodges and Lehmann also consider many other examples.

A second alternative type of efficiency was introduced by Bahadur (1960a, 1960b, 1967). The basic difference between the Bahadur and the Pitman approaches to efficiency corresponds to what is being held fixed and what is converging as the sample size(s) become large. The Pitman approach deals with fixed significance level and power and sequences of alternatives converging to the null hypothesis. The efficiency then involves evaluations of the relative rates of convergence for the two tests under comparison. On the other hand, Bahadur considers fixed alternative and power values and studies the relative rates at which the attained significance levels for the two tests converge to zero. For a given distributional form the Bahadur efficiency is, in many cases, not a single number—as is the case for Pitman efficiency—but is instead a function of the fixed alternative under consideration. The Pitman efficiency value can often be obtained from the Bahadur efficiency expression by letting the alternative, say, θ, converge to the null hypothesis value, say, θ_0. For examples of Bahadur efficiency calculations and associated approximations, see Bahadur (1960a, 1960b, 1967). Additional concepts of asymptotic efficiency have been proposed by Cochran (1952) and Anderson and Goodman (1957), among others.

6 CONFIDENCE INTERVALS AND BOUNDS

In this chapter we discuss some of the techniques for obtaining confidence intervals or bounds for parameters of interest in a particular setting. We demonstrate how to derive a distribution-free confidence interval from a distribution-free test of hypothesis, and consider large sample properties of the resulting intervals. In addition, asymptotically distribution-free confidence intervals for certain parameters are obtained using the U-statistics results of Chapter 3.

6.1. Distribution-Free Confidence Intervals

Let $X_1,\ldots,X_n$ be a random sample from a distribution with c.d.f. $F_\theta(\cdot)$, where $\theta\in\Theta$ is some unknown parameter. Let $L(X_1,\ldots,X_n)$ and $U(X_1,\ldots,X_n)$ be statistics such that

$$P_\theta(L(X_1,\ldots,X_n)<\theta<U(X_1,\ldots,X_n)) = 1-\gamma,$$

for all $\theta\in\Theta$, where $0<\gamma<1$ is arbitrary and P_θ indicates that the probability is computed under $F_\theta(\cdot)$. Then we say that $(L(X_1,\ldots,X_n), U(X_1,\ldots,X_n))$ is a **$100(1-\gamma)$ percent confidence interval for θ**, and $1-\gamma$ is called the **confidence coefficient**. For example, if

$$F_\theta(x) = \Phi\left(\frac{x-\theta}{\sigma}\right),\ -\infty<x<\infty,\ 0<\sigma^2<\infty \text{ and } \Theta=(-\infty,\infty),$$

then

$$(L(X_1,\ldots,X_n), U(X_1,\ldots,X_n)) = \left(\bar{X}-t_{(\gamma/2,n-1)}\frac{S}{\sqrt{n}}, \bar{X}+t_{(\gamma/2,n-1)}\frac{S}{\sqrt{n}}\right)$$

is a $100(1-\gamma)$ percent confidence interval for θ, where

$$\bar{X} = \frac{1}{n}\sum_{i=1}^{n} X_i, \qquad S^2 = \frac{\sum_{i=1}^{n}(X_i-\bar{X})^2}{(n-1)},$$

and $t_{(\gamma/2,n-1)}$ is the upper $100(\gamma/2)$th percentile point for the t-distribution with $n-1$ degrees of freedom.

In this confidence interval for the mean of a normal distribution, the coverage probability $1-\gamma$ is maintained exactly when $F_\theta(\cdot)$ is normal with unknown mean θ and variance σ^2. We therefore say that this confidence interval is **distribution-free** over the class of i.i.d. normal variates. However, for a different distribution with mean θ, the probability

$$P_\theta\left(\bar{X} - t_{(\gamma/2,n-1)}\frac{S}{\sqrt{n}} < \theta < \bar{X} + t_{(\gamma/2,n-1)}\frac{S}{\sqrt{n}}\right)$$

will depend on the form of $F_\theta(\cdot)$, potentially varying either above or below the nominal value $1-\gamma$. Thus, for distributions other than the normal, the interval

$$\left(\bar{X} - t_{(\gamma/2,n-1)}\frac{S}{\sqrt{n}}, \bar{X} + t_{(\gamma/2,n-1)}\frac{S}{\sqrt{n}}\right)$$

is not a $100(1-\gamma)$ percent confidence interval for the mean θ, and consequently the class for which this exact coverage probability is valid is not nonparametric in nature. If a confidence interval maintains the designated confidence coefficient over a class of distributions containing more than one parametric form, we term it **nonparametric distribution-free**. To obtain confidence intervals that maintain their coverage probabilities over a broad class of distributions, we utilize nonparametric distribution-free test statistics. We consider a technique for obtaining such nonparametric distribution-free confidence intervals for the one- and two-sample location parameter settings. Actually, the same approach can be used to obtain nonparametric distribution-free confidence intervals for parameters in other settings as well. Although such usage is not discussed in text, this more general approach is illustrated in Exercises 6.1.15–6.1.18.

Some further remarks are necessary before we proceed with the general methodology. Although the interval $(L(X_1,\ldots,X_n), U(X_1,\ldots,X_n))$ used in our description of a $100(1-\gamma)$ percent confidence interval for a parameter θ is open, this is not essential. A confidence interval could also be either totally closed or half-open, half-closed (on either end). For example, if we obtain a probability statement such as

$$P_\theta(L(X_1,\ldots,X_n)\leq\theta<U(X_1,\ldots,X_n)) = 1-\gamma$$

for all $\theta\in\Theta$, then the half-open, half-closed interval $[L(X_1,\ldots,X_n), U(X_1,\ldots,X_n))$ is a $100(1-\gamma)$ percent confidence interval for θ. As we shall see, many of the nonparametric distribution-free confidence intervals to be discussed in this chapter are of this half-open, half-closed type. On the other hand, when the statistics $L(X_1,\ldots,X_n)$ and $U(X_1,\ldots,X_n)$ have continuous distributions, we can freely open or close either or both of the endpoints of the confidence interval $[L(X_1,\ldots,X_n), U(X_1,\ldots,X_n))$ without affecting the associated confidence coefficient $1-\gamma$. Although the endpoints for all of the nonparametric distribution-free confidence intervals to be discussed in this text have continuous distributions, we discuss each such interval as half-open, half-closed. This will be more naturally in agreement with the way they are derived, and will more readily enable us to later discuss the conservative nature of many of these confidence intervals when the data arise from certain underlying discrete distribution(s).

In addition, since nonparametric distribution-free tests usually have associated null distributions that are discrete, there are only certain **natural confidence coefficient values** available for confidence intervals (or bounds) associated with such procedures, just as there are only certain natural α-levels for the tests. Hence throughout this chapter our discussion of confidence coefficients for nonparametric distribution-free confidence intervals (or bounds) is restricted to the natural confidence coefficient values of the associated two-sided (one-sided) tests.

For the one-sample location setting, let $X_1,\ldots,X_n$ be a random sample from a continuous distribution with c.d.f. $F(x-\theta)$, where $\theta\in\Theta=(-\infty,\infty)$ and $F(0)=1/2$; that is, the Xs are i.i.d. with median θ. Let $T(X_1,\ldots,X_n)$ be a test statistic for testing H_0: $\theta=0$ such that $T(X_1,\ldots,X_n)$ is distribution-free over $F\in\mathcal{F}$ when H_0: $\theta=0$ is true. Thus, for appropriate γ, we can select constants d_1 and d_2 to be possible values of $T(X_1,\ldots,X_n)$ such that

$$P_0(d_1<T(X_1,\ldots,X_n)<d_2) = 1-\gamma, \qquad \textbf{(6.1.1)}$$

with this probability holding for every $F(\cdot)\in\mathcal{F}$. On the other hand, in view of our location model, we know that when θ is the true parameter value, $(X_1-\theta)$ has the same distribution as does X_1 when 0 is the true value of the parameter. We express this fact notationally by $\{(X_1-\theta)|_\theta\} \stackrel{d}{=} \{X_1|_{\theta=0}\}$. Since the X_is are i.i.d., this implies that

$$\{T(X_1-\theta,\ldots,X_n-\theta)|_\theta\} \stackrel{d}{=} \{T(X_1,\ldots,X_n)|_{\theta=0}\}. \qquad \textbf{(6.1.2)}$$

Combining the properties in (6.1.1) and (6.1.2) we obtain

$$P_\theta(d_1 < T(X_1-\theta,\ldots,X_n-\theta) < d_2) = 1-\gamma. \tag{6.1.3}$$

Hence if for a given setting we can show that

$$\{d_1 < T(X_1-\theta,\ldots,X_n-\theta) < d_2\} \text{ iff } \{L_T(X_1,\ldots,X_n) \leqslant \theta < U_T(X_1,\ldots,X_n)\}, \tag{6.1.4}$$

for some statistics $L_T(X_1,\ldots,X_n)$ and $U_T(X_1,\ldots,X_n)$ and every $\theta \in \Theta$, then (6.1.3) and (6.1.4) would imply that $[L_T(X_1,\ldots,X_n), U_T(X_1,\ldots,X_n))$ is a $100(1-\gamma)$ percent distribution-free (over $\mathfrak{F}$) confidence interval for θ. The equivalence in (6.1.4) is often quite easy to establish as is demonstrated in the following series of examples.

Example 6.1.5. Let $\mathfrak{F}_1 = \{F(\cdot)$: $F(\cdot)$ is a c.d.f. for a normal distribution with mean 0 and variance $0<\sigma^2<\infty\}$ and take $T(X_1,\ldots,X_n) = \bar{X}/(S/\sqrt{n}\,)$ to be the one-sample t-statistic for testing H_0: $\theta=0$. Then $T(X_1,\ldots,X_n)$ is distribution-free over the parametric class $\mathfrak{F}_1$, and if $t_{(\gamma/2,n-1)}$ is the upper $100(\gamma/2)$th percentile point for the t-distribution with $n-1$ degrees of freedom, we have that (6.1.1) is satisfied with $d_2 = -d_1 = t_{(\gamma/2,n-1)}$. In addition, we note that, with probability one, the equivalence in (6.1.4) holds with $L_T(X_1,\ldots,X_n) = \bar{X} - t_{(\gamma/2,n-1)}S/\sqrt{n}$ and $U_T(X_1,\ldots,X_n) = \bar{X} + t_{(\gamma/2,n-1)}S/\sqrt{n}$. Thus the interval $[L_T(X_1,\ldots,X_n), U_T(X_1,\ldots,X_n))$ [or, as previously mentioned, the more commonly quoted open interval $(L(X_1,\ldots,X_n), U(X_1,\ldots,X_n))$] is a $100(1-\gamma)$ percent distribution-free (over $\mathfrak{F}_1$) confidence interval for θ.

The next two examples deal with nonparametric distribution-free statistics and illustrate a useful technique for establishing the equivalence relation (6.1.4) for such settings.

Example 6.1.6. Let $\mathfrak{F}_2 = \{F(\cdot)$: $F(\cdot)$ is a c.d.f. for a continuous distribution with zero median$\}$. Let $B(X_1,\ldots,X_n)$ = (number of positive Xs) be the sign statistic of (2.2.3) for testing H_0: $\theta=0$. Then $B(X_1,\ldots,X_n)$ is nonparametric distribution-free over $\mathfrak{F}_2$ when H_0: $\theta=0$ is true, and if $b(\gamma/2,n,\frac{1}{2})$ is the upper $100(\gamma/2)$th percentile point for the binomial $(n,\frac{1}{2})$ distribution, we see that (6.1.1) holds with $d_2 = b(\gamma/2,n,\frac{1}{2})$ and $d_1 = n - b(\gamma/2,n,\frac{1}{2})$. For (6.1.4) we note that $B(X_1-\theta,\ldots,X_n-\theta)$ = (number of Xs greater than θ). Thus, letting $X_{(1)} \leqslant \cdots \leqslant X_{(n)}$ be the ordered sample observations, we see that $\{B(X_1-\theta,\ldots,X_n-\theta) < b(\gamma/2,n,\frac{1}{2})\}$ is equivalent to

$$\{X_{(n+1-b(\gamma/2,n,1/2))} \leqslant \theta\},$$

and $\{n-b(\gamma/2,n,\frac{1}{2})<B(X_1-\theta,\ldots,X_n-\theta)\}$ is equivalent to

$$\{\theta<X_{(b(\gamma/2,n,1/2))}\}.$$

Hence (6.1.4) is satisfied with

$$L_B(X_1,\ldots,X_n) = X_{(n+1-b(\gamma/2,n,1/2))}$$

and

$$U_B(X_1,\ldots,X_n) = X_{(b(\gamma/2,n,1/2))};$$

consequently the interval

$$\left[X_{(n+1-b(\gamma/2,n,1/2))}, X_{(b(\gamma/2,n,1/2))}\right)$$

is a $100(1-\gamma)$ percent nonparametric distribution-free (over $\mathcal{F}_2$) confidence interval for θ.

Example 6.1.7. Let $\mathcal{F}_3=\{F(\cdot)\colon F(\cdot)$ is a c.d.f. for a continuous distribution that is symmetric about 0$\}$. If $W^+(X_1,\ldots,X_n)$ is the Wilcoxon signed rank statistic given in (2.4.11) for testing H_0: $\theta=0$, then $W^+(X_1,\ldots,X_n)$ is nonparametric distribution-free over $\mathcal{F}_3$ when H_0 is true. Thus letting $w^+(\gamma/2,n)$ be the upper $100(\gamma/2)$th percentile point for the null (H_0) distribution of W^+, we see that (6.1.1) is satisfied by setting $d_2=w^+(\gamma/2,n)$ and $d_1=(n(n+1)/2)-w^+(\gamma/2,n)$, since the null distribution of W^+ is symmetric about $(n(n+1)/4)$ (see Theorem 2.4.17). Now, let $W_{(1)}\leqslant\cdots\leqslant W_{(M)}$ be the ordered values for the $M=(n(n+1)/2)$ Walsh averages $(X_i+X_j)/2$, $i\leqslant j=1,2,\ldots,n$. Then it can be shown (Exercise 6.1.1) that (6.1.4) is satisfied with $L_{W^+}(X_1,\ldots,X_n)=W_{(M+1-w^+(\gamma/2,n))}$ and $U_{W^+}(X_1,\ldots,X_n)=W_{(w^+(\gamma/2,n))}$; thus $[W_{(M+1-w^+(\gamma/2,n))}, W_{(w^+(\gamma/2,n))})$ is a $100(1-\gamma)$ percent nonparametric distribution-free (over $\mathcal{F}_3$) confidence interval for θ.

The next example deals with the two-sample location setting. Although the basic approach is similar to that for the one-sample setting, we require some slight modifications. Let $X_1,\ldots,X_m$ and $Y_1,\ldots,Y_n$ be independent random samples from continuous distributions with c.d.f.'s $F(x)$ and $F(x-\Delta)$, respectively, where $-\infty<\Delta<\infty$. Let $S(X_1,\ldots,X_m;Y_1,\ldots,Y_n)$ be a test statistic for testing the null hypothesis H_0: $\Delta=0$ such that $S(X_1,\ldots,X_m;Y_1,\ldots,Y_n)$ is distribution-free over $F(\cdot)\in\mathcal{F}$ when H_0 is true. Since $\{(Y_1-\Delta)|_\Delta\}\stackrel{d}{=}\{Y_1|_{\Delta=0}\}$, we have that

$$\{S(X_1,\ldots,X_m;Y_1-\Delta,\ldots,Y_n-\Delta)|_\Delta\}\stackrel{d}{=}\{S(X_1,\ldots,X_m;Y_1,\ldots,Y_n)|_{\Delta=0}\}. \tag{6.1.8}$$

Combining (6.1.8) with the fact that $S(X_1,\ldots,X_m; Y_1,\ldots,Y_n)$ is distribution-free over $F(\cdot)\in\mathcal{F}$ we see that

$$P_\Delta(c_1 < S(X_1,\ldots,X_m; Y_1-\Delta,\ldots,Y_n-\Delta) < c_2) = 1-\gamma \qquad \textbf{(6.1.9)}$$

holds for every $F(\cdot)\in\mathcal{F}$ if c_1 and c_2 can be selected so that

$$P_{\Delta=0}(c_1 < S(X_1,\ldots,X_m; Y_1,\ldots,Y_n) < c_2) = 1-\gamma. \qquad \textbf{(6.1.10)}$$

Thus, as in the one-sample situation, if we can show that

$$\{c_1 < S(X_1,\ldots,X_m; Y_1-\Delta,\ldots,Y_n-\Delta) < c_2\}$$

$$\text{iff } \{L_S(X_1,\ldots,X_m; Y_1,\ldots,Y_n) \leqslant \Delta < U_S(X_1,\ldots,X_m; Y_1,\ldots,Y_n)\} \qquad \textbf{(6.1.11)}$$

for some statistics $L_S(X_1,\ldots,X_m; Y_1,\ldots,Y_n)$ and $U_S(X_1,\ldots,X_m; Y_1,\ldots,Y_n)$, then $[L_S(X_1,\ldots,X_m; Y_1,\ldots,Y_n), U_S(X_1,\ldots,X_m; Y_1,\ldots,Y_n))$ is a $100(1-\gamma)$ percent distribution-free (over $\mathcal{F}$) confidence interval for Δ.

Example 6.1.12. For this two-sample location setting, let $\mathcal{F}_4=\{F(\cdot)$: $F(\cdot)$ is a c.d.f. for a continuous distribution$\}$ and let $V(X_1,\ldots,X_m; Y_1,\ldots,Y_n)$ be the Mann–Whitney form [see (2.3.8)] of the Mann–Whitney–Wilcoxon statistic. For $i=1,\ldots,m$ and $j=1,\ldots,n$, define $D_{ij}=Y_j-X_i$. Thus,

$$V(X_1,\ldots,X_m; Y_1,\ldots,Y_n) = (\text{number of positive } D\text{s}).$$

Letting $D_{(1)}\leqslant\cdots\leqslant D_{(mn)}$ be the ordered D_{ij} differences, we see that (6.1.11) holds for appropriate c_1 and c_2 if we set $L_V(X_1,\ldots,X_m; Y_1,\ldots,Y_n) = D_{(mn+1-c_2)}$ and $U_V(X_1,\ldots,X_m; Y_1,\ldots,Y_n) = D_{(mn-c_1)}$. Since $V(X_1,\ldots,X_m; Y_1,\ldots,Y_n)$ is distribution-free over $\mathcal{F}_4$ under H_0: $\Delta=0$, then taking c_2 to be the upper $100(\gamma/2)$th percentile for the null distribution of V, say, $v((\gamma/2),m,n)$, and using the symmetry of the null distribution of V about the point $mn/2$, we have that

$$[D_{(mn+1-v(\gamma/2,m,n))}, D_{(v(\gamma/2,m,n))})$$

is a $100(1-\gamma)$ percent nonparametric distribution-free (over $\mathcal{F}_4$) confidence interval for Δ.

The technique used in Examples 6.1.6, 6.1.7 and 6.1.12 can be used to construct nonparametric distribution-free confidence intervals for a general class of problems, provided the associated test statistics have certain counting representations. We now state the general result as a theorem, leaving the proof as an exercise.

Theorem 6.1.13. *Let $Z_1,\ldots,Z_k$ be a set of statistics based on random sample(s) from a continuous distribution(s) in a class $\mathcal{F}$ for which η is a parameter of interest. Let $S(Z_1,\ldots,Z_k)=\sum_{i=1}^{k}\Psi(Z_i)$, where $\Psi(t)=1,0$ as $t>,\leqslant 0$. Suppose that $S(Z_1,\ldots,Z_k)$ is distribution-free over $\mathcal{F}$ when $\eta=0$. If the joint distribution of $(Z_1,\ldots,Z_k)$ is such that*

$$\left\{(Z_1-\eta,\ldots,Z_k-\eta)|_{\eta}\right\} \stackrel{d}{=} \left\{(Z_1,\ldots,Z_k)|_{\eta=0}\right\},$$

then $[Z_{(k+1-c_2)},Z_{(k-c_1)})$ is a $100(1-\gamma)$ percent distribution-free (over $\mathcal{F}$) confidence interval for η, where $Z_{(1)}\leqslant\cdots\leqslant Z_{(k)}$ are the ordered Zs and c_1 and c_2 are selected so that

$$P_{\eta=0}(c_1<S(Z_1,\ldots,Z_k)<c_2) = 1-\gamma.$$

Proof: Exercise 6.1.2. ■

Thus, for example, the distribution-free confidence intervals considered in Example 6.1.12 follow from Theorem 6.1.13 by taking the identifications $\mathcal{F}=\mathcal{F}_4$, $k=mn$, $Z_{(i)}=D_{(i)}$, $i=1,\ldots,mn$, $\eta=\Delta$, and $S(Z_1,\ldots,Z_k)=V(X_1,\ldots,X_m;Y_1,\ldots,Y_n)$. The intervals in Examples 6.1.6 and 6.1.7 follow similarly, and additional applications of Theorem 6.1.13 are given in some of the exercises.

In our examples we considered confidence intervals that are half-open, half-closed random intervals. However, for these examples the coverage probability $1-\gamma$ would not change if we closed both ends of the intervals, since in each case the involved variables have continuous distributions. However, for many nonparametric distribution-free confidence intervals the simple closing of the intervals results in conservative confidence intervals, even when the underlying distributions are discrete. That is, if $(1-\gamma)$ is the coverage probability for such a confidence interval when the appropriate distributions are continuous, then the corresponding closed confidence interval will have coverage probability *at least* $(1-\gamma)$ when the underlying distributions are discrete and have finite numbers of positive probability masses on any bounded interval of the real line. A technique for establishing this property is due to Noether (1967), and we illustrate its use on the confidence intervals of Example 6.1.12. To do so we need the following lemma.

Lemma 6.1.14. *Let $F^*(\cdot)$ be the c.d.f. for a discrete distribution having positive probability on a set of (finite or countably infinite) points $\{u_j\}$. In addition, assume that any bounded interval of the real line contains at most a finite number of the u_j's. Let $F(\cdot)$ be the c.d.f. for a continuous distribution such that $F(\cdot)$ agrees with $F^*(\cdot)$ at the latter's points of discontinuity. If X has distribution $F(\cdot)$, then*

$$X^* = \underset{\{j:\, X \leqslant u_j\}}{\text{minimum}}\, u_j \tag{6.1.15}$$

will have distribution $F^*(\cdot)$.

Proof: Exercise 6.1.6. ■

Example 6.1.16. Let $F^*(\cdot)$ be the c.d.f. for an arbitrary discrete distribution having positive probability on the set of points $\{u_j\}$. Let $X_1,\ldots,X_m$ and $Y_1,\ldots,Y_n$ be independent random samples from $F(x)$ and $F(x-\Delta)$, respectively, where $-\infty<\Delta<\infty$ and $F(x)$ is the c.d.f. for a continuous distribution such that $F(x)$ agrees with $F^*(x)$ at all points of discontinuity of $F^*(x)$. [In Exercise 6.1.7 the reader is asked to construct such an $F(\cdot)$ for a particular $F^*(\cdot)$.] Define

$$X_s^* = \underset{\{i:\, X_s \leqslant u_i\}}{\text{minimum}}\, u_i, \qquad s = 1,\ldots,m, \tag{6.1.17}$$

and

$$Y_t^* = \underset{\{i:\, Y_t \leqslant (u_i+\Delta)\}}{\text{minimum}}\, (u_i+\Delta), \qquad t = 1,\ldots,n. \tag{6.1.18}$$

Then, from Lemma 6.1.14, we see that $X_1^*,\ldots,X_m^*$ and $Y_1^*,\ldots,Y_n^*$ are independent random samples from $F^*(x)$ and $F^*(x-\Delta)$, respectively. Now, for any $i=1,\ldots,m$ and $j=1,\ldots,n$, we note that

$$(Y_j - X_i) > \Delta \text{ implies } (Y_j^* - X_i^*) \geqslant \Delta$$

and

$$(Y_j - X_i) < \Delta \text{ implies } (Y_j^* - X_i^*) \leqslant \Delta. \tag{6.1.19}$$

[The proof of (6.1.19) is left as Exercise 6.1.8.] Letting I_Δ and I_Δ^* be the closed forms of the $100(1-\gamma)$ percent confidence intervals for Δ (as discussed in Example 6.1.12) based on the $(Y_j - X_i)$ and $(Y_j^* - X_i^*)$ differences, respectively, we see from (6.1.19) that $\Delta \in I_\Delta$ implies $\Delta \in I_\Delta^*$, but

$\Delta \in I_\Delta^*$ does not necessarily imply $\Delta \in I_\Delta$. Thus

$$P_{F^*}(\Delta \in I_\Delta^*) \geqslant P_F(\Delta \in I_\Delta) = 1 - \gamma, \qquad \textbf{(6.1.20)}$$

and the conservative nature of the closed interval I_Δ for these discrete distributions follows from the arbitrariness of $F^*(\cdot)$.

Noether (1967) demonstrated this conservative nature of the confidence intervals based on the Wilcoxon signed rank statistic, as discussed in Example 6.1.7. In addition, you are asked to show in Exercise 6.1.9 that the confidence intervals based on the sign statistic (see Example 6.1.6) also have a similar conservative property for discrete distributions. (See also Exercise 6.1.10.)

Often in practice we do not want a confidence *interval* for a parameter θ, but rather we are interested in obtaining one-sided confidence bounds for θ. (This would, of course, correspond to the desire for a one-sided hypothesis test instead of a two-sided procedure.) For the technique illustrated in this section, the switch from confidence intervals to confidence bounds is essentially accomplished by simply using γ instead of $\gamma/2$ in the appropriate calculations. Thus, for example, using the notations of Examples 6.1.6 and 6.1.7, we have that

$$(-\infty, X_{(b(\gamma, n, 1/2))})$$

is a $100(1-\gamma)$ percent nonparametric distribution-free (over $\mathcal{F}_2$) upper confidence bound for θ, and $[W_{(M+1-w^+(\gamma,n))}, \infty)$ is a $100(1-\gamma)$ percent nonparametric distribution-free (over $\mathcal{F}_3$) lower confidence bound for θ. (We note that the closed forms of these confidence bounds possess the same conservative nature for certain discrete distributions as do the corresponding intervals.)

Exercises 6.1.

6.1.1. For the setting of Example 6.1.7, show that

$$\left[W_{(M+1-w^+(\gamma/2,n))}, W_{(w^+(\gamma/2,n))}\right]$$

is a $100(1-\gamma)$ percent nonparametric distribution-free (over $\mathcal{F}_3$) confidence interval for θ.

6.1.2. Prove Theorem 6.1.13.

6.1.3. For the Lamp (1976) data of Exercise 2.2.3, find an 89.06% confidence interval for θ using the method in Example 6.1.6.

6.1.4. For the Bick, Adams, and Schmalhorst (1976) data of Example 2.4.15, find a 97.5% upper confidence bound for θ using the method in Example 6.1.7.

6.1.5. For the Zelzano, Zelzano, and Kolb (1972) data of Example 2.3.14, find a 97.9% lower confidence bound for Δ using the method in Example 6.1.12.

6.1.6. Prove Lemma 6.1.14.

6.1.7. Let X^* have a Poisson distribution with probability function

$$P(X^*=x^*) = \frac{\lambda^{x^*}e^{-\lambda}}{(x^*)!}, \qquad x^* = 0, 1, \ldots$$
$$= 0, \qquad \text{elsewhere.}$$

Construct a continuous random variable that is related to X^* in the sense of Lemma 6.1.14.

6.1.8. Verify the implications in (6.1.19) and establish (6.1.20).

6.1.9. Show that the closed forms of the confidence intervals for θ discussed in Example 6.1.6 are conservative for discrete distributions having unique median θ and finite numbers of positive probability masses on any bounded interval of the real line.

6.1.10. Consider the $100(1-\gamma)$ percent confidence interval for the median θ in the setting of Example 6.1.6. For the test of H_0: $\theta=\theta_0$ against H_1: $\theta\neq\theta_0$ that rejects H_0 iff $|B(X_1-\theta_0,\ldots,X_n-\theta_0)-n/2| \geqslant b(\gamma/2,n,\frac{1}{2})-n/2$, show that the collection of parameter values θ_0 that would not be rejected as null hypothesis values is precisely

$$A = \left\{\theta_0\colon \theta_0 \in \left[X_{(n+1-b(\gamma/2,n,1/2))}, X_{(b(\gamma/2,n,1/2))}\right)\right\}.$$

Does an analogous statement apply to the settings of Examples 6.1.7 and 6.1.12? What would be the corresponding property for one-sided confidence bounds?

6.1.11. Let $X_1,\ldots,X_n$ be a random sample from a distribution having c.d.f. $F(x-\theta)$, with $F(\cdot)\in\mathfrak{F}$. If $S_1(X_1,\ldots,X_n)$ and $S_2(X_1,\ldots,X_n)$ are statistics such that

(i) $S_1(X_1,\ldots,X_n)$ is a translation statistic in the Xs,

(ii) $S_2(X_1,\ldots,X_n)$ is a translation-invariant statistic in the Xs,

and

(iii) $T=\dfrac{S_1(X_1,\ldots,X_n)}{S_2(X_1,\ldots,X_n)}$ is distribution-free over $F\in\mathfrak{F}$ when H_0: $\theta=0$ is true,

then show that $(S_1(X_1,\ldots,X_n)-c_2S_2(X_1,\ldots,X_n), S_1(X_1,\ldots,X_n)-c_1S_2(X_1,\ldots,X_n))$ is a $100(1-\gamma)$ percent distribution-free (over $\mathfrak{F}$) confidence interval for θ, provided that c_1 and c_2 are the $100(\gamma/2)$th and upper $100(\gamma/2)$th percentiles, respectively, for the null H_0: $\theta=0$ distribution of T. Discuss how this result can be used as an alternate derivation of the confidence intervals of Example 6.1.5.

6.1.12. Let $X_1,\ldots,X_m$ and $Y_1,\ldots,Y_n$ be independent random samples from distributions with c.d.f.'s $F(x)$ and $F(x-\Delta)$, respectively, with $F(\cdot)\in\mathfrak{F}$. Discuss conditions similar to those in Exercise 6.1.11 that would lead to distribution-free (over $\mathfrak{F}$) confidence intervals for Δ. Apply these conditions to the case where $\mathfrak{F}=\{$all normal distributions with mean 0 and nonzero variance σ^2 (unknown)$\}$ to obtain confidence intervals for Δ that are based on the two-sample t-statistic.

6.1.13. Let $X_{(1)}\leqslant\cdots\leqslant X_{(n)}$ be the order statistics for a random sample from a continuous distribution that is symmetric about θ, and set

$$W_{ij}=\frac{X_{(i)}+X_{(j)}}{2}, 1\leqslant i\leqslant j\leqslant n.$$

Thus, $A=\{W_{ij}\colon 1\leqslant i\leqslant j\leqslant n\}$ is the set of $M=(n(n+1))/2$ Walsh averages for the X observations. Let $A_C\subset A$ be defined by

$$A_C=\{W_{ij}\colon (i,j)\in C\},$$

where C is an arbitrary subset of $\{(i,j)\colon 1\leqslant i\leqslant j\leqslant n\}$ containing at least two such (i,j) pairs. Set $S(X_1,\ldots,X_n)=\Sigma_{(i,j)\in C}\Psi(W_{ij})$, where $\Psi(t)=1,0$ as $t>0, \leqslant 0$. [Thus $S(X_1,\ldots,X_n)$ is the number of positive Walsh averages in A_C.]

(a) Argue that $S(X_1,\ldots,X_n)$ is distribution-free over the class $\mathfrak{F}$ of all continuous distributions that are symmetric about zero.

(b) Find the form of a $100(1-\gamma)$ percent distribution-free (over $\mathfrak{F}$) confidence interval for θ based on $S(X_1,\ldots,X_n)$.

(c) In the definition of A_C, what sets C_1 and C_2 yield the $100(1-\gamma)$ percent distribution-free (over $\mathfrak{F}$) confidence intervals for θ based on the sign and the Wilcoxon signed rank statistics, respectively, as presented in Examples 6.1.6 and 6.1.7? [Note: Noether (1973) and

Policello and Hettmansperger (1976) discussed the use of C sets other than C_1 and C_2.]

6.1.14. Let $X_{(1)} \leqslant \cdots \leqslant X_{(m)}$ and $Y_{(1)} \leqslant \cdots \leqslant Y_{(n)}$ be the order statistics for independent random samples from continuous distributions with c.d.f.'s $F(x)$ and $F(x-\Delta)$, respectively. Set $D_{ij} = Y_{(j)} - X_{(i)}$ and let $A_C = \{D_{ij} | (i,j) \in C\}$, where C is an arbitrary subset of $\{(i,j) | i=1,\ldots,m$ and $j=1,\ldots,n\}$ containing at least two such (i,j) pairs. Let $S(X_1,\ldots,X_m; Y_1,\ldots,Y_n) = \Sigma_{(i,j)\in C} \Psi(D_{ij})$, where $\Psi(t) = 1,0$ as $t >, \leqslant 0$.

(a) Argue that $S(X_1,\ldots,X_m; Y_1,\ldots,Y_n)$ is distribution-free over the class $\mathfrak{F}$ of all continuous distributions $F(\cdot)$ when H_0: $\Delta = 0$ is true.

(b) Find the form of a $100(1-\gamma)$ percent distribution-free (over $\mathfrak{F}$) confidence interval for Δ based on $S(X_1,\ldots,X_m; Y_1,\ldots,Y_n)$.

(c) In the definition of A_C, what set C_1 will yield the $100(1-\gamma)$ percent distribution-free (over $\mathfrak{F}$) confidence interval for Δ based on the Mann–Whitney–Wilcoxon statistic, as discussed in Example 6.1.12?

6.1.15. Let $X_1,\ldots,X_n$ be a random sample from a distribution with c.d.f. $F_\theta(x)$, where $\theta \in \Theta$ is some unknown parameter. Let $T(X_1,\ldots,X_n)$ be a test statistic for testing H_0: $\theta = \theta_0$ such that $T(X_1,\ldots,X_n)$ is distribution-free over $F_\theta(\cdot) \in \mathfrak{F}$ when H_0: $\theta = \theta_0$ is true. Suppose that:

(i) There exists a function $g(\cdot)$ defined on the real line and possibly depending on θ such that

$$\{g(X_1)|_\theta\} \stackrel{d}{=} \{X_1|_{\theta=\theta_0}\}. \tag{6.1.21}$$

(ii) For any possible $0 < \gamma < 1$ and constants d_1 and d_2 such that

$$P_{\theta_0}(d_1 < T(X_1,\ldots,X_n) < d_2) = 1 - \gamma \tag{6.1.22}$$

there exist statistics $L_T(X_1,\ldots,X_n)$ and $U_T(X_1,\ldots,X_n)$ such that

$$\{d_1 < T(g(X_1),\ldots,g(X_n)) < d_2\} \quad \text{iff} \quad \{L_T(X_1,\ldots,X_n) \leqslant \theta < U_T(X_1,\ldots,X_n)\}. \tag{6.1.23}$$

Show that these conditions imply that $[L_T(X_1,\ldots,X_n), U_T(X_1,\ldots,X_n))$ is a $100(1-\gamma)$ percent distribution-free (over $\mathfrak{F}$) confidence interval for θ. [Note that the distribution-free confidence intervals for a one-sample location parameter (see, for example, expression (6.1.4)) correspond to the special case of this result where $g(x) = x - \theta$ and $\theta_0 = 0$.]

6.1.16. Let $X_1,\ldots,X_n$ be a random sample from a $n(\mu,\theta)$ distribution, where $-\infty<\mu<\infty$ and $0<\theta<\infty$ are both unknown, and set $V(X_1,\ldots,X_n) = \sum_{i=1}^n (X_i-\bar{X})^2$, with $\bar{X}=\sum_{i=1}^n X_i/n$. If $\chi^2_{(\delta,n-1)}$ denotes the upper 100δth percentile point for the chi-square distribution with $n-1$ degrees of freedom, use the result in Exercise 6.1.15 to show that

$$\left(\frac{V(X_1,\ldots,X_n)}{\chi^2_{(\gamma/2,n-1)}},\ \frac{V(X_1,\ldots,X_n)}{\chi^2_{(1-\gamma/2,n-1)}}\right)$$

is a $100(1-\gamma)$ percent distribution-free (over the class $\mathfrak{F}$ of all normal distributions) confidence interval for θ. [Hint: Apply Exercise 6.1.15 to the variables $X_i-\mu$, $i=1,\ldots,n$.]

6.1.17. Let $X_1,\ldots,X_n$ be a random sample from a continuous distribution with c.d.f. $F_\theta(x)=F(x/\theta)$, where $\theta>0$ and $F(\cdot)$ is completely known and such that $F(x)+F(-x)=1$ for every x (thus the Xs have a distribution that is symmetric about zero). Set $T(X_1,\ldots,X_n)=$(number of $|X_i|$'s greater than x_0), where $F(x_0)=\frac{3}{4}$. Consider the test of H_0: $\theta=1$ against H_1: $\theta\neq 1$ that rejects H_0 for large or small values of $T(X_1,\ldots,X_n)$.

- **(a)** What is the distribution of $T(X_1,\ldots,X_n)$ when H_0: $\theta=1$ is true?
- **(b)** Use the results in **(a)** and Exercise 6.1.15 to find the form of a $100(1-\gamma)$ percent nonparametric distribution-free (over $\mathfrak{F}=\{F(\cdot)$: $F(\cdot)$ is the c.d.f. for a continuous distribution that is symmetric about $0\}$) confidence interval for θ.

6.1.18. Let $X_1,\ldots,X_m$ and $Y_1,\ldots,Y_n$ be independent random samples from continuous distributions with c.d.f.'s $F(x-\mu_X)$ and $F((y-\mu_Y)/\eta)$, respectively, where both μ_X and μ_Y are known and $\eta>0$. Let $V(X_1^*,\ldots,X_m^*;\,Y_1^*,\ldots,Y_n^*)$ be the Mann–Whitney–Wilcoxon statistic based on $X_i^*=|X_i-\mu_X|$ and $Y_j^*=|Y_j-\mu_Y|$, as discussed in Exercise 3.4.11. Using the test statistic V and the general approach of Exercise 6.1.15, modified as necessary for this two-sample setting, find the form of a $100(1-\gamma)$ percent nonparametric distribution-free (over $\mathfrak{F}=\{F(\cdot)$: $F(\cdot)$ is the c.d.f. for a continuous distribution$\}$) confidence interval for the scale parameter η.

6.2. Large-Sample Properties and Asymptotically Distribution-Free Confidence Intervals

In Section 6.1 we discussed how to obtain a distribution-free confidence interval for a parameter θ by essentially inverting a distribution-free statistic, say T, used for testing H_0: $\theta=0$. The confidence coefficient for the resulting confidence interval is obtained from the tables for the null (H_0) distribution of T. However, such null distribution tables are generally available for small or moderate sample size(s) only. Thus we must consider how to use a large-sample approximation to the null distribution of T to obtain approximate $100(1-\gamma)$ percent confidence intervals for θ when the sample size(s) exceed those for which exact null distribution tables are available.

Let $X_1,\ldots,X_n$ be a random sample from a continuous distribution having c.d.f. $F(x-\theta)$, with $-\infty<\theta<\infty$ and $F(0)=1/2$, and let $T(X_1,\ldots,X_n)$ be a test statistic for testing H_0: $\theta=0$. Consider the following conditions on $T(X_1,\ldots,X_n)$ and $F(\cdot)$.

B1. $T(X_1,\ldots,X_n)$ is distribution-free over $F(\cdot)\in\mathfrak{F}$ when H_0: $\theta=0$ is true,

and

B2. For every pair of constants d_1 and d_2 such that

$$P_0(d_1<T(X_1,\ldots,X_n)<d_2) = 1-\gamma,$$

there exist statistics $L_T(X_1,\ldots,X_n)$ and $U_T(X_1,\ldots,X_n)$ satisfying (6.1.4) for every $\theta\in(-\infty,\infty)$.

Theorem 6.2.1. *Let $T(X_1,\ldots,X_n)$ be a statistic for testing H_0: $\theta=0$ such that conditions* B1 *and* B2 *are satisfied. Let $\mu_0(T)$ and $\sigma_0^2(T)$ be the null (H_0) mean and variance, respectively, for T. If $[T(X_1,\ldots,X_n)-\mu_0(T)]/\sigma_0(T)$ has a limiting ($n\to\infty$) continuous distribution with c.d.f. $H(t)$ when H_0 is true, then $[L_T(X_1,\ldots,X_n), U_T(X_1,\ldots,X_n))$ is an approximate* $100(1-\gamma)$ *percent distribution-free (over $\mathfrak{F}$) confidence interval for θ, where $L_T(X_1,\ldots,X_n)$ and $U_T(X_1,\ldots,X_n)$ correspond in the sense of* (6.1.4) *to the constants $d_1=\mu_0(T)+t_1\sigma_0(T)$ and $d_2=\mu_0(T)+t_2\sigma_0(T)$, with t_1 and t_2 equal to the* $100(\gamma/2)$*th percentile and upper* $100(\gamma/2)$*th percentile, respectively, for the $H(t)$ distribution.*

Proof: From the limiting behavior of $[T(X_1,\ldots,X_n)-\mu_0(T)]/\sigma_0(T)$ when H_0 is true and n is large, we know that

$$P_0\left(t_1<\frac{T(X_1,\ldots,X_n)-\mu_0(T)}{\sigma_0(T)}<t_2\right)\approx 1-\gamma,$$

or, equivalently,

$$P_0(\mu_0(T)+t_1\sigma_0(T)<T(X_1,\ldots,X_n)<\mu_0(T)+t_2\sigma_0(T))\approx 1-\gamma. \quad \textbf{(6.2.2)}$$

Combining (6.2.2) with condition B2, we obtain (as in Section 6.1)

$$P_\theta(L_T(X_1,\ldots,X_n) \leqslant \theta < U_T(X_1,\ldots,X_n)) \approx 1-\gamma, \qquad \textbf{(6.2.3)}$$

for every $\theta \in (-\infty,\infty)$, where $L_T(X_1,\ldots,X_n)$ and $U_T(X_1,\ldots,X_n)$ correspond to $d_1 = \mu_0(T) + t_1\sigma_0(T)$ and $d_2 = \mu_0(T) + t_2\sigma_0(T)$ in (6.1.4), and the theorem is established. ■

We now consider two applications of Theorem 6.2.1.

Example 6.2.4. Let $B(X_1,\ldots,X_n)$ be the sign statistic for testing H_0: $\theta=0$ for the setting of Example 6.1.6. Then conditions B1 and B2 are satisfied (see Example 6.1.6) for $T=B$ with $\mathcal{F}=\mathcal{F}_2$ and $L_B(X_1,\ldots,X_n)= X_{(n+1-d_2)}$ and $U_B(X_1,\ldots,X_n)=X_{(n-d_1)}$ for $d_1<d_2$ in condition B2. Since $B(X_1,\ldots,X_n)$ has a distribution that is binomial $(n,\frac{1}{2})$ when H_0: $\theta=0$ is true, we also know that

$$\frac{B(X_1,\ldots,X_n)-\dfrac{n}{2}}{\sqrt{\dfrac{n}{4}}}$$

has a limiting $(n\to\infty)$ distribution that is standard normal. Hence the probability statement in condition B2 holds approximately (for large n) with $(n/2)-z_{(\gamma/2)}\sqrt{n/4}$ and $(n/2)+z_{(\gamma/2)}\sqrt{n/4}$ substituted for d_1 and d_2, respectively, where $z_{(\gamma/2)}$ is the upper $100(\gamma/2)$th percentile for the standard normal distribution. Thus, if we let l be the integer closest to $(n/2)-z_{(\gamma/2)}\sqrt{n/4}$, then Theorem 6.2.1 implies that $[X_{(l+1)},X_{(n-l)})$ is an approximate $100(1-\gamma)$ percent confidence interval for θ.

Example 6.2.5. Let $W^+(X_1,\ldots,X_n)$ be the Wilcoxon signed rank statistic for testing H_0: $\theta=0$ for the setting of Example 6.1.7. Then conditions B1 and B2 are satisfied (see Example 6.1.7) for $T=W^+$ with $\mathcal{F}=\mathcal{F}_3$, $L_{W^+}(X_1,\ldots,X_n)=W_{(M+1-d_2)}$ and $U_{W^+}(X_1,\ldots,X_n)=W_{(M-d_1)}$, where $d_1<d_2$ satisfy condition B2, $M=(n(n+1))/2$ and $W_{(1)}\leqslant\cdots\leqslant W_{(M)}$ are the ordered Walsh averages. From Example 3.3.17 we know that under H_0: $\theta=0$,

$$\frac{W^+-\dfrac{n(n+1)}{4}}{\sqrt{\dfrac{n(n+1)(2n+1)}{24}}}$$

has a limiting standard normal distribution. Thus for the signed rank statistic we can take d_1 and d_2 in condition B2 to be

$$\frac{n(n+1)}{4} - z_{(\gamma/2)}\sqrt{\frac{n(n+1)(2n+1)}{24}}$$

and

$$\frac{n(n+1)}{4} + z_{(\gamma/2)}\sqrt{\frac{n(n+1)(2n+1)}{24}},$$

respectively. Hence, from Theorem 6.2.1 we see that $[W_{(q+1)}, W_{(M-q)})$ is an approximate $100(1-\gamma)$ percent confidence interval for θ, where q is the integer closest to

$$\frac{n(n+1)}{4} - z_{(\gamma/2)}\sqrt{\frac{n(n+1)(2n+1)}{24}}.$$

For the two-sample location problem, a result similar to Theorem 6.2.1 can be established (see Exercise 6.2.1) and used to obtain approximate confidence intervals for the shift parameter Δ.

Example 6.2.6. Let $V(X_1,\ldots,X_m; Y_1,\ldots,Y_n)$ be the Mann–Whitney form of the Mann–Whitney–Wilcoxon statistic for testing H_0: $\Delta=0$ in the setting of Example 6.1.12. Then it can be shown (Exercise 6.2.2) that $[D_{(s+1)}, D_{(mn-s)})$ is an approximate $100(1-\gamma)$ percent confidence interval for Δ, where s is the integer closest to

$$\frac{mn}{2} - z_{(\gamma/2)}\sqrt{\frac{mn(m+n+1)}{12}}.$$

We note that all these examples have dealt with obtaining approximate confidence intervals for a parameter. However, the same approach can be used to obtain approximate upper or lower confidence bounds. For example, for the setting of Example 6.2.5 you are asked to show in Exercise 6.2.3 that $(-\infty, W_{(M-c)})$ is an approximate $100(1-\gamma)$ percent upper confidence bound for θ, where c is the integer closest to

$$\frac{n(n+1)}{4} - z_{(\gamma)}\sqrt{\frac{n(n+1)(2n+1)}{24}}.$$

We have just discussed how to obtain large-sample approximations to nonparametric distribution-free confidence intervals. However, we are not able to obtain nonparametric distribution-free confidence intervals for many important parameters. In such situations the best we can hope for is an asymptotically nonparametric distribution-free confidence interval, in the same vein as the asymptotically nonparametric distribution-free test procedures of Section 3.5. We consider now a method for obtaining such asymptotically nonparametric distribution-free confidence intervals for the class of parameters associated with one- or two-sample U-statistics, as presented in Chapter 3.

Let $U(X_1,\ldots,X_n)$ be a one-sample U-statistic for the estimable parameter η of degree r for the family of distributions $\mathcal{F}$, and let ζ_1 be given by (3.1.14) for this U-statistic. Then if the conditions of Theorem 3.3.13 are satisfied, we know that $\sqrt{n}\,(U(X_1,\ldots,X_n)-\eta)/\sqrt{r^2\zeta_1}$ has a limiting distribution that is standard normal for any value of η. This fact leads us to the following result for obtaining an asymptotically distribution-free confidence interval for η.

Theorem 6.2.7. *Let $U(X_1,\ldots,X_n)$ be a one-sample U-statistic for the estimable parameter η of degree r for the family of distributions $\mathcal{F}$ and let ζ_1 be given by* (3.1.14) *for this U-statistic. If $\hat{\zeta}_1(X_1,\ldots,X_n)$ is a consistent estimator for ζ_1 and the conditions of Theorem* 3.3.13 *are satisfied, then*

$$\left(U(X_1,\ldots,X_n)-z_{(\gamma/2)}r\sqrt{\frac{\hat{\zeta}_1(X_1,\ldots,X_n)}{n}}\,,\right.$$

$$\left.U(X_1,\ldots,X_n)+z_{(\gamma/2)}r\sqrt{\frac{\hat{\zeta}_1(X_1,\ldots,X_n)}{n}}\right)$$

is an asymptotically nonparametric distribution-free (over $\mathcal{F}$) $100(1-\gamma)$ percent confidence interval for η, where $z_{(\gamma/2)}$ is the upper $100(\gamma/2)$th percentile for the standard normal distribution.

Proof: Since $\hat{\zeta}_1(X_1,\ldots,X_n)$ is a consistent estimator of ζ_1, we see from Theorem 3.3.13 and Slutsky's Theorem 3.2.8 that $\sqrt{n}\,(U(X_1,\ldots,X_n)-\eta)/\sqrt{r^2\hat{\zeta}_1(X_1,\ldots,X_n)}$ has a limiting standard normal distribution. Thus, for large n,

$$P_\eta\left[-z_{(\gamma/2)}<\frac{\sqrt{n}\,(U(X_1,\ldots,X_n)-\eta)}{\sqrt{r^2\hat{\zeta}_1(X_1,\ldots,X_n)}}<z_{(\gamma/2)}\right]\approx 1-\gamma.$$

The result then follows from isolating η in the middle of the inequalities in this probability expression. ■

As applications of Theorem 6.2.7, we consider the parameters $\eta_1 = P(X_1 > 0)$ and $\eta_2 = P(X_1 + X_2 > 0)$.

Example 6.2.8. Let $X_1, \ldots, X_n$ be a random sample from a distribution with c.d.f. $F \in \mathcal{F} = \{\text{all univariate distribution functions}\}$ and set $\eta_1 = P(X_1 > 0) = 1 - F(0)$. If $U_3(X_1, \ldots, X_n)$ is the U-statistic of Example 3.1.8 (i.e., U_3 is $(1/n)$ times the sign statistic), then η_1 is of degree $r = 1$ and $\zeta_1 = \eta_1(1 - \eta_1)$. Thus since U_3 is a consistent estimator of η_1, it follows (see Theorem 3.2.6) that $U_3(1 - U_3)$ is a consistent estimator of ζ_1. This, together with the fact that the conditions of Theorem 3.3.13 are satisfied for U_3, enables us to apply Theorem 6.2.7 and infer that

$$\left(U_3 - z_{(\gamma/2)} \sqrt{\frac{U_3(1 - U_3)}{n}}, U_3 + z_{(\gamma/2)} \sqrt{\frac{U_3(1 - U_3)}{n}} \right)$$

is an asymptotically nonparametric distribution-free (over $\mathcal{F}$) $100(1 - \gamma)$ percent confidence interval for $\eta_1 = P(X_1 > 0)$.

Example 6.2.9. Let $X_1, \ldots, X_n$ be a random sample from a distribution with c.d.f. $F \in \mathcal{F}$, where $\mathcal{F}$ is as given in Example 6.2.8, and set $\eta_2 = P(X_1 + X_2 > 0)$. The corresponding U-statistic for η_2 is $U_4(X_1, \ldots, X_n)$ as presented in Example 3.1.11, and we have $r = 2$ and ζ_1 is given in Example 3.1.20. Defining

$$V(X_1, \ldots, X_n) = \left[\sum_{i=1}^{n} \sum_{\substack{1 \leqslant j < k \leqslant n \\ j \neq i \neq k}} \Psi(X_i + X_j)\Psi(X_i + X_k) \right] \Big/ \left[n(n-1)(n-2)/2 \right], \tag{6.2.10}$$

where $\Psi(t) = 1, 0$ as $t >, \leqslant 0$, we note that $V(X_1, \ldots, X_n)$ is a consistent estimator of $P(X_1 + X_2 > 0, X_1 + X_3 > 0)$. (See Exercise 6.2.6.) In addition, $U_4(X_1, \ldots, X_n)$ is a consistent estimator for η_2 (see Corollary 3.2.5). Thus from Theorem 3.2.6 it follows that $V(X_1, \ldots, X_n) - [U_4(X_1, \ldots, X_n)]^2$ is a consistent estimator of ζ_1. Since the conditions of Theorem 3.3.13 are satisfied for U_4, we can now apply Theorem 6.2.7 to establish that

$$\left(U_4 - 2z_{(\gamma/2)} \sqrt{\frac{V - U_4^2}{n}}, U_4 + 2z_{(\gamma/2)} \sqrt{\frac{V - U_4^2}{n}} \right)$$

is an asymptotically nonparametric distribution-free (over $\mathcal{F}$) $100(1-\gamma)$ percent confidence interval for $\eta_2 = P(X_1 + X_2 > 0)$.

To obtain asymptotically nonparametric distribution-free confidence intervals for certain parameters in a two-sample setting, we can construct a theorem analogous to Theorem 6.2.7, but for two-sample U-statistics. This problem is considered in Exercise 6.2.7, and you are asked in Exercise 6.2.8 to use this two-sample result to obtain asymptotically nonparametric distribution-free confidence intervals for $P(X < Y)$, where X and Y have distributions with c.d.f.'s $F(\cdot)$ and $G(\cdot)$, respectively, both members of $\mathcal{F}$ as given in Example 6.2.8.

Finally, as with the large-sample approximations for the nonparametric distribution-free confidence intervals or bounds, these techniques can be used to obtain asymptotically nonparametric distribution-free upper or lower confidence bounds for appropriate parameters. Thus, for example,

$$\left(-\infty, U_4 + 2z_{(\gamma)}\sqrt{\frac{V - U_4^2}{n}}\,\right)$$

is an asymptotically nonparametric distribution-free $100(1-\gamma)$ percent upper confidence bound for $\eta_2 = P(X_1 + X_2 > 0)$, as discussed in Example 6.2.9.

Exercises 6.2.

6.2.1. State (including all the necessary conditions) and prove a theorem similar to Theorem 6.2.1, but for the two-sample location setting.

6.2.2. In Example 6.2.6 verify that $[D_{(s+1)}, D_{(mn-s)})$ is an approximate $100(1-\gamma)$ percent confidence interval for Δ.

6.2.3. For the setting of Example 6.2.5, show that $(-\infty, W_{(M-c)})$ is an approximate $100(1-\gamma)$ percent upper confidence bound for θ, where c is the integer closest to

$$\frac{n(n+1)}{4} - z_{(\gamma)}\sqrt{\frac{n(n+1)(2n+1)}{24}}\,.$$

6.2.4. For the Lamp (1976) data of Exercise 2.2.3, find an approximate 89.06% confidence interval for θ using the method of Example 6.2.4. Compare this interval with the exact interval of Exercise 6.1.3.

6.2.5. For the Bick, Adams, and Schmalhorst (1976) data of Example 2.4.15, find an approximate 97.5% upper confidence bound for θ using the method of Example 6.2.5. Compare this bound with the exact bound of Exercise 6.1.4.

6.2.6. Show that $V(X_1,\ldots,X_n)$ in Example 6.2.9 is a consistent estimator of $P(X_1+X_2>0, X_1+X_3>0)$.

6.2.7. State and prove a theorem analogous to Theorem 6.2.7, but for two-sample U-statistics and associated estimable parameters. [Hint: Consider Theorem 3.4.13.]

6.2.8. Let $X_1,\ldots,X_m$ and $Y_1,\ldots,Y_n$ be independent random samples from distributions with c.d.f.'s $F(\cdot)$ and $G(\cdot)$, respectively. Use the result of Exercise 6.2.7 to construct an asymptotically nonparametric distribution-free (over the class of all pairs of univariate distributions $\mathscr{F}$) $100(1-\gamma)$ percent confidence interval for $\eta = P(X_1 < Y_1)$. [For a more detailed discussion of this problem, see, for example, Govindarajulu (1968).]

6.2.9. Use the result of Exercise 6.2.8 on the Zelzano, Zelzano, and Kolb (1972) data of Example 2.3.14 and find an approximate 90% confidence interval for $P(X<Y)$, where X and Y represent random walking-age observations from the active-exercise and no-exercise groups, respectively.

6.2.10. Find the form of an asymptotically nonparametric distribution-free $100(1-\gamma)$ percent confidence interval for the asymmetry parameter γ of Example 3.5.4.

6.2.11. For the Lamp data of Exercise 2.2.3, find an approximate 90% confidence interval for $P(X>22$ micrometer divisions$)$, where X represents a random head width observation for a mayfly in habitat A.

6.2.12. For the Bick, Adams, and Schmalhorst (1976) data of Example 2.4.15, find an approximate 97.5% lower confidence bound for $P(Z_1+Z_2 > 0)$, where Z_1 and Z_2 are independent differences in bleeding times observations.

6.2.13. Find the form of an asymptotically nonparametric distribution-free $100(1-\gamma)$ percent upper confidence bound for the scale parameter θ of Examples 3.4.18 and 3.5.3.

6.2.14. Let $\bar{X}$ and S^2 be the sample mean and variance, respectively, for a random sample of size n from a distribution with mean θ and variance $0<\sigma^2<\infty$. Show that

$$\left(\bar{X} - z_{(\gamma/2)}\frac{S}{\sqrt{n}},\ \bar{X} + z_{(\gamma/2)}\frac{S}{\sqrt{n}}\right)$$

is an asymptotically distribution-free (over the class of distributions with positive, finite variance) $100(1-\gamma)$ percent confidence interval for θ. [See the discussion following equation (5.4.1).]

6.2.15. Let $\bar{X}$, $\bar{Y}$, S_1^2, and S_2^2 be the sample means and variances, respectively, for independent random samples of sizes m and n, respectively, from distributions with means θ and $\theta+\Delta$, respectively, and common variance $0<\sigma^2<\infty$. Show how to use the two-sample t-statistic to construct an asymptotically distribution-free (over the class of distributions with common finite variance) $100(1-\gamma)$ percent confidence interval for Δ. [Note the discussion following equation (5.4.1).]

6.3. Asymptotic Relative Efficiency of Confidence Intervals and Bounds

In this section we consider the problem of measuring the relative effectiveness of two competitive confidence intervals (or bounds) for the same parameter. As was the situation in Chapter 5 (where we compared test procedures), such comparisons for small sample sizes are neither definitive nor practical, and so we turn to asymptotics to provide the answers. We shall study the large-sample properties of confidence intervals and bounds under sequences of real numbers converging to the true value of the parameter, as considered for the tests in Section 5.2. For such sequences, two different criteria considered by Lehmann (1963b) are often used to evaluate the performance of a given confidence interval or bound. The first of these criteria takes direct advantage of the link between hypothesis tests and confidence intervals and bounds, and the second uses a separate approach. Since we have already derived the ARE results for tests under similar parameter sequences (Chapter 5), we use those known properties and the first criterion to provide detailed development of ARE properties for confidence intervals and bounds. We then briefly outline the second criterion for comparison of confidence intervals and bounds and simply note that in many settings the two criteria result in the same asymptotic relative efficiency values.

We assume throughout this section that all confidence intervals under discussion are totally open intervals. This is presumed only for the sake of simplicity, since the arguments of this section would apply as well to half-open, half-closed, or totally closed intervals. Consequently, this formal restriction to open intervals poses no constraints on the application of the results in this section to the confidence intervals discussed in Sections 6.1 and 6.2.

Definition 6.3.1. Let $\{I_n\}$ and $\{J_{n'}\}$ be two sequences of confidence intervals (or bounds with one endpoint either $-\infty$ or ∞) for a parameter θ

with true, but unknown, value θ^*. Assume that these two sequences of confidence intervals achieve a common confidence coefficient of $(1-\gamma)$ in the limit; that is,

$$\lim_{n\to\infty} P_{\theta^*}[\theta^* \in I_n] = \lim_{n'\to\infty} P_{\theta^*}[\theta^* \in J_{n'}] = 1-\gamma.$$

Consider a sequence of false parameter values $\{\theta_i\}$ such that $\lim_{i\to\infty}\theta_i = \theta^*$, and let $\xi_{I_n}(\theta_i)$ and $\xi_{J_{n'}}(\theta_i)$ be the respective probabilities of covering the false value θ_i; that is,

$$\xi_{I_n}(\theta_i) = P_{\theta^*}(\theta_i \in I_n)$$

and

$$\xi_{J_{n'}}(\theta_i) = P_{\theta^*}(\theta_i \in J_{n'}). \tag{6.3.2}$$

Let $\{n_i\}$ and $\{n_i'\}$ be increasing sequences of positive integers such that

$$0 < \lim_{i\to\infty} \xi_{I_{n_i}}(\theta_i) = \lim_{i\to\infty} \xi_{J_{n_i'}}(\theta_i) < (1-\gamma). \tag{6.3.3}$$

Then the **asymptotic relative efficiency of** $\{I_{n_i}\}$ **relative to** $\{J_{n_i'}\}$ (or simply of I to J) is

$$\text{ARE}(I,J) = \lim_{i\to\infty} \frac{n_i'}{n_i}, \tag{6.3.4}$$

provided that this limit is the same for all such sequences $\{n_i\}$ and $\{n_i'\}$, and independent of the choice of the $\{\theta_i\}$ sequence and the value of θ^*.

Thus our definition of ARE for confidence intervals and bounds is very similar to that used for ARE of test procedures (see Definition 5.1.5). In essence, ARE(I,J) is the limiting ratio of sample sizes required to achieve the same limiting probability of incorrectly covering a sequence of false parameter values (converging to the true value of the parameter) when the limiting coverage probabilities for the true parameter value are the same (i.e., both $(1-\gamma)$). [This concept of asymptotic efficiency for confidence intervals and bounds was considered by Lehmann (1963b).] Thus, similar to the situation for hypothesis tests, a value of ARE$(I,J)=1.4$ means roughly that we would need approximately 1.4 times as many observations for the J interval as for the I interval to achieve the same asymptotic probability of incorrectly covering the sequence of false parameter values $\{\theta_i\}$.

In many settings the ARE of one confidence interval (or bound) relative to another confidence interval (or bound) is identical with the ARE between the two-sided (or one-sided) hypothesis tests. Since we presented the details of the ARE results for one-sided hypothesis tests that rejected for large values of the test statistics (the corresponding asymptotic efficiency properties for two-sided tests were left as Exercise 5.2.2), we discuss the details of the ARE properties for the analogous lower confidence *bounds* for our parameter and leave the similar connection between confidence *intervals* and two-sided tests as Exercise 6.3.1. We consider the one- and two-sample location problems separately and in that order.

Let X have a continuous distribution with c.d.f $F(x-\theta)$, where $F(x)+F(-x)=1$ for all x, and let $\{L_{I_n}\}$ and $\{L_{J_{n'}}\}$ be two sequences of limiting $100(1-\gamma)$ percent lower confidence bounds for θ^*, the true value for θ. Thus, $I_n=(L_{I_n},\infty)$ and $J_{n'}=(L_{J_{n'}},\infty)$ are confidence intervals for θ^* with limiting confidence coefficient $(1-\gamma)$, as discussed in Definition 6.3.1. In addition, let $\{\theta_i\}$ be a sequence of false parameter values such that $\lim_{i\to\infty}\theta_i=\theta^*$, and let n_i and n_i' be defined by (6.3.3). (For this case of lower confidence bounds, we note that a false parameter value θ_i satisfies $\theta_i<\theta^*$.)

Theorem 6.3.5. *Let $\{I_{n_i}\}$, $\{J_{n_i'}\}$, $\{L_{I_{n_i}}\}$, $\{L_{J_{n_i'}}\}$, $\{\theta_i\}$, θ and θ^* be as just defined for the one-sample location problem. Let $X_{I1},\ldots,X_{In_i}$ and $X_{J1},\ldots,X_{Jn_i'}$ be independent random samples from $F(x-\theta)$. If there exist statistics $S(X_{I1},\ldots,X_{In_i})$ and $T(X_{J1},\ldots,X_{Jn_i'})$ and sequences of constants $\{c_{n_i}\}$ and $\{d_{n_i'}\}$ such that*

$$S(X_{I1}-v,\ldots,X_{In_i}-v)<c_{n_i} \quad \text{iff} \quad L_{I_{n_i}}(X_{I1},\ldots,X_{In_i})<v \tag{6.3.6}$$

and

$$T(X_{J1}-v,\ldots,X_{Jn_i'}-v)<d_{n_i'} \quad \text{iff} \quad L_{J_{n_i'}}(X_{J1},\ldots,X_{Jn_i'})<v, \tag{6.3.7}$$

for every number v and all $i=1,2,\ldots,$ then

$$\text{ARE}(I,J)=\text{ARE}(S,T), \tag{6.3.8}$$

where $S(X_{I1},\ldots,X_{In_i})$ and $T(X_{J1},\ldots,X_{Jn_i'})$ are viewed as test statistics for the null hypothesis H_0: $\theta=0$ against the alternative H_1: $\theta>0$ that reject H_0 for large values.

Proof: From (6.3.6) we see that (using the notation of Definition 6.3.1)

$$\xi_{I_{n_i}}(\theta_i)=P_{\theta^*}\big(L_{I_{n_i}}(X_{I1},\ldots,X_{In_i})<\theta_i\big)=P_{\theta^*}\big(S(X_{I1}-\theta_i,\ldots,X_{In_i}-\theta_i)<c_{n_i}\big),$$

which, in view of our location model and the independence of the Xs, implies

$$\xi_{I_{n_i}}(\theta_i) = P_{\theta^*-\theta_i}\big(S(X_{I1},\ldots,X_{In_i})<c_{n_i}\big). \tag{6.3.9}$$

Similarly, for $\xi_{J_{n_i'}}(\theta_i)$ we obtain

$$\xi_{J_{n_i'}}(\theta_i) = P_{\theta^*-\theta_i}\big(T(X_{J1},\ldots,X_{Jn_i'})<d_{n_i'}\big). \tag{6.3.10}$$

Thus the limit equality in (6.3.3) is equivalent to

$$\lim_{i\to\infty} P_{\theta^*-\theta_i}\big(S(X_{I1},\ldots,X_{In_i})<c_{n_i}\big) = \lim_{i\to\infty} P_{\theta^*-\theta_i}\big(T(X_{J1},\ldots,X_{Jn_i'})<d_{n_i'}\big),$$

or

$$\lim_{i\to\infty} P_{\theta^*-\theta_i}\big(S(X_{I1},\ldots,X_{In_i})\geqslant c_{n_i}\big) = \lim_{i\to\infty} P_{\theta^*-\theta_i}\big(T(X_{J1},\ldots,X_{Jn_i'})\geqslant d_{n_i'}\big). \tag{6.3.11}$$

Since both sequences of confidence bounds have limiting confidence coefficient $(1-\gamma)$, we have

$$\begin{aligned}(1-\gamma) &= \lim_{i\to\infty} P_{\theta^*}\big[L_{I_{n_i}}(X_{I1},\ldots,X_{In_i})<\theta^*\big]\\ &= \lim_{i\to\infty} P_{\theta^*}\big[S(X_{I1}-\theta^*,\ldots,X_{In_i}-\theta^*)<c_{n_i}\big]\\ &= 1-\lim_{i\to\infty} P_0\big[S(X_{I1},\ldots,X_{In_i})\geqslant c_{n_i}\big]\end{aligned}$$

and, similarly,

$$\begin{aligned}(1-\gamma) &= \lim_{i\to\infty} P_{\theta^*}\big[L_{J_{n_i'}}(X_{J1},\ldots,X_{Jn_i'})<\theta^*\big]\\ &= 1-\lim_{i\to\infty} P_0\big[T(X_{J1},\ldots,X_{Jn_i'})\geqslant d_{n_i'}\big].\end{aligned}$$

Thus we see that the tests of hypotheses defined by

$$\text{reject } H_0\colon \theta=0 \quad \text{iff} \quad S(X_{I1},\ldots,X_{In_i}) \geqslant c_{n_i} \tag{6.3.12}$$

and

$$\text{reject } H_0\colon \theta=0 \quad \text{iff} \quad T(X_{J1},\ldots,X_{Jn_i'}) \geqslant d_{n_i'}, \tag{6.3.13}$$

are both limiting γ level hypothesis tests of H_0: $\theta=0$ against the alternative

H_1: $\theta > 0$. In addition, we note that $\{\theta^* - \theta_i\}$ is a sequence of alternatives (i.e., $\theta^* - \theta_i > 0$, $i = 1, 2, \ldots$) such that $\lim_{i\to\infty}(\theta^* - \theta_i) = 0$, and that equation (6.3.11) represents the condition that the two sequences of tests $\{S_{n_i}\}$ and $\{T_{n_i'}\}$ have the same limiting power against the sequence of alternatives $\{\theta^* - \theta_i\}$ converging to the null hypothesis value $\theta = 0$. Thus all the conditions for the definition of asymptotic relative efficiency (Definition 5.1.5) of two tests are satisfied for the sequences $\{S_{n_i}\}$, $\{T_{n_i'}\}$ and $\{\theta^* - \theta_i\}$, with $\theta_0 = 0$ and $\alpha = \gamma$. Consequently, we have from (5.1.7) that

$$\lim_{i\to\infty} \frac{n_i'}{n_i} = \text{ARE}(S, T), \qquad \textbf{(6.3.14)}$$

provided that the limit is the same for all such sequences $\{n_i\}$ and $\{n_i'\}$, and independent of the $\{\theta^* - \theta_i\}$ sequence. Coupling (6.3.14) with Definition 6.3.1 [in particular, equation (6.3.4)], we see that

$$\text{ARE}(S, T) = \text{ARE}(I, J),$$

as desired. ■

As an application of Theorem 6.3.5 and the corresponding result for confidence intervals and two-sided hypothesis tests (Exercise 6.3.1), we consider the ARE for a confidence interval based on the Wilcoxon signed rank statistic relative to one based on the one-sample t-statistic.

Example 6.3.15. Let $(L_T(X_1, \ldots, X_n), U_T(X_1, \ldots, X_n))$ be the $100(1-\gamma)$ percent confidence interval for θ based on the one-sample t-statistic as discussed in Example 6.1.5, and let $(L_{W^+}(X_1, \ldots, X_n), U_{W^+}(X_1, \ldots, X_n))$ be the corresponding $100(1-\gamma)$ percent confidence interval for θ based on the Wilcoxon signed rank statistic and developed in Example 6.1.7. In addition, assume $0 < \sigma^2 = \text{Var}(X_1) < \infty$. Then conditions analogous to (6.3.6) and (6.3.7), but for confidence intervals, are satisfied for the t and W^+ confidence intervals by using the test statistics $\bar{X}/(S/\sqrt{n})$ and $W^+ =$ (number of positive Walsh averages), respectively. (See also the discussions in Examples 6.1.5 and 6.1.7.) Hence the asymptotic relative efficiency of the signed rank confidence interval for θ relative to the analogous t confidence interval for θ is identical with that given in (5.4.6) for the corresponding tests.

Consider now the two-sample location problem. Let X and Y have continuous distributions with c.d.f.'s $F(x)$ and $F(x - \Delta)$, respectively, $-\infty < \Delta < \infty$, and let $\{I_N\}$ and $\{J_{N'}\}$ be two sequences of confidence intervals for Δ^* (the true value of Δ) which achieve a confidence coefficient $(1-\gamma)$

in the limit; that is,

$$(1-\gamma) = \lim_{N\to\infty} P_{\Delta^*}[\Delta^* \in I_N] = \lim_{N'\to\infty} P_{\Delta^*}[\Delta^* \in J_{N'}].$$

In addition, let $\{\Delta_i\}$ denote a sequence of false parameter values such that $\lim_{i\to\infty}\Delta_i = \Delta^*$. Define $\{N_i\}$ and $\{N_i'\}$ to be any two increasing sequences of sample sizes satisfying

$$0 < \lim_{i\to\infty} P_{\Delta^*}[\Delta_i \in I_{N_i}] = \lim_{i\to\infty} P_{\Delta^*}[\Delta_i \in J_{N_i'}] < (1-\gamma).$$

Theorem 6.3.16. *Let $\{I_{N_i}\}$, $\{J_{N_i'}\}$, $\{\Delta_i\}$, Δ, and Δ^* be as just defined for the two-sample location problem. Let $(X_{I1},\ldots,X_{Im_i}),(X_{J1},\ldots,X_{Jm_i'})$ and $(Y_{I1},\ldots,Y_{In_i}),(Y_{J1},\ldots,Y_{Jn_i'})$ be independent random samples from $F(x)$ and $F(x-\Delta)$, respectively, with $N_i = m_i + n_i$ and $N_i' = m_i' + n_i'$. If there exist statistics $S(X_{I1},\ldots,X_{Im_i};Y_{I1},\ldots,Y_{In_i})$ and $T(X_{J1},\ldots,X_{Jm_i'};Y_{J1},\ldots,Y_{Jn_i'})$ and sequences of constants $\{c_{1N_i}\}$, $\{c_{2N_i}\}$, $\{d_{1N_i'}\}$ and $\{d_{2N_i'}\}$ such that*

$$c_{1N_i} < S(X_{I1},\ldots,X_{Im_i};Y_{I1}-v,\ldots,Y_{In_i}-v) < c_{2N_i} \quad \text{iff} \quad v \in I_{N_i} \tag{6.3.17}$$

and

$$d_{1N_i'} < T(X_{J1},\ldots,X_{Jm_i'};Y_{J1}-v,\ldots,Y_{Jn_i'}-v) < d_{2N_i'} \quad \text{iff} \quad v \in J_{N_i'}, \tag{6.3.18}$$

holds for every number v and all $i = 1,2,\ldots,$ then

$$\text{ARE}(I,J) = \text{ARE}(S,T),$$

where $S(X_{I1},\ldots,X_{Im_i};Y_{I1},\ldots,Y_{In_i})$ and $T(X_{J1},\ldots,X_{Jm_i'};Y_{J1},\ldots,Y_{Jn_i'})$ are viewed as test statistics for the null hypothesis H_0: $\Delta=0$ against the alternative H_1: $\Delta>0$ that reject H_0 for large values of $S(\cdot)$ and $T(\cdot)$, respectively.

Proof: Exercise 6.3.3. ■

The conditions of Theorem 6.3.16 and the corresponding assumptions for two-sample confidence bounds and one-sided tests (see Exercise 6.3.4) are very similar to those for the one-sample location problem (Exercise 6.3.1 and Theorem 6.3.5), and hence we do not provide an example illustrating a two-sample application of these results. However, we do note (and you are asked to prove this in Exercise 6.3.5) that the necessary conditions are satisfied for both the confidence intervals and bounds for Δ based on the two-sample t-statistic (Exercise 6.2.15) and those based on the Mann–Whitney–Wilcoxon statistic (Example 6.1.12). Thus the asymptotic relative efficiency of the Mann–Whitney–Wilcoxon confidence interval (or bound) for Δ relative to the two-sample t confidence interval (or bound) for Δ is identical with that given in (5.4.16) for the corresponding tests.

An alternative way to measure the asymptotic relative efficiency of two confidence intervals for a parameter is to compare the asymptotic behavior of the lengths of the intervals. Clearly, the one with the smaller asymptotic length is preferred.

Definition 6.3.19. Let $\{I_n\}$ and $\{J_n\}$ be two sequences of $100(1-\gamma)$ percent confidence intervals for a parameter θ. Let D_{I_n} and D_{J_n} be the lengths of the respective intervals. Then the **length-criterion asymptotic relative efficiency** of $\{I_n\}$ relative to $\{J_n\}$ (or of I relative to J) is denoted by L-ARE(I,J) and represents the value to which $D_{J_n}^2/D_{I_n}^2$ converges in probability.

Lehmann (1963b) has shown that for many confidence interval settings, the two Definitions 6.3.1 and 6.3.19 of asymptotic relative efficiency provide the same answer; that is, ARE$(I,J)=L$-ARE(I,J). In particular, this is the case for a Wilcoxon signed rank confidence interval for the location parameter θ relative to an interval for θ based on the one-sample t-statistic and likewise for Mann–Whitney–Wilcoxon and two-sample t confidence intervals for the shift parameter Δ. As an illustration, we consider the one-sample location setting, providing the details for the one-sample t-statistic, while simply stating the necessary result for the Wilcoxon signed rank statistic.

Example 6.3.20. Consider the one-sample location setting of Example 6.1.7, with the additional assumption that $0<\sigma^2=\text{Var}(X_1)<\infty$. Then the approximate $100(1-\gamma)$ percent confidence interval for θ based on the one-sample t-statistic T_n^+ has form

$$\left(\overline{X}-z_{(\gamma/2)}\frac{S}{\sqrt{n}},\overline{X}+z_{(\gamma/2)}\frac{S}{\sqrt{n}}\right).$$

The squared length for this interval is $D_{T_n^+}^2=4z_{(\gamma/2)}^2(S^2/n)$. Thus, from Khintchin's weak law of large numbers [see, for example, page 463 of Bickel and Doksum (1977)], we have that

$$nD_{T_n^+}^2 \text{ converges in probability to } 4z_{(\gamma/2)}^2\sigma^2, \text{ as } n\to\infty. \quad \textbf{(6.3.21)}$$

For the corresponding result for the Wilcoxon signed rank confidence intervals, we state without proof the following theorem due to Lehmann (1963b).

Theorem 6.3.22. *Consider the one-sample location setting of Example 6.1.7 and let $D^2_{W_n^+}$ be the squared length of the Wilcoxon signed rank $100(1-\gamma)$ percent confidence interval for θ. Then*

$$nD^2_{W_n^+} \text{ converges in probability to } z^2_{(\gamma/2)}/3\left[\int_{-\infty}^{\infty} f^2(x)\,dx\right]^2, \text{ as } n \to \infty, \tag{6.3.23}$$

where $f(x)$ is the density corresponding to $F(x)$.

Proof: See Lehmann (1963b). ■

Utilizing these results in the framework of Definition 6.3.19, we obtain the equivalence between the ARE and L-ARE values for the one-sample t and Wilcoxon signed rank confidence intervals.

Theorem 6.3.24. *Let $I_{W_n^+}$ and $J_{T_n^+}$ be the $100(1-\gamma)$ percent confidence intervals for θ considered in Theorem 6.3.22 and Example 6.3.20, respectively. Then*

$$\text{ARE}(I_{W^+},J_{T^+}) = L\text{-ARE}(I_{W^+},J_{T^+}),$$

with the common expression given in (5.4.6).

Proof: It follows from Exercise 3.2.5 that $(D^2_{T_n^+}/D^2_{W_n^+})$ converges in probability to the same value as does $(nD^2_{T_n^+}/nD^2_{W_n^+})$, which by (6.3.21) and (6.3.23) converges in probability to

$$12\sigma^2\left[\int_{-\infty}^{\infty} f^2(x)\,dx\right]^2,$$

as given in (5.4.6) and Example 6.3.15. ■

Results similar to those of Theorems 6.3.22 and 6.3.24 hold for the Mann–Whitney–Wilcoxon confidence interval for the two-sample shift parameter Δ relative to one based on the two-sample t-statistic.

One final point should be made regarding these asymptotic properties of confidence intervals. Theorem 6.3.22 and Theorem 3.2.6 show that

$$\frac{\sqrt{3n}\, D_{W_n^+}}{z_{(\gamma/2)}}$$

converges in probability to the quantity $[\int_{-\infty}^{\infty} f^2(x)\,dx]^{-1}$. Thus for a symmetric distribution with c.d.f. $F(x-\theta)$, $\sqrt{3n}\,/z_{(\gamma/2)}$ times the length of a Wilcoxon signed rank $100(1-\gamma)$ percent confidence interval for θ is a consistent estimator of $[\int_{-\infty}^{\infty} f^2(x)\,dx]^{-1}$. (A similar property pertains for the Mann–Whitney–Wilcoxon confidence interval for the location shift Δ.) This fact can be used to estimate the asymptotic relative efficiency of the Wilcoxon signed rank test relative to the one-sample t-test. (See Exercise 6.3.6.)

Exercises 6.3.

6.3.1. Establish the result analogous to Theorem 6.3.5, but for confidence intervals and two-sided tests. (See Exercise 5.2.2.)

6.3.2. Show that the asymptotic relative efficiency of the sign confidence interval or bound (see Example 6.1.6) for the one-sample location parameter θ with respect to the analogous one-sample t confidence interval or bound for θ is given by (5.4.9).

6.3.3. Prove Theorem 6.3.16.

6.3.4. State and prove the result analogous to Theorem 6.3.16, but for confidence bounds and one-sided tests.

6.3.5. Show that the asymptotic relative efficiency of the Mann–Whitney–Wilcoxon confidence interval or bound for the two-sample location parameter Δ with respect to the analogous two-sample t confidence interval or bound for Δ is given by (5.4.16).

6.3.6. Let W^+ and T^+ be the Wilcoxon signed rank statistic and the one-sample t-statistic, respectively, for testing H_0: $\theta=0$, where θ is the point of symmetry for a continuous distribution with positive and finite variance. If we do not know the form of this underlying distribution, show how the results of Section 6.3 can be used to obtain a consistent estimator for $\text{ARE}(W^+,T^+)$ (and, in view of Example 6.3.15, for the asymptotic relative efficiency of the corresponding intervals and bounds).

7 POINT ESTIMATION

In Chapter 3 we studied properties of U-statistic estimators for estimable parameters. In particular, for several of the parameters that are probability statements the resulting U-statistic estimators lead to nonparametric distribution-free test procedures. For example, the Mann–Whitney form (divided by mn) of the Mann–Whitney–Wilcoxon statistic is the U-statistic estimator for the two-sample probability parameter $P(X<Y)$. In the first part of this chapter we concern ourselves with essentially the converse method; that is, how do we use a given nonparametric distribution-free test statistic to construct a point estimator for a related parameter? Unlike Chapter 3, however, we concentrate here on important parameters that are not probability statements. In the latter part of this chapter we discuss M-estimators and influence curves.

7.1. Estimators Associated with Distribution-Free Test Statistics

In this section we consider a general procedure for deriving a point estimator for a parameter θ from a test statistic that is distribution-free under an appropriate null hypothesis about θ. This important technique was first proposed by Hodges and Lehmann (1963). We introduce the ideas via the one-sample location setting and then consider the direct extension to the two-sample location problem.

Let $X_1,\ldots,X_n$ be a random sample from a continuous distribution with c.d.f. $F(x-\theta)$, where $F(\cdot)$ is the c.d.f. for a distribution that is symmetric about 0. Let $V(X_1,\ldots,X_n)$ be a test statistic for testing H_0: $\theta=0$ against H_1: $\theta>0$ that satisfies the following three conditions:

C1. H_0: $\theta=0$ is rejected for large values of $V(X_1,\ldots,X_n)$. **(7.1.1)**

C2. $V(x_1+h,\ldots,x_n+h)$ is a nondecreasing function of h for each $(x_1,\ldots,x_n)$. **(7.1.2)**

C3. When H_0: $\theta=0$ is true, the distribution of $V(X_1,\ldots,X_n)$ is symmetric about some value ξ for every continuous distribution $F(\cdot)$ that is symmetric about zero. **(7.1.3)**

For such a setting we motivate a Hodges–Lehmann estimator of θ as follows. The random variables $X_1-\theta,\ldots,X_n-\theta$ are independent, and each has the same distribution that is symmetric about zero. Thus it would be desirable for an estimator of θ, say, $\hat{\theta}$, to possess the property that the variables $X_1-\hat{\theta},\ldots,X_n-\hat{\theta}$ "look as close as possible" to being symmetrically distributed about 0. However, we must have some way to better define the criterion "look as close as possible," and that is the point at which the test statistic $V(X_1,\ldots,X_n)$ enters the problem. Since $V(X_1,\ldots,X_n)$ is used to test H_0: $\theta=0$, one intuitive way to evaluate this "closeness" property would be to choose $\hat{\theta}$ so that $V(X_1-\hat{\theta},\ldots,X_n-\hat{\theta})$ assumes a value as near as possible to the median of the null H_0: $\theta=0$ distribution of $V(X_1,\ldots,X_n)$. In view of condition C3, this implies that we choose $\hat{\theta}$ so that $V(X_1-\hat{\theta},\ldots,X_n-\hat{\theta})$ is as close as possible to ξ, the point of symmetry for the null distribution of $V(X_1,\ldots,X_n)$; that is, so that $X_1-\hat{\theta},\ldots,X_n-\hat{\theta}$ "look as close as possible" to being symmetrically distributed about 0, when "viewed" through the $V(X_1,\ldots,X_n)$ statistic.

Formally, the **Hodges–Lehmann estimator** for θ based on a test statistic $V(X_1,\ldots,X_n)$ satisfying (7.1.1), (7.1.2), and (7.1.3) is given by

$$\hat{\theta} = \hat{\theta}(X_1,\ldots,X_n) = \frac{\theta^* + \theta^{**}}{2}, \tag{7.1.4}$$

where

$$\theta^* = \theta^*(X_1,\ldots,X_n) = \text{supremum } \{\theta\colon V(X_1-\theta,\ldots,X_n-\theta)>\xi\} \tag{7.1.5}$$

and

$$\theta^{**} = \theta^{**}(X_1,\ldots,X_n) = \text{infimum } \{\theta\colon V(X_1-\theta,\ldots,X_n-\theta)<\xi\}. \tag{7.1.6}$$

In cases where the statistic $V(X_1-\theta,\ldots,X_n-\theta)$ is continuous and strictly decreasing in θ, we have that $\hat{\theta}=\theta^*=\theta^{**}$ satisfies $V(X_1-\hat{\theta},\ldots,X_n-\hat{\theta})=\xi$. However, for most nonparametric settings the statistic $V(X_1-\theta,\ldots,X_n-\theta)$ is not continuous in θ, and, in fact, it is often impossible for $V(X_1-\theta,\ldots,X_n-\theta)$ to assume the value ξ. For such cases, $\theta^*\neq\theta^{**}$ and the estimator $\hat{\theta}$ (7.1.4) will force $V(X_1-\hat{\theta},\ldots,X_n-\hat{\theta})$ to be "as close as possible," in the sense of (7.1.5) and (7.1.6), to ξ.

As a first example, we consider the one-sample t-statistic. Although the associated t-test is distribution-free under H_0 only over the class of all normal distributions (and thus is not nonparametric distribution-free), the

Hodges–Lehmann technique still applies and provides a simple introduction to this method of estimation.

Example 7.1.7. Let $V_1(X_1,\ldots,X_n)=n^{1/2}\overline{X}/S$ be the one-sample t-statistic for testing H_0: $\theta=0$ against H_1: $\theta>0$. Condition C1 is clearly satisfied, and condition C2 follows from the fact that $\overline{X}$ and S are translation and translation invariant statistics (see Definition 1.3.14), respectively, and the fact that S is nonnegative. In addition, for every continuous distribution $F(\cdot)$ that is symmetric about 0, condition C3 with $\xi=0$ follows from Theorem 1.3.16 and the fact that V is an odd statistic (see Definition 1.3.13). Thus the Hodges–Lehmann conditions are satisfied. Moreover, since $V(X_1-\theta,\ldots,X_n-\theta)=n^{1/2}(\overline{X}-\theta)/S$ is continuous and strictly decreasing in θ, we know that $\hat{\theta}_1$ is such that $n^{1/2}(\overline{X}-\hat{\theta}_1)/S=\xi=0$. Thus the Hodges–Lehmann estimator for θ, using the one-sample t-statistic to evaluate the symmetry of the $(X_i-\hat{\theta}_1)$'s, is $\hat{\theta}_1=\overline{X}$, a not too surprising result.

In the next example we consider the estimator of θ that is associated with the signed rank statistic $W^+(X_1,\ldots,X_n)$.

Example 7.1.8. Let $V_2(X_1,\ldots,X_n)=W^+(X_1,\ldots,X_n)$ be the Wilcoxon signed rank statistic, and let $W_{(1)}\leqslant\cdots\leqslant W_{(M)}$ be the ordered values for the $M=(n(n+1))/2$ Walsh averages $(X_i+X_j)/2$, $i\leqslant j=1,2,\ldots,n$. Then, $V_2(X_1,\ldots,X_n)=$[number of positive Walsh averages], and we have that $V_2(x_1+h,\ldots,x_n+h)=$[number of $((x_i+x_j)/2)$ averages greater than $-h$] is a nondecreasing function of h for each $(x_1,\ldots,x_n)$. Thus conditions C1(7.1.1) and C2(7.1.2) are satisfied for $V_2(X_1,\ldots,X_n)$. Moreover, C3(7.1.3) is also satisfied, since the null distribution of $V_2(X_1,\ldots,X_n)$ is symmetric about $\xi=(n(n+1))/4$ for every underlying continuous distribution that is symmetric about 0 (see Theorem 2.4.17). Now, to get an expression for the Hodges–Lehmann estimator for θ based on $V_2(X_1,\ldots,X_n)$, we consider two cases.

Case 1: M odd

Let $M=2k+1$, where k is an integer. Then $\xi=(n(n+1))/4=M/2=k+\frac{1}{2}$ is not a possible value for $V_2(X_1,\ldots,X_n)$. Thus we have

$$\begin{aligned}\theta_2^* &= \text{supremum}\left\{\theta\text{: } V_2(X_1-\theta,\ldots,X_n-\theta)>k+\tfrac{1}{2}\right\}\\ &= \text{supremum}\left\{\theta\text{: more than } \left(k+\tfrac{1}{2}\right) \text{ Walsh averages exceed } \theta\right\}\\ &= \text{supremum}\left\{\theta\text{: } W_{(k+1)}>\theta\right\} = W_{(k+1)}. \qquad \textbf{(7.1.9)}\end{aligned}$$

On the other hand,

$$\begin{aligned}\theta_2^{**} &= \text{infimum}\left\{\theta\colon V_2(X_1-\theta,\ldots,X_n-\theta)<k+\tfrac{1}{2}\right\}\\ &= \text{infimum}\left\{\theta\colon \text{fewer than } \left(k+\tfrac{1}{2}\right) \text{ Walsh averages exceed } \theta\right\}\\ &= \text{infimum}\left\{\theta\colon W_{(k+1)}\leqslant\theta\right\} = W_{(k+1)}.\end{aligned} \tag{7.1.10}$$

Using (7.1.9) and (7.1.10) in (7.1.4) we see that

$$\hat{\theta}_2 = \frac{W_{(k+1)}+W_{(k+1)}}{2} = W_{(k+1)}. \tag{7.1.11}$$

Case 2: M even

Let $M=2k$, where k is an integer. Then $\xi=(n(n+1))/4=k$, a possible value for $V_2(X_1,\ldots,X_n)$, and

$$\begin{aligned}\theta_2^{*} &= \text{supremum}\left\{\theta\colon V_2(X_1-\theta,\ldots,X_n-\theta)>k\right\}\\ &= \text{supremum}\left\{\theta\colon \text{more than } k \text{ of the Walsh averages exceed } \theta\right\}\\ &= \text{supremum}\left\{\theta\colon W_{(k)}>\theta\right\} = W_{(k)}.\end{aligned} \tag{7.1.12}$$

Similarly,

$$\begin{aligned}\theta_2^{**} &= \text{infimum}\left\{\theta\colon V_2(X_1-\theta,\ldots,X_n-\theta)<k\right\}\\ &= \text{infimum}\left\{\theta\colon \text{fewer than } k \text{ of the Walsh averages exceed } \theta\right\}\\ &= \text{infimum}\left\{\theta\colon W_{(k+1)}\leqslant\theta\right\} = W_{(k+1)}.\end{aligned} \tag{7.1.13}$$

Using (7.1.12) and (7.1.13) in (7.1.4) we see that for M even

$$\hat{\theta}_2 = \frac{W_{(k)}+W_{(k+1)}}{2}. \tag{7.1.14}$$

Thus, with the usual definition of the median of an even number of variables, we have that the Hodges–Lehmann estimator for θ, using the Wilcoxon signed rank statistic to evaluate the symmetry of the $(X_i-\hat{\theta}_2)$'s, is

$$\hat{\theta}_2 = \text{median}\left\{\frac{X_i+X_j}{2},\qquad i\leqslant j=1,\ldots,n\right\}. \tag{7.1.15}$$

Our final example for this one-sample setting deals with the sign statistic.

Example 7.1.16. Let $V_3(X_1,\ldots,X_n)=B(X_1,\ldots,X_n)$ be the sign statistic. Then the Hodges–Lehmann estimator for θ, using the sign statistic to evaluate the symmetry of the $(X_i-\hat{\theta}_3)$'s, is given by

$$\hat{\theta}_3 = \text{median}\,\{X_i, \qquad i=1,\ldots,n\}. \tag{7.1.17}$$

(The proof of this fact is left as Exercise 7.1.1.)

For the Hodges–Lehmann estimation technique in the two-sample location setting, we consider the independent random samples $X_1,\ldots,X_m$ and $Y_1,\ldots,Y_n$ from continuous distributions with c.d.f.'s $F(x)$ and $F(x-\Delta)$, respectively. Let $U(X_1,\ldots,X_m;Y_1,\ldots,Y_n)$ be a test statistic for testing H_0: $\Delta=0$ against H_1: $\Delta>0$ that satisfies the conditions:

D1. H_0: $\Delta=0$ is rejected for large values of $U(X_1,\ldots,X_m;Y_1,\ldots,Y_n)$. **(7.1.18)**

D2. $U(x_1,\ldots,x_m;y_1+h,\ldots,y_n+h)$ is a nondecreasing function of h for each $(x_1,\ldots,x_m;y_1,\ldots,y_n)$. **(7.1.19)**

D3. When $\Delta=0$, the distribution of $U(X_1,\ldots,X_m;Y_1,\ldots,Y_n)$ is symmetric about some value ξ for every continuous $F(\cdot)$. **(7.1.20)**

[For some applications, we have to place an additional symmetry requirement on $F(\cdot)$ to guarantee that U is symmetrically distributed as required in D3. See, for example, Exercises 7.1.6, 7.1.7, and 7.1.8.] In direct analogy to the one-sample scheme, the Hodges–Lehmann estimator for Δ based on $U(X_1,\ldots,X_m;Y_1,\ldots,Y_n)$ is that statistic $\hat{\Delta}=\hat{\Delta}(X_1,\ldots,X_m;Y_1,\ldots,Y_n)$ such that $X_1,\ldots,X_m$ and $Y_1-\hat{\Delta},\ldots,Y_n-\hat{\Delta}$ look, when viewed by the statistic U, as much as possible as if they have the same locations. Thus for U satisfying (7.1.18)–(7.1.20) we have

$$\hat{\Delta} = \hat{\Delta}(X_1,\ldots,X_m;Y_1,\ldots,Y_n) = \frac{\Delta^*+\Delta^{**}}{2}, \tag{7.1.21}$$

where

$$\begin{aligned}\Delta^* &= \Delta^*(X_1,\ldots,X_m;Y_1,\ldots,Y_n)\\ &= \text{supremum}\,\{\Delta\colon U(X_1,\ldots,X_m;Y_1-\Delta,\ldots,Y_n-\Delta)>\xi\}\end{aligned} \tag{7.1.22}$$

and

$$\begin{aligned}\Delta^{**} &= \Delta^{**}(X_1,\ldots,X_m;Y_1,\ldots,Y_n)\\ &= \text{infimum}\,\{\Delta\colon U(X_1,\ldots,X_m;Y_1-\Delta,\ldots,Y_n-\Delta)<\xi\}.\end{aligned} \tag{7.1.23}$$

[We note, as with the one-sample estimators, that if $U(X_1,\ldots,X_m; Y_1 - \Delta,\ldots, Y_n - \Delta)$ is continuous and strictly decreasing in Δ, then $\hat{\Delta} = \Delta^* = \Delta^{**}$ satisfies $U(X_1,\ldots,X_m; Y_1 - \hat{\Delta},\ldots, Y_n - \hat{\Delta}) = \xi$.]

Example 7.1.24. Let $U(X_1,\ldots,X_m; Y_1,\ldots, Y_n)$ be the Mann–Whitney form of the Mann–Whitney–Wilcoxon statistic. Thus U=[number of positive Ds], where $D_{(1)} \leqslant \cdots \leqslant D_{(mn)}$ are the ordered values of the differences $Y_j - X_i$, $i = 1,\ldots,m$ and $j = 1,\ldots,n$, and we see that $U(x_1,\ldots,x_m; y_1 + h,\ldots,y_n + h)$=[number of Ds greater than $-h$] is a nondecreasing function of h for each $(x_1,\ldots,x_m;y_1,\ldots,y_n)$. In addition, the distribution of $U(X_1,\ldots,X_m; Y_1,\ldots, Y_n)$ is symmetric about $\xi = mn/2$ whenever $\Delta = 0$ and $F(\cdot)$ is a continuous c.d.f. [See, for example, Theorem 2.3.16 and equation (2.3.9).] Hence conditions D1–D3 are satisfied for the Mann–Whitney statistic U and, with arguments similar to those in Example 7.1.8, it follows (Exercise 7.1.5) that the Hodges–Lehmann estimator for Δ, using the Mann–Whitney statistic to evaluate the difference in locations, is

$$\hat{\Delta} = \text{median}\left\{ Y_j - X_i : i = 1,\ldots,m \quad \text{and} \quad j = 1,\ldots,n \right\}. \qquad \textbf{(7.1.25)}$$

Exercises 7.1.

7.1.1. Show that conditions C1–C3 in (7.1.1)–(7.1.3) are satisfied for the sign statistic $B(X_1,\ldots,X_n)$. Verify the form (7.1.17) of the Hodges–Lehmann estimator of θ based on $B(X_1,\ldots,X_n)$.

7.1.2. Let $X_1,\ldots,X_n$ be a random sample from a continuous distribution with c.d.f. $F(x-\theta)$, where $F(\cdot)$ corresponds to a distribution that is symmetric about 0. Let $U_i = U_i(X_1,\ldots,X_n)$, $i = 1,\ldots,k$, be odd translation statistics (see Definitions 1.3.13 and 1.3.14) and set $V(X_1,\ldots,X_n) = \sum_{i=1}^{k} \Psi(U_i)$, where $\Psi(t) = 1, 0$ as $t >, \leqslant 0$. Consider the test of H_0: $\theta = 0$ against H_1: $\theta > 0$ that rejects for large values of V.

(a) Show that conditions C2 and C3 in (7.1.2) and (7.1.3) are satisfied for this V. (You may use without proof the fact that when the underlying population is continuous, any odd translation statistic has a continuous distribution.)

(b) Find the form of the Hodges–Lehmann estimator $\hat{\theta}$ (7.1.4) for θ, when using the V statistic to evaluate the symmetry of the $(X_i - \hat{\theta})$'s.

(c) How does this general result apply to the estimators associated with the signed rank and sign statistics in Examples 7.1.8 and 7.1.16, respectively?

7.1.3. Let $X_{(1)} \leqslant \cdots \leqslant X_{(n)}$ be the order statistics for a random sample from a continuous distribution that is symmetric about θ, and set $W_{ij} = (X_{(i)} + X_{(j)})/2$, $1 \leqslant i \leqslant j \leqslant n$. Thus $A = \{W_{ij}: 1 \leqslant i \leqslant j \leqslant n\}$ is the set of $M = (n(n+1))/2$ Walsh averages for the X sample. Let $A_C \subset A$ be defined by

$$A_C = \{W_{ij}: (i,j) \in C\},$$

where C is a subset of $\{(i,j): 1 \leqslant i \leqslant j \leqslant n\}$ such that $(i,j) \in C$ if and only if $(n+1-j, n+1-i) \in C$. Consider the test of $H_0: \theta = 0$ against $H_1: \theta > 0$ that rejects for large values of $V(X_1, \ldots, X_n) = \Sigma_{(i,j) \in C} \Psi(W_{ij})$, where $\Psi(t) = 1, 0$ as $t >, \leqslant 0$. [Thus $V(X_1, \ldots, X_n)$ is the number of positive Walsh averages in A_C.]

(a) Show that conditions C2 and C3 in (7.1.2) and (7.1.3) are satisfied for this V.

(b) Find the form of the Hodges–Lehmann estimator $\hat{\theta}$ (7.1.4) for θ, when using this V statistic to evaluate the symmetry of the $(X_i - \hat{\theta})$'s.

(c) In the definition of A_C, what sets C_1 and C_2 result in $V(X_1, \ldots, X_n)$ being the sign and Wilcoxon signed rank statistics, respectively? [Note: Noether (1973) and Policello and Hettmansperger (1976) considered criteria that lead to the selection of C sets other than C_1 and C_2.]

7.1.4. Let $X_1, \ldots, X_n$ be a random sample from a continuous distribution that is symmetric about θ, and set $W_0 = W^+ - W^-$, where W^+ is the Wilcoxon signed rank statistic and W^- is the sum of the absolute ranks for those Xs that are nonpositive (i.e., $W_0 = 2W^+ - (n(n+1))/2$). Let $R_i^+(\theta)$ denote the rank of $|X_i - \theta|$ among $|X_1 - \theta|, \ldots, |X_n - \theta|$ and define $\text{sign}(t) = 1, 0, -1$ as $t >, =, < 0$.

(a) If $\hat{\theta}$ denotes the estimator of θ associated with W^+ (see Example 7.1.8), show that $\hat{\theta}$ satisfies

$$\sum_{\{i: X_i \neq \hat{\theta}\}}^{n} R_i^+(\hat{\theta}) \, \text{sign}(X_i - \hat{\theta}) \approx 0. \tag{7.1.26}$$

(b) Show that (7.1.26) is equivalent to

$$\sum_{\{i: X_i \neq \hat{\theta}\}}^{n} Q_i(\hat{\theta})(X_i - \hat{\theta}) \approx 0, \tag{7.1.27}$$

where

$$Q_i(\hat{\theta}) = R_i^+(\hat{\theta})/|X_i - \hat{\theta}|.$$

(c) Discuss how equation (7.1.27) can be used to develop an iterative scheme for obtaining the value of $\hat{\theta}$. In particular, if $\hat{\theta}_0$ is an initial estimator for θ (e.g., $\hat{\theta}_0 = \text{median}\ \{X_1,\ldots,X_n\}$) and $\hat{\theta}_1$ is the first iterate of $\hat{\theta}_0$, find the function $g(\hat{\theta}_0)$ so that

$$\hat{\theta}_1 = g(\hat{\theta}_0). \qquad \textbf{(7.1.28)}$$

[Note: For even moderate sample size n the number of calculations and operations necessary to find the value of $\hat{\theta}$ directly is relatively large. To help alleviate this problem, the simple iteration scheme in (7.1.28) was suggested by Hettmansperger and Utts (1977), who noted that two or three iterations beginning with $\hat{\theta}_0 = \text{median}\ \{X_1,\ldots,X_n\}$ are usually sufficient to yield a good approximate solution.]

7.1.5. Show that the estimator $\hat{\Delta}$ associated with the Mann–Whitney–Wilcoxon statistic is indeed given by (7.1.25).

7.1.6. Consider the two-sample location setting with $F(\cdot)$ symmetric about some point η. Obtain the Hodges–Lehmann estimator of the shift parameter Δ, say $\hat{\Delta}$, that is associated with the two-sample t-statistic. Be sure to check all the necessary conditions.

7.1.7. Let $X_1,\ldots,X_m$ and $Y_1,\ldots,Y_n$ be independent random samples from distributions with c.d.f.'s $F(x)$ and $F(x-\Delta)$, respectively, where $F(\cdot)$ is symmetric about some point η. Let $\tilde{Y} = \text{median}\ \{Y_1,\ldots,Y_n\}$ and set

$$U(X_1,\ldots,X_m; Y_1,\ldots,Y_n) = \sum_{i=1}^{m} \Psi(\tilde{Y} - X_i),$$

where $\Psi(t) = 1, 0$ as $t >, \leqslant 0$. Find the Hodges–Lehmann estimator of Δ, say, $\hat{\Delta}$, that is associated with U. Be sure to check all the necessary conditions. [Note: In Chapter 9 we study properties of rank statistics that enable us to drop the symmetry requirement on $F(\cdot)$ for this problem.]

7.1.8. Let $X_{(1)} \leqslant \cdots \leqslant X_{(m)}$ and $Y_{(1)} \leqslant \cdots \leqslant Y_{(n)}$ be the order statistics for independent random samples from continuous distributions with c.d.f.'s $F(x)$ and $F(x-\Delta)$, respectively, where $F(\cdot)$ is symmetric about some point η. Set $D_{ij} = Y_{(j)} - X_{(i)}$ and let $A_C = \{D_{ij}\colon (i,j) \in C\}$, where C is a subset of $\{(i,j): i=1,\ldots,m \text{ and } j=1,\ldots,n\}$ such that $(i,j)\in C$ if and only if $(m+1-i, n+1-j)\in C$. Let $U(X_1,\ldots,X_m; Y_1,\ldots,Y_n)=\Sigma_{(i,j)\in C}\Psi(D_{ij})$, where $\Psi(t)=1,0$ as $t>, \leqslant 0$. Consider the test of H_0: $\Delta=0$ against H_1: $\Delta>0$ that rejects for large values of U.

(a) Show that conditions D2 and D3 in (7.1.19) and (7.1.20) are satisfied for this U. [Note: In Chapter 9 we study properties of rank statistics that enable us to drop the symmetry requirement on $F(\cdot)$ for this problem.]

(b) Find the form of the Hodges–Lehmann estimator $\hat{\Delta}$ (7.1.21) for Δ, when using the statistic U to evaluate the difference in locations.

(c) In the definition of A_C, what set C_1 will yield the estimator $\hat{\Delta}$ associated with the Mann–Whitney–Wilcoxon statistic and discussed in Example 7.1.24?

7.1.9. Consider a situation in which we observe $Z_1,\ldots,Z_n$ such that

$$Z_i = \beta c_i + E_i,$$

where $c_1,\ldots,c_n$ are known positive numbers and $E_1,\ldots,E_n$ are i.i.d. continuous random variables with c.d.f. $F(\cdot)$ satisfying $F(0)=\frac{1}{2}$. Suppose that we wish to test

$$H_0\colon \beta = \beta_0 \text{ versus } H_1\colon \beta > \beta_0$$

by rejecting for large values of

$$V(Z_1-\beta_0 c_1,\ldots,Z_n-\beta_0 c_n) = \sum_{i=1}^{n} \Psi(Z_i-\beta_0 c_i),$$

with $\Psi(t)=1,0$ as $t>, \leqslant 0$.

(a) Argue that this test has a nonparametric distribution-free property for the specified problem.

(b) Under H_0, argue that $V(Z_1-\beta_0 c_1,\ldots,Z_n-\beta_0 c_n)$ is symmetrically distributed about some number ξ. What is ξ?

(c) Show that $V(z_1-\beta c_1,\ldots,z_n-\beta c_n)$ is a nonincreasing function of β.

(d) Define

$$\beta^* = \text{supremum}\{\beta \mid V(Z_1-\beta c_1,\ldots,Z_n-\beta c_n)>\xi\},$$
$$\beta^{**} = \text{infimum}\{\beta \mid V(Z_1-\beta c_1,\ldots,Z_n-\beta c_n)<\xi\},$$

and set

$$\hat{\beta} = \tfrac{1}{2}(\beta^* + \beta^{**}).$$

Find the form of this Hodges–Lehmann estimator $\hat{\beta}$, expressing it in terms of $(Z_1/c_1), \ldots, (Z_n/c_n)$.

7.2. Exact Small Sample Properties of the Hodges–Lehmann Location Estimators

In this section we develop some of the small sample properties of the Hodges–Lehmann location estimators. In particular, we consider conditions under which these estimators will (i) have distributions that are symmetric about the values of the estimated parameters and (ii) be median unbiased. We consider first the one-sample location problem.

Let $X_1, \ldots, X_n$ be a random sample from a continuous distribution with c.d.f. $F(x-\theta)$, where $F(\cdot)$ corresponds to a distribution that is symmetric about 0.

Lemma 7.2.1. *Let $\hat{\theta}(X_1, \ldots, X_n)$ be the Hodges–Lehmann estimator* (7.1.4) *of θ based on a test statistic $V(X_1, \ldots, X_n)$ satisfying* (7.1.1), (7.1.2), *and* (7.1.3). *Then $\hat{\theta}(X_1, \ldots, X_n)$ is a translation statistic, that is, $\hat{\theta}(x_1 + k, \ldots, x_n + k) = k + \hat{\theta}(x_1, \ldots, x_n)$ for every k and $x_1, \ldots, x_n$.*

Proof: For any constant k and any $(x_1, \ldots, x_n)$ we have

$$\begin{aligned}
\theta^*(x_1 + k, \ldots, x_n + k) &= \text{supremum}\{\theta \mid V(x_1 - (\theta - k), \ldots, x_n - (\theta - k)) > \xi\} \\
&= k + \text{supremum}\{\theta \mid V(x_1 - \theta, \ldots, x_n - \theta) > \xi\} \\
&= \theta^*(x_1, \ldots, x_n) + k.
\end{aligned}$$

Similarly,

$$\begin{aligned}
\theta^{**}(x_1 + k, \ldots, x_n + k) &= \text{infimum}\{\theta \mid V(x_1 - (\theta - k), \ldots, x_n - (\theta - k)) < \xi\} \\
&= \theta^{**}(x_1, \ldots, x_n) + k.
\end{aligned}$$

The translation property for $\hat{\theta}(X_1, \ldots, X_n)$ follows from these two facts and its form (7.1.4). ■

The next lemma demonstrates that when studying distributional properties of a Hodges–Lehmann estimator $\hat{\theta}(X_1,\ldots,X_n)$, we can assume without loss of generality that $\theta=0$. The corresponding properties for arbitrary θ then follow from a simple translation. Since this result is valid for the entire class of translation statistics, we establish the lemma for this more general (in view of Lemma 7.2.1) setting. In this form we are also able to apply the result to M-estimators in Section 7.4.

Lemma 7.2.2. *Let $X_1,\ldots,X_n$ be a random sample from a distribution with c.d.f. $F(x-\theta)$, where $F(\cdot)$ corresponds to a distribution with median 0. If $T(X_1,\ldots,X_n)$ is any translation statistic (in the sense of Definition 1.3.14), then*

$$P_\theta((T(X_1,\ldots,X_n)-\theta)\leqslant t) = P_0(T(X_1,\ldots,X_n)\leqslant t) \qquad \textit{for all } t, \tag{7.2.3}$$

where the notation P_θ indicates that the probability is computed when θ is the true value of the point of symmetry.

Proof: From the translation property, we see that

$$\begin{aligned}\{T(X_1,\ldots,X_n)-\theta|_\theta\} &= \{T(X_1-\theta,\ldots,X_n-\theta)|_\theta\}\\ &\stackrel{d}{=} \{T(X_1,\ldots,X_n)|_{\theta=0}\},\end{aligned}$$

the last distributional equality following from the nature of $F(x-\theta)$. ■

We are now ready to establish the conditions which lead to a symmetric distribution for a Hodges–Lehmann estimator $\hat{\theta}(X_1,\ldots,X_n)$.

Theorem 7.2.4. *Let $X_1,\ldots,X_n$ be a random sample from a continuous distribution that is symmetric about θ, and let $\hat{\theta}(X_1,\ldots,X_n)$ be the Hodges–Lehmann estimator (7.1.4) associated with a test statistic $V(X_1,\ldots,X_n)$ satisfying (7.1.1), (7.1.2), and (7.1.3). If, for every $(x_1,\ldots,x_n)$,*

$$V(x_1,\ldots,x_n)+V(-x_1,\ldots,-x_n)=2\xi, \tag{7.2.5}$$

where ξ is the point of symmetry in condition C3 of (7.1.3), then $\hat{\theta}(X_1,\ldots,X_n)$ is an odd statistic (in the sense of Definition 1.3.13) and is therefore symmetrically distributed about θ.

Proof: From (7.1.5) and (7.1.6) we see that for any $(x_1,\ldots,x_n)$

$$\begin{aligned}\theta^*(-x_1,\ldots,-x_n) &= \text{supremum}\{\theta\,|\,V(-x_1-\theta,\ldots,-x_n-\theta)>\xi\}\\ &= \text{supremum}\{\theta\,|(2\xi-V(x_1+\theta,\ldots,x_n+\theta))>\xi\},\\ &\quad\text{from (7.2.5)},\\ &= \text{supremum}\{-\theta\,|\,V(x_1-\theta,\ldots,x_n-\theta)<\xi\}\\ &= -\text{infimum}\{\theta\,|\,V(x_1-\theta,\ldots,x_n-\theta)<\xi\}\\ &= -\theta^{**}(x_1,\ldots,x_n).\end{aligned}$$

Since this equality holds for all $(x_1,\ldots,x_n)$, we also have

$$\theta^*(x_1,\ldots,x_n) = \theta^*(-(-x_1),\ldots,-(-x_n)) = -\theta^{**}(-x_1,\ldots,-x_n).$$

Thus

$$\begin{aligned}\hat{\theta}(-x_1,\ldots,-x_n) &= \frac{\theta^*(-x_1,\ldots,-x_n)+\theta^{**}(-x_1,\ldots,-x_n)}{2}\\ &= -\hat{\theta}(x_1,\ldots,x_n)\end{aligned}$$

for every $(x_1,\ldots,x_n)$, and we have that $\hat{\theta}(X_1,\ldots,X_n)$ is an odd statistic. The symmetry of the distribution of $\hat{\theta}(X_1,\ldots,X_n)$ follows from this fact, Lemma 7.2.1 and Corollary 1.3.19. ■

Before considering an application of Theorem 7.2.4 we note that condition (7.2.5) need only hold with probability one for the conclusion of the theorem to be valid. Indeed, this is often how the result is used in practice, as is illustrated in the next example.

Example 7.2.6. Let $\hat{\theta}_2(X_1,\ldots,X_n)$ be the Hodges–Lehmann estimator for θ associated (see Example 7.1.8) with the Wilcoxon signed rank statistic $W^+(X_1,\ldots,X_n) =$ (number of positive Walsh averages). Since $W^+(-X_1,\ldots,-X_n)=$(number of negative Walsh averages), we see that

$$\begin{aligned}W^+(X_1,\ldots,X_n) + W^+(-X_1,\ldots,-X_n) &= \text{(total number of Walsh averages)}\\ &= \frac{n(n+1)}{2} = 2\left(\frac{n(n+1)}{4}\right) \qquad (7.2.7)\end{aligned}$$

with probability one, since the probability of a zero Walsh average is zero with an underlying continuous X distribution. Hence, from Theorem 7.2.4 we see that

$$\hat{\theta}_2(X_1,\ldots,X_n) = \text{median}\left\{\frac{X_i+X_j}{2}, \qquad i \leqslant j = 1,2,\ldots,n\right\}$$

has a distribution that is symmetric about θ whenever the underlying continuous population is symmetric about θ.

We note that when a Hodges–Lehmann estimator $\hat{\theta}(X_1,\ldots,X_n)$ can be written in closed form it is often as simple to verify the odd property of the estimator directly rather than establishing (7.2.5) in order to apply Theorem 7.2.4. In fact, this is actually the case with the estimator $\hat{\theta}_2(X_1,\ldots,X_n)$ in Example 7.2.6.

One particular implication of Theorem 7.2.4 is that under the stated conditions, $E[\hat{\theta}(X_1,\ldots,X_n)] = \theta$, provided this expectation exists. Thus under those conditions $\hat{\theta}(X_1,\ldots,X_n)$ is an unbiased estimator for θ. In general, however, when the distribution of $\hat{\theta}(X_1,\ldots,X_n)$ is not symmetric [e.g., either the underlying X distribution is not symmetric or $V(X_1,\ldots,X_n)$ does not satisfy (7.2.5)], the estimator need not be unbiased. For such situations the concept of median unbiasedness is useful.

Definition 7.2.8. Let $\hat{\eta}$ be an estimator for η. We say that $\hat{\eta}$ is **median unbiased** for η if

$$P_\eta(\hat{\eta} \leqslant \eta) = \tfrac{1}{2}; \tag{7.2.9}$$

that is, if η is a median for the distribution of $\hat{\eta}$.

For some of the situations where a Hodges–Lehmann estimator $\hat{\theta}(X_1,\ldots,X_n)$ is not symmetrically distributed (and thus not necessarily unbiased for θ), it is still median unbiased for θ. For example, the assumption of symmetry for the X distribution can be dropped and still have median unbiasedness for certain Hodges–Lehmann estimators. We illustrate this fact with an example, but first we establish two necessary lemmas. For these two results, let $X_1,\ldots,X_n$ be a random sample from a continuous, but not necessarily symmetric, distribution with c.d.f. $F(x-\theta)$, where $F(\cdot)$ has median zero.

Lemma 7.2.10. *Let $\hat{\theta}(X_1,\ldots,X_n)$ be the Hodges–Lehmann estimator associated with a statistic $V(X_1,\ldots,X_n)$ through equations* (7.1.4), (7.1.5), *and*

(7.1.6). *For* ξ *as in* (7.1.3) *and any real number* a, *we have the inequalities*

$$P_\theta(V(X_1-a,\ldots,X_n-a)<\xi) \leqslant P_\theta(\hat{\theta}(X_1,\ldots,X_n)<a)$$
$$\leqslant P_\theta(V(X_1-a,\ldots,X_n-a)\leqslant\xi). \tag{7.2.11}$$

Proof: From the definition of $\theta^*(X_1,\ldots,X_n)$ and $\theta^{**}(X_1,\ldots,X_n)$ we have

$$\theta^* > a \Rightarrow V(X_1-a,\ldots,X_n-a) > \xi \Rightarrow \theta^* \geqslant a \tag{7.2.12}$$

and

$$\theta^{**} < a \Rightarrow V(X_1-a,\ldots,X_n-a) < \xi \Rightarrow \theta^{**} \leqslant a. \tag{7.2.13}$$

Since our underlying X distribution is continuous, it follows [see Theorem 6.1 of Hodges and Lehmann (1963)] that the distributions of θ^* and θ^{**} are both continuous also. Thus (7.2.12) and (7.2.13) imply that

$$P_\theta(\theta^*(X_1,\ldots,X_n)<a) = P_\theta(V(X_1-a,\ldots,X_n-a)\leqslant\xi)$$

and

$$P_\theta(\theta^{**}(X_1,\ldots,X_n)<a) = P_\theta(V(X_1-a,\ldots,X_n-a)<\xi),$$

Since $\hat{\theta}(X_1,\ldots,X_n)=\frac{1}{2}[\theta^*(X_1,\ldots,X_n)+\theta^{**}(X_1,\ldots,X_n)]$ and $\theta^*(X_1,\ldots,X_n) \leqslant\theta^{**}(X_1,\ldots,X_n)$, these relations imply the desired result. ■

Lemma 7.2.14. *Let* $\hat{\theta}(X_1,\ldots,X_n)$ *be the Hodges–Lehmann estimator associated with a statistic* $V(X_1,\ldots,X_n)$ *through equations* (7.1.4)–(7.1.6). *If* $V(X_1,\ldots,X_n)$ *has a distribution that is symmetric about some point* ξ *whenever* $\theta=0$, *then*

$$\frac{1-\epsilon}{2} \leqslant P_\theta(\hat{\theta}(X_1,\ldots,X_n)\leqslant\theta) \leqslant \frac{1+\epsilon}{2}, \tag{7.2.15}$$

where $\epsilon=P_0(V(X_1,\ldots,X_n)=\xi)$.

Proof: From the definition of ϵ and the symmetry of the distribution of $V(X_1,\ldots,X_n)$ when $\theta=0$, we have

$$P_0(V(X_1,\ldots,X_n)<\xi) = \frac{1-\epsilon}{2}$$

and

$$P_0(V(X_1,\ldots,X_n)\leqslant\xi)=\frac{1+\epsilon}{2}.$$

Combining these facts with (7.2.11) we obtain

$$\frac{1-\epsilon}{2}\leqslant P_0(\hat{\theta}(X_1,\ldots,X_n)\leqslant 0)\leqslant\frac{1+\epsilon}{2},$$

and (7.2.15) follows from Lemmas 7.2.1 and 7.2.2. ■

Corollary 7.2.16. For the setting of Lemma 7.2.14, if $\epsilon=0$ then $\hat{\theta}(X_1,\ldots,X_n)$ is median unbiased for θ.

Proof: Immediate from (7.2.15) and the definition of median unbiased. ■

Example 7.2.17. Let $X_1,\ldots,X_n$ be a random sample from a continuous distribution with c.d.f. $F(x-\theta)$, where $F(\cdot)$ has median zero. The sign statistic $B(X_1,\ldots,X_n)$ has a binomial $(n,\frac{1}{2})$ distribution when $\theta=0$, and the conditions of Lemma 7.2.14 are satisfied for $V(X_1,\ldots,X_n)=B(X_1,\ldots,X_n)$ and $\xi=n/2$. Thus if n is an odd integer, we have $\epsilon=P_0(V(X_1,\ldots,X_n)=n/2)=0$, and Corollary 7.2.16 implies that the associated (see Example 7.1.16) Hodges–Lehmann estimator $\hat{\theta}_3(X_1,\ldots,X_n)=\text{median}\{X_1,\ldots,X_n\}$ is median unbiased for θ. (We note that even if n is an even integer, the estimator $\hat{\theta}_3$ will not be far from median unbiasedness, since the ϵ in Lemma 7.2.14 will still be close to zero, especially for moderate or large sample sizes.)

We now turn our attention to the two-sample location problem. The distributional properties for two-sample Hodges–Lehmann estimators are established via proofs very similar to those for the analogous one-sample properties. Consequently, we simply state the results and leave the proofs as exercises. Let $X_1,\ldots,X_m$ and $Y_1,\ldots,Y_n$ be independent random samples from continuous distributions with c.d.f.'s $F(x)$ and $F(x-\Delta)$, respectively. We begin by studying the behavior of a two-sample Hodges–Lehmann estimator for the shift parameter Δ.

Lemma 7.2.18. *Let $\hat{\Delta}(X_1,\ldots,X_m;Y_1,\ldots,Y_n)$ be the Hodges–Lehmann estimator* (7.1.21) *associated with a test statistic $U(X_1,\ldots,X_m;Y_1,\ldots,Y_n)$ satisfying* (7.1.18), (7.1.19), *and* (7.1.20). *Then $\hat{\Delta}$ is a shift statistic; that is, it* *satisfies*

$$\hat{\Delta}(x_1,\ldots,x_m;y_1+k,\ldots,y_n+k) = \hat{\Delta}(x_1,\ldots,x_m;y_1,\ldots,y_n) + k$$

for all $(x_1,\ldots,x_m;y_1,\ldots,y_n)$ *and* k.

Proof: Exercise 7.2.6. ■

Lemma 7.2.19. *Let* $X_1,\ldots,X_m$ *and* $Y_1,\ldots,Y_n$ *be independent random samples from distributions with c.d.f.'s* $F(x)$ *and* $F(x-\Delta)$, *respectively. If* $T(X_1,\ldots,X_m;\,Y_1,\ldots,Y_n)$ *is any shift statistic, that is, if it satisfies*

$$T(x_1,\ldots,x_m;y_1+k,\ldots,y_n+k) = k + T(x_1,\ldots,x_m;y_1,\ldots,y_n)$$

for all $(x_1,\ldots,x_m;y_1,\ldots,y_n)$ *and all* k, *then*

$$P_\Delta\big[(T(X_1,\ldots,X_m;\,Y_1,\ldots,Y_n)-\Delta)\leqslant t\big]$$

$$= P_0\big[T(X_1,\ldots,X_m;\,Y_1,\ldots,Y_n)\leqslant t\big] \qquad \text{for all } t, \qquad \textbf{(7.2.20)}$$

where P_Δ *indicates that the probability is computed assuming* Δ *to be the true value of the shift parameter.*

Proof: Exercise 7.2.7. ■

In direct analogy to the one-sample Hodges–Lehmann estimators, we see from Lemmas 7.2.18 and 7.2.19 that for distributional properties of the estimator $\hat{\Delta}$ we can assume, without loss of generality, that $\Delta=0$, with the corresponding properties for arbitrary Δ following from (7.2.20). We now state a distributional symmetry result for a two-sample $\hat{\Delta}$ estimator.

Theorem 7.2.21. *Let* $\hat{\Delta}(X_1,\ldots,X_m;\,Y_1,\ldots,Y_n)$ *be as in Lemma* 7.2.18. *If*

$$\text{(i)}\quad U(x_1,\ldots,x_m;y_1,\ldots,y_n) + U(-x_1,\ldots,-x_m;\,-y_1,\ldots,-y_n) = 2\xi \qquad \textbf{(7.2.22)}$$

and

$$\text{(ii)}\quad U(x_1+k,\ldots,x_m+k;y_1+k,\ldots,y_n+k) = U(x_1,\ldots,x_m;y_1,\ldots,y_n), \qquad \textbf{(7.2.23)}$$

for every k *and* $(x_1,\ldots,x_m;y_1,\ldots,y_n)$,

and

(iii) $F(\cdot)$ *corresponds to a distribution that is symmetric about some value* η, **(7.2.24)**

then $\hat{\Delta}(X_1,\ldots,X_m; Y_1,\ldots,Y_n)$ *is an odd shift statistic (in the sense of Definitions* 1.3.13 *and* 1.3.23) *and is therefore symmetrically distributed about* Δ.

Proof: Exercise 7.2.8. ■

As with the application of Theorem 7.2.4, we note that condition (7.2.22) need only be satisfied with probability one in order for the conclusion of Theorem 7.2.21 to be valid. We illustrate this in the next example.

Example 7.2.25. Let $\hat{\Delta}(X_1,\ldots,X_m; Y_1,\ldots,Y_n)$ be the Hodges–Lehmann shift estimator associated with the Mann–Whitney form of the Mann–Whitney–Wilcoxon statistic, namely, $U(X_1,\ldots,X_m; Y_1,\ldots,Y_n)=$(number of positive Y_j-X_i differences, $i=1,\ldots,m$ and $j=1,\ldots,n$). We see that

$$U(-X_1,\ldots,-X_m;-Y_1,\ldots,-Y_n)+U(X_1,\ldots,X_m;Y_1,\ldots,Y_n)$$
$$=(\text{total number of } Y_j-X_i \text{ differences}) = 2\left(\frac{mn}{2}\right) \quad \textbf{(7.2.26)}$$

with probability one (since the underlying distributions are continuous and hence the probability of a difference equaling zero is zero.) Also we note that

$$U(x_1+k,\ldots,x_m+k;y_1+k,\ldots,y_n+k)$$
$$=\left(\text{number of positive } (y_j+k-x_i-k)\text{'s}\right)$$
$$=U(x_1,\ldots,x_m;y_1,\ldots,y_n)$$

for every k and $(x_1,\ldots,x_m;y_1,\ldots,y_n)$. Thus from Theorem 7.2.21 and Example 7.1.24 we have that $\hat{\Delta}(X_1,\ldots,X_m;Y_1,\ldots,Y_n)=\text{median}\{Y_j-X_i| i=1,\ldots,m \text{ and } j=1,\ldots,n\}$ has a distribution that is symmetric about Δ, provided $F(\cdot)$ corresponds to a continuous distribution that is symmetric about some point η.

For those Hodges–Lehmann shift estimators $\hat{\Delta}(X_1,\ldots,X_m;Y_1,\ldots,Y_n)$ that have closed form representations it is often as simple to verify the odd shift property of the estimator directly rather than establishing (7.2.22) and (7.2.23) so that Theorem 7.2.21 can be applied. This is the case for the estimator $\hat{\Delta}(X_1,\ldots,X_m;Y_1,\ldots,Y_n)$ in the previous Example 7.2.25.

When the underlying distribution $F(\cdot)$ is not symmetric, a Hodges–Lehmann estimator $\hat{\Delta}$ does not, in general, have a distribution that is symmetric about Δ, even if conditions (7.2.22) and (7.2.23) are satisfied. In fact, without this underlying $F(\cdot)$ symmetry, the estimator $\hat{\Delta}$ is often not even unbiased for Δ. (See Exercise 7.2.12.) However, under certain regularity conditions we can be sure that $\hat{\Delta}$ is median unbiased for Δ.

Lemma 7.2.27. *Let $\hat{\Delta}(X_1,\ldots,X_m; Y_1,\ldots,Y_n)$ be as in Lemma* 7.2.18. *For ξ as in* (7.1.20) *and any real number a, we have the inequalities*

$$P_\Delta(U(X_1,\ldots,X_m; Y_1-a,\ldots,Y_n-a)<\xi)$$
$$\leqslant P_\Delta\big(\hat{\Delta}(X_1,\ldots,X_m; Y_1,\ldots,Y_n)<a\big)$$
$$\leqslant P_\Delta(U(X_1,\ldots,X_m; Y_1-a,\ldots,Y_n-a)\leqslant\xi). \tag{7.2.28}$$

Proof: Exercise 7.2.9. ■

Lemma 7.2.29. *Let $\hat{\Delta}(X_1,\ldots,X_m; Y_1,\ldots,Y_n)$ be as in Lemma* 7.2.18. *If $\delta=P_0(U(X_1,\ldots,X_m; Y_1,\ldots,Y_n)=\xi)$, then*

$$\frac{1-\delta}{2}\leqslant P_\Delta\big(\hat{\Delta}(X_1,\ldots,X_m; Y_1,\ldots,Y_n)\leqslant\Delta\big)\leqslant\frac{1+\delta}{2}. \tag{7.2.30}$$

Proof: Exercise 7.2.10. ■

Corollary 7.2.31. For the setting of Lemma 7.2.29, if $\delta=0$, then $\hat{\Delta}(X_1,\ldots,X_m; Y_1,\ldots,Y_n)$ is median unbiased for Δ.

Proof: Immediate from (7.2.30). ■

Thus median unbiasedness for $\hat{\Delta}$ can be obtained without requiring the three conditions of Theorem 7.2.21. In particular, the assumption that the underlying $F(\cdot)$ is symmetric is not necessary to have $\hat{\Delta}$ be median unbiased for Δ.

Example 7.2.32. Let $\hat{\Delta}$ be the Hodges–Lehmann estimator associated with the Mann–Whitney–Wilcoxon statistic and discussed in Example 7.2.25. If mn is an odd integer, then $\delta=P_0(U(X_1,\ldots,X_m; Y_1,\ldots,Y_n)=mn/2)=0$ and $\hat{\Delta}$ is median unbiased with no additional assumptions on $F(\cdot)$. (As with the one-sample estimators, the estimator $\hat{\Delta}$ will not be far from median unbiasedness for even mn values either, since δ will typically be small, especially for moderate or large sample sizes m or n.)

Exercises 7.2.

7.2.1. Let $\hat{\theta}_3(X_1,\ldots,X_n)=\text{median}\{X_1,\ldots,X_n\}$ be the Hodges–Lehmann one-sample estimator for θ associated with the sign statistic (see Example 7.1.16). Show that $\hat{\theta}_3(X_1,\ldots,X_n)$ has a distribution that is symmetric about θ whenever the underlying X distribution is symmetric about θ.

7.2.2. Let $\hat{\theta}(X_1,\ldots,X_n)$ be the Hodges–Lehmann one-sample estimator associated with the statistic $V(X_1,\ldots,X_n)=\Sigma_{(i,j)\in C}\Psi(W_{ij})$ of Exercise 7.1.3. Show that $\hat{\theta}(X_1,\ldots,X_n)$ has a distribution that is symmetric about θ whenever the underlying X distribution is symmetric about θ and continuous.

7.2.3. Let $X_1,\ldots,X_n$ be a random sample from a continuous distribution that is symmetric about θ. Use Theorem 7.2.4 to show that the sample mean $\bar{X}=\frac{1}{n}\Sigma_{i=1}^{n}X_i$ has a distribution that is symmetric about θ.

7.2.4. Let $\hat{\Delta}(X_1,\ldots,X_m;Y_1,\ldots,Y_n)$ be the Hodges–Lehmann two-sample estimator associated with the test statistic $U(X_1,\ldots,X_m;Y_1,\ldots,Y_n)$ of Exercise 7.1.7. Show that $\hat{\Delta}$ has a distribution that is symmetric about Δ whenever $F(\cdot)$ is symmetric about some point η.

7.2.5. Let $\hat{\Delta}(X_1,\ldots,X_m;Y_1,\ldots,Y_n)$ be the Hodges–Lehmann two-sample estimator associated with the statistic $U(X_1,\ldots,X_m;Y_1,\ldots,Y_n)=\Sigma_{(i,j)\in C}\Psi(D_{ij})$ of Exercise 7.1.8. Show that $\hat{\Delta}$ is symmetrically distributed about Δ whenever the underlying distribution $F(\cdot)$ is symmetric about some point η.

7.2.6. Prove Lemma 7.2.18.

7.2.7. Prove Lemma 7.2.19.

7.2.8. Prove Theorem 7.2.21. [Hint: See the proof of Theorem 7.2.4.]

7.2.9. Prove Lemma 7.2.27. [Hint: You may use without proof the fact that both Δ^* and Δ^{**} have continuous distributions for continuous $F(\cdot)$. See Hodges and Lehmann (1963) for a proof of this fact.]

7.2.10. Prove Lemma 7.2.29.

7.2.11. Let $X_1,\ldots,X_m$ and $Y_1,\ldots,Y_n$ be independent random samples from continuous distributions with c.d.f.'s $F(x)$ and $F(x-\Delta)$, respectively, where $F(\cdot)$ corresponds to a distribution that is symmetric about some point η. Use Theorem 7.2.21 to show that the difference in sample means $\bar{Y}-\bar{X}$ is symmetrically distributed about Δ.

7.2.12. In Example 7.2.25 we showed that the Hodges–Lehmann two-sample estimator $\hat{\Delta}(X_1,\ldots,X_m; Y_1,\ldots,Y_n)=\text{median}\{Y_j-X_i: i=1,\ldots,m$ and $j=1,\ldots,n\}$ has a distribution that is symmetric about Δ, provided $F(\cdot)$ corresponds to a distribution that is symmetric about some point η. Argue or show by example that $\hat{\Delta}$ need not even be unbiased for Δ if the $F(\cdot)$ symmetry assumption is dropped. [Hint: Consider $m=1$ and n large.]

7.3. Asymptotic Properties of the Hodges–Lehmann Location Estimators

In this section we consider some large sample (asymptotic) properties for the Hodges–Lehmann location estimators. Of particular interest will be to obtain an expression for the asymptotic relative efficiency between two such competing estimators. As with the small sample properties in Section 7.2, we provide the details for the one-sample estimators and simply state the analogous results for the two-sample estimators, leaving the necessary proofs as exercises.

Let $X_1,\ldots,X_n$ be a random sample from a continuous distribution with c.d.f. $F(x-\theta)$, where $F(\cdot)$ corresponds to a distribution that is symmetric about 0. Let $\hat{\theta}(X_1,\ldots,X_n)$ be the Hodges–Lehmann estimator (7.1.4) of θ based on a test statistic $V(X_1,\ldots,X_n)$ satisfying (7.1.1), (7.1.2), and (7.1.3). We first establish a theorem linking the asymptotic distributional property of a (properly normed) $\hat{\theta}$ with that of the associated test statistic V.

Lemma 7.3.1. *Let a denote an arbitrary constant and define the sequence of parameter values $\{\theta_n\}$ by $\theta_n=(-a/\sqrt{n}\,)$, for each positive integer n. Let $G(\cdot)$ be the c.d.f. for a continuous random variable with zero mean and unit variance, and suppose that*

$$\lim_{n\to\infty} P_{\theta_n}\{V(X_1,\ldots,X_n)\leqslant \xi_n\} = G\left(\frac{aB}{A}\right), \tag{7.3.2}$$

where A and B are constants, P_{θ_n} indicates that the probability is computed under the parameter value θ_n, and ξ_n is the value of ξ in (7.1.3) for sample size n. Then for any fixed θ we have

$$\lim_{n\to\infty} P_{\theta}\left\{\sqrt{n}\,\left(\hat{\theta}(X_1,\ldots,X_n)-\theta\right)\leqslant a\right\} = G\left(\frac{aB}{A}\right), \tag{7.3.3}$$

where P_θ indicates a probability computed under the parameter value θ.

Proof: In view of Lemmas 7.2.1 and 7.2.2, it is sufficient to establish (7.3.3) when $\theta=0$. Since $\hat{\theta}$ has a continuous distribution (see the proof of Lemma 7.2.10) and $G(\cdot)$ corresponds to a continuous distribution, we apply Lemma 7.2.10 (with $\theta=0$) to obtain

$$\lim_{n\to\infty} P_0\left(\sqrt{n}\,\hat{\theta}(X_1,\ldots,X_n)\leqslant a\right)$$

$$= \lim_{n\to\infty} P_0\left(\hat{\theta}(X_1,\ldots,X_n)\leqslant \frac{a}{\sqrt{n}}\right)$$

$$= \lim_{n\to\infty} P_0\left(V\left(X_1-\frac{a}{\sqrt{n}},\ldots,X_n-\frac{a}{\sqrt{n}}\right)\leqslant \xi_n\right),$$

which, since $X_i-(a/\sqrt{n}\,)$ has c.d.f. $F(x+(a/\sqrt{n}\,))=F(x-\theta_n)$, yields

$$= \lim_{n\to\infty} P_{\theta_n}(V(X_1,\ldots,X_n)\leqslant \xi_n) = G\left(\frac{aB}{A}\right),$$

the last equality following from assumption (7.3.2). ■

Thus if we can find a sequence of parameter values $\{\theta_n\}=\{-(a/\sqrt{n}\,)\}$ such that $V(X_1,\ldots,X_n)$ has the limiting property of (7.3.2), then $\sqrt{n}\,(\hat{\theta}(X_1,\ldots,X_n)-\theta)$ will have the limiting distribution property of (7.3.3). We illustrate the application of Lemma 7.3.1 with an example.

Example 7.3.4. Let $\hat{\theta}_2(X_1,\ldots,X_n)=\text{median}\{(X_i+X_j)/2:\ i\leqslant j=1,2,\ldots,n\}$ be the Hodges–Lehmann estimator for θ associated (see Example 7.1.8) with the Wilcoxon signed rank statistic $W^+(X_1,\ldots,X_n)$. Then, ξ_n in (7.1.3) is $(n(n+1))/4$, and we see that the sequence $\{\theta_n\}=\{-(a/\sqrt{n}\,)\}$ is simply a sequence of one-sample Pitman translation alternatives for the H_0 value $\theta_0=0$, as was considered in Chapter 5 when developing the asymptotic efficiencies of test procedures. (We note that we have not required $-a$ to be positive as was done for the Pitman alternatives in Chapter 5. However, the positivity requirement at that time was due to the one-sided nature of the alternatives under consideration there. Such is not the case here, and the results we use from Chapter 5 do not depend upon the positivity of $-a$.)

Letting $\mu_n(\theta)=2(n-1)^{-1}P_\theta(X_1>0)+P_\theta(X_1>-X_2)$ be the sequence of constants used in Section 5.4 for developing the testing efficacy for the signed rank test, we see from those arguments that

$$\sqrt{n}\left[\frac{W^+(X_1,\ldots,X_n)}{\binom{n}{2}} - \mu_n(\theta_n)\right]$$

has a limiting ($n\to\infty$) distribution under θ_n that is normal with mean zero and variance $1/3$. Now, with $\xi_n=(n(n+1))/4$, we write

$$\frac{\sqrt{n}}{\binom{n}{2}}[W^+(X_1,\ldots,X_n)-\xi_n] = \sqrt{n}\left[\frac{W^+(X_1,\ldots,X_n)}{\binom{n}{2}} - \mu_n(\theta_n)\right] + \sqrt{n}\left[\mu_n(\theta_n) - \frac{\xi_n}{\binom{n}{2}}\right].$$

Since $\xi_n=\binom{n}{2}\mu_n(0)$, we see that

$$\begin{aligned}
\lim_{n\to\infty}\sqrt{n}\left[\mu_n(\theta_n)-\frac{\xi_n}{\binom{n}{2}}\right] &= \lim_{n\to\infty}\sqrt{n}\,[\mu_n(\theta_n)-\mu_n(0)] \\
&= -a\lim_{n\to\infty}\left[\frac{\mu_n\left(-\frac{a}{\sqrt{n}}\right)-\mu_n(0)}{\left(-\frac{a}{\sqrt{n}}\right)}\right] \\
&= -a\lim_{n\to\infty}\mu_n'(0) \\
&= -a\lim_{n\to\infty}2\left[\frac{f(0)}{(n-1)}+\int_{-\infty}^{\infty}f^2(y)\,dy\right] \\
&= -2a\int_{-\infty}^{\infty}f^2(y)\,dy,
\end{aligned}$$

where we have used the expression for $\mu_n'(0)$ given in the Examples of Section 5.4, and we have now assumed that $f(x)=dF(x)/dx$ is bounded by some positive number M. Combining this limit with the limiting normality of

$$\sqrt{n}\left[\frac{W^+(X_1,\ldots,X_n)}{\binom{n}{2}} - \mu_n(\theta_n)\right],$$

we see from Slutsky's Theorem 3.2.8 that $[\sqrt{n}/\binom{n}{2}](W^+(X_1,\ldots,X_n) - \xi_n)$ has a limiting distribution (under θ_n) that is normal with mean $-2a\int_{-\infty}^{\infty} f^2(x)\,dx$ and variance $1/3$. This implies that (7.3.2) is satisfied with the identifications:

(i) $G(\cdot) = \Phi(\cdot)$, the standard normal c.d.f.
(ii) $A = \sqrt{1/3}$
(iii) $B = -2\int_{-\infty}^{\infty} f^2(x)\,dx$.

Hence, applying Lemma 7.3.1, we have that $\sqrt{n}\,(\hat{\theta}_2(X_1,\ldots,X_n) - \theta)$ has a limiting distribution (under θ) that is normal with mean 0 and variance

$$\frac{A^2}{B^2} = \frac{1}{12\left[\int_{-\infty}^{\infty} f^2(x)\,dx\right]^2}.$$

In Example 7.3.4 we appealed to the properties of the Wilcoxon signed rank test statistic under Pitman alternatives to establish the form of the limiting distribution of $\hat{\theta}_2(X_1,\ldots,X_n)$. This technique can be used as a general method for any Hodges–Lehmann estimator, so long as the necessary information is available concerning the limiting properties of the associated test statistic under Pitman alternatives. We state this fact as a theorem, and leave the proof as an exercise.

Theorem 7.3.5. *Let $\hat{\theta}(X_1,\ldots,X_n)$ be the Hodges–Lehmann estimator* (7.1.4) *associated with the test statistic $V(X_1,\ldots,X_n)$ satisfying* (7.1.1), (7.1.2), *and* (7.1.3) *for some point of symmetry ξ_n. Assume that $V(X_1,\ldots,X_n)$ also satisfies Assumptions A1–A6 of Theorem* 5.2.7 *for sequences $\{\mu_n(\theta)\}$ and $\{\sigma_n^2(\theta)\}$ and some c.d.f. $H(\cdot)$ with mean zero and variance one. If*

$$\lim_{n\to\infty} \frac{\mu_n(0) - \xi_n}{\sigma_n(0)} = 0,$$

then $\sqrt{n}\,(\hat{\theta}(X_1,\ldots,X_n) - \theta)$ has a limiting distribution with distributional form $H(\cdot)$, mean zero, and variance $[1/K_V^2]$, where K_V is the efficacy of the test based on the statistic V (see Definition 5.2.14).

Proof: Exercise 7.3.4. ■

Example 7.3.6. Let $\hat{\theta}_3(X_1,\ldots,X_n) = \text{median}\{X_1,\ldots,X_n\}$ be the Hodges–Lehmann estimator for θ associated (see Example 7.1.16) with the sign statistic $B(X_1,\ldots,X_n)$. Then applying Theorem 7.3.5 (see Exercise 7.3.2), we see that $\sqrt{n}\,(\hat{\theta}_3(X_1,\ldots,X_n)-\theta)$ has a limiting distribution (under θ) that is normal with mean 0 and variance $[4f^2(0)]^{-1}$, provided we assume that the density $f(x)=F'(x)$ is positive and continuous at 0.

We note that Theorem 7.3.5 implies that a Hodges–Lehmann estimator $\hat{\theta}(X_1,\ldots,X_n)$, satisfying the stated conditions, will be asymptotically unbiased for the parameter θ. When this is the case for two competing estimators, we are naturally led to the following definition of the asymptotic relative efficiency of one point estimator with respect to another.

Definition 7.3.7. Let $\hat{\eta}_1$ and $\hat{\eta}_2$ be asymptotically unbiased estimators for a parameter η in the sense that both $\sqrt{n}\,(\hat{\eta}_1-\eta)$ and $\sqrt{n}\,(\hat{\eta}_2-\eta)$ have asymptotic distributions with zero means. We define the **asymptotic relative efficiency of $\hat{\eta}_1$ with respect to $\hat{\eta}_2$**, denoted by $\mathrm{ARE}(\hat{\eta}_1,\hat{\eta}_2)$, to be

$$\mathrm{ARE}(\hat{\eta}_1,\hat{\eta}_2)=\frac{\sigma^2_{\hat{\eta}_2}}{\sigma^2_{\hat{\eta}_1}}, \tag{7.3.8}$$

where $\sigma^2_{\hat{\eta}_i}=\lim_{n\to\infty}\mathrm{Var}(\sqrt{n}\,\hat{\eta}_i)$, $i=1,2$.

Thus between two asymptotically unbiased estimators for a parameter, the one with the smaller variance limit is to be preferred. We now establish a theorem relating the asymptotic relative efficiency properties for Hodges–Lehmann one-sample estimators to the asymptotic relative efficiency properties of the associated test statistics.

Theorem 7.3.9. *Let $\hat{\theta}_a(X_1,\ldots,X_n)$ and $\hat{\theta}_b(X_1,\ldots,X_n)$ be two Hodges–Lehmann one-sample estimators associated with test statistics $V_a(X_1,\ldots,X_n)$ and $V_b(X_1,\ldots,X_n)$, respectively, satisfying the assumptions of Theorem* 7.3.5. *Then*

$$\mathrm{ARE}(\hat{\theta}_a,\hat{\theta}_b)=\frac{K^2_{V_a}}{K^2_{V_b}}=\mathrm{ARE}(V_a,V_b). \tag{7.3.10}$$

Proof: The result follows immediately from Definition 7.3.7, Theorem 7.3.5 and Theorem 5.2.7. ■

Example 7.3.11. Let

$$\hat{\theta}_1(X_1,\ldots,X_n)=\bar{X},\ \hat{\theta}_2(X_1,\ldots,X_n)=\text{median}\left\{\frac{X_i+X_j}{2}:i\leqslant j=1,\ldots,n\right\}$$

and

$$\hat{\theta}_3(X_1,\ldots,X_n) = \text{median}\{X_1,\ldots,X_n\}$$

be the Hodges–Lehmann one-sample estimators associated with the one-sample t test statistic, Wilcoxon signed rank statistic, and sign statistic, respectively. (See Examples 7.1.7, 7.1.8, and 7.1.16). Then, from Theorem 7.3.9, Examples 7.3.4 and 7.3.6, Exercise 7.3.3, and expressions (5.4.6), (5.4.9), and (5.4.10), we immediately obtain the following estimator efficiency forms:

(i)
$$\text{ARE}(\hat{\theta}_2,\hat{\theta}_1) = 12\sigma^2\left[\int_{-\infty}^{\infty} f^2(x)\,dx\right]^2, \tag{7.3.12}$$

provided that the density $f(x)=F'(x)$ is bounded, has finite variance σ^2, and is continuous at 0 and at all but a countable number of other x-values. Also,

(ii)
$$\text{ARE}(\hat{\theta}_3,\hat{\theta}_1) = 4\sigma^2 f^2(0), \tag{7.3.13}$$

provided the density $f(x)=F'(x)$ is positive and continuous at 0 and has finite variance σ^2. Likewise,

(iii)
$$\text{ARE}(\hat{\theta}_3,\hat{\theta}_2) = f^2(0)\Big/\left[3\left\{\int_{-\infty}^{\infty} f^2(x)\,dx\right\}^2\right], \tag{7.3.14}$$

provided the density $f(x)=F'(x)$ is bounded, is continuous and positive at 0, and is continuous at all but a countable number of other x-values. Numerical values for these ARE expressions for several different underlying distributions $F(x)$ can be found in Tables 5.4.7 and 5.4.11 and the solution to Exercise 5.4.10.

The remainder of this section is devoted to a brief description of the analogous large-sample properties of the Hodges–Lehmann two-sample estimators $\hat{\Delta}(X_1,\ldots,X_m; Y_1,\ldots, Y_n)$. Because of the similarity of the proofs of the large-sample results for the one- and two-sample estimators, we simply state the necessary properties for $\hat{\Delta}$ and leave the proofs as exercises.

Let $X_1,\ldots,X_m$ and $Y_1,\ldots, Y_n$ be independent random samples from continuous distributions with c.d.f.'s $F(x)$ and $F(x-\Delta)$, respectively, and set $N=m+n$. Let $\hat{\Delta}(X_1,\ldots,X_m; Y_1,\ldots, Y_n)$ be the Hodges–Lehmann estimator (7.1.21) of Δ based on a test statistic $U(X_1,\ldots,X_m; Y_1,\ldots, Y_n)$ satisfying (7.1.18), (7.1.19), and (7.1.20), and assume that $\lim_{N\to\infty}(m/N)=\lambda$ and $\lim_{N\to\infty}(n/N)=(1-\lambda)$, where $0<\lambda<1$.

Lemma 7.3.15. *Let b denote an arbitrary constant and define the sequence of parameter values $\{\Delta_N\}$ by $\Delta_N = -(b/\sqrt{N}\,)$, for each positive integer N. Let $H(\cdot)$ be the c.d.f. for a continuous random variable with mean zero and unit variance, and suppose that*

$$\lim_{N\to\infty} P_{\Delta_N}\{U(X_1,\ldots,X_m; Y_1,\ldots,Y_n) \leqslant \xi_N\} = H\left(\frac{bD}{C}\right), \tag{7.3.16}$$

where C and D are constants, P_{Δ_N} indicates that the probability is computed under the parameter value Δ_N, and ξ_N is the value of ξ in (7.1.20) for the combined sample size N. Then, for any fixed Δ we have

$$\lim_{N\to\infty} P_{\Delta}\left\{\sqrt{n}\left(\hat{\Delta}(X_1,\ldots,X_m; Y_1,\ldots,Y_n)-\Delta\right)\leqslant b\right\} = H\left(\frac{bD}{C}\right), \tag{7.3.17}$$

where P_Δ indicates a probability computed under the parameter value Δ.

Proof: Exercise 7.3.5. ■

Theorem 7.3.18. *Let $\hat{\Delta}(X_1,\ldots,X_m; Y_1,\ldots,Y_n)$ be the Hodges–Lehmann estimator of (7.1.21) associated with the test statistic $U(X_1,\ldots,X_m; Y_1,\ldots,Y_n)$ satisfying (7.1.18), (7.1.19), and (7.1.20) for some point of symmetry ξ_N. Assume that $U(X_1,\ldots,X_m; Y_1,\ldots,Y_n)$ also satisfies Assumptions A1–A6 of Theorem 5.2.7 for sequences $\{\mu_N(\theta)\}$ and $\{\sigma_N^2(\theta)\}$ and some c.d.f. $H(\cdot)$ with mean zero and variance one. If*

$$\lim_{n\to\infty}\left[\frac{\mu_N(0)-\xi_N}{\sigma_N(0)}\right] = 0,$$

then $\sqrt{N}\,(\hat{\Delta}(X_1,\ldots,X_m; Y_1,\ldots,Y_n)-\Delta)$ has a limiting distribution with distributional form $H(\cdot)$, mean zero, and variance $[1/K_U^2]$, where K_U is the efficacy of the test based on U.

Proof: Exercise 7.3.6. ■

Example 7.3.19. Let $\hat{\Delta}(X_1,\ldots,X_m; Y_1,\ldots,Y_n)=\text{median}\{Y_j-X_i\colon\ i=1,\ldots,m \text{ and } j=1,\ldots,n\}$ be the Hodges–Lehmann estimator associated (see Example 7.1.24) with the Mann–Whitney–Wilcoxon statistic $U(X_1,\ldots,X_m; Y_1,\ldots,Y_n)$. Then from Theorem 7.3.18 and the asymptotic relative efficiency calculations for the hypothesis tests based on U (see Section 5.4), we have that $\sqrt{N}\,(\hat{\Delta}(X_1,\ldots,X_m;Y_1,\ldots,Y_n)-\Delta)$ has a limiting distribution (under Δ) that is normal with zero mean and variance

$$\left\{12\lambda(1-\lambda)\left[\int_{-\infty}^{\infty} f^2(x)\,dx\right]^2\right\}^{-1},$$

where we assume that the density $f(x)=F'(x)$ is bounded and continuous at all but a countable number of x-values.

Theorem 7.3.18 implies that a Hodges–Lehmann estimator $\hat{\Delta}(X_1,\ldots,X_m;Y_1,\ldots,Y_n)$, satisfying the stated conditions, will be asymptotically unbiased for the parameter Δ. This fact leads us to the theorem for the asymptotic relative efficiency of one two-sample Hodges–Lehmann estimator with respect to a second competing estimator.

Theorem 7.3.20. *Let* $\hat{\Delta}_a(X_1,\ldots,X_m;Y_1,\ldots,Y_n)$ *and* $\hat{\Delta}_b(X_1,\ldots,X_m;Y_1,\ldots,Y_n)$ *be two Hodges–Lehmann two-sample estimators associated with test statistics* $U_a(X_1,\ldots,X_m;Y_1,\ldots,Y_n)$ *and* $U_b(X_1,\ldots,X_m;Y_1,\ldots,Y_n)$, *respectively, satisfying the conditions of Theorem* 7.3.18. *Then*

$$\mathrm{ARE}(\hat{\Delta}_a,\hat{\Delta}_b) = \frac{K_{U_a}^2}{K_{U_b}^2} = \mathrm{ARE}(U_a,U_b). \tag{7.3.21}$$

Proof: The result follows at once from Definition 7.3.7, Theorem 7.3.18, and Theorem 5.2.7. ■

Example 7.3.22. Let $\hat{\Delta}_1(X_1,\ldots,X_m;Y_1,\ldots,Y_n)=\bar{Y}-\bar{X}$ and $\hat{\Delta}_2(X_1,\ldots,X_m;Y_1,\ldots,Y_n)=\text{median}\{Y_j-X_i\!: i=1,\ldots,m \text{ and } j=1,\ldots,n\}$. Then, from Theorem 7.3.20, Exercise 7.3.8 and expressions (5.4.13) and (5.4.14), we have

$$\mathrm{ARE}(\hat{\Delta}_2,\hat{\Delta}_1) = 12\sigma^2\left[\int_{-\infty}^{\infty} f^2(x)\,dx\right]^2, \tag{7.3.23}$$

provided that the density $f(x)=F'(x)$ is bounded, has finite variance σ^2, and is continuous at all but a countable number of x-values.

Exercises 7.3.

7.3.1. Let $\hat{\theta}(X_1,\ldots,X_n)$ be a Hodges–Lehmann estimator for which the conditions of Lemma 7.3.1 are satisfied. Show that $\hat{\theta}$ is a consistent estimator for θ.

7.3.2. Verify the necessary conditions to apply Theorem 7.3.5 to obtain the limiting distribution of $\sqrt{n}\,(\hat{\theta}_3(X_1,\ldots,X_n)-\theta)$ in Example 7.3.6.

7.3.3. Show that the Hodges–Lehmann one-sample estimator $\hat{\theta}_1(X_1,\ldots,X_n)=\bar{X}$ and the associated one-sample t-statistic satisfy the conditions of Theorem 7.3.5. Do we require any conditions on the density $f(x)=(dF(x)/dx)$?

7.3.4. Prove Lemma 7.3.5. [Hint: See Example 7.3.4 and Lemma A.3.12 in the Appendix.]

7.3.5. Prove Lemma 7.3.15.

7.3.6. Prove Theorem 7.3.18. [Hint: See Example 7.3.4 and Lemma A.3.12 in the Appendix.]

7.3.7. Let $\hat{\Delta}(X_1,\ldots,X_m;Y_1,\ldots,Y_n)$ be a Hodges–Lehmann estimator for which the conditions of Lemma 7.3.15 are satisfied. Show that $\hat{\Delta}$ is a consistent estimator of Δ.

7.3.8. Show that the Hodges–Lehmann two-sample estimator $\hat{\Delta}(X_1,\ldots,X_m;Y_1,\ldots,Y_n)=\bar{Y}-\bar{X}$ and the associated two-sample t-statistic satisfy the assumptions of Theorem 7.3.18. Do we require any conditions on the density $f(x)=(dF(x)/dx)$?

7.3.9. Let

$$\hat{\theta}_1(X_1,\ldots,X_n) = \bar{X} \text{ and } \hat{\theta}_2(X_1,\ldots,X_n) = \text{median}\left\{\frac{X_i+X_j}{2} : i \leqslant j=1,\ldots,n\right\}$$

be the Hodges–Lehmann estimators associated with the one-sample t-statistic and the Wilcoxon signed rank statistic, respectively. Then from expression (7.3.12) we see that $\text{ARE}(\hat{\theta}_2,\hat{\theta}_1)=.955$ when the underlying distribution $F(\cdot)$ is normal. To compare this value with the corresponding variance ratio values for small n, consider the three cases $n=1,2,3$.

(a) For $n=1$ or 2, show that

$$\frac{\text{Var}(\hat{\theta}_1)}{\text{Var}(\hat{\theta}_2)} = 1.$$

(b) For $n=3$, show that

$$\frac{\text{Var}(\hat{\theta}_1)}{\text{Var}(\hat{\theta}_2)} = .979.$$

[Hint: With $n=3$, show that $\hat{\theta}_2 = \frac{1}{4}(X_{(1)} + 2X_{(2)} + X_{(3)})$, where $X_{(1)} \leqslant X_{(2)} \leqslant X_{(3)}$ are the order statistics for the Xs. Then, use tables of covariances of normal order statistics (see Godwin (1949), for example) to compute $\text{Var}(\hat{\theta}_2)$.]

7.3.10. Carry out the same comparisons as in Exercise 7.3.9 for an underlying exponential distribution with density

$$f(x) = \begin{cases} \dfrac{1}{\beta} e^{-x/\beta}, & x > 0, \\ 0, & \text{elsewhere,} \end{cases}$$

where $\beta > 0$. [Hint: Consider the results of Exercise 1.2.7.]

7.4. *M*-Estimators and Influence Curves

This section deals with two important topics in robust estimation. To concentrate on the basic principles and avoid complexities, we confine our description of M-estimators to the problem of estimating the center of a symmetric, univariate, continuous population. Let $X_1, \ldots, X_n$ denote a random sample from a continuous population that is symmetric about the unknown parameter θ and has c.d.f. $F(x-\theta)$. A classical approach to the problem of estimating θ would be to estimate the center with the θ-value that is closest to the sample data $X_1, \ldots, X_n$, where closeness is measured by the sum of the squared deviations

$$\sum_{i=1}^{n} (X_i - \theta)^2. \tag{7.4.1}$$

Choosing the estimator to be the θ-value that minimizes (7.4.1) is the well-known **least squares** criterion for estimation. Taking the derivative of (7.4.1) with respect to θ, setting it equal to zero and multiplying by -1 shows that this estimator $\hat{\theta}$ must satisfy

$$\sum_{i=1}^{n} 2(X_i - \hat{\theta}) = 0, \tag{7.4.2}$$

which, of course, yields the sample mean

$$\hat{\theta} = \bar{X}.$$

Let us generalize this solution process by taking $\rho(\cdot)$ to be a distance function satisfying

(i) $\rho(t) \geqslant 0$ for all t, and $\rho(0)=0$,
(ii) $\rho(t)=\rho(-t)$ for all t,

and

(iii) The derivative of $\rho(\cdot)$, namely $\rho'(t) \equiv \Psi(t)$, exists for all t and is a nondecreasing function of t. [Please note that in this section the notation $\Psi(t)$ is not used to denote a specific indicator function as it is elsewhere in this text.]

We measure the distance of any θ-value from the sample by

$$\sum_{i=1}^{n} \rho(X_i - \theta), \tag{7.4.3}$$

and then seek the θ-value that minimizes this quantity. As before, taking the derivative of (7.4.3), setting it equal to zero, and multiplying by -1 shows that such an estimator $\hat{\theta}$ must satisfy

$$\sum_{i=1}^{n} \Psi(X_i - \hat{\theta}) = 0, \tag{7.4.4}$$

where $\Psi(t)=\rho'(t)$. [This equation corresponds to (7.4.2).] Clearly, the previous example fits into this formulation with the identifications $\rho(t)=t^2$ and $\Psi(t)=2t$.

Now let us consider a second important example that almost (but not quite) fits into this general estimation scheme. Suppose we measure the distance of each θ from the sample by

$$\sum_{i=1}^{n} |X_i - \theta|. \tag{7.4.5}$$

The $\rho(\cdot)$ function here is $\rho(t)=|t|$, which does not have a derivative at $t=0$. Suppose n is odd, say, $n=2k+1$. Let θ be an arbitrary real number, and let $X_{(1)} < \cdots < X_{(n)}$ represent the ordered X_i-values. Note that for $i=1,\ldots,k$,

$$\begin{aligned} |X_{(i)} - X_{(k+1)}| + |X_{(k+1)} - X_{(n+1-i)}| &= |X_{(i)} - X_{(n+1-i)}| \\ &\leqslant |X_{(i)} - \theta| + |\theta - X_{(n+1-i)}|. \end{aligned}$$

Summing this inequality over $i=1,\ldots,k$ and including the two sides of

$$|X_{(k+1)} - X_{(k+1)}| \leqslant |X_{(k+1)} - \theta|$$

in the respective sums yields

$$\sum_{i=1}^{n} |X_{(i)} - X_{(k+1)}| \leqslant \sum_{i=1}^{n} |X_{(i)} - \theta|$$

or

$$\sum_{i=1}^{n} |X_i - X_{(k+1)}| \leqslant \sum_{i=1}^{n} |X_i - \theta|. \tag{7.4.6}$$

When n is an even integer a similar result (see Exercise 7.4.1) holds. These two inequalities yield the fact that for any n the value of θ that minimizes (7.4.5) is $\hat{\theta} = \text{median}(X_1, \ldots, X_n)$.

With $\text{sign}(t) = -1, 0, 1$ as $t <, =, > 0$, a pseudosolution to this problem can be obtained by finding the value $\hat{\theta}$ that satisfies

$$\sum_{i=1}^{n} \text{sign}(X_i - \hat{\theta}) = 0. \tag{7.4.7}$$

This is analogous to (7.4.4), but is not a proper solution, since $\text{sign}(t)$ is not quite the derivative function of $|t|$. Moreover, (7.4.7) might not be satisfied by any $\hat{\theta}$, or it might be satisfied by an interval of θ-values. Hence (7.4.7) does not produce an adequate description of $\hat{\theta}$, and we replace it by

$$\hat{\theta} = \hat{\theta}(X_1, \ldots, X_n) = \frac{\theta^* + \theta^{**}}{2} = \text{median}(X_1, \ldots, X_n), \tag{7.4.8}$$

where

$$\theta^* = \theta^*(X_1, \ldots, X_n) = \sup\left\{ \theta \,\middle|\, \sum_{i=1}^{n} \text{sign}(X_i - \theta) > 0 \right\}$$

and

$$\theta^{**} = \theta^{**}(X_1, \ldots, X_n) = \inf\left\{ \theta \,\middle|\, \sum_{i=1}^{n} \text{sign}(X_i - \theta) < 0 \right\}.$$

Since the sample median is derived via an estimation scheme that is very similar to one producing the sample mean, we consider a general formulation that includes both.

Definition 7.4.9. Let $\Psi(\cdot)$ be a function with the following properties:

(i) $\Psi(t)$ is defined over the real line and is nondecreasing and nonconstant,

(ii) $\Psi(t) = -\Psi(-t)$, for all t.

The ***M*-estimator** corresponding to $\Psi(\cdot)$ is then

$$\hat{\theta} = \hat{\theta}(X_1, \dots, X_n) = \frac{\theta^* + \theta^{**}}{2},$$

where

$$\theta^* = \theta^*(X_1, \dots, X_n) = \sup\left\{ \theta \mid \sum_{i=1}^{n} \Psi(X_i - \theta) > 0 \right\}$$

and

$$\theta^{**} = \theta^{**}(X_1, \dots, X_n) = \inf\left\{ \theta \mid \sum_{i=1}^{n} \Psi(X_i - \theta) < 0 \right\}.$$

For any $\Psi(\cdot)$ satisfying (i) and (ii) there exists a $t_0 > 0$ such that $\Psi(t) > 0$ whenever $t > t_0$ and $\Psi(t) < 0$ whenever $t < -t_0$. Thus if $\theta < X_{(1)} - t_0$, where $X_{(1)}$ is the smallest X_i value, then $\Sigma_{i=1}^{n} \Psi(X_i - \theta) > 0$. Likewise, if $\theta > X_{(n)} + t_0$, where $X_{(n)}$ is the largest X_i value, then $\Sigma_{i=1}^{n} \Psi(X_i - \theta) < 0$. These facts, together with the fact that $\Sigma_{i=1}^{n} \Psi(X_i - \theta)$ is a nonincreasing function of θ, show that both θ^* and θ^{**} are well defined and finite. Hence Definition 7.4.9 produces a unique $\hat{\theta}$-value for any such $\Psi(\cdot)$. If $\Psi(\cdot)$ happens to be continuous and strictly increasing, then the *M*-estimator is the unique value satisfying

$$\sum_{i=1}^{n} \Psi\left(X_i - \hat{\theta}\right) = 0. \tag{7.4.10}$$

The concept of *M*-estimators was introduced by Huber (1964). The name derives from the fact that most maximum likelihood estimators are included in this general class. To see this, suppose the underlying distribution is continuous with density function $f(x - \theta)$. We find the maximum likelihood estimator of θ by maximizing

$$\ln\left[\prod_{i=1}^{n} f(X_i - \theta) \right].$$

For those distributions for which this can be accomplished by solving

$$\sum_{i=1}^{n} -\frac{f'\left(X_i - \hat{\theta}\right)}{f\left(X_i - \hat{\theta}\right)} = 0,$$

we set

$$\Psi_0(t) = -\frac{f'(t)}{f(t)}. \tag{7.4.11}$$

Then, under conditions (i) and (ii), the corresponding maximum likelihood estimator is an M-estimator. Insisting that $\Psi_0(t)$ have properties (i) and (ii) is the same as requiring $f(x)$ to be strongly unimodal $[-f'(t)/f(t)$ nondecreasing] and symmetric about zero.

In his paper, Huber (1964) was particularly interested in M-estimators derived from a $\Psi(\cdot)$ function of the form

$$\begin{aligned} \Psi_k(t) &= t \qquad \text{if } |t| \leqslant k, \\ &= k\,\text{sign}(t) \qquad \text{if } |t| > k, \end{aligned} \tag{7.4.12}$$

for some $k>0$, which corresponds to a $\rho(\cdot)$ function

$$\begin{aligned} \rho_k(t) &= \tfrac{1}{2}t^2 \qquad \text{if } |t| \leqslant k, \\ &= k|t| - \tfrac{1}{2}k^2 \qquad \text{if } |t| > k. \end{aligned}$$

For this $\rho(\cdot)$ function we see that the distance measure is like the square function in the middle but like the absolute value function in the extremes. Hence the corresponding M-estimator will tend to treat central observations much as does $\overline{X}$ but will de-emphasize the extreme observations in the fashion of median $(X_1,\ldots,X_n)$. (See Exercise 7.4.7.)

Note the similarity between the form of the M-estimator as given in Definition 7.4.9 and that of the Hodges–Lehmann estimator shown in (7.1.4). However, in the Hodges–Lehmann approach the estimator is derived from a *test statistic*, whereas the $\Psi(\cdot)$ function for an M-estimator is motivated by *distance measures* and the *maximum likelihood* estimation criterion. Nevertheless, the similarity of these two estimators carries over to the following result.

Theorem 7.4.13. *Let $\hat{\theta}(X_1,\ldots,X_n)$ denote an M-estimator as given in Definition* 7.4.9. *Then,*

(i) *$\hat{\theta}$ is an odd translation statistic,*

and

(ii) *if the underlying distribution is symmetric about θ, then $\hat{\theta}$ is symmetrically distributed about θ.*

Proof: Exercise 7.4.3. ■

In addition to the translation property of an estimator, as discussed in the previous theorem, another important consideration is how the estimator adjusts to changes in the scale of the underlying distribution. Specifically, suppose $Y=cX+d$, where $c>0$. Then, $E[Y]=cE[X]+d$, provided

$E[X]$ exists. It also follows that $\text{median}[Y] = c\,\text{median}[X] + d$. Thus, in general, we would want our location estimator $\hat{\theta}(X_1, \ldots, X_n)$ to have the property

$$\hat{\theta}(cX_1 + d, \ldots, cX_n + d) = c\hat{\theta}(X_1, \ldots, X_n) + d. \tag{7.4.14}$$

Both the sample mean and sample median have this property, but it is not possessed by all M-estimators as described in Definition 7.4.9. In particular, using Huber's $\Psi_k(\cdot)$, with $0 < k < \infty$, yields an M-estimator that does not satisfy (7.4.14). To ensure this desirable property, we must modify the working definition of an M-estimator for a location parameter to allow for a scale factor. For example, in the case where $\Psi(\cdot)$ is continuous and strictly increasing, we take $\hat{\theta}$ to satisfy

$$\sum_{i=1}^{n} \Psi\left(\frac{X_i - \hat{\theta}}{S}\right) = 0, \tag{7.4.15}$$

where S is a scale estimator satisfying

$$S(cx_1 + d, \ldots, cx_n + d) = cS(x_1, \ldots, x_n),$$

for all $x_1, \ldots, x_n$, d, and $c > 0$. (A similar modification is necessary for an arbitrary M-estimator.) The role of S could be played, for example, by the sample standard deviation. Another choice that has received much attention is

$$S^*(X_1, \ldots, X_n) = \frac{\text{median}\left[|X_1 - M|, \ldots, |X_n - M|\right]}{.6745}, \tag{7.4.16}$$

where $M = \text{median}\,(X_1, \ldots, X_n)$. The numerator of S^* is often called the **median absolute deviation** (MAD) estimator of scale. The number .6745 is included in the denominator so that S^* will be a consistent estimator of σ when the underlying distribution is normal.

Solving (7.4.15) may be difficult for certain choices of $\Psi(\cdot)$. Iterative schemes are often used to approximate the exact solution. One such iterative scheme uses (7.4.15) and rewrites it as

$$\sum_{i=1}^{n} \left[\frac{\Psi\left(\dfrac{X_i - \hat{\theta}}{S}\right)}{\left(\dfrac{X_i - \hat{\theta}}{S}\right)} \right] \left(\frac{X_i - \hat{\theta}}{S}\right) = 0.$$

If we let

$$w_i = \frac{\Psi\left(\dfrac{X_i - \hat{\theta}}{S}\right)}{\left(\dfrac{X_i - \hat{\theta}}{S}\right)},$$

this becomes

$$\sum_{i=1}^{n} w_i\left(\frac{X_i - \hat{\theta}}{S}\right) = 0$$

which is easily solved for $\hat{\theta}$, yielding

$$\hat{\theta} = \frac{\sum_{i=1}^{n} w_i X_i}{\sum_{i=1}^{n} w_i},$$

a weighted average of the X_is. However, this is not a proper solution, since the weights depend on $\hat{\theta}$. Thus the solution process must proceed iteratively, with the estimator of θ on the kth iteration being

$$\hat{\theta}_k = \frac{\sum_{i=1}^{n} w_{i,k-1} X_i}{\sum_{i=1}^{n} w_{i,k-1}}, \tag{7.4.17}$$

where the weights depend on the value of the location estimator in the previous iteration; specifically,

$$w_{i,k-1} = \frac{\Psi\left(\dfrac{X_i - \hat{\theta}_{k-1}}{S}\right)}{\left(\dfrac{X_i - \hat{\theta}_{k-1}}{S}\right)}, \qquad i = 1,\ldots,n.$$

Thus this solution method requires the specification of an initial estimator of θ, namely, $\hat{\theta}_0$. (The sample median is often used for this purpose.) However, we must specify the value of the weight when $X_i = \hat{\theta}_{k-1}$, particularly when using a starting value like the sample median. In such cases it is natural to set $w_{i,k-1}=\lim_{t\to 0}[\Psi(t)/t]$ whenever $X_i=\hat{\theta}_{k-1}$. (If this limit does not exist, the weight $w_{i,k-1}$ is taken to be 1, and all other weights are set equal to zero.) Usually, only a few iterations of the scheme described in (7.4.17) are sufficient to provide a good estimator. In particular, one estimator which has received considerable attention results from (7.4.17) by using the scale estimator S^* in (7.4.16), the starting estimator $\hat{\theta}_0=$ median$(X_1,\ldots,X_n)$ and

$$\Psi_{1.5}(t) = \min(1.5, \max(-1.5, t)),$$

which is merely (7.4.12) with $k=1.5$.

Since Huber's (1964) paper a great deal of attention has been focused on *M*-estimators, popularized, in part, by Andrews, Bickel, Hampel, Huber, Rogers, and Tukey (1972) and the survey paper by Huber (1972). Other types of $\Psi(\cdot)$ functions, including some that are not monotone, have been studied. See, for example, Andrews, et al. (1972), Beaton and Tukey (1974), and Andrews (1974). A theory for estimators based on a broad class of $\Psi(\cdot)$ functions is given by Collins (1976). In addition, *M*-estimation techniques have been developed for use in a variety of other problems, including regression (Huber (1973) and Andrews (1974)) and multivariate settings (Maronna (1976) and Huber (1977)). Interesting comparisons between several *M*-estimators of location parameters are given by Andrews, et al. (1972), Stigler (1976), and Boos and Serfling (1976).

We now make a slight transition to discuss the influence curve, an important concept in describing the robustness (or lack of it) of estimators. This tool plays a central role in the development of properties of *M*-estimators. To understand the purpose and nature of the influence curve we must first begin to think of estimators and their corresponding parameters as functionals. Let $X_1,\ldots,X_n$ be a random sample from some population with c.d.f. $F(x)$. The **empirical distribution** of these observations is often defined in terms of its c.d.f.

$$F_n(t) = \frac{1}{n}\left[\text{number of } X_i\text{s} \leqslant t\right], \qquad -\infty < t < \infty, \tag{7.4.18}$$

and corresponds to the discrete distribution that associates probability $1/n$ with each of the sample values $X_1,\ldots,X_n$. Let $T(\cdot)$ denote a real-valued functional defined on some subset of the set of all probability distribution c.d.f.'s such that the parameter of interest is

$$\gamma = T(F).$$

That is, the parameter of interest, namely, γ, is some function, namely, $T(\cdot)$, of the underlying distribution. One reasonable estimator of γ would then be obtained by substituting the empirical c.d.f. into this functional, yielding

$$\hat{\gamma} = T(F_n).$$

One of the simplest examples of this estimation technique concerns the **mean functional** defined by

$$\gamma = T_1(F) = \int_{-\infty}^{\infty} t\,dF(t) \tag{7.4.19}$$

for all distributions with finite first moments. The estimator corresponding to this functional is then

$$\begin{aligned}\hat{\gamma} = T_1(F_n) &= \int_{-\infty}^{\infty} t\,dF_n(t)\\ &= \sum_{i=1}^{n} \frac{X_i}{n} = \overline{X}.\end{aligned}$$

The **variance functional** is defined by

$$\gamma = T_2(F) = \int_{-\infty}^{\infty} t^2\,dF(t) - \left[\int_{-\infty}^{\infty} t\,dF(t)\right]^2, \tag{7.4.20}$$

and its estimator is

$$\begin{aligned}\hat{\gamma} = T_2(F_n) &= \int_{-\infty}^{\infty} t^2\,dF_n(x) - \left[\int_{-\infty}^{\infty} t\,dF_n(t)\right]^2\\ &= \frac{1}{n}\sum_{i=1}^{n} X_i^2 - \overline{X}^2 = \frac{1}{n}\sum_{i=1}^{n}\left(X_i - \overline{X}\right)^2,\end{aligned}$$

which is simply the sample variance with an n in the denominator instead of an $(n-1)$.

The **median functional** is more difficult, because it must be expressed implicitly. We seek the value $\gamma = T_3(F)$ that satisfies

$$F(T_3(F)) = \frac{1}{2}, \tag{7.4.21}$$

which has a solution (possibly nonunique) as long as F is continuous. To include discrete distributions and to give a unique value for all continuous

Fs, the definition must be extended to

$$\gamma = T_3(F) = \frac{1}{2}(\theta^* + \theta^{**}),$$

where $\theta^* = \sup\{x|F(x) < 1/2\}$ and $\theta^{**} = \inf\{x|F(x) > 1/2\}$. With this definition, we note that

$$\hat{\gamma} = T_3(F_n) = \text{median}(X_1, \ldots, X_n).$$

The ***M*-estimator functional** must also be expressed implicitly. For the location parameter problem described earlier in this section, $\gamma = T_4(F)$ represents a value that satisfies

$$\int_{-\infty}^{\infty} \Psi(t - T_4(F))\, dF(t) = 0, \qquad \textbf{(7.4.22)}$$

as long as either $\Psi(\cdot)$ or $F(\cdot)$ is continuous. Substituting $F_n(\cdot)$ into this equation shows that the estimator $\hat{\gamma} = T_4(F_n)$ should satisfy

$$\int_{-\infty}^{\infty} \Psi(t - T_4(F_n))\, dF_n(t) = \frac{1}{n} \sum_{i=1}^{n} \Psi(X_i - \hat{\gamma}) = 0,$$

which is the same as (7.4.10). (Definition 7.4.9 is merely an extension of this description of an M-estimator to provide a unique solution for $\hat{\gamma}$ over a broader range of $\Psi(\cdot)$ and $F(\cdot)$ combinations.)

Note that if the underlying distribution is symmetric about θ, we may write its c.d.f. as $G(x - \theta)$ with $G(x) = 1 - G(-x)$ for all x. Then

$$\int_{-\infty}^{\infty} \Psi(t - \theta)\, dF(t) = \int_{-\infty}^{\infty} \Psi(t - \theta)\, dG(t - \theta) = \int_{-\infty}^{\infty} \Psi(s)\, dG(s) = 0,$$

by the symmetry of the underlying distribution and the fact that $\Psi(s) = -\Psi(-s)$. Thus, from (7.4.22), we see that the associated M-estimator is estimating $T_4(F) = \gamma = \theta$, the point of symmetry of the underlying distribution.

With this view of estimators as functionals, we now proceed to define the influence curve of such a functional.

Definition 7.4.23. Let $T(\cdot)$ be a functional defined on some subset of probability distribution c.d.f.'s. Let x be an arbitrary real number and take $D_x(t)$ to be the c.d.f. of the distribution that puts probability 1 on the value x. The **influence curve (IC) of $T(\cdot)$ at $F(\cdot)$** is defined for each x by

$$\text{IC}_{T,F}(x) = \lim_{\epsilon \downarrow 0} \left[\frac{T[(1-\epsilon)F(\cdot) + \epsilon D_x(\cdot)] - T[F(\cdot)]}{\epsilon} \right].$$

We recognize that the influence curve is simply a type of derivative of the functional $T(\cdot)$ (more explicitly, it is a Volterra-type derivative). Since $(1-\epsilon)F(\cdot)+\epsilon D_x(\cdot)$ is a c.d.f. iff $0 \leqslant \epsilon \leqslant 1$, we see that the influence curve is essentially the right-hand derivative of $T_\epsilon \equiv T[(1-\epsilon)F(\cdot)+\epsilon D_x(\cdot)]$ at $\epsilon=0$. It represents the rate of change in the functional that results from adding to the distribution $F(\cdot)$ an infinitesimal point mass at the value x. Note that the derivative is evaluated at $F(\cdot)$ and not $F_n(\cdot)$. Thus since $F_n(\cdot)$ is a consistent estimator of $F(\cdot)$, it represents an *asymptotic* rate of change in the corresponding estimator.

In constructing influence curves we often use the fact that

$$\int k(t)d[(1-\epsilon)F(t)+\epsilon G(t)] = (1-\epsilon)\int k(t)dF(t) + \epsilon\int k(t)dG(t)$$

for any $0 \leqslant \epsilon \leqslant 1$ and any c.d.f.'s $F(\cdot)$ and $G(\cdot)$ for which the integrals on the right-hand side exist.

Example 7.4.24. Let $T_1(\cdot)$ be the mean functional defined in (7.4.19). Then

$$\begin{aligned} \mathrm{IC}_{\text{mean}, F}(x) &= \lim_{\epsilon \downarrow 0}\left[\frac{\int_{-\infty}^{\infty} td[(1-\epsilon)F(t)+\epsilon D_x(t)] - \int_{-\infty}^{\infty} tdF(t)}{\epsilon}\right] \\ &= \lim_{\epsilon \downarrow 0}\left[\frac{(1-\epsilon)\int_{-\infty}^{\infty} tdF(t)+\epsilon x - \int_{-\infty}^{\infty} tdF(t)}{\epsilon}\right] \\ &= x - \mu, \end{aligned}$$

where $\mu = \int_{-\infty}^{\infty} tdF(t)$ is the mean of $F(\cdot)$. Thus, as shown in Figure 7.4.25, the influence of adding to the distribution $F(\cdot)$ an infinitesimal point mass

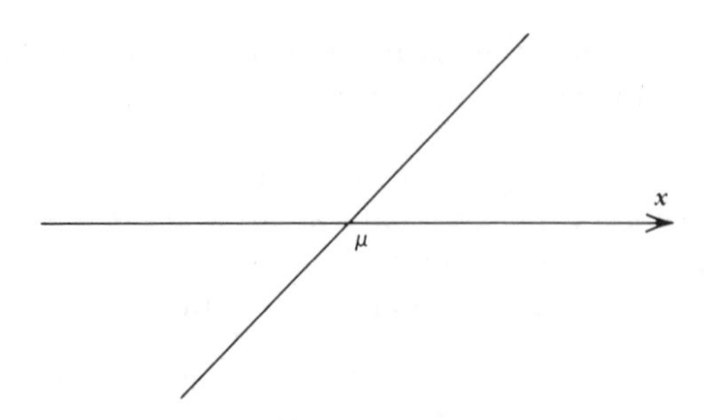

FIGURE 7.4.25.

at x is directly proportional to the distance between x and the mean of $F(\cdot)$. In particular, it follows that the influence of an extremely large or small x is great.

Before proceeding with the next example we first note something that will often simplify the derivation of an influence curve. Although $(1-\epsilon)F(\cdot)+\epsilon D_x(\cdot)$ is a c.d.f. only when $0 \leqslant \epsilon \leqslant 1$, the equation or expression defining $T[(1-\epsilon)F(\cdot)+\epsilon D_x(\cdot)]$ may still be valid even when $\epsilon<0$. As an illustration, consider the previous example in which

$$T_\epsilon \equiv T[(1-\epsilon)F(\cdot)+\epsilon D_x(\cdot)] = (1-\epsilon)\int_{-\infty}^{\infty} t\,dF(t) + \epsilon\int_{-\infty}^{\infty} t\,dD_x(t).$$

This is merely a linear function of ϵ and, *as an expression*, it still makes sense when $\epsilon<0$. In such cases the influence curve is given by

$$\mathrm{IC}_{T,F}(x) = \left.\frac{\partial T_\epsilon}{\partial \epsilon}\right|_{\epsilon=0} \equiv \frac{\partial}{\partial \epsilon} T[(1-\epsilon)F(\cdot)+\epsilon D_x(\cdot)]|_{\epsilon=0},$$

since, if this partial derivative exists at zero, it will be equal to the right-hand derivative at zero (which is the definition of the influence curve).

Example 7.4.26. Let $\Psi(t)$ satisfy conditions (i) and (ii) of Definition 7.4.9. In addition, assume that $\Psi'(t)$ exists at all but at most a countable number of t-values and that, when necessary, we can interchange differentiation and integration operations. (See, for example, the conditions in Theorem A.2.4.) The M-estimator functional is defined implicitly by equation (7.4.22). Hence, to find the influence curve of the M-estimator associated with $\Psi(t)$ we differentiate (with respect to ϵ) the equation

$$\int_{-\infty}^{\infty} \Psi(t-T_\epsilon)\,d[(1-\epsilon)F(t)+\epsilon D_x(t)] = 0,$$

which is the same as

$$(1-\epsilon)\int_{-\infty}^{\infty} \Psi(t-T_\epsilon)\,dF(t) + \epsilon\Psi(x-T_\epsilon) = 0.$$

Here T_ϵ again denotes $T[(1-\epsilon)F(\cdot)+\epsilon D_x(\cdot)]$. This differentiation yields

$$-\int_{-\infty}^{\infty} \Psi(t-T_\epsilon)\,dF(t) - (1-\epsilon)\left[\frac{\partial T_\epsilon}{\partial \epsilon}\right]\int_{-\infty}^{\infty} \Psi'(t-T_\epsilon)\,dF(t)$$
$$+\Psi(x-T_\epsilon) - \epsilon\left[\frac{\partial T_\epsilon}{\partial \epsilon}\right]\Psi'(x-T_\epsilon) = 0.$$

Setting $\epsilon=0$, using (7.4.22), and solving for $(\partial T_\epsilon/\partial\epsilon)|_{\epsilon=0}$ yields

$$\mathrm{IC}_{\Psi,F}(x)=\frac{\Psi(x-\theta)}{\int_{-\infty}^{\infty}\Psi'(t-\theta)\,dF(t)}.$$

Thus, the influence curve of such an M-estimator is proportional to its $\Psi(\cdot)$ function. To have an influence curve with designated properties, we merely choose a $\Psi(\cdot)$ with those properties. One particular choice, namely Huber's $\Psi_k(\cdot)$ defined in (7.4.12), has an influence curve as depicted in Figure 7.4.27.

Although the embryo of the influence curve can be found in earlier work by von Mises (1947) and others, a clear statement of this tool and its importance as a unifying concept of statistical robustness was first considered by Hampel (1968). In Hampel (1974) the development of this topic, including its history, is discussed in detail. Numerous important measures of robustness have been proposed by Hampel (1971, 1974), some of which can be observed from the influence curve. One such measure is the **gross error sensitivity**, denoted by $\gamma^*(T,F)$ and defined as

$$\gamma^*(T,F)=\sup_{-\infty<x<\infty}|\mathrm{IC}_{T,F}(x)|.$$

This is interpreted as the worst approximate influence that the addition of an infinitesimal point mass can have on the value of the associated estimator. A second measure of robustness is **local shift sensitivity**, the definition of which is

$$\lambda^*(T,F)=\sup_{-\infty<x\neq y<\infty}\left|\frac{\mathrm{IC}_{T,F}(x)-\mathrm{IC}_{T,F}(y)}{x-y}\right|.$$

It represents the worst possible effect on the estimator of slight shifts in the data, something that is particularly important in applications where there may be rounding or grouping of the observed values.

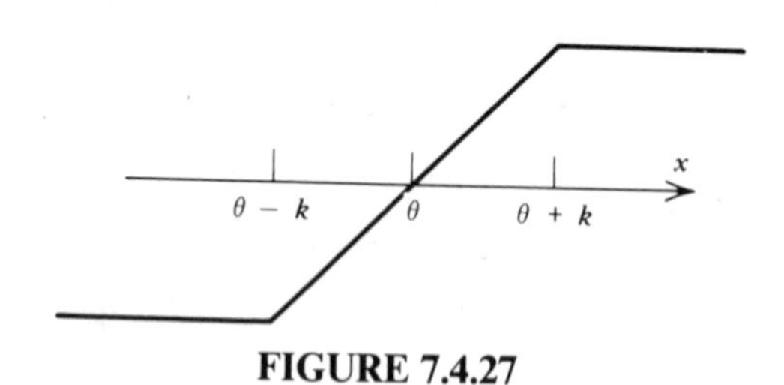

FIGURE 7.4.27

From Examples 7.4.24 and 7.4.26 we see that the local shift sensitivities for the sample mean and Huber's $\Psi_k(\cdot)$ M-estimator are $\lambda^*(\text{mean}, F) \equiv 1$ and $\lambda^*(H, F) = [F(k) - F(-k)]^{-1}$, respectively. This would indicate that both estimators (provided k for the M-estimator is reasonably large) are relatively insensitive to slight shifts in the data. Since $[F(k) - F(-k)] \leqslant 1$ for any $k > 0$ and c.d.f. $F(\cdot)$, the Huber estimator is always at least as sensitive to slight shifts as is the sample mean. The gross error sensitivity of Huber's $\Psi_k(\cdot)$ M-estimator is $\gamma^*(H, F) = k/[F(k) - F(-k)]$, while the corresponding value for the sample mean is $\gamma^*(\text{mean}, F) = +\infty$. This re-emphasizes the point that the sample mean is drastically influenced by very extreme observations. With a judicious choice of k (depending on the scale of the underlying distribution), we see that Huber's $\Psi_k(\cdot)$ M-estimator is not greatly affected by extreme values among the data.

Verifying the asymptotic normality of certain standardized estimators is another important use of the influence curve. In particular, subject to certain restrictions on $\Psi(\cdot)$, an M-estimator $\hat{\theta}$ of the parameter θ will be such that $\sqrt{n}\,(\hat{\theta} - \theta)$ can be approximated by the random variable

$$V_n = \frac{1}{\sqrt{n}} \sum_{i=1}^{n} \mathrm{IC}_{\Psi,F}(X_i),$$

in the sense that $\sqrt{n}\,(\hat{\theta} - \theta) - V_n$ goes to zero in probability as $n \to \infty$. [See Filippova (1962).] The Central Limit Theorem and Theorem 3.2.12 then show that $\sqrt{n}\,(\hat{\theta} - \theta)$ has a limiting normal distribution with mean 0 and variance equal to

$$\mathrm{Var}[\mathrm{IC}_{\Psi,F}(X)] = \frac{\int_{-\infty}^{\infty} \Psi^2(t-\theta)\, dF(t)}{\left[\int_{-\infty}^{\infty} \Psi'(t-\theta)\, dF(t)\right]^2},$$

since, from (7.4.22),

$$E[\mathrm{IC}_{\Psi,F}(X)] = \frac{\int_{-\infty}^{\infty} \Psi(t-\theta)\, dF(t)}{\int_{-\infty}^{\infty} \Psi'(t-\theta)\, dF(t)} = 0.$$

A direct proof of the asymptotic normality of M-estimators can be found in Huber (1964).

A heuristic justification for the fact that $[n(\hat{\theta}-\theta)-V_n]$ goes to zero in probability can be seen as follows. Note that the equation

$$k(\hat{\theta}) \equiv \sum_{i=1}^{n} \Psi(X_i-\hat{\theta}) = 0$$

can be re-expressed as

$$\sum_{i=1}^{n} \Psi(X_i-\theta) - (\hat{\theta}-\theta) \sum_{i=1}^{n} \Psi'(X_i-\theta^*) = 0$$

by taking a Taylor series expansion of $k(\cdot)$ (with a remainder term) around θ. Here θ^* denotes some value between $\hat{\theta}$ and θ. It follows that

$$\sqrt{n}\,(\hat{\theta}-\theta) = \frac{1}{\sqrt{n}} \frac{\sum_{i=1}^{n} \Psi(X_i-\theta)}{\left[\frac{1}{n}\sum_{i=1}^{n} \Psi'(X_i-\theta^*)\right]}.$$

Since the denominator, $(1/n)\Sigma_{i=1}^{n}\Psi'(X_i-\theta^*)$, converges in probability to $E[\Psi'(X-\theta)]$, it follows that $\sqrt{n}\,(\hat{\theta}-\theta)$ has the same limiting distribution as

$$\frac{1}{\sqrt{n}} \sum_{i=1}^{n} \frac{\Psi(X_i-\theta)}{E[\Psi'(X-\theta)]} = \frac{1}{\sqrt{n}} \sum_{i=1}^{n} \mathrm{IC}_{\Psi,F}(X_i) = V_n.$$

Exercises 7.4.

7.4.1. Derive a result similar to (7.4.6) which shows that whenever n is even, say, $n=2k$, any value M such that $X_{(k)} \leqslant M \leqslant X_{(k+1)}$ satisfies

$$\sum_{i=1}^{n} |X_i - M| \leqslant \sum_{i=1}^{n} |X_i - \theta|.$$

7.4.2. Assume that $F(\cdot)$ has a continuous, unimodal density that is symmetric about 0. Hampel (1968) shows the following two results:

(I) Let $k^* = [\![n\alpha]\!]$, for $0<\alpha<\frac{1}{2}$, and form

$$V(X_1,\ldots,X_n) = \frac{X_{(k^*+1)} + X_{(k^*+2)} + \cdots + X_{(n-k^*)}}{n-2k^*}.$$

$V(X_1,\ldots,X_n)$ is called the **α-trimmed mean**. Its associated influence curve is

$$\begin{aligned}
\mathrm{IC}_{V,F}(x) &= \frac{F^{-1}(\alpha)}{1-2\alpha}, \qquad \text{if } x < F^{-1}(\alpha),\\
&= \frac{x}{1-2\alpha}, \qquad \text{if } F^{-1}(\alpha) \leqslant x \leqslant F^{-1}(1-\alpha),\\
&= \frac{F^{-1}(1-\alpha)}{1-2\alpha}, \qquad \text{if } F^{-1}(1-\alpha) < x.
\end{aligned}$$

(II) Let $k^* = [\![n\alpha]\!]$, for $0<\alpha<1/2$, and form

$$W(X_1,\ldots,X_n) = \frac{(k^*+1)X_{(k^*+1)} + X_{(k^*+2)} + \cdots + X_{(n-k^*-1)} + (k^*+1)X_{(n-k^*)}}{n}.$$

This estimator is known as the **α-Winsorized mean**. Its associated influence curve is

$$\begin{aligned}
\mathrm{IC}_{W,F}(x) &= x \qquad \text{if } |x| \leqslant F^{-1}(1-\alpha),\\
&= \left[F^{-1}(1-\alpha) + \frac{\alpha}{f[F^{-1}(1-\alpha)]}\right]\mathrm{sign}(x) \qquad \text{if } |x| > F^{-1}(1-\alpha).
\end{aligned}$$

Suppose that the underlying distribution is a double exponential with density $f(x)=\frac{1}{2}e^{-|x|}$. Draw the influence curves associated with the following estimators on the same graph and determine the gross error sensitivity, γ^*, and the local shift sensitivity, λ^*, for each of these cases.

(i) an α-trimmed mean with $\alpha=1/4$,
(ii) an α-Windsorized mean with $\alpha=1/4$,
(iii) Huber's $\Psi_k(\cdot)$ M-estimator with $k=\ln(2)$.

7.4.3. Prove Theorem 7.4.13.

7.4.4. Let $x_1,\ldots,x_n$ denote the observed values of a random sample and let $F_n(t)$ be the empirical c.d.f. based on these n values. Consider the $n+1$ values $x_1,\ldots,x_n$, x and let $F^*_{n+1}(t)$ represent the empirical c.d.f. of these $n+1$ values. The finite sample size version of the influence curve for a functional $T(\cdot)$ (sometimes called a **sensitivity curve**) is defined to be

$$S(x) = T(F^*_{n+1}) - T(F_n).$$

For $n=5$, consider the observed sample data $x_1=1.2$, $x_2=4.5$, $x_3=-1.8$, $x_4=3.6$, and $x_5=2.2$. Construct and draw a graph of the sensitivity curves for (i) the sample mean and (ii) the sample median.

7.4.5. Suppose $X_1,\ldots,X_n$ is a random sample from a continuous population with c.d.f. $F(x)$. The relationship of the underlying distribution with respect to zero is sometimes assessed by means of the parameter $\gamma=P[X_1+X_2>0]$.

(a) Describe this parameter as a functional of $F(\cdot)$.

(b) Substitute the empirical c.d.f. $F_n(\cdot)$ into the functional in **(a)** and express the resulting estimator $\hat{\theta}$ in simple terms. [Hint: The estimator may be expressed in terms of W^+, the Wilcoxon signed rank test statistic.]

7.4.6. Let $X_1,\ldots,X_n$ denote a random sample from a continuous population with c.d.f. $F(x)$. One characteristic of this population is its spread around zero, which could be measured with the parameter

$$\gamma = E\big[|X_1|\big].$$

(a) Describe this parameter in terms of the c.d.f. $F(\cdot)$.

(b) What is the estimator of γ that corresponds to this functional? Describe it as simply as possible.

(c) Derive the influence curve of this functional.

(d) Draw a typical graph of this influence curve.

(e) What are the gross error sensitivity γ^* and the local shift sensitivity λ^* of this functional?

7.4.7. Consider the data points $x_1=0$, $x_2=2.5$, $x_3=1$, $x_4=1$, and $x_5=40$. Compute

(a) Sample mean.

(b) Sample median.

(c) Huber's M-estimator using $\Psi_k(\cdot)$ as in (7.4.12) with $k=1.5$ and the scale estimator S^* as defined in (7.4.16). Evaluate this Huber estimator by performing two iterations of the scheme described in (7.4.17) using $\hat{\theta}_0=\text{median}(X_1,\ldots,X_5)$ as the initial estimator.

7.4.8. Let $X_1,\ldots,X_n$ denote a random sample from a population with c.d.f. $F(x)$. A common measure of spread is

$$\gamma = \text{Var}(X_1) = E\big[(X_1-\mu)^2\big],$$

where $\mu = E(X_1)$. The variance functional and the corresponding estimator, $\hat{\gamma}$, were discussed in text [see (7.4.20)].

(a) Derive the influence curve of this functional.

(b) Draw a typical graph of this influence curve.

(c) What are the gross error sensitivity γ^* and the local shift sensitivity λ^* of this functional?

7.4.9. Suppose that the median functional is defined by equation (7.4.21). Use this to verify that the influence curve for the median functional is given by

$$\mathrm{IC}_{\text{median}, F}(x) = \frac{\frac{1}{2}\operatorname{sign}(x)}{f(0)},$$

when $F(\cdot)$ is the c.d.f. of a continuous distribution with density $f(\cdot)$ that is positive for all x and symmetric about 0. Determine the gross error sensitivity γ^* and the local shift sensitivity λ^* of this estimator.

7.4.10. With an M-estimator defined by (7.4.15), show that the location and scale property (7.4.14) is satisfied.

7.4.11. Let $\rho(t) = t^4$ and find an expression for the corresponding M-estimator of a location parameter. [See (7.4.4).]

7.4.12. Let $X_1, \ldots, X_n$ denote a random sample from a continuous population that is symmetrically distributed about 0. Suppose γ, a measure of the spread of the distribution about 0, is implicitly defined by

$$E\left[\Psi\left(\frac{X_1}{\gamma}\right)\right] = 1,$$

where $\Psi(\cdot)$ is a function satisfying

(i) $\Psi(t) = \Psi(-t)$ for all t

(ii) $\Psi(t)$ is continuous and nondecreasing for all $t \geqslant 0$

(iii) $\Psi(0) = 0$ and $\Psi(t_0) > 1$ for some $t_0 > 0$.

(a) Express $\gamma = T(F)$ as a functional that is implicitly defined in terms of $F(\cdot)$.

(b) Give the implicit definition of the estimator that corresponds to this functional.

(c) Derive the influence curve for this functional.

7.5. Discussion

This chapter includes only a few of the possible approaches and techniques for robust estimation. Indeed, a whole book (possibly several) could be devoted to this topic. In particular, we have omitted the important class of estimators based on linear functions of order statistics, for which general theory has been developed by Chernoff, Gastwirth, and Johns (1967), Bickel (1967) and Stigler (1973). In addition, we did not include a discussion of Bayesian nonparametric methods based on Dirichlet priors. This very elegant theory, begun by Ferguson (1973), has spawned a number of interesting solutions to estimation problems. See, for example, Doksum (1972), Ferguson (1973, 1974), Antoniak (1974), Goldstein (1975), Korwar and Hollander (1976), and Campbell and Hollander (1978).

Another important class of robust estimators are the adaptive ones. Examples are given by Hogg (1967), Birnbaum and Miké (1970), van Eeden (1970), Jaeckel (1971), and Takeuchi (1971). See Hogg (1974) for a survey of this approach.

8 LINEAR RANK STATISTICS UNDER THE NULL HYPOTHESIS

This chapter introduces a general class of linear rank statistics that includes the Mann–Whitney–Wilcoxon test statistic discussed in Section 2.3. Such statistics play a prominent role in distribution-free statistical methods, as is demonstrated in succeeding chapters. In this chapter we develop distributional properties of these linear rank statistics for assumptions that commonly hold under various null hypotheses. Such properties as their mean, variance, and symmetry of their distribution are considered. In addition, we prove an important asymptotic normality theorem due to Hájek (1961). In later chapters we study rank procedures for several important problems using statistics that are members of the general class considered in this chapter.

8.1. Linear Rank Statistics

In Section 2.3 we introduced the rank sum statistic

$$W = \sum_{i=1}^{n} R_i,$$

where R_i is the rank of Y_i among $X_1, \ldots, X_m, Y_1, \ldots, Y_n$. There are two primary reasons for the importance of this test statistic. Under the associated null hypothesis H_0 the random variables $X_1, \ldots, X_m, Y_1, \ldots, Y_n$ are independent and identically distributed. Since W is a rank statistic, Corollary 2.3.6 shows that it is distribution-free (under H_0) over the class of continuous populations. Thus one important property of W as a test statistic is that it maintains the desired α-level over a very broad class of distributional models. A second important property of the statistic W is that its power is excellent for detecting a shift in a medium-tailed distribution like the normal or the logistic. In addition, if a broad range of underlying distributions is possible, we noted in Sections 4.1 and 5.4 that this test has good power to detect a shift alternative.

We now describe a general class of rank statistics which includes W and other statistics that share many of the positive attributes of W. Let $\mathbf{R}^* = (R_1^*, \ldots, R_N^*)$ denote a vector of ranks; that is, $\mathbf{R}^*$ assumes only values in the collection of all $N!$ permutations of the integers $(1, \ldots, N)$.

Definition 8.1.1. Let $a(1), \ldots, a(N)$ and $c(1), \ldots, c(N)$ be two sets of N constants such that the numbers within each set are not all the same. A statistic of the form

$$S = \sum_{i=1}^{N} c(i) a(R_i^*)$$

is called a **linear rank statistic**. The constants $a(1), \ldots, a(N)$ are called the **scores**, and $c(1), \ldots, c(N)$ are termed the **regression constants**.

The choice of regression constants for a linear rank statistic is usually dictated by the nature of the particular testing problem and thus may not be under our control. On the other hand, we are frequently at liberty to choose the scores so as to achieve desirable power properties, which is discussed in later chapters. For the moment, let us consider a few examples of linear rank statistics.

In two-sample problems $\mathbf{R}^*$ is the rank vector of the variables $X_1, \ldots, X_m, Y_1, \ldots, Y_n$, with $N = m + n$. That is, $R_1^*, \ldots, R_m^*$ denote the joint ranks of $X_1, \ldots, X_m$, respectively, and $R_{m+1}^*, \ldots, R_{m+n}^*$ denote the joint ranks of $Y_1, \ldots, Y_n$, respectively. If

$$\begin{aligned} c(i) &= 0, \qquad i = 1, \ldots, m, \\ &= 1, \qquad i = m+1, \ldots, N, \end{aligned} \tag{8.1.2}$$

then

$$S = \sum_{i=1}^{N} c(i) a(R_i^*) = \sum_{j=1}^{n} a(R_{m+j}^*)$$

is the sum of the scores associated with the ranks of the Ys. The constants $c(i)$ defined in (8.1.2) are referred to as the **two-sample regression constants**.

If, in addition, we let

$$a(i) = i, \qquad i = 1,\ldots,N, \tag{8.1.3}$$

S becomes

$$\sum_{j=1}^{n} R^*_{m+j}, \tag{8.1.4}$$

which is the rank sum statistic W of the Mann–Whitney–Wilcoxon test. The scores $a(1),\ldots,a(N)$ defined in (8.1.3) are called the **Wilcoxon scores**.

A different choice of $a(1),\ldots,a(N)$ in the two-sample problem will yield a test statistic with different properties. For example, let

$$\begin{aligned} a(i) &= 0 \qquad \text{if } i \leqslant \frac{N+1}{2}, \\ &= 1 \qquad \text{if } i > \frac{N+1}{2}. \end{aligned} \tag{8.1.5}$$

Using these $a(i)$'s together with the two-sample regression constants produces

$$\begin{aligned} S &= \sum_{j=1}^{n} a(R^*_{m+j}) \\ &= [\text{the number of } Y_j\text{s greater than the combined sample median}]. \end{aligned} \tag{8.1.6}$$

This is the test statistic for a two-sample **median test** attributed to Mood (1950) and Westenberg (1948), and the scores in (8.1.5) are appropriately called the **median scores**.

In other settings variations on the regression constants are also important. For example, suppose that $Z_1,\ldots,Z_N$ are N independent, continuous random variables and that R^*_i is the rank of Z_i among $Z_1,\ldots,Z_N$. A problem that is often of interest is to test

$$H_0\colon Z_1,\ldots,Z_N \text{ are i.i.d.}$$

versus

$$H_1\colon Z_i \text{ is stochastically smaller than } Z_{i+1},\ i = 1,\ldots,N-1.$$

To detect this "increasing trend" alternative, an intuitive test statistic is created by correlating the ranks of $Z_1,\ldots,Z_N$ with the integers $1,2,\ldots,N$. Thus, by forming the product moment correlation coefficient on the pairs $(1,R_1^*),\ldots,(N,R_N^*)$, we obtain

$$\hat{\rho} = \frac{\sum_{i=1}^{N}\left(R_i^* - \frac{N+1}{2}\right)\left(i - \frac{N+1}{2}\right)}{\sqrt{\sum_{j=1}^{N}\left(R_j^* - \frac{N+1}{2}\right)^2 \sum_{i=1}^{N}\left(i - \frac{N+1}{2}\right)^2}},$$

since $(1/N)\Sigma_{i=1}^{N}R_i^* = (1/N)\Sigma_{i=1}^{N}i = (N+1)/2$. Noting that

$$\sum_{i=1}^{N}\left(R_i^* - \frac{N+1}{2}\right)^2 = \sum_{i=1}^{N}\left(i - \frac{N+1}{2}\right)^2 = \frac{N(N+1)(N-1)}{12},$$

we see that

$$\hat{\rho} = \frac{12}{N(N+1)(N-1)}\left[\sum_{i=1}^{N} iR_i^* - \frac{N(N+1)^2}{4}\right]. \qquad \textbf{(8.1.7)}$$

Thus the rank test of H_0 based on $\hat{\rho}$ is equivalent to one based on the linear rank statistic

$$S = \sum_{i=1}^{N} iR_i^*, \qquad \textbf{(8.1.8)}$$

in which the regression constants are $c(i) = i$ for $i = 1,\ldots,N$, and the scores are the Wilcoxon ones given in 8.1.3.

In this chapter we do not present a detailed account of the use of linear rank statistics in any particular testing problem. Such examples of their use are explored in later chapters. Instead, we develop here the general properties of linear rank statistics under their null hypothesis. We use the expression "**null hypothesis**" to refer to any set of assumptions that will result in the rank vector, $\mathbf{R}^*$, having a uniform distribution over $\mathcal{R}$, the set of permutations of the integers $1,\ldots,N$. Corollary 2.3.4 showed that under such a null hypothesis we have

$$P[R_i^* = r] = \frac{1}{N}, \qquad r = 1,\ldots,N, \qquad \textbf{(8.1.9)}$$

and if $i \neq j$,

$$\begin{aligned} P[R_i^* = r, R_j^* = s] &= \frac{1}{N(N-1)} \qquad r \neq s = 1,\ldots,N, \\ &= 0, \qquad \text{otherwise}. \end{aligned} \qquad \textbf{(8.1.10)}$$

These facts yield the following results.

Lemma 8.1.11. *Let $a(1),\ldots,a(N)$ denote a set of N constants. Then, if $\mathbf{R}^*$ is uniformly distributed over $\mathcal{R}$,*

(i)
$$E[a(R_i^*)] = \bar{a}, \qquad \text{for } i = 1,\ldots,N,$$

(ii)
$$\operatorname{Var}[a(R_i^*)] = \frac{1}{N}\sum_{k=1}^{N}(a(k)-\bar{a})^2, \qquad \text{for } i = 1,\ldots,N,$$

and

(iii)
$$\operatorname{Cov}[a(R_i^*),a(R_j^*)] = \frac{-1}{N(N-1)}\sum_{k=1}^{N}(a(k)-\bar{a})^2, \quad \text{for } i \neq j = 1,\ldots,N,$$

where $\bar{a}=(1/N)\Sigma_{i=1}^{N}a(i)$.

Proof:

(i)
$$E[a(R_i^*)] = \sum_{k=1}^{N} a(k)P[R_i^*=k] = \bar{a}$$

(ii)
$$\operatorname{Var}[a(R_i^*)] = \sum_{k=1}^{N}(a(k)-E[a(R_i^*)])^2 P[R_i^*=k]$$
$$= (1/N)\sum_{k=1}^{N}(a(k)-\bar{a})^2$$

(iii) If $i \neq j$,

$$\operatorname{Cov}[a(R_i^*),a(R_j^*)] = \sum_{h\neq k}^{N}\sum^{N}\{a(h)-\bar{a}\}\{a(k)-\bar{a}\}P[R_i^*=h,R_j^*=k]$$
$$= \frac{1}{N(N-1)}\sum_{h\neq k}^{N}\sum^{N}\{a(h)-\bar{a}\}\{a(k)-\bar{a}\}$$
$$= \frac{1}{N(N-1)}\left\{\left[\sum_{k=1}^{N}(a(k)-\bar{a})\right]^2 - \sum_{k=1}^{N}(a(k)-\bar{a})^2\right\}$$
$$= \frac{-1}{N(N-1)}\sum_{k=1}^{N}(a(k)-\bar{a})^2,$$
$$\text{since } \sum_{k=1}^{N}(a(k)-\bar{a}) = 0. \quad \blacksquare$$

Using this lemma we now establish the null mean and variance of a linear rank statistic.

Theorem 8.1.12. *Let S denote a linear rank statistic with regression constants $c(1),\dots,c(N)$ and scores $a(1),\dots,a(N)$. If $\mathbf{R}^*$ is uniformly distributed over $\mathcal{R}$, then*

$$\text{(i)} \qquad E[S] = N\bar{c}\bar{a}$$

and

$$\text{(ii)} \qquad \operatorname{Var}[S] = \frac{1}{(N-1)}\left[\sum_{i=1}^{N}(c(i)-\bar{c})^2\right]\left[\sum_{k=1}^{N}(a(k)-\bar{a})^2\right],$$

where $\bar{a}=(1/N)\Sigma_{i=1}^{N}a(i)$ and $\bar{c}=(1/N)\Sigma_{i=1}^{N}c(i)$.

Proof:

$$\text{(i)} \qquad E[S] = \sum_{i=1}^{N} c(i)E[a(R_i^*)] = N\bar{c}\bar{a}$$

$$\begin{aligned}
\text{(ii)} \qquad \operatorname{Var}[S] &= \sum_{i=1}^{N} c^2(i)\operatorname{Var}[a(R_i^*)] \\
&\quad + \sum_{i\neq j}^{N}\sum^{N} c(i)c(j)\operatorname{Cov}[a(R_i^*),a(R_j^*)] \\
&= \sum_{i=1}^{N}\frac{c^2(i)}{N}\left\{\sum_{k=1}^{N}(a(k)-\bar{a})^2\right\} \\
&\quad - \sum_{i\neq j}^{N}\sum^{N}\frac{c(i)c(j)}{N(N-1)}\left\{\sum_{k=1}^{N}(a(k)-\bar{a})^2\right\} \\
&= \frac{1}{N(N-1)}\left[(N-1)\sum_{i=1}^{N}c^2(i)\right. \\
&\quad \left. - \sum_{i\neq j}^{N}\sum^{N}c(i)c(j)\right]\left[\sum_{k=1}^{N}(a(k)-\bar{a})^2\right] \\
&= \frac{1}{(N-1)}\left[\sum_{i=1}^{N}(c(i)-\bar{c})^2\right]\left[\sum_{k=1}^{N}(a(k)-\bar{a})^2\right],
\end{aligned}$$

where the last step follows because

$$(N-1)\sum_{i=1}^{N} c^2(i) - \sum_{i\neq j}^{N}\sum^{N} c(i)c(j) = N\sum_{i=1}^{N} c^2(i) - \left(\sum_{i=1}^{N} c(i)\right)^2$$
$$= N\left[\sum_{i=1}^{N} c^2(i) - N\bar{c}^2\right]$$
$$= \left[N\sum_{i=1}^{N} (c(i)-\bar{c})^2\right]. \quad \blacksquare$$

Note that, as used in the preceding proof, the traditional computation formula for variances yields

$$\sum_{i=1}^{N} (c(i)-\bar{c})^2 = \sum_{i=1}^{N} c^2(i) - N\bar{c}^2,$$

with a similar computation formula for the $a(\cdot)$'s.

Remark 8.1.13. Let S denote a linear rank statistic with regression constants $c(1),\ldots,c(N)$ and scores $a(1),\ldots,a(N)$. Let $b_N>0$ and $-\infty<d_N<\infty$ be constants that (possibly) depend on N. Consider new scores defined by

$$a'(i) = b_N a(i) + d_N, \qquad i = 1,\ldots,N,$$

and the associated linear rank statistic

$$S' = \sum_{i=1}^{N} c(i)a'(R_i^*).$$

Noting the relationship $S'=b_N S+N\bar{c}d_N$, it follows that

$$E[S'] = b_N E[S] + N\bar{c}d_N$$

and

$$\operatorname{Var}[S'] = b_N^2 \operatorname{Var}[S].$$

Moreover, for any test of hypothesis based on S there is an equivalent test based on S', and vice versa. This observation points out a degree of flexibility in the specification of scores for conducting a test of hypothesis based on a linear rank statistic. Thus, as previously noted, basing a test of hypothesis on the estimated correlation $\hat{\rho}$ in equation (8.1.7) is equivalent to basing the test on the linear rank statistic S in (8.1.8).

Exercises 8.1.

8.1.1. Find the null mean and variance of the linear rank statistic with regression constants $c(i)=i$ and scores $a(i)=i^2$, $i=1,\ldots,N$. [Hint: See A.4.6 of the Appendix.]

8.1.2. Consider the linear rank statistic S with the two-sample regression constants defined in (8.1.2) and the **Savage scores** $a(i)=\Sigma_{j=N+1-i}^{N}(1/j)$. Show that under the null hypothesis H_0,

$$E[S] = n \text{ and } \operatorname{Var}[S] = \frac{mn}{(N-1)}\left[1-\frac{1}{N}\sum_{j=1}^{N}(1/j)\right].$$

8.1.3. Find the null mean and variance of the linear rank statistic $S=\Sigma_{i=1}^{N}c(i)a(R_i^*)$, where

$$\begin{aligned} c(i) &= i-1-q && \text{if } i=1,\ldots,q \\ &= 0 && \text{if } i=q+1,\ldots,N \end{aligned}$$

for $q=[\![(N+1)/2]\!]$ and where

$$\begin{aligned} a(i) &= -1 && \text{if } i \leqslant \frac{N+1}{3} \\ &= 0 && \text{if } \frac{N+1}{3} < i < \frac{2}{3}(N+1) \\ &= 1 && \text{if } i \geqslant \frac{2}{3}(N+1). \end{aligned}$$

8.1.4. Consider linear rank statistics $S=\Sigma_{i=1}^{N}c(i)a(R_i^*)$ and $S'=\Sigma_{i=1}^{N}c'(i)a'(R_i^*)$, where $a'(i)=b_N a(i)+d_N$ and $c'(i)=b_N^* c(i)+d_N^*$, $i=1,\ldots,N$, with $b_N>0$ and $b_N^*>0$.

(a) Express S' as a function of S and find the mean and variance of S' in terms of the mean and variance of S.

(b) If you conduct a test that rejects for $S>12$, what is the equivalent rejection region based on S'?

8.1.5. Let $S=\sum_{i=1}^{N}c(i)a(R_i^*)$ and $S'=\sum_{i=1}^{N}c'(i)a'(R_i^*)$ be any two linear rank statistics. Show that when $\mathbf{R}^*$ is uniformly distributed over $\mathcal{R}$,

$$\operatorname{Cov}(S,S') = \frac{1}{(N-1)}\sum_{i=1}^{N}(c(i)-\bar{c})(c'(i)-\bar{c}')\sum_{j=1}^{N}(a(j)-\bar{a})(a'(j)-\bar{a}').$$

8.1.6. Use the result of Exercise 8.1.5 to find an expression for the null hypothesis correlation coefficient between S and S' when $c(i)=c'(i)=i$ and $a(i)=i$ for $i=1,\ldots,N$, and $a'(i)=-1$, 0 or 1 as $i<$, $=$ or $>(N+1)/2$. Evaluate this expression when $N=10$.

8.1.7. Show that if

$$a(i)+a(N+1-i)=k \qquad i=1,\ldots,N$$

for some constant k and

$$a'(i)=a'(N+1-i) \qquad i=1,\ldots,N,$$

then the two linear rank statistics S and S' as defined in Exercise 8.1.5 are uncorrelated under H_0.

8.1.8. Let $S=\sum_{i=1}^{N}c(i)a(R_i^*)$. If $\mathbf{R}^*$ is uniformly distributed over $\mathcal{R}$, show that

$$E[\{S-E(S)\}^3] = \frac{N}{(N-1)(N-2)}\sum_{i=1}^{N}(c(i)-\bar{c})^3\sum_{j=1}^{N}(a(j)-\bar{a})^3.$$

8.2. Distributional Properties

Under null hypothesis assumptions $\mathbf{R}^*$ assumes each of the $N!$ permutations of the integers $(1,\ldots,N)$ with probability $1/N!$; that is, it has a uniform distribution over $\mathcal{R}$. Thus it is natural in such a setting to use the structure of these permutations in describing properties of rank statistics. We define the **composition** of two permutations as follows: If $\mathbf{r}=(r_1,\ldots,r_N)\in\mathcal{R}$ and $\mathbf{s}=(s_1,\ldots,s_N)\in\mathcal{R}$, then

$$\mathbf{r}\circ\mathbf{s}=(r_{s_1},r_{s_2},\ldots,r_{s_N}). \qquad \textbf{(8.2.1)}$$

It specifies a rearrangement of $\mathbf{r}$ that begins with the s_1th element of $\mathbf{r}$. For example, if $N=4$,

$$(2,4,3,1)\circ(1,3,4,2)=(2,3,1,4)$$

and

$$(1,3,4,2)\mathbf{o}(2,4,3,1) = (3,2,4,1).$$

(As this example shows, **ros** is not in general equal to **sor**.)

In what follows we discuss functions $k(\mathbf{r})$ which map $\mathfrak{R}$ onto $\mathfrak{R}$ in a one-to-one fashion. That is,

(i) For every $\mathbf{r} \in \mathfrak{R}$, $k(\mathbf{r}) \in \mathfrak{R}$

and **(8.2.2)**

(ii) For every $\mathbf{s} \in \mathfrak{R}$ there is an $\mathbf{r}$ in $\mathfrak{R}$, such that $k(\mathbf{r}) = \mathbf{s}$.

Since $\mathfrak{R}$ is finite, we can picture such a function as taking each $\mathbf{r}$ into a unique element $\mathbf{s}$ of $\mathfrak{R}$. (Any $\mathbf{r}$ may be mapped into itself.) One function of this type for $N=3$ is shown in Figure 8.2.3. Each such one-to-one function has an inverse function which merely reverses the direction of the mapping and is also one-to-one.

Theorem 8.2.4. *If $k(\cdot)$ is a one to one function mapping $\mathfrak{R}$ onto $\mathfrak{R}$ and if* $\mathbf{R}^*$ *is uniformly distributed over $\mathfrak{R}$, then*

$$\mathbf{S}^* = k(\mathbf{R}^*)$$

also has a uniform distribution over $\mathfrak{R}$.

Proof: If $k^{-1}(\cdot)$ denotes the inverse function of $k(\cdot)$, then $\{\mathbf{S}^*=\mathbf{s}\}$ is equivalent to $\{\mathbf{R}^*=k^{-1}(\mathbf{s})\}$. Thus

$$P[\mathbf{S}^*=\mathbf{s}] = P[\mathbf{R}^*=k^{-1}(\mathbf{s})] = \frac{1}{N!}.$$

Since this holds for each $\mathbf{s}\in\mathfrak{R}$, the proof is complete. ■

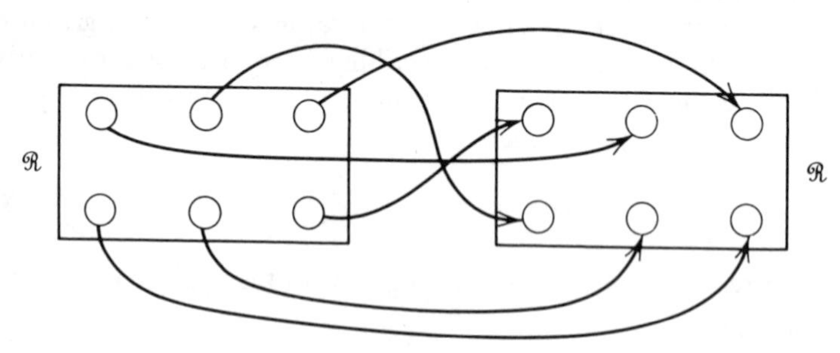

FIGURE 8.2.3.

The importance of the preceding theorem can be illustrated by considering a particular one-to-one function. For each $\mathbf{r}=(r_1,r_2,\ldots,r_N)\in\mathfrak{R}$ define a vector

$$\mathbf{d} = k_1(\mathbf{r}) \tag{8.2.5}$$

such that d_i is the position of i in the vector $\mathbf{r}$, for $i=1,\ldots,N$. For example, let $N=6$ and $\mathbf{r}=(4,1,6,2,3,5)$. Then $d=(2,4,5,1,6,3)$, since 1 is in the second position in $\mathbf{r}$, 2 is in the fourth position, and so on. To show that associating such a $\mathbf{d}$ with each $\mathbf{r}$ in $\mathfrak{R}$ defines a one-to-one function on $\mathfrak{R}$, note first that $\mathbf{d}\in\mathfrak{R}$ since each integer in $\{1,2,\ldots,N\}$ is in one and only one position in $\mathbf{d}$. Second, if $\mathbf{d}^\circ=(d_1^\circ,\ldots,d_N^\circ)$ is an arbitrary element in $\mathfrak{R}$, define $\mathbf{r}^\circ$ so that the integer i is in position d_i° in $\mathbf{r}^\circ$, for $i=1,\ldots,N$. Then $k_1(\mathbf{r}^\circ)=\mathbf{d}^\circ$ and clearly $\mathbf{r}^\circ$ is unique. The vector $\mathbf{d}=k_1(\mathbf{r})$ is called the **antirank** of the rank vector $\mathbf{r}$. Theorem 8.2.4 shows that if $\mathbf{R}^*$ is uniformly distributed over $\mathfrak{R}$, so is $\mathbf{D}^*=k_1(\mathbf{R}^*)$. Since D_j^* is the position of j in $\mathbf{R}^*$ for $j=1,\ldots,N$, we see that

$$S = \sum_{i=1}^{N} c(i)a(R_i^*) = \sum_{j=1}^{N} a(j)c(D_j^*).$$

This equation yields an important duality between the roles of the regression constants and the scores. Since $\mathbf{R}^*$ and $\mathbf{D}^*$ are both uniformly distributed over $\mathfrak{R}$, we see that if we reverse the roles of the $a(\cdot)$'s and the $c(\cdot)$'s, the distribution of S under H_0 is the same. That is, under H_0

$$\sum_{i=1}^{N} c(i)a(R_i^*) = \sum_{i=1}^{N} a(i)c(D_i^*) \stackrel{d}{=} \sum_{i=1}^{N} a(i)c(R_i^*). \tag{8.2.6}$$

This distributional equality will not usually hold under alternatives, H_1, but it is useful in proving certain properties of the H_0 distribution of S, as we see later in this section.

Other important one-to-one functions include the **reverse ranking** function defined by

$$k_2(\mathbf{r}) = (N+1-r_1,\ldots,N+1-r_N). \tag{8.2.7}$$

The value $k_2(\mathbf{r})$ is the result of ranking from largest to smallest. That is, the rank 1 is replaced by N, 2 by $N-1$, and so on. Thus

$$k_2(4,6,2,1,3,5) = (3,1,5,6,4,2).$$

Another useful one-to-one function is the **reverse sequencing** one,

$$k_3(\mathbf{r}) = (r_N, r_{N-1}, \ldots, r_1), \tag{8.2.8}$$

which, for example, yields

$$k_3(4,6,2,1,3,5) = (5,3,1,2,6,4).$$

The two **composition functions**

$$k_4(\mathbf{r}) = \mathbf{r o r^*} \tag{8.2.9}$$

and

$$k_5(\mathbf{r}) = \mathbf{r^* o r}, \tag{8.2.10}$$

where $\mathbf{r}^*$ is some fixed permutation in $\mathfrak{R}$, are also one-to-one. Uses for $k_2(\cdot)$ through $k_5(\cdot)$ are demonstrated later in this section. The verification of the one-to-one nature of these functions is left as an exercise.

Theorem 8.2.11. *Let $c'(1),\ldots,c'(N)$ and $a'(1),\ldots,a'(N)$ be fixed permutations of $c(1),\ldots,c(N)$ and $a(1),\ldots,a(N)$, respectively. If $\mathbf{R}^*$ is uniformly distributed over $\mathfrak{R}$, then*

$$S = \sum_{i=1}^{N} c(i)a(R_i^*) \stackrel{d}{=} S' = \sum_{i=1}^{N} c'(i)a'(R_i^*).$$

Proof: Let α_i be the position of $c'(i)$ in $(c(1),\ldots,c(N))$, for $i=1,\ldots,N$. Thus $(c(\alpha_1),\ldots,c(\alpha_N))=(c'(1),\ldots,c'(N))$. Similarly, let β_i be the position of $a'(i)$ in $(a(1),\ldots,a(N))$; hence, $(a(\beta_1),\ldots,a(\beta_N))=(a'(1),\ldots,a'(N))$. Then it follows that

$$\begin{aligned} S' &= \sum_{i=1}^{N} c'(i)a'(R_i^*) = \sum_{i=1}^{N} c(\alpha_i)a(\beta_{R_i^*}) \\ &= \sum_{i=1}^{N} c(\alpha_i)a(B_i), \end{aligned}$$

where $\boldsymbol{\beta}=(\beta_1,\ldots,\beta_N)$ and B_i is the ith component of $\mathbf{B}=\boldsymbol{\beta}\mathbf{o}\mathbf{R}^*$. Using the one-to-one function $k_5(\cdot)$ in (8.2.10) and Theorem 8.2.4, we see that $\mathbf{B}$ is also uniformly distributed over $\mathfrak{R}$; that is, $\mathbf{B} \stackrel{d}{=} \mathbf{R}^*$. Therefore,

$$S' = \sum_{i=1}^{N} c(\alpha_i)a(B_i) \stackrel{d}{=} \sum_{i=1}^{N} c(\alpha_i)a(R_i^*). \tag{8.2.12}$$

Let γ_j denote the position of j in $(\alpha_1,\ldots,\alpha_N)$. Then

$$\sum_{i=1}^{N} c(\alpha_i)a(R_i^*) = \sum_{j=1}^{N} c(j)a(R_{\gamma_j}^*) = \sum_{j=1}^{N} c(j)a(A_j),$$

where $\boldsymbol{\gamma}=(\gamma_1,\ldots,\gamma_N)$ and A_j is the jth component of $\mathbf{A}=\mathbf{R}^*\mathbf{o}\boldsymbol{\gamma}$. Using $k_4(\cdot)$ in (8.2.9) and Theorem 8.2.4, we see that $\mathbf{A} \overset{d}{=} \mathbf{R}^*$ and thus

$$\sum_{i=1}^{N} c(\alpha_i)a(R_i^*) = \sum_{i=1}^{N} c(i)a(A_i) \overset{d}{=} \sum_{i=1}^{N} c(i)a(R_i^*) = S.$$

Combining this with (8.2.12) completes the proof. ■

The conclusion of Theorem 8.2.11 is not surprising. It says that when H_0 holds the original, individual arrangements of the $a(\cdot)$'s and the $c(\cdot)$'s do not affect the distribution of S, since all $N!$ of the possible simultaneous pairings of every $c(j)$ with an $a(i)$ are included in the S distribution and they are all equally likely. We should also note that, in general, this result holds only under the null hypothesis condition that $\mathbf{R}^*$ is uniformly distributed over $\mathfrak{R}$. To illustrate the use of this theorem consider a two-sample problem with equal sample sizes; that is, $m=n$. Let $c(1),\ldots,c(N)$ denote the two-sample regression constants defined in (8.1.2) and let

$$\begin{aligned} c'(i) &= 1 \qquad \text{for } i = 1,\ldots,m, \\ &= 0 \qquad \text{for } i = m+1,\ldots,N. \end{aligned}$$

Since $m=n$, the $c'(\cdot)$'s are simply a rearrangement of the $c(\cdot)$'s. Theorem 8.2.11 then shows that $\mathbf{R}^*$ being uniformly distributed over $\mathfrak{R}$ implies

$$\sum_{i=1}^{N} c(i)a(R_i^*) = \sum_{j=1}^{n} a(R_{m+j}^*) \overset{d}{=} \sum_{i=1}^{m} a(R_i^*) = \sum_{i=1}^{N} c'(i)a(R_i^*).$$

Thus, when $m=n$, whether we add the scores of the ranks of the Y_j's or the scores of the ranks of the X_i's, the two statistics have the same null hypothesis distribution.

Theorem 8.2.11 is also useful in proving the following theorem due to Hájek (1969) dealing with conditions under which a linear rank statistic has a symmetric null distribution. This symmetry implies that it is necessary to tabulate only one tail of the null distribution. Since the point of symmetry will be known, the characteristics of the opposite tail will be automatic.

Theorem 8.2.13. *Consider the linear rank statistic*

$$S = \sum_{i=1}^{N} c(i)a(R_i^*).$$

Let $c^*(1) \leqslant \cdots \leqslant c^*(N)$ and $a^*(1) \leqslant \cdots \leqslant a^*(N)$ denote the ordered values of $c(1),\ldots,c(N)$ and $a(1),\ldots,a(N)$, respectively. Assume $\mathbf{R}^*$ is uniformly distributed over $\mathfrak{R}$. If either

(i) $$a^*(i) + a^*(N+1-i) = \text{constant}, \qquad i = 1,\ldots,N,$$

or

(ii) $$c^*(i) + c^*(N+1-i) = \text{constant}, \qquad i = 1,\ldots,N,$$

then the distribution of S is symmetric about its mean $N\bar{c}\bar{a}$.

Proof: Assume

$$a^*(i) + a^*(N+1-i) = k$$

for $i=1,\ldots,N$ and some constant k. Summing and dividing by N yields

$$\frac{1}{N}\sum_{i=1}^{N} a^*(i) + \frac{1}{N}\sum_{i=1}^{N} a^*(N+1-i) = k$$

or

$$k = 2\bar{a}^* = 2\bar{a}.$$

Therefore,

$$a^*(i) - \bar{a} = \bar{a} - a^*(N+1-i),$$

for $i=1,\ldots,N$. Define

$$S^* = \sum_{i=1}^{N} c(i)a^*(R_i^*).$$

Theorem 8.2.11 shows that, under H_0, $S^* \stackrel{d}{=} S$, so that it suffices to show that S^* is symmetrically distributed about $N\bar{c}\bar{a}$. Now,

$$\begin{aligned} S^* - N\bar{c}\bar{a} &= \sum_{i=1}^{N} c(i)\left[a^*(R_i^*) - \bar{a}\right] \\ &= \sum_{i=1}^{N} c(i)\left[\bar{a} - a^*(N+1-R_i^*)\right]. \end{aligned} \qquad \textbf{(8.2.14)}$$

Using $k_2(\cdot)$ in (8.2.7) and Theorem 8.2.4, we see that

$$(N+1-R_1^*,\ldots,N+1-R_N^*) \stackrel{d}{=} (R_1^*,\ldots,R_N^*).$$

Substituting into (8.2.14) yields

$$\begin{aligned} S^* - N\bar{c}\bar{a} &= \sum_{i=1}^{N} c(i)\left[\bar{a} - a^*(N+1-R_i^*)\right] \\ &\stackrel{d}{=} \sum_{i=1}^{N} c(i)\left[\bar{a} - a^*(R_i^*)\right] = N\bar{c}\bar{a} - S^*, \end{aligned}$$

and S^* is symmetrically distributed about $N\bar{c}\bar{a}$. The proof for (ii) follows by interchanging the roles of the $a(\cdot)$'s and $c(\cdot)$'s, as shown in (8.2.6). ∎

Example 8.2.15. The Wilcoxon scores, $a(i)=i$ for $i=1,\ldots,N$, satisfy the conditions of Theorem 8.2.13, since $a^*(i)=a(i)$ and

$$a(i) + a(N+1-i) = i + N + 1 - i = N + 1$$

for $i=1,\ldots,N$. This yields, for example, the symmetry under H_0 of the Mann–Whitney–Wilcoxon rank sum statistic, as was shown earlier in Theorem 2.3.16. It also demonstrates that a linear rank statistic with Wilcoxon scores will have a symmetric null distribution no matter what regression constants are used.

Example 8.2.16. When $m=n$ the two-sample regression constants defined in (8.1.2) satisfy the conditions of Theorem 8.2.13, since the $c(i)$'s are ordered and

$$c(i) + c(N+1-i) = 1$$

for $i=1,\ldots,N$. Thus for any choice of scores a linear rank statistic with these regression constants has a symmetric null distribution when $m=n$.

Example 8.2.17. When $m+n$ is even, the median scores defined in (8.1.5) satisfy the conditions of Theorem 8.2.13. However, when $m+n$ is odd, they do not. A slight variation on these scores is required to ensure these conditions for all $(m+n)$ values. Let

$$\begin{aligned} a(i) &= 1 \qquad \text{if } i > (N+1)/2, \\ &= \tfrac{1}{2} \qquad \text{if } i = (N+1)/2, \\ &= 0 \qquad \text{if } i < (N+1)/2. \end{aligned}$$

These **symmetric median scores** are ordered and satisfy

$$a(i)+a(N+1-i)=1$$

for $i=1,\ldots,N$, and linear rank statistics using these scores will have symmetric null distributions. In particular, when combined with the two-sample regression constants in (8.1.2) we get the two-sample median type test statistic

$$S=\left[\text{number of } Y_j\text{'s greater than the combined sample median}\right]+\Delta^*,$$

where Δ^* is $\frac{1}{2}$ if the combined sample median is a Y_j, and is 0, otherwise.

The following general theorem due to Randles and Hogg (1971) provides a necessary and sufficient condition for a rank statistic to have a symmetric null distribution. It is applicable to linear rank statistics but also to those rank statistics that are not linear in form.

Theorem 8.2.18. *Let $V(\mathbf{R}^*)$ be any rank statistic and assume that R^* is uniformly distributed over $\mathfrak{R}$. Then $V(\mathbf{R}^*)$ is symmetrically distributed iff there exists a one-to-one function $k(\cdot)$ mapping $\mathfrak{R}$ onto $\mathfrak{R}$ such that*

$$V(\mathbf{r})+V(k(\mathbf{r}))=c \qquad \textbf{(8.2.19)}$$

for every $\mathbf{r}\in\mathfrak{R}$ and some constant c. The distribution of $V(\mathbf{R}^*)$ is then symmetric about $c/2$.

Proof: Suppose there exists a one-to-one function satisfying (8.2.19). Then by Theorem 8.2.4

$$k(\mathbf{R}^*)\overset{d}{=}\mathbf{R}^*.$$

Therefore

$$\begin{aligned}V(\mathbf{R}^*)-\frac{c}{2}&=\frac{c}{2}-V(k(\mathbf{R}^*))\\&\overset{d}{=}\frac{c}{2}-V(\mathbf{R}^*),\end{aligned}$$

and $V(\mathbf{R}^*)$ is symmetrically distributed about $c/2$.

Assume $V(\mathbf{R}^*)$ is symmetrically distributed about the value μ. Recall that $\mathfrak{R}$ is finite and that all $N!$ points in it are equally likely. Let d be any nonnegative number such that $P[V(\mathbf{R}^*)=\mu+d]>0$. Let $A_d=\{\mathbf{r}|\mathbf{r}\in\mathfrak{R}$ and $V(\mathbf{r})=\mu+d\}$ and $B_d=\{\mathbf{r}|\mathbf{r}\in\mathfrak{R}$ and $V(r)=\mu-d\}$. The symmetry of the distribution of $V(\mathbf{R}^*)$ (when $\mathbf{R}^*$ is uniformly distributed) implies that both A_d and B_d contain the same number of rank vectors. Pair up the rank vectors in A_d and B_d so that each $\mathbf{r}$ in A_d is associated with one and only one $\mathbf{r}'$ in B_d. Define $k(\cdot)$ to be such that $k(\mathbf{r})=\mathbf{r}'$ and $k(\mathbf{r}')=\mathbf{r}$ for each such pair. Doing this for all nonnegative d's defines a one-to-one mapping, $k(\cdot)$, of $\mathfrak{R}$ onto $\mathfrak{R}$ such that

$$V(\mathbf{r}) + V(k(\mathbf{r})) = 2\mu \equiv c$$

for every $\mathbf{r} \in \mathfrak{R}$.

Example 8.2.20. Consider the two-sample location problem where we measure the location difference by

$$V_1(\mathbf{R}^*) = \underset{1 \leqslant j \leqslant n}{\text{median}}\,(R^*_{m+j}) - \underset{1 \leqslant i \leqslant m}{\text{median}}\,(R^*_i).$$

Letting $k_2(\cdot)$ be as defined in (8.2.7), we have

$$\begin{aligned} V_1(k_2(\mathbf{r})) &= \underset{1 \leqslant j \leqslant n}{\text{median}}\,(N+1-r_{m+j}) - \underset{1 \leqslant i \leqslant m}{\text{median}}\,(N+1-r_i) \\ &= (N+1) - \underset{1 \leqslant j \leqslant n}{\text{median}}\,(r_{m+j}) - \left[N+1 - \underset{1 \leqslant i \leqslant m}{\text{median}}\,(r_i)\right] \\ &= -V_1(\mathbf{r}), \qquad \text{for every } \mathbf{r} \in \mathfrak{R}. \end{aligned}$$

Thus (8.2.19) is satisfied with $c=0$, and $V_1(\mathbf{R}^*)$ is symmetrically distributed about 0 whenever $\mathbf{R}^*$ has a uniform distribution over $\mathfrak{R}$.

Example 8.2.21. Suppose $N \geqslant 3$ and set

$$V_2(\mathbf{R}^*) = \frac{R_1^* - R_2^*}{R_3^*}.$$

Then, considering the special case of (8.2.9) corresponding to

$$k(\mathbf{r}) = (r_2, r_1, r_3, r_4, \ldots, r_N),$$

we have

$$V_2(\mathbf{r}) + V_2(k(\mathbf{r})) = 0, \qquad \text{for every } \mathbf{r} \in \mathfrak{R}.$$

Thus if $\mathbf{R}^*$ is uniformly distributed over $\mathfrak{R}$, $V_2(\mathbf{R}^*)$ is symmetrically distributed about 0.

Exercises 8.2.

8.2.1. Let S be the linear rank statistic with two-sample regression constants and scores

$$
\begin{aligned}
a(i) &= \frac{i}{N+1} - \frac{1}{4} && \text{if } i \leqslant \frac{N+1}{4} \\
&= 0 && \text{if } \frac{N+1}{4} < i < \frac{3}{4}(N+1) \\
&= \frac{i}{N+1} - \frac{3}{4} && \text{if } i \geqslant \frac{3}{4}(N+1).
\end{aligned}
$$

Show that, under H_0, S is symmetrically distributed about its mean. Find the value of this null mean.

8.2.2. Consider the linear rank statistic S with regression constants $c(i) = i$ and scores $a(i) = i^2$, $i = 1, \ldots, N$. Show that, under H_0, S is symmetrically distributed about its mean. What is the value of this null hypothesis mean?

8.2.3. Consider the scores

$$a(i) = G^{-1}\left(\frac{i}{N+1}\right), \qquad i = 1, \ldots, N,$$

where $G(x)$ is the c.d.f. of a continuous random variable symmetrically distributed about zero with support set equal to the interval $[-a, a]$ for some $0 < a \leqslant \infty$. Show that for any set of regression constants, the H_0 distribution of a linear rank statistic with these scores is symmetric about its mean. What is the value of this null hypothesis mean?

8.2.4. Consider the scores $a(1) = 1$, $a(2) = N$, $a(3) = 2$, $a(4) = N - 1$, $a(5) = 3, \ldots$.

(a) Show that, under H_0, a linear rank statistic S with these scores and with the two-sample regression constants is symmetrically distributed about its mean. What is the value of this null hypothesis mean?

(b) The H_0 distribution of S is the same as that of what well-known statistic? Why?

8.2.5. Using the definition of one-to-one given in (8.2.2), show that the following mappings are one-to-one:

(a) $k_2(\mathbf{r}) = (N+1-r_1, \ldots, N+1-r_N)$,

(b) $k_3(\mathbf{r}) = (r_N, r_{N-1}, \ldots, r_1)$,

(c) $k_4(\mathbf{r})=\mathbf{ror^*}$, where $\mathbf{r}^*$ is a fixed member of $\mathcal{R}$,

(d) $k_5(\mathbf{r})=\mathbf{r^*or}$, where $\mathbf{r}^*$ is a fixed member of $\mathcal{R}$.

8.2.6. Show that in a two-sample problem with $N=m+n$, the Haga (1960) statistic

$$\left[\left(\max_{1\leqslant j\leqslant n} R_{m+j}^* + \min_{1\leqslant j\leqslant n} R_{m+j}^*\right)/2\right] - \left[\left(\max_{1\leqslant i\leqslant m} R_i^* + \min_{1\leqslant i\leqslant m} R_i^*\right)/2\right]$$

is symmetrically distributed under H_0. What is the value of the point of symmetry of its null distribution?

8.2.7. Show that if $N \geqslant 4$, the rank statistic

$$C(\mathbf{R}^*) = \ln\left[(R_1^* \cdot R_4^*)/(R_2^* \cdot R_3^*)\right]$$

is symmetrically distributed under H_0. What is the value of this point of symmetry?

8.2.8. In a two-sample setting the Kamat (1956) statistic is

$$\operatorname*{range}_{1\leqslant j\leqslant n} R_{m+j}^* - \operatorname*{range}_{1\leqslant i\leqslant m} R_i^*.$$

Show that when $m=n$ this statistic is symmetrically distributed under H_0, and find the value of the point of symmetry.

8.2.9. Show that the scores

$$a(i) = \operatorname{sine}\left(\frac{2\pi i}{N+1}\right), \qquad i = 1,\ldots,N,$$

satisfy condition (i) of Theorem 8.2.13.

8.2.10. **Theorem 8.2.22.** Consider two rank statistics $V(\mathbf{R}^*)$ and $W(\mathbf{R}^*)$. Suppose there exists a one-to-one mapping $k(\mathbf{r})$ of $\mathcal{R}$ onto $\mathcal{R}$ such that

$$V(\mathbf{r}) + V(k(\mathbf{r})) = c$$

and

$$W(\mathbf{r}) = W(k(\mathbf{r}))$$

for some constant c and every $\mathbf{r}\in\mathcal{R}$. Then if $\mathbf{R}^*$ is uniformly distributed over $\mathcal{R}$,

$$\left[V(\mathbf{R}^*)-\mu, W(\mathbf{R}^*)\right] \stackrel{d}{=} \left[\mu - V(\mathbf{R}^*), W(\mathbf{R}^*)\right]. \qquad \textbf{(8.2.23)}$$

Prove this result and show that (8.2.23) implies $V(\mathbf{R}^*)$ and $W(\mathbf{R}^*)$ are uncorrelated.

8.2.11. Use the results of Exercise 8.2.10 to show that under H_0 in a two-sample problem with $m=n$ the rank sum statistic, $\Sigma_{j=1}^{n} R_{m+j}^*$, is uncorrelated with the two-sided Kolmogorov (1933)–Smirnov (1939) statistic

$$U(\mathbf{R}^*) = \left(\frac{mn}{m+n}\right)^{\frac{1}{2}} \max_{1 \leqslant k \leqslant N} \left| \frac{1}{m} (\text{number of } R_i^* \leqslant k,\, i=1,\ldots,m) - \frac{1}{n} (\text{number of } R_{m+j}^* \leqslant k,\, j=1,\ldots,n) \right|.$$

8.2.12. Show that $\mathcal{R}$ and composition, as defined in (8.2.1), form a *group*. That is, show: (i) closure with respect to composition, (ii) composition is associative, (iii) existence of an identity, and (iv) existence of an inverse element for each $\mathbf{r} \in \mathcal{R}$.

8.3. Some Preliminaries for Asymptotics

The next two sections deal with the asymptotic behavior of linear rank statistics under H_0. We are building toward Theorem 8.4.9 in the next section which yields the asymptotic normality under H_0 of a very large class of properly standardized linear rank statistics. Consider statistics of the form

$$S_N = \sum_{i=1}^{N} c_N(i) a_N(R_i^*). \tag{8.3.1}$$

Note that both the regression constants and the scores are indexed by N; that is, they may change as N changes.

Since the regression constants $c_N(1),\ldots,c_N(N)$ are often dictated by the structure of the problem and, as such, are not under our control, we place a very weak restriction on these constants. This restriction essentially requires that (asymptotically) no individual constant should dominate (be much larger or smaller than) the others. Mathematically, this restriction is written as

$$\frac{\sum_{i=1}^{N} \left(c_N(i) - \bar{c}_N\right)^2}{\max_{1 \leqslant i \leqslant N} \left(c_N(i) - \bar{c}_N\right)^2} \to \infty \tag{8.3.2}$$

as $N \to \infty$, where $\bar{c}_N = (1/N)\Sigma_{i=1}^{N} c_N(i)$. We refer to (8.3.2) as **Noether's condition** (1949) on the regression constants.

Example 8.3.3. Consider the two-sample regression constants defined in (8.1.2). Here $\bar{c}_N = n/N$ and

$$\sum_{i=1}^{N} (c_N(i) - \bar{c}_N)^2 = m\bar{c}_N^2 + n(1-\bar{c}_N)^2$$
$$= \frac{mn}{N}.$$

Also

$$\max_{1 \leqslant i \leqslant N} (c_N(i) - \bar{c}_N)^2 = \max\left[\bar{c}_N^2, (1-\bar{c}_N)^2\right]$$
$$= \frac{\left[\max(m,n)\right]^2}{N^2}.$$

Noether's condition (8.3.2) then says that

$$\frac{Nmn}{\left[\max(m,n)\right]^2} = \frac{N\min(m,n)}{\max(m,n)} = \left[1 + \frac{\min(m,n)}{\max(m,n)}\right]\min(m,n)$$

must go to infinity as $N \to \infty$. This, clearly, will hold iff $\min(m,n) \to \infty$. Thus with two-sample regression constants, condition (8.3.2) merely requires that both sample sizes approach infinity simultaneously.

Example 8.3.4. Let us examine the regression constants $c_N(i) = i$ for $i = 1, \ldots, N$, as used in (8.1.8). Here $\bar{c}_N = (N+1)/2$,

$$\sum_{i=1}^{N} \left(i - \frac{N+1}{2}\right)^2 = \frac{N(N+1)(N-1)}{12},$$

and

$$\max_{1 \leqslant i \leqslant N} \left(i - \frac{N+1}{2}\right)^2 = \frac{(N-1)^2}{4}.$$

Since $[N(N+1)(N-1)/12]/[(N-1)^2/4] = N(N+1)/(3N-3)$, goes to infinity as $N \to \infty$, Noether's condition (8.3.2) is satisfied for these regression constants.

The scores $a_N(1),\ldots,a_N(N)$ are usually subject to our control and are often chosen on considerations of power. (See, for example, Section 9.1.) Hence we place slightly stronger restrictions on them. Let

$$a_N'(i) = b_N\phi\left(\frac{i}{N+1}\right) + d_N, \tag{8.3.5}$$

where $\{b_N\}$ and $\{d_N\}$ are sequences of constants with $b_N > 0$ for each N. We assume that the function $\phi(u)$ is defined on $(0,1)$, does not depend on N, and can be written as the difference of two nondecreasing functions. In addition, it must satisfy

$$0 < \int_0^1 \{\phi(u) - \bar{\phi}\}^2 du < \infty, \tag{8.3.6}$$

with $\bar{\phi} = \int_0^1 \phi(u)\,du$. Throughout this chapter we consider only $\phi(\cdot)$ functions with these properties, and we refer to them as **square integrable score functions**. Since the expression in the middle of (8.3.6) is the variance of $\phi(U)$, when U is uniform $(0,1)$, it may be computed using

$$\int_0^1 \{\phi(u) - \bar{\phi}\}^2 du = \int_0^1 \phi^2(u)\,du - (\bar{\phi})^2.$$

Note that if we define $a_N(i) = \phi(i/(N+1))$, $i = 1,\ldots,N$, then (8.3.5) and Remark 8.1.13 show that

$$S_N' = b_N S_N + N\bar{c}_N d_N,$$

where S_N' and S_N are linear rank statistics that use the scores $a_N'(i)$ of (8.3.5) and $a_N(i)$, respectively, and the same regression constants $c_N(1),\ldots,c_N(N)$. Hence we see that

$$\frac{S_N' - E(S_N')}{\sqrt{\operatorname{Var}(S_N')}} = \frac{S_N - E(S_N)}{\sqrt{\operatorname{Var}(S_N)}}. \tag{8.3.7}$$

Thus while the mean and variance of S_N' depend on b_N and d_N, these two sequences of constants play no essential role in determining whether the standardized linear rank statistic is asymptotically normal. This determination depends solely on the properties of $\phi(\cdot)$, which provides the smoothness of the scores. Thus, without loss of generality, we emphasize **function scores** of the form

$$a_N(i) = \phi\left(\frac{i}{N+1}\right), \qquad i = 1,\ldots,N, \tag{8.3.8}$$

where $\phi(\cdot)$ is a square integrable score function; that is, we set $b_N = 1$ and $d_N = 0$.

Example 8.3.9. The symmetric median scores of Example 8.2.17 are obtained by setting $a_N(i)=\phi(i/(N+1))$, where

$$\begin{aligned}\phi(u) &= 1 && \text{if } \tfrac{1}{2}<u<1\\ &= \tfrac{1}{2} && \text{if } u=\tfrac{1}{2}\\ &= 0 && \text{if } 0<u<\tfrac{1}{2}.\end{aligned}$$

Note that $\phi(u)$ is nondecreasing and satisfies $\bar{\phi}=\frac{1}{2}$ and $\int_0^1\{\phi(u)-\bar{\phi}\}^2\,du=\frac{1}{4}$.

Example 8.3.10. The Wilcoxon scores $a'_N(i)=i$, $i=1,\ldots,N$, satisfy

$$a'_N(i) = (N+1)\phi\left(\frac{i}{N+1}\right),$$

where $\phi(u)=u$ for $0<u<1$. Thus the $\phi(\cdot)$ function associated with the Wilcoxon scores has $\int_0^1\phi(u)\,du=\frac{1}{2}$, $\int_0^1\{\phi(u)-\bar{\phi}\}^2\,du=\frac{1}{12}$, and is monotone increasing.

The remainder of this section is devoted to results that pertain to proving the asymptotic normality under H_0 of standardized linear rank statistics with scores and regression constants restricted as just described. We begin with a fundamental lemma that is used in later chapters as well.

Lemma 8.3.11. *Let $Z_1,\ldots,Z_N$ be* i.i.d. *continuous random variables. Let $\mathbf{R}^*$ denote the rank vector of these observations; that is, R_i^* is the rank of Z_i among $Z_1,\ldots,Z_N$. Also, let $Z_{(1)}<\cdots<Z_{(N)}$ be the order statistics of $Z_1,\ldots,Z_N$. Then $\mathbf{R}^*$ and $Z_{(1)}<\cdots<Z_{(N)}$ are independent.*

Proof: It suffices to show that the conditional distribution of $Z_{(1)}<\cdots<Z_{(N)}$, given $\mathbf{R}^*=\mathbf{r}^*$, is equal to the marginal distribution of $Z_{(1)}<\cdots<Z_{(N)}$, for each $\mathbf{r}^*\in\mathfrak{R}$. Consider $\mathbf{r}^*=(1,2,\ldots,N)$, the argument being similar for the other $\mathbf{r}^*$ vectors in $\mathfrak{R}$. If $\mathbf{R}^*=(1,2,\ldots,N)$, then $Z_{(1)}=Z_1,\ldots,Z_{(N)}=Z_N$, and the joint density of $Z_{(1)},\ldots,Z_{(N)}$, given $\mathbf{R}^*=\mathbf{r}^*$, is

$$\begin{aligned}g(z_{(1)},\ldots,z_{(N)}|\mathbf{R}^*=\mathbf{r}^*) &= \left[\prod_{i=1}^{N} f(z_{(i)})\right]\Big/P[\mathbf{R}^*=(1,\ldots,N)],\\ &\qquad\qquad -\infty<z_{(1)}<\cdots<z_{(N)}<\infty\\ &= N!\prod_{i=1}^{N} f(z_{(i)}), \qquad -\infty<z_{(1)}<\cdots<z_{(N)}<\infty,\end{aligned}$$

which is also the joint unconditional distribution of $Z_{(1)}<\cdots<Z_{(N)}$. ■

Lemma 8.3.12. *Let* $\{V_k\}$ *and* $\{W_k\}$ *be sequences of random variables such that*

$$\lim_{k\to\infty} E\left[\{V_k - W_k\}^2\right] = 0.$$

Then

(i)
$$\lim_{k\to\infty} E[V_k] = \lim_{k\to\infty} E[W_k]$$

and

(ii)
$$\lim_{k\to\infty} E[V_k^2] = \lim_{k\to\infty} E[W_k^2].$$

Proof: Since

$$\mathrm{Var}[(V_k - W_k)] \leqslant E\left[(V_k - W_k)^2\right],$$

Lemma 3.2.14 shows that

$$\lim_{k\to\infty} \mathrm{Var}[V_k] = \lim_{k\to\infty} \mathrm{Var}[W_k]. \tag{8.3.13}$$

Jensen's inequality A.3.3 in the Appendix shows that

$$\{E[V_k] - E[W_k]\}^2 \leqslant E\left[\{V_k - W_k\}^2\right]$$

and thus

$$\lim_{k\to\infty} E[V_k] = \lim_{k\to\infty} E[W_k].$$

Combining this with (8.3.13) yields part (ii). ■

Theorem 8.3.14. *Let* $\phi(u)$ *be a square integrable score function and set* $a_N(i) = \phi(i/(N+1))$ *for* $i = 1,\ldots,N$. *If* R_1^* *equals each of the integers* $1,\ldots,N$ *with probability* $1/N$, *then*

$$\lim_{N\to\infty} E\left[\phi^2\left(\frac{R_1^*}{N+1}\right)\right] = \lim_{N\to\infty} \frac{1}{N}\sum_{i=1}^{N} a_N^2(i) = \int_0^1 \phi^2(u)\,du. \tag{8.3.15}$$

Proof: Define

$$q_N(u) = \phi\left(\frac{i}{N+1}\right) \text{ for } \frac{i-1}{N} \leqslant u < \frac{i}{N}, \qquad i = 1,\ldots,N.$$

Then

$$\int_0^1 q_N^2(u)\,du = \frac{1}{N}\sum_{i=1}^{N}\phi^2\left(\frac{i}{N+1}\right) = \frac{1}{N}\sum_{i=1}^{N} a_N^2(i).$$

Taking $V_N = q_N(U)$ and $W_N = \phi(U)$, where U is a uniform $(0,1)$ random variable, we see from Lemma 8.3.12 that to verify (8.3.15) we need only show

$$\lim_{N\to\infty}\int_0^1 \{q_N(u)-\phi(u)\}^2\,du = 0. \qquad \textbf{(8.3.16)}$$

To check (8.3.16) it suffices to consider $\phi(\cdot)$'s that are nonnegative and nondecreasing. The proof for $\phi(\cdot)$'s that are nonnegative and nonincreasing proceeds similarly. Moreover, an arbitrary $\phi(\cdot)$ can be written as

$$\phi(u) = \phi_1(u) - \phi_2(u) - \phi_3(u) + \phi_4(u),$$

where $\phi_1(\cdot)$ and $\phi_3(\cdot)$ are nonnegative and nondecreasing while $\phi_2(\cdot)$ and $\phi_4(\cdot)$ are nonnegative and nonincreasing. Define $q_{Ni}(\cdot)$ corresponding to $\phi_i(\cdot)$ and apply the inequality $(a+b)^2 \leqslant 2a^2 + 2b^2$ two times to obtain the expression

$$\begin{aligned}\int_0^1 \{q_N(u)-\phi(u)\}^2\,du \leqslant\ & 4\int_0^1 \{q_{N1}(u)-\phi_1(u)\}^2\,du \\ & +4\int_0^1 \{q_{N2}(u)-\phi_2(u)\}^2\,du \\ & +4\int_0^1 \{q_{N3}(u)-\phi_3(u)\}^2\,du \\ & +4\int_0^1 \{q_{N4}(u)-\phi_4(u)\}^2\,du.\end{aligned}$$

This shows that the result will then follow for an arbitrary $\phi(\cdot)$.

For a $\phi(\cdot)$ that is both nonnegative and nondecreasing,

$$\frac{1}{N}\phi^2\left(\frac{i}{N+1}\right) \leqslant \int_{\frac{i}{N}}^{\frac{i+1}{N}} \phi^2(u)\,du, \qquad 1 \leqslant i \leqslant (N-1)$$

and

$$\frac{1}{N+1}\phi^2\left(\frac{N}{N+1}\right) \leqslant \int_{\frac{N}{N+1}}^{1} \phi^2(u)\,du.$$

Hence

$$\int_0^1 q_N^2(u)\,du \leqslant \int_{\frac{1}{N}}^1 \phi^2(u)\,du + \frac{N+1}{N}\int_{\frac{N}{N+1}}^1 \phi^2(u)\,du$$

and thus

$$\limsup_{N\to\infty} \int_0^1 q_N^2(u)\,du \leqslant \int_0^1 \phi^2(u)\,du.$$

Since $\lim_{N\to\infty} q_N(u) = \phi(u)$ at all continuity points of $\phi(u)$ in $(0,1)$ (there are at most a countable number of discontinuity points), equation (8.3.16) then follows from part (iii) of Theorem A.4.4 in the Appendix. ■

The following two results are used in the proof of Lemma 8.3.19.

Lemma 8.3.17. *Let* $f_N(i) = \binom{N}{i} p^i(1-p)^{N-i}$, *for* $i = 1, \ldots, N$, *zero elsewhere, denote the probability function for a binomial random variable with parameters* N *and* p. *Then*

(i) $$f_N(i) > f_N(i+1) \qquad \text{if } i > (N+1)p - 1$$

and

(ii) $$f_N(i-1) < f_N(i) \qquad \text{if } i < (N+1)p.$$

Proof: Exercise 8.3.5. ■

Lemma 8.3.18. *Suppose* $b(1) \geqslant \cdots \geqslant b(k)$ *and* $d(1) \leqslant \cdots \leqslant d(k)$ *are constants. Then*

$$\sum_{i=1}^{k} b(i)d(i) \leqslant k\bar{b}\bar{d},$$

where

$$\bar{b} = \left(\frac{1}{k}\right)\sum_{i=1}^{k} b(i) \text{ and } \bar{d} = \left(\frac{1}{k}\right)\sum_{i=1}^{k} d(i).$$

Proof: Exercise 8.3.6 ■

Lemma 8.3.19. *Let* $\phi(u)$ *be a square integrable score function, and let* Y_N *denote a binomial random variable with parameters* N *and* p. *Then*

$$\lim_{N\to\infty} E\left[\phi\left(\frac{Y_N+1}{N+2}\right)\right] = \phi(p)$$

for all but at most a countable number of p*-values in* $(0,1)$.

Proof: Without loss of generality we assume that $\phi(u)$ is nondecreasing, since an arbitrary $\phi(\cdot)$ can be written as a difference of two such functions. Since $\phi(u)$ is monotone, it has at most a countable number of discontinuities. Let $0<p<1$ be a value at which $\phi(\cdot)$ is continuous. Also, since $\phi(\cdot)$ is square integrable, set $K^* = \int_0^1\{\phi(u)-\phi(p)\}^2\,du<\infty$. Let $\epsilon>0$ be arbitrary but fixed. Then, since $\phi(\cdot)$ is continuous at p, there exists an a_ϵ satisfying $\min(p,1-p)>a_\epsilon>0$ such that $\{u|[\phi(u)-\phi(p)]^2>\epsilon\}\subset\{u||u-p|>a_\epsilon\}$. Let $F_N(\cdot)$ denote the c.d.f. of Y_N. Using Jensen's inequality A.3.3 we write

$$\left\{E\left[\phi\left(\frac{Y_N+1}{N+2}\right)-\phi(p)\right]\right\}^2 \leqslant E\left[\left\{\phi\left(\frac{Y_N+1}{N+2}\right)-\phi(p)\right\}^2\right]$$

$$= \int_0^1\left[\phi\left(\frac{y+1}{N+2}\right)-\phi(p)\right]^2 dF_N(y)$$

$$\leqslant \epsilon + \int_{\left\{y:\left[\phi\left(\frac{y+1}{N+2}\right)-\phi(p)\right]^2>\epsilon\right\}}\left[\phi\left(\frac{y+1}{N+2}\right)-\phi(p)\right]^2 dF_N(y)$$

$$\leqslant \epsilon + \int_{\left\{y:\left|\frac{y+1}{N+2}-p\right|>a_\epsilon\right\}}\left[\phi\left(\frac{y+1}{N+2}\right)-\phi(p)\right]^2 dF_N(y). \quad \textbf{(8.3.20)}$$

Consider the portion of the integral in (8.3.20) corresponding to $y>[(N+2)(p+a_\epsilon)-1]$. Since $[(N+2)(p+a_\epsilon)-1]>[(N+1)p-1]$, we see from Lemma 8.3.17 that $f_N(y)$, the probability function of Y_N, is decreasing over $\{y: y>[(N+2)(p+a_\epsilon)-1]\}$. Combining this with the monotonicity of $\phi(\cdot)$ and using Lemma 8.3.18, we obtain

$$\int_{\{y:y>[(N+2)(p+a_\epsilon)-1]\}}\left[\phi\left(\frac{y+1}{N+2}\right)-\phi(p)\right]^2 dF_N(y)$$

$$= \sum_{i=k_N}^{N}\left[\phi\left(\frac{i+1}{N+2}\right)-\phi(p)\right]^2 f_N(i)$$

$$\leqslant \bar{f}_N(k_N,N)\sum_{i=k_N}^{N}\left[\phi\left(\frac{i+1}{N+2}\right)-\phi(p)\right]^2, \quad \textbf{(8.3.21)}$$

where $k_N = [\![(N+2)(p+a_\epsilon)]\!]$, $\bar{f}_N(i,j)=\sum_{h=i}^{j} f_N(h)/(j-i+1)$, and N is large enough so that $k_N\leqslant N$. Since $\phi(\cdot)$ is nondecreasing, the right side of (8.3.21) is bounded above by

$$(N+2)\bar{f}_N(k_N,N)\int_{\frac{k_N+1}{N+2}}^{1}[\phi(u)-\phi(p)]^2\,du \leqslant (N+2)\bar{f}_N(k_N,N)K^*.$$

This fact, combined with a similar argument applied to the other part of the integral in (8.3.20), shows that for N sufficiently large,

$$\left\{E\left[\phi\left(\frac{Y_N+1}{N+2}\right)-\phi(p)\right]\right\}^2 \leqslant \epsilon + K^*(N+2)\left\{\bar{f}_N(0,k'_N)+\bar{f}_N(k_N,N)\right\}, \tag{8.3.22}$$

where $k'_N = [\![(N+2)(p-a_\epsilon)]\!]$. Now,

$$(N+2)\bar{f}_N(k_N,N) = \frac{N+2}{N-[\![(N+2)(p+a_\epsilon)]\!]+1}P\left[\frac{Y_N}{N} > \frac{k_N-1}{N}\right].$$

Since

$$\frac{k_N}{N} = \frac{[\![(N+2)(p+a_\epsilon)]\!]}{N} \to (p+a_\epsilon),$$

as $N\to\infty$, and Y_N/N converges in probability to p as $N\to\infty$ (see Example 3.2.4), it follows that $\lim_{N\to\infty} P[Y_N/N > (k_N-1)/N]=0$. In addition, $a_\epsilon < 1-p$ implies that

$$\frac{N+2}{N-[\![(N+2)(p+a_\epsilon)]\!]+1} \to \frac{1}{1-(p+a_\epsilon)} > 0, \qquad \text{as } N\to\infty.$$

Combining these facts yields

$$\lim_{N\to\infty} (N+2)\bar{f}_N(k_N,N) = 0.$$

Similarly,

$$\lim_{N\to\infty} (N+2)\bar{f}_N(0,k'_N) = 0$$

and, since ϵ was arbitrarily chosen, the desired result follows from (8.3.22). ■

The preceding series of lemmas leads to the following one, which provides the key step in the proof of Theorem 8.4.9, establishing the asymptotic normality under H_0 of a large class of standardized linear rank statistics.

Lemma 8.3.23. *Let $\phi(\cdot)$ denote a square integrable score function, and let* $\mathbf{R}^*$ *be the rank vector corresponding to $U_1,\ldots,U_N$, i.i.d. uniform* $(0,1)$

variates. Then

$$\lim_{N\to\infty} E\left[\left\{\phi\left(\frac{R_1^*}{N+1}\right)-\phi(U_1)\right\}^2\right]=0.$$

Proof: From Theorem 8.3.14 we note that

$$\lim_{N\to\infty} E\left[\left\{\phi\left(\frac{R_1^*}{N+1}\right)-\phi(U_1)\right\}^2\right]$$

$$=\lim_{N\to\infty}\left\{E\left[\phi^2\left(\frac{R_1^*}{N+1}\right)\right]+E[\phi^2(U_1)]-2E\left[\phi\left(\frac{R_1^*}{N+1}\right)\phi(U_1)\right]\right\}$$

$$=2\left\{E[\phi^2(U_1)]-\lim_{N\to\infty}E\left[\phi\left(\frac{R_1^*}{N+1}\right)\phi(U_1)\right]\right\}.$$

Hence it remains to show that

$$\lim_{N\to\infty} E\left[\phi\left(\frac{R_1^*}{N+1}\right)\phi(U_1)\right]=\int_0^1\phi^2(u)\,du. \qquad \textbf{(8.3.24)}$$

Now,

$$\phi\left(\frac{R_1^*}{N+1}\right)=\phi\left(\left[1+\sum_{i=2}^{N}\Psi(U_1-U_i)\right]/(N+1)\right),$$

where $\Psi(t)=1,0$ as $t>,\leqslant 0$. Note that, given $U_1=u$, $\Sigma_{i=2}^N\Psi(u-U_i)\equiv Y(u,N-1)$ is a binomial random variable with parameters $N-1$ and p. Thus, defining

$$b_N(u)=E\left[\phi\left(\frac{1+Y(u,N-1)}{N+1}\right)\right],$$

we see that

$$E\left[\phi\left(\frac{R_1^*}{N+1}\right)\phi(U_1)\right]=\int_0^1 b_N(u)\phi(u)\,du.$$

From Lemma 8.3.19 we see that $b_N(u)\to\phi(u)$ for all but at most a countable number of u-values in $(0,1)$. In addition, since $b_N(u)$ is an expected value, we apply Jensen's inequality A.3.3 in the Appendix to obtain

$$\int_0^1 b_N^2(u)\,du = \int_0^1 \left\{ E\left[\phi\left(\frac{1+Y(u,N-1)}{N+1}\right)\right]\right\}^2 du$$
$$\leqslant \int_0^1 E\left[\phi^2\left(\frac{1+Y(u,N-1)}{N+1}\right)\right] du$$
$$= E\left[\phi^2\left[\frac{1+\sum_{i=2}^{N}\Psi(U_1-U_i)}{N+1}\right]\right]$$
$$= E\left[\phi^2\left(\frac{R_1^*}{N+1}\right)\right].$$

Lemma 8.3.14 then shows that

$$\limsup_{N\to\infty} \int_0^1 b_N^2(u)\,du \leqslant \int_0^1 \phi^2(u)\,du,$$

and so by part (ii) of Theorem A.4.4 in the Appendix, we have

$$\lim_{N\to\infty} \int_0^1 b_N(u)\phi(u)\,du = \int_0^1 \phi^2(u)\,du. \quad \blacksquare$$

Exercises 8.3.

8.3.1. Consider the regression constants

$$c_N(i) = 0 \qquad \text{if } i < (N+1)/4$$
$$= \frac{i}{N+1} - \frac{1}{4} \qquad \text{if } (N+1)/4 \leqslant i \leqslant 3(N+1)/4$$
$$= \frac{1}{2} \qquad \text{if } i > 3(N+1)/4.$$

Show that they satisfy Noether's condition (8.3.2).

8.3.2. Verify Noether's condition (8.3.2) for the regression constants

$$c_N(i) = \left(i - \frac{N+1}{2}\right)^2, \; i = 1,\ldots,N.$$

8.3.3. If $\phi(u)=G^{-1}(u)$, where $G(x)$ is the c.d.f. of a continuous random variable with a positive, finite variance, show that $\phi(\cdot)$ is a square integrable score function.

8.3.4. Verify that

$$\begin{aligned}\phi(u) &= u-\tfrac{1}{4} && \text{if } u<\tfrac{1}{4}\\ &=0 && \text{if } \tfrac{1}{4}\leqslant u\leqslant\tfrac{3}{4}\\ &=u-\tfrac{3}{4} && \text{if } u>\tfrac{3}{4}\end{aligned}$$

is a square integrable score function.

8.3.5. Prove Lemma 8.3.17.

8.3.6. Prove Lemma 8.3.18.

8.3.7. Consider the scores $a_N(i)=\{i-(N+1)/2\}^2$, $i=1,\ldots,N$.

(a) With what $\phi(\cdot)$ function are these scores associated?

(b) Show that this $\phi(\cdot)$ function is a square integrable score function.

8.3.8. Define a set of scores as follows:

$$\begin{aligned}a_N(i) &= \ln(i)+\ln\left(\frac{2}{N+1}\right), && i<\frac{N+1}{2},\\ &= -\ln(N+1-i)-\ln\left(\frac{2}{N+1}\right), && i\geqslant\frac{N+1}{2}.\end{aligned}$$

(a) With what $\phi(\cdot)$ function are these scores associated?

(b) Show that this $\phi(\cdot)$ function is a square integrable score function.

8.3.9. Find the $\phi(\cdot)$ associated with the scores

$$\begin{aligned}a_N(i) &= 1\ - && \text{if } i<\frac{N+1}{2}\\ &= 2i-N && \text{if } i\geqslant\frac{N+1}{2},\end{aligned}$$

and show that it is a square integrable score function.

8.4. Asymptotic Normality under H_0

In this section we prove an important limit theorem, due to Hájek (1961), establishing the asymptotic normality of certain properly normed linear rank statistics when $\mathbf{R}^*$, the rank vector, is uniformly distributed over $\mathcal{R}$ for every N. The nature of the proof is somewhat similar to that of the U-statistic Theorem 3.3.13, in that we find a second random variable that (i) is easier to establish as asymptotically normal and (ii) is such that the square of the difference between it and the linear rank statistic has an expected value that (after being standardized) goes to zero as $N \to \infty$. The cleverness of Hájek's proof stems from assuming that the rank vector $\mathbf{R}^*$ is generated by $U_1, \ldots, U_N$, i.i.d. uniform $(0,1)$ random variables. That is, R_i^* is assumed to be the rank of U_i among $U_1, \ldots, U_N$. This assumption leads to the ingenious construction of an approximating random variable by replacing $\phi(R_i^*/(N+1))$ with $\phi(U_i)$. Thus if the scores satisfy $a_N(i) = \phi(i/(N+1))$ for $i = 1, \ldots, N$, then the linear rank statistic

$$S_N = \sum_{i=1}^{N} (c_N(i) - \bar{c}_N) a_N(R_i^*) + N\bar{c}_N \bar{a}_N$$

$$= \sum_{i=1}^{N} (c_N(i) - \bar{c}_N) \phi\left(\frac{R_i^*}{N+1}\right) + N\bar{c}_N \bar{a}_N \tag{8.4.1}$$

is approximated by

$$V_N = \sum_{i=1}^{N} (c_N(i) - \bar{c}_N) \phi(U_i) + N\bar{c}_N \bar{a}_N. \tag{8.4.2}$$

An important step in the proof of Hájek's result is to verify that V_N is indeed asymptotically normal when appropriately standardized.

Theorem 8.4.3. *Let V_N be defined by* (8.4.2), *where $c_N(1), \ldots, c_N(N)$ satisfy* (8.3.2), *$\phi(\cdot)$ is a square integrable score function, and $a_N(i) = \phi(i/(N+1))$, $i = 1, \ldots, N$. Then $(V_N - \mu_N)/\sigma_{Na}$ has a limiting standard normal distribution with*

$$\mu_N = N\bar{c}_N \bar{a}_N \tag{8.4.4}$$

and

$$\sigma_{Na}^2 = \sum_{i=1}^{N} (c_N(i) - \bar{c}_N)^2 \int_0^1 \{\phi(u) - \bar{\phi}\}^2 \, du. \tag{8.4.5}$$

Proof: Note that

$$E[V_N] = \left[\sum_{i=1}^{N} (c_N(i) - \bar{c}_N) \int_0^1 \phi(u)\,du\right] + N\bar{c}_N\bar{a}_N$$
$$= N\bar{c}_N\bar{a}_N = \mu_N$$

and

$$\operatorname{Var}[V_N] = \sum_{i=1}^{N} (c_N(i) - \bar{c}_N)^2 \operatorname{Var}[\phi(U_i)] = \sigma_{Na}^2.$$

We apply the Lindeberg central limit theorem (Theorem A.3.8 in the Appendix) to

$$V_N - \mu_N = \sum_{i=1}^{N} (c_N(i) - \bar{c}_N)\phi(U_i)$$
$$= \sum_{i=1}^{N} \{(c_N(i) - \bar{c}_N)\phi(U_i) - (c_N(i) - \bar{c}_N)\bar{\phi}\}. \qquad \textbf{(8.4.6)}$$

In analogy to the notation of Theorem A.3.8, we let

$$X_{i:N} = (c_N(i) - \bar{c}_N)\phi(U_i)$$

$$\mu_{i:N} = (c_N(i) - \bar{c}_N)\bar{\phi}$$

and

$$\sigma_{i:N}^2 = (c_N(i) - \bar{c}_N)^2 \int_0^1 \{\phi(u) - \bar{\phi}\}^2\,du,$$

for $i = 1, \ldots, N$. Then condition (A.3.9) of that theorem amounts to showing

$$\lim_{N\to\infty} \frac{1}{\sigma_{Na}^2} \sum_{i=1}^{N} \left[\int_{|x-\mu_{i:N}|>\epsilon\sigma_{Na}} (x - \mu_{i:N})^2\,dF_{i:N}(x)\right] = 0 \qquad \textbf{(8.4.7)}$$

for every $\epsilon > 0$, where $F_{i:N}(\cdot)$ is the c.d.f. of $X_{i:N}$. Before taking the limit, the left side of (8.4.7) is

$$\frac{1}{\sigma_{Na}^2}\sum_{i=1}^{N}\left[\int_{\{u:\ |c_N(i)-\bar{c}_N||\phi(u)-\bar{\phi}|>\epsilon\sigma_{Na}\}}(c_N(i)-\bar{c}_N)^2(\phi(u)-\bar{\phi})^2\,du\right]$$

$$\leqslant \frac{1}{\sigma_{Na}^2}\sum_{i=1}^{N}\left[c_N(i)-\bar{c}_N\right]^2\int_{\{u:\ |\phi(u)-\bar{\phi}|>\epsilon\delta_N\}}\left[\phi(u)-\bar{\phi}\right]^2du$$

$$=\left[\int_0^1\{\phi(u)-\bar{\phi}\}^2du\right]^{-1}\int_{\{u:\ |\phi(u)-\bar{\phi}|>\epsilon\delta_N\}}\left[\phi(u)-\bar{\phi}\right]^2du, \qquad \textbf{(8.4.8)}$$

where

$$\delta_N \equiv \frac{\epsilon\sigma_{Na}}{\max_{1\leqslant j\leqslant N}|c_N(j)-\bar{c}_N|}$$

$$=\epsilon\left[\int_0^1\{\phi(u)-\bar{\phi}\}^2du\frac{\sum_{i=1}^{N}(c_N(i)-\bar{c}_N)^2}{\max_{1\leqslant j\leqslant N}(c_N(j)-\bar{c}_N)^2}\right]^{\frac{1}{2}}.$$

Noether's condition (8.3.2) implies that $\delta_N\to\infty$ as $N\to\infty$, and hence the upper bound (8.4.8) goes to zero. Therefore, (8.4.7) is satisfied and $(V_N-\mu_N)/\sigma_{Na}$ has a limiting standard normal distribution. ■

We now consider the main result of this section, the important theorem that establishes the asymptotic normality of a large class of properly normed linear rank statistics under a null hypothesis condition.

Theorem 8.4.9. [Hájek (1961)] *Let S_N denote a linear rank statistic with regression constants $c_N(1),\ldots,c_N(N)$ satisfying Noether's condition* (8.3.2) *and with scores $a_N(i)=\phi(i/(N+1))$ for $i=1,\ldots,N$, where $\phi(\cdot)$ is a square integrable score function. Let*

$$\mu_N = N\bar{c}_N\bar{a}_N \qquad \textbf{(8.4.10)}$$

and

$$\sigma_N^2=\frac{1}{(N-1)}\left[\sum_{i=1}^{N}(c_N(i)-\bar{c}_N)^2\right]\left[\sum_{j=1}^{N}(a_N(j)-\bar{a}_N)^2\right]. \qquad \textbf{(8.4.11)}$$

Then, if $\mathbf{R}^*$ *is uniformly distributed over* $\mathcal{R}$ *for each* N, *the statistic*

$$\frac{S_N - \mu_N}{\sigma_N}$$

has a limiting standard normal distribution.

Proof: Assume $\mathbf{R}^*$ is the rank vector of $U_1, \ldots, U_N$, i.i.d. uniform $(0,1)$ variables and let V_N be defined by (8.4.2). Recall from Lemma 8.3.11 that the conditional distribution of $\mathbf{R}^*$, given the order statistics $\mathbf{U}_{(\,)} = (U_{(1)}, \ldots, U_{(N)})$, is still uniform over $\mathcal{R}$. Thus taking a conditional expected value given these order statistics, we obtain

$$E\left[(S_N - V_N)^2 | \mathbf{U}_{(\,)} = \mathbf{u}_{(\,)}\right] = E\left[\left\{\sum_{i=1}^{N} (c_N(i) - \bar{c}_N)\left[a_N(R_i^*) - \phi(u_{(R_i^*)})\right]\right\}^2\right]$$

$$= \frac{1}{(N-1)} \sum_{i=1}^{N} (c_N(i) - \bar{c}_N)^2 \sum_{j=1}^{N} \left[a_N(j) - \phi(u_{(j)}) - \bar{a}_N + \bar{\phi}_u\right]^2, \tag{8.4.12}$$

where this last step follows from part (ii) of Theorem 8.1.12 with $\bar{\phi}_u = (1/N)\sum_{j=1}^{N}\phi(u_{(j)})$. Since $\sum_{i=1}^{N}(x_i - \bar{x})^2 \leqslant \sum_{i=1}^{N} x_i^2$, it follows that expression (8.4.12) is bounded above by

$$\frac{N}{N-1} \sum_{i=1}^{N} (c_N(i) - \bar{c}_N)^2 \left\{\frac{1}{N} \sum_{j=1}^{N} (a_N(j) - \phi(u_{(j)}))^2\right\}$$

$$= \frac{N}{N-1} \sum_{i=1}^{N} (c_N(i) - \bar{c}_N)^2 E\left[\{a(R_1^*) - \phi(U_1)\}^2 | \mathbf{U}_{(\,)} = \mathbf{u}_{(\,)}\right].$$

Let $\mathbf{U}_{(\,)}$ be the random variable in both (8.4.12) and this upper bound. Then taking expected values of both sides yields

$$E\left[(S_N - V_N)^2\right] \leqslant \frac{N}{N-1} \sum_{i=1}^{N} (c_N(i) - \bar{c}_N)^2 E\left[\{a(R_1^*) - \phi(U_1)\}^2\right].$$

Using σ_{Na}^2 as defined in (8.4.5), we have

$$\lim_{N\to\infty} E\left[\left(\frac{S_N - V_N}{\sigma_{Na}}\right)^2\right]$$

$$\leqslant \left\{\int_0^1 \{\phi(u) - \bar{\phi}\}^2 du\right\}^{-1} \lim_{N\to\infty} E\left[\{a(R_1^*) - \phi(U_1)\}^2\right] = 0$$

by Lemma 8.3.23. Theorem 3.2.15 then shows that $(S_N - \mu_N)/\sigma_N$ has the same limiting distribution as $(V_N - \mu_N)/\sigma_{Na}$, and the result follows from Theorem 8.4.3. ■

This theorem provides, for example, the asymptotic normality of the standardized Mann–Whitney–Wilcoxon test statistic W under H_0. In (8.1.4), W was shown to correspond to the two-sample regression constants (8.1.2) and the Wilcoxon scores (8.1.3). Example 8.3.3 demonstrated that the two-sample regression constants satisfy Noether's condition, provided that both sample sizes simultaneously go to infinity. In Example 8.3.10 the Wilcoxon scores were shown to be generated by a square integrable score function. Thus the conditions of Theorem 8.4.9 are satisfied, and

$$\frac{W - E_0(W)}{\sqrt{\text{Var}_0(W)}}$$

has a limiting standard normal distribution under H_0. Additional applications of this theorem are presented in Chapter 9.

Remark 8.4.13. As applied to the setting in Theorem 8.4.9, the proof of Theorem 3.2.15 shows that

$$\lim_{N\to\infty} \frac{1}{N} \sum_{i=1}^{N} \{a_N(i) - \bar{a}_N\}^2 = \int_0^1 \{\phi(u) - \bar{\phi}\}^2 du. \qquad \textbf{(8.4.14)}$$

Hence, in view of Slutsky's Theorem 3.2.8, we can standardize $(S_N - \mu_N)$ with either the exact standard deviation of S_N, namely, σ_N as defined in (8.4.11), or the asymptotic standard deviation of S_N, namely, σ_{Na} as defined in (8.4.5).

Remark 8.4.15. In Hájek (1961) it was shown that regression constants $c_N(1),\ldots,c_N(N)$ satisfying Noether's condition (8.3.2) and scores $a_N(1),\ldots,$ $a_N(N)$ satisfying a similar but slightly stronger Noether condition provides a necessary and sufficient criterion for the asymptotic normality of a properly standardized linear rank statistic. The use of scores produced by a square integrable score function simplifies the proof of asymptotic normality, and such scores were shown by Hájek (in that same paper) to satisfy the necessary strengthened Noether condition.

Some scores of interest are not directly determined from a square integrable score function by the relation $a_N(i) = \phi(i/(N+1))$. (We study a few such scores in the next chapter.) The asymptotic normality of standardized linear rank statistics with these scores can be established if the scores are asymptotically equivalent to ones which are produced by such a $\phi(\cdot)$. The formal statement of the required equivalence is given in the following theorem.

Theorem 8.4.16. *Let S_N denote a linear rank statistic with regression constants $c_N(1),\ldots,c_N(N)$ satisfying Noether's condition* (8.3.2) *and scores $a_N(i)=\phi(i/(N+1))$, $i=1,\ldots,N$, where $\phi(\cdot)$ is a square integrable score function. Let S_N^* denote another linear rank statistic with the same regression constants but scores $a_N^*(1),\ldots,a_N^*(N)$. Assume $\mathbf{R}^*$ is uniformly distributed over $\mathfrak{R}$ for each N. If*

$$\lim_{N\to\infty}\frac{1}{N}\sum_{i=1}^{N}\left[a_N(i)-a_N^*(i)\right]^2=0, \tag{8.4.17}$$

then $(S_N^*-N\bar{c}_N\bar{a}_N^*)/\sigma_N$ has, under H_0, a standard normal limiting distribution, where σ_N^2 is defined in (8.4.11). *That is, identically standardized forms of $(S_N-N\bar{c}_N\bar{a}_N)$ and $(S_N^*-N\bar{c}_N\bar{a}_N^*)$ have the same limiting distribution under H_0.*

Proof: Exercise 8.4.4. ■

One method for generating scores that are of interest corresponds to setting

$$a_N^*(i)=E[\phi(U_{(i)})], \qquad i=1,\ldots,N, \tag{8.4.18}$$

where $U_{(1)}<\cdots<U_{(N)}$ are the order statistics for a random sample of size N from a uniform $(0,1)$ distribution and $\phi(\cdot)$ is a square integrable score function. Scores of this type are referred to as **expected value scores** and are clearly related to the function scores

$$a_N(i)=\phi\left(\frac{i}{N+1}\right)=\phi(E[U_{(i)}]), \qquad i=1,\ldots,N.$$

The asymptotic equivalence under H_0 of these two sets of scores is given in the following theorem.

Theorem 8.4.19. *Let $a_N^*(1),\ldots,a_N^*(N)$ be defined by* (8.4.18). *If $a_N(1),\ldots,a_N(N)$ are determined directly from the same square integrable score function by $a_N(i)=\phi(i/(N+1))$, $i=1,\ldots,N$, then*

$$\lim_{N\to\infty}\frac{1}{N}\sum_{i=1}^{N}\{a_N(i)-a_N^*(i)\}^2=0.$$

Proof: See Theorem V.1.4.b and Lemma V.1.6.a of Hájek and Šidák (1967). A proof under more restrictive conditions is suggested in Exercise 8.4.5. ■

Remark 8.4.20. The choice of V_N in the proof of Theorem 8.4.9 is motivated by the **projection principle**, as expounded in Section 3.3. There, when dealing with U-statistics, we obtained the projection of interest by taking a conditional expected value of the U-statistic. That projection was the closest point (in the sense of quadratic mean) to the U-statistic in the space of sums of i.i.d. random variables. (See Lemma 3.3.11.) In the present setting, for any regression constants $c_N(1),\ldots,c_N(N)$ consider the random variable

$$V_N^* = \sum_{i=1}^{N} (c_N(i) - \bar{c}_N)\phi(U_i), \tag{8.4.21}$$

where $U_1,\ldots,U_N$ are i.i.d. uniform (0, 1) variables. Determining the conditional expected value of V_N^* given the rank vector $\mathbf{R}^*$, we obtain

$$\begin{aligned} E[V_N^*|\mathbf{R}^*=\mathbf{r}^*] &= \sum_{i=1}^{N} (c_N(i)-\bar{c}_N)E[\phi(U_i)|R_i^*=r_i^*] \\ &= \sum_{i=1}^{N} (c_N(i)-\bar{c}_N)a_N^*(r_i^*), \end{aligned} \tag{8.4.22}$$

where $a_N^*(1),\ldots,a_N^*(N)$ represent the expected value scores defined in (8.4.18). Thus the linear rank statistic with these scores is the projection of V_N^* onto the space of rank statistics. Notice that in the present setting the projection process is employed in an opposite direction from its usage with U-statistics. Here the projection is the statistic of interest. Thus V_N was used in Theorem 8.4.9 because V_N^* projects into a linear rank statistic with the expected value scores $a_N^*(i)$ defined in (8.4.18) and because this linear rank statistic is asymptotically equivalent to the corresponding one with function scores $a_N(i)=\phi(i/(N+1))$, as shown in Theorems 8.4.16 and 8.4.19.

Exercises 8.4.

8.4.1. Consider the linear rank statistic S_N that uses the regression constants $c_N(i)=i$, $i=1,\ldots,N$, and the scores $a_N(i)=\Phi^{-1}(i/(N+1))$, where $\Phi(x)$ is the c.d.f. of the standard normal distribution. Verify the conditions necessary to ensure that $(S_N - \mu_N)/\sigma_{Na}$ has a limiting standard normal distribution under H_0 and give expressions for both μ_N and the asymptotic variance σ_{Na}^2.

8.4.2. Let $S_N = \sum_{i=1}^{N} i^2 \ln(R_i^*/(N+1))$. Show that this linear rank statistic satisfies the conditions necessary to ensure that $(S_N - \mu_N)/\sigma_{Na}$ is asymptotically standard normal under H_0. Find expressions for both μ_N and the asymptotic variance σ_{Na}^2.

8.4.3. Let $a_N(i)$ and $a_N^*(i)$ denote two sets of scores that satisfy condition (8.4.17). Assuming these limits exist, show that

(a)
$$\lim_{N\to\infty} \bar{a}_N = \lim_{N\to\infty} \bar{a}_N^*$$

and

(b)
$$\lim_{N\to\infty} \frac{1}{N} \sum_{i=1}^{N} (a_N(i) - \bar{a}_N)^2 = \lim_{N\to\infty} \frac{1}{N} \sum_{j=1}^{N} (a_N^*(j) - \bar{a}_N^*)^2$$

[Hint: Consider Lemma 8.3.12.]

8.4.4. Using an approach similar to that employed in the proof of Theorem 8.4.9, prove Theorem 8.4.16.

8.4.5. Prove Theorem 8.4.19 for the special case in which $\phi(\cdot)$ has first and second derivatives on $(0, 1)$ and the second derivative is bounded in absolute value. [Hint: Use a Taylor series expansion of $\phi(U_{(i)})$ around $i/(N+1)$, and the form of the variance of the ith uniform order statistic.]

8.4.6. Use the approach of the proof of Lemma 3.3.11 to verify that the statistic given in (8.4.22) is the closest rank statistic to the statistic in (8.4.21) if distance is measured by the expected value of the square of their difference.

8.4.7. Apply the Lindeberg central limit theorem (A.3.8 in the Appendix) to show that $\sum_{i=1}^{n} X_i$, when appropriately standardized, has a limiting standard normal distribution when $X_1, \ldots, X_n$ are i.i.d. continuous random variables with a finite and positive variance.

9 TWO-SAMPLE LOCATION AND SCALE PROBLEMS

9.1. The Two-Sample Location Problem

One of the fundamental problems of statistics, often encountered in applications, is the two-sample location problem. The model for this setting assumes that random samples are drawn from each of two univariate populations that differ only in their locations. Specifically, we let $X_1,\ldots,X_m$ and $Y_1,\ldots,Y_n$ be independent random samples from continuous distributions with c.d.f.'s $F(x)$ and $F(x-\Delta)$, respectively. The parameter Δ is termed a **shift parameter**, and if $\Delta>0$ the Y distribution is shifted to the right; that is, in that case the Ys would tend to be larger than the Xs.

The null hypothesis of interest is H_0: $\Delta=0$, and under H_0, $X_1,\ldots,X_m$, $Y_1,\ldots,Y_n$ represent a single random sample of size $N=(m+n)$ from a continuous population with c.d.f. $F(x)$. If $Q_i(R_j)$ denotes the rank of $X_i(Y_j)$ among all N observations, then we know from Theorem 2.3.3 that under H_0 the rank vector $\mathbf{R}^*=(Q_1,\ldots,Q_m,R_1,\ldots,R_n)$ is uniformly distributed over $\mathcal{R}$. Hence all rank statistics (functions only of $\mathbf{R}^*$) will be nonparametric distribution-free under H_0, since $\mathbf{R}^*$ will have this same uniform distribution for any $F(\cdot)$, so long as the population is continuous.

For simplicity, we consider only the alternative H_1: $\Delta>0$. (Appropriate modifications for the cases $\Delta<0$ and $\Delta\neq 0$ should be clear.) To detect $\Delta>0$ we want to reject H_0 when the ranks of $Y_1,\ldots,Y_n$ are in some sense large. Suppose we consider linear rank statistics with the two-sample regression constants

$$c(i)=0, \qquad i=1,\ldots,m,$$

$$=1, \qquad i=m+1,\ldots,m+n.$$

A two-sample linear rank statistic is then of the form

$$S = \sum_{j=1}^{n} a(R_j). \tag{9.1.1}$$

If the scores satisfy a nondecreasing and nonconstant condition, namely,

$$a(1) \leqslant \cdots \leqslant a(N), \qquad a(1) \neq a(N), \tag{9.1.2}$$

then we would reject H_0 in favor of H_1 for large values of S. For any such choice of scores this provides a nonparametric distribution-free test of H_0 versus H_1.

As noted earlier, if we use the **Wilcoxon scores**, $a_W(i)=i$, for $i=1,\ldots,N$, then S becomes the rank sum statistic

$$W = \sum_{j=1}^{n} R_j, \tag{9.1.3}$$

of the Mann–Whitney–Wilcoxon test. If the **median scores** are used, namely, $a_M(i)=1,\ 0$ as $i>, \leqslant (N+1)/2$, then the linear rank statistic becomes

$$\begin{aligned} M &= \sum_{j=1}^{n} a_M(R_j) \\ &= [\text{number of } Y_j\text{'s greater than the combined sample median}], \end{aligned} \tag{9.1.4}$$

which is the test statistic for a **two-sample median test** attributed to Mood (1950) and Westenberg (1948). This median test is a popular two-sample location procedure, due in part to the simplicity of the test statistic, which, in turn, yields a simple H_0 distribution. In Exercise 9.1.9 you are asked to show that, under H_0,

$$P_0[M=k] = \frac{\binom{n}{k}\binom{m}{[\![N/2]\!]-k}}{\binom{N}{[\![N/2]\!]}}$$

for all values of k for which $P_0[M=k]$ is positive. The median test is particularly effective in detecting a shift in a distribution that is symmetric and heavy-tailed (this is demonstrated in Table 9.2.21).

Another type of scores that is often mentioned in the literature is one that mimics a normal distribution; that is, the scores assume values like an ideal sample from a standard normal distribution. One version of these **normal scores** is the function type

$$a_{NS}(i) = \Phi^{-1}\left(\frac{i}{N+1}\right), \qquad i = 1,\ldots,N, \qquad \textbf{(9.1.5)}$$

where $\Phi(\cdot)$ is the c.d.f. of a standard normal distribution. These are referred to as the **quantile normal scores**, since an inverse c.d.f. is often called a quantile function. The related **expected value normal scores** are

$$a_{NS^*}(i) = E\left[\Phi^{-1}(U_{(i)})\right], \qquad i = 1,\ldots,N, \qquad \textbf{(9.1.6)}$$

where $U_{(1)} < \cdots < U_{(N)}$ are the order statistics for a random sample of size N from a uniform $(0,1)$ distribution. These, of course, may be written

$$a_{NS^*}(i) = E\left[Z_{(i)}\right], \qquad i = 1,\ldots,N,$$

where $Z_{(1)} < \cdots < Z_{(N)}$ are the order statistics for a random sample of size N from a standard normal distribution. (See Theorem 1.2.15.) The use of these expected value normal scores in the two-sample problem was first proposed by Fisher and Yates (1938) and later studied by Terry (1952). The quantile normal scores were developed by van der Waerden (1952, 1953a, 1953b). It can be shown that these two versions of normal scores are asymptotically equivalent in the sense of Theorem 8.4.19. We develop some of the properties of these normal scores tests later in this chapter and show that these scores are particularly effective in detecting a shift in a normal distribution. (See Table 9.2.21.)

A wide variety of scores satisfy (9.1.2). Suppose that $F_0(x)$ is the c.d.f. for any nondegenerate distribution. We can then define the **quantile F_0 scores** by

$$a_{F_0}(i) = F_0^{-1}\left(\frac{i}{N+1}\right), \qquad i = 1,\ldots,N, \qquad \textbf{(9.1.7)}$$

and the **expected value F_0 scores** by

$$a_{F_0}^*(i) = E_{F_0}\left[V_{(i)}\right] = E\left[F_0^{-1}(U_{(i)})\right], \qquad i = 1,\ldots,N, \qquad \textbf{(9.1.8)}$$

where $V_{(1)} < \cdots < V_{(N)}$ $[U_{(1)} < \cdots < U_{(N)}]$ denotes the order statistics for a random sample of size N from the $F_0(\cdot)$ [uniform $(0,1)$] distribution. Clearly, the two normal scores are special cases of these, as are the Savage scores defined in Exercise 8.1.2, since they are the expected value F_0 scores corresponding to an exponential distribution. (See Exercise 1.2.7.) A set of scores that is simple in form and which emphasizes the extreme ranks (large and small) is

$$a_L(i) = \begin{cases} i-[(N+1)/4] & \text{if } i \leqslant (N+1)/4 \\ 0 & \text{if } (N+1)/4 < i < 3(N+1)/4 \\ i-[3(N+1)/4] & \text{if } i \geqslant 3(N+1)/4. \end{cases} \quad \textbf{(9.1.9)}$$

These scores are effective in detecting a shift in light-tailed distributions, as we see in Table 9.2.21, and are among those proposed by Gastwirth (1965). A set of scores that de-emphasizes the extremes is

$$\begin{aligned} a_H(i) &= -(N+1)/4 && \text{if } i < (N+1)/4 \\ &= i-[(N+1)/2] && \text{if } (N+1)/4 \leqslant i \leqslant 3(N+1)/4 \\ &= (N+1)/4 && \text{if } i > 3(N+1)/4. \end{aligned} \quad \textbf{(9.1.10)}$$

They are effective for detecting shifts in moderately heavy-tailed distributions. (See Table 9.2.21.) Hogg, Fisher, and Randles (1975) considered scores that emphasize the small ranks such as

$$a_{\mathrm{RS}}(i) = \begin{cases} i-[(N+1)/2] & i \leqslant (N+1)/2 \\ 0 & i > (N+1)/2, \end{cases} \quad \textbf{(9.1.11)}$$

or the large ranks, such as

$$a_{\mathrm{LS}}(i) = \begin{cases} 0 & i \leqslant (N+1)/2 \\ i-[(N+1)/2] & i > (N+1)/2. \end{cases} \quad \textbf{(9.1.12)}$$

They are effective for detecting shifts in distributions that are skewed to the right or left, respectively, as is evidenced in Tables 9.2.21 and 11.6.6.

Before we broach the subject of how to select an appropriate set of scores, we first develop a few properties of two-sample linear rank statistics for the location problem. Theorem 8.1.12 and Example 8.3.3 show that under H_0 the mean and variance of the two-sample linear rank statistic

$$S = \sum_{i=1}^{n} a(R_i)$$

are

$$E_0[S] = n\bar{a} \quad \text{and} \quad \text{Var}_0[S] = \frac{mn}{N(N-1)} \sum_{i=1}^{N} (a(i) - \bar{a})^2. \quad \textbf{(9.1.13)}$$

In addition, the null hypothesis results of Chapter 8 yield the following theorem.

Theorem 9.1.14. *Let $S = \Sigma_{i=1}^{n} a(R_i)$ denote a linear rank statistic for the two-sample location problem such that the scores are nondecreasing and nonconstant. Then, under H_0, S has a distribution that is symmetric about the value $n\bar{a}$ provided either*

(i) $$m = n$$

or

(ii) $$a(i) + a(N+1-i) = \text{constant}, \qquad i = 1, \ldots, N.$$

Proof: This is a direct consequence of Theorem 8.2.13 and Example 8.2.16. ■

Our discussion of power function properties in Chapter 4 provides the following result.

Theorem 9.1.15. *Let $X_1, \ldots, X_m$ and $Y_1, \ldots, Y_n$ be independent random samples from continuous populations with c.d.f.'s $F(x)$ and $F(x-\Delta)$, respectively. For testing H_0: $\Delta = 0$ versus H_1: $\Delta > 0$ suppose that we reject for large values of $S = \Sigma_{i=1}^{n} a(R_i)$, where R_i is the rank of Y_i among all $m+n$ observations and the scores satisfy* (9.1.2). *Then, for any such $F(x)$, the power function of this test is a nondecreasing function of Δ, and the test is unbiased at all its natural α-levels.*

Proof: Exercise 9.1.4. ■

Next we consider the question of how to select an appropriate set of scores. In the process we develop an important property of certain linear rank statistics. Our approach is to search among all rank tests for the one that is most powerful, in some sense, for a particular problem. Suppose $X_1, \ldots, X_m$ and $Y_1, \ldots, Y_n$ are independent random samples from continuous populations with c.d.f.'s $F(x)$ and $F(x-\Delta)$, respectively. We wish to test H_0: $\Delta = 0$ versus H_1: $\Delta > 0$ for a setting in which $F(\cdot)$ is known. Thus the power of the test will be a function only of Δ. We would like to find the rank test that is most powerful for all alternatives for which $\Delta > 0$. Unfortunately, there is no such uniformly most powerful procedure, so we instead seek the one that is locally most powerful.

Definition 9.1.16. Suppose that our model has a fixed distributional form $F(\cdot)$ and is indexed only by a shift parameter Δ. We wish to test H_0: $\Delta=0$ versus H_1: $\Delta>0$. A particular rank test is termed the **locally most powerful rank test** if there exists some $\epsilon>0$ such that this test is the uniformly most powerful rank test for detecting $0<\Delta<\epsilon$ at every natural α-level of the test.

The idea is simply we choose the test that does well for the local alternatives, since they are the hardest to detect. If a test has good power for the most difficult alternatives, then, although it may not be best, it surely will have good power for the alternatives that are easier to detect.

There are $(m+n)!$ total rankings of the Xs and Ys, and each rank test determines an *order* in which these rank configurations enter the rejection region. For example, suppose $m=2$, $n=3$ and we reject H_0 for large values of a linear rank statistic. The values of three linear rank statistics for different rank configurations and the subsequent order in which the configurations go into the rejection region are displayed in Table 9.1.17. The tabled rank configurations are indexed only by the Y ranks, since any two configurations with the same Y ranks yield the same value for a linear rank statistic. The three rank statistics included are the rank sum statistic W in (9.1.3), the median statistic M in (9.1.4), and the linear rank statistic L that uses the $a_L(\cdot)$ scores in (9.1.9). We see that the associated orderings do indeed differ. The actual number of rank configurations to be placed in a rejection region will, of course, depend on the chosen α-level.

TABLE 9.1.17

Y RANKS	W-VALUE	W ORDERING	M-VALUE	M ORDERING	L-VALUE	L ORDERING
5,4,3	12	1	2	1	1/2	1
5,4,2	11	2	2	1	1/2	1
5,4,1	10	3	2	1	0	2
5,3,2	10	3	1	2	1/2	1
5,3,1	9	4	1	2	0	2
5,2,1	8	5	1	2	0	2
4,3,2	9	4	1	2	0	2
4,3,1	8	5	1	2	−1/2	3
4,2,1	7	6	1	2	−1/2	3
3,2,1	6	7	0	3	−1/2	3

Recall that under H_0 each rank configuration in $\mathcal{R}$ is equally likely. A critical region for an $\alpha = k/N!$ level rank test will thus consist of k of the $N!$ equally likely rank configurations. To make such a test the most powerful against a particular alternative $\Delta > 0$, the critical region should consist of the k rank configurations with the largest probabilities for the particular Δ. Since the most powerful rank test for detecting $\Delta > 0$ must have this property for *all* of its natural α-levels, the order in which it puts the rank configurations into the critical region must coincide with ordering the rank configurations by their probabilities under the alternative $\Delta > 0$. That is, the rank configuration with largest probability under $\Delta > 0$ must be placed first in the critical region, and so on.

Denote the probability of an arbitrary rank configuration $\mathbf{r}^*$ at the parameter value Δ by

$$\mathcal{K}_{\mathbf{r}^*}(\Delta) = P_{\Delta}[\mathbf{R}^* = \mathbf{r}^*],$$

where we view this probability as a function of Δ.

Assumption 9.1.18. Suppose there exists a $\delta > 0$ such that $\mathcal{K}'_{\mathbf{r}^*}(\Delta)$ exists and is continuous on $(-\delta, \delta)$ for each $\mathbf{r}^* \in \mathcal{R}$.

This assumption enables us to use Taylor's formula, with remainder, at $\Delta = 0$, yielding

$$\begin{aligned} \mathcal{K}_{\mathbf{r}^*}(\Delta) &= \mathcal{K}_{\mathbf{r}^*}(0) + \Delta \mathcal{K}'_{\mathbf{r}^*}(\Delta^*) \\ &= \frac{1}{N!} + \Delta \mathcal{K}'_{\mathbf{r}^*}(\Delta^*), \end{aligned}$$

where $0 < \Delta^* < \Delta < \delta$. Using such an expansion for each of two different rank configurations in $\mathcal{R}$, namely $\mathbf{r}^*$ and $\mathbf{r}^\circ$, we see that

$$\lim_{\Delta \downarrow 0} \frac{1}{\Delta} [\mathcal{K}_{\mathbf{r}^*}(\Delta) - \mathcal{K}_{\mathbf{r}^\circ}(\Delta)] = \mathcal{K}'_{\mathbf{r}^*}(0) - \mathcal{K}'_{\mathbf{r}^\circ}(0).$$

Thus

$$\mathcal{K}'_{\mathbf{r}^*}(0) - \mathcal{K}'_{\mathbf{r}^\circ}(0) > 0$$

implies

$$\mathcal{K}_{\mathbf{r}^*}(\Delta) - \mathcal{K}_{\mathbf{r}^\circ}(\Delta) > 0$$

for Δ sufficiently small. Consequently, the locally most powerful rank test will order the rank configurations according to the ordered values of

$$\mathcal{K}'_{\mathbf{r}^*}(0) = \frac{\partial}{\partial \Delta} P_\Delta[\mathbf{R}^* = \mathbf{r}^*]\Big|_{\Delta=0}, \tag{9.1.19}$$

with the configuration yielding the largest $\mathcal{K}'_{\mathbf{r}^*}(0)$ value going into the rejection region first, and so on.

Theorem 9.1.20. *Let $X_1,\ldots,X_m$ and $Y_1,\ldots,Y_n$ be independent random samples from continuous populations with respective c.d.f.'s $F(x)$ and $F(x-\Delta)$. Assume that the support of $F(\cdot)$ is the whole real line and that the density function $f(x)$ is the indefinite integral of a function $f'(x)$, which is continuous at all but at most a countable number of x-values. Here $f'(x)$ denotes the derivative of $f(\cdot)$, where it exists, and is defined to be zero otherwise. Assume also that for some $\delta > 0$ there exist functions $g_1(\cdot)$ and $g_2(\cdot)$ satisfying*

$$|f'(x-\Delta)| \leqslant g_1(x)$$

and **(9.1.21)**

$$f(x-\Delta) \leqslant g_2(x)$$

for all $-\delta < \Delta < \delta$ and all x and, in addition,

$$E_F\left[\frac{g_1(W_1)\prod_{j=2}^{n} g_2(W_j)}{\prod_{i=1}^{n} f(W_i)}\right] < \infty, \tag{9.1.22}$$

where $N = m+n$ and $W_1,\ldots,W_N$ are i.i.d. with c.d.f. $F(\cdot)$. Then the locally most powerful rank test for H_0: $\Delta = 0$ versus H_1: $\Delta > 0$ rejects for large values of the linear rank statistic

$$S = \sum_{i=1}^{n} a^*(R_i),$$

where

$$a^*(i) = E\left[\frac{-f'(F^{-1}(U_{(i)}))}{f(F^{-1}(U_{(i)}))}\right], \qquad i = 1,\ldots,N, \tag{9.1.23}$$

and $U_{(1)} < \cdots < U_{(N)}$ are the order statistics for a random sample of size N from a uniform $(0,1)$ distribution.

Proof: From Hoeffding's Lemma 4.3.12 we see that

$$P_\Delta[\mathbf{R}^*=\mathbf{r}^*]=\frac{1}{N!}E\left[\frac{\prod_{j=1}^{n} f(W_{(r^*_{m+j})}-\Delta)}{\prod_{i=1}^{n} f(W_{(r^*_{m+i})})}\right].$$

Note that

$$E\left[\frac{g_1(W_1)\prod_{j=2}^{n} g_2(W_j)}{\prod_{i=1}^{n} f(W_i)}\right]=\frac{1}{N!}\sum_{\mathbf{r}^*\in\mathfrak{R}} E\left[\frac{g_1(W_{(r_1^*)})\prod_{j=2}^{n} g_2(W_{(r_j^*)})}{\prod_{i=1}^{n} f(W_{(r_i^*)})}\right],$$

so that (9.1.22) implies

$$E\left[\frac{g_1(W_{(r_1^*)})\prod_{j=2}^{n} g_2(W_{(r_j^*)})}{\prod_{i=1}^{n} f(W_{(r_i^*)})}\right]<\infty \tag{9.1.24}$$

for every $\mathbf{r}^*\in\mathfrak{R}$, where $W_{(1)}<\cdots<W_{(N)}$ are the ordered values of $W_1,\ldots,W_N$. Using Theorems A.2.3 and A.2.4 of the Appendix we see that (9.1.21) and (9.1.22) are what are necessary to differentiate with respect to Δ inside the expected value expression for $P_\Delta(\mathbf{R}^*=\mathbf{r}^*)$ and for the resulting derivative to be a continuous function of Δ for $-\delta<\Delta<\delta$ (Assumption 9.1.18). Thus

$$\begin{aligned}\frac{\partial}{\partial\Delta}P_\Delta[\mathbf{R}^*=\mathbf{r}^*]\Big|_{\Delta=0}&=\frac{1}{N!}\sum_{j=1}^{n} E\left[\frac{-f'(W_{(r^*_{m+j})})}{f(W_{(r^*_{m+j})})}\right]\\&=\frac{1}{N!}\sum_{j=1}^{n} E\left[\frac{-f'(F^{-1}(U_{(r_j)}))}{f(F^{-1}(U_{(r_j)}))}\right],\end{aligned}$$

where r_j is the rank of Y_j among all $m+n$ observations. According to (9.1.19), ordering the rank configurations by the ordered values of this expression yields the locally most powerful rank test for detecting a shift in the distribution $F(\cdot)$. ■

Let $F(\cdot)$ denote the c.d.f. of a continuous distribution and assume $f'(x)=d^2F(x)/dx^2$ exists at all but at most a countable number of x-values. We refer to

$$\phi(u,f) = \frac{-f'(F^{-1}(u))}{f(F^{-1}(u))} \tag{9.1.25}$$

as the **optimal score function** for the two-sample location problem with underlying distribution $F(\cdot)$. Then

$$a_{\mathrm{OF}}(i) = \frac{-f'\left(F^{-1}\left(\frac{i}{N+1}\right)\right)}{f\left(F^{-1}\left(\frac{i}{N+1}\right)\right)}, \qquad i = 1,\ldots,N, \tag{9.1.26}$$

are called the **optimal function scores** (OF) and

$$a_{\mathrm{OF}^*}(i) = E\left[\frac{-f'(F^{-1}(U_{(i)}))}{f(F^{-1}(U_{(i)}))}\right], \qquad i = 1,\ldots,N, \tag{9.1.27}$$

are referred to as the **optimal expected value scores (OF*)** for the two-sample location problem.

Theorem 9.1.20 shows that when certain assumptions are satisfied the optimal expected value scores produce a test that is locally most powerful among rank tests for detecting a shift in the distribution $F(\cdot)$. This property holds for any finite sample sizes m and n. An asymptotic optimality property of both the scores in (9.1.26) and (9.1.27) is discussed in Section 9.2. The following lemma shows that the test based on the optimal scores does not depend on either the location or scale parameter values of the distribution.

Lemma 9.1.28. *Suppose $F(x)=G((x-\mu)/\sigma)$, for some $-\infty<\mu<\infty$ and $\sigma>0$. Then the corresponding optimal score functions satisfy*

$$\phi(u,f) = \frac{1}{\sigma}\phi(u,g)$$

for all $0<u<1$.

Proof: Exercise 9.1.2. ■

Multiplying the scores by the positive constant $1/\sigma$ does not change the test. (See Remark 8.1.13.) Thus the optimal scores for F and G produce equivalent rank tests.

In the examples which follow we do not check the conditions (9.1.21) and (9.1.22). These are technical details that, while not difficult, would tend to distract from the purpose of the examples.

Example 9.1.29. Let $f(x)$ denote the density of a standard normal distribution. Then

$$\begin{aligned}\frac{-f'(x)}{f(x)} &= \frac{-d}{dx}\left[\ln(f(x))\right]\\ &= \frac{-d}{dx}\left[\frac{-1}{2}\ln(2\pi)-\frac{x^2}{2}\right]\\ &= x.\end{aligned}$$

Substituting into (9.1.23), we see that the associated locally most powerful rank test rejects for large values of

$$\sum_{j=1}^{n} E\left[\Phi^{-1}(U_{(r_j)})\right],$$

where $U_{(1)} < \cdots < U_{(N)}$ are the order statistics for a random sample of size N from a uniform $(0,1)$ distribution and $\Phi^{-1}(u)$ is the inverse c.d.f. of a standard normal. Thus the locally most powerful rank test for detecting a shift in a standard normal distribution rejects for large values of the two-sample linear rank statistic with expected value normal scores (9.1.6). The corresponding optimal score function $\phi(\cdot)$ is

$$\phi(u,\text{normal}) = \Phi^{-1}(u).$$

Lemma 9.1.28 shows that this is the locally most powerful rank test for detecting a shift in any underlying normal distribution regardless of the mean and variance.

Example 9.1.30. Consider a logistic distribution with c.d.f.

$$F(x) = \frac{1}{1+e^{-x}}$$

and density function

$$f(x) = \frac{e^{-x}}{(1+e^{-x})^2}, \qquad -\infty < x < \infty.$$

It follows that

$$\frac{-f'(x)}{f(x)} = 2F(x) - 1,$$

and, using (9.1.23), we see that the locally most powerful rank test rejects for large values of

$$2 \sum_{j=1}^{n} E\left[U_{(r_j)} \right] - n,$$

where $U_{(1)} < \cdots < U_{(N)}$ are the order statistics for a random sample of size N from a uniform $(0, 1)$ distribution. Since $E[U_{(r_j)}] = r_j/(N+1)$, the locally most powerful rank test is equivalent to the Mann–Whitney–Wilcoxon test that rejects for large values of

$$\sum_{j=1}^{n} r_j.$$

Lemma 9.1.28 also implies that the Mann-Whitney-Wilcoxon test is the locally most powerful rank test for detecting a shift in any logistic distribution, regardless of location or scale parameter values.

Exercises 9.1.

9.1.1. Find the null mean and variance of the two-sample linear rank statistic

$$S = \sum_{i=1}^{n} a(R_i)$$

for the following choices of scores:

(a) The median scores (see (9.1.4)),

(b) The light-tailed distribution scores given in (9.1.9). [Hint: Note that $a(i) + a(N+1-i) = k$, for $i = 1, \ldots, N$.]

9.1.2. Prove Lemma 9.1.28.

9.1.3. Find the optimal score function for the continuous distribution with density

$$f(x) = \exp\left[x - e^{x} \right], \qquad -\infty < x < \infty.$$

9.1.4. Prove Theorem 9.1.15.

9.1.5. Consider the RST-Lambda distributions defined by the inverse c.d.f.

$$F^{-1}(u) = \lambda_1 + \left[u^{\lambda_3} - (1-u)^{\lambda_4}\right]/\lambda_2,$$

(see (A.1.8) in the Appendix). Find the form of the optimal score function for this family of distributions. Assume λ_3 and λ_4 are neither 0 nor 1.

9.1.6. Find the scores for the locally most powerful rank test for a shift in a double exponential distribution. Express them in terms of the c.d.f.'s of Beta random variables. [You need not verify conditions (9.1.21) and (9.1.22).]

9.1.7. Show that the Mann–Whitney–Wilcoxon test is the locally most powerful rank test of

$$H_0: \Delta = 0 \text{ versus } H_1: 0 < \Delta < 1,$$

when $X_1, \ldots, X_m$ and $Y_1, \ldots, Y_n$ represent independent random samples from continuous distributions with respective c.d.f.'s $F(x)$ and $(1-\Delta)F(x) + \Delta F^2(x)$. [Hint: Proceed in a fashion similar to that in the proof of Theorem 9.1.20. You need not verify the conditions necessary to differentiate under the integral or for the resulting differentiated integral to be a continuous function of Δ.]

9.1.8. Find the optimal score function for detecting a shift in a Cauchy distribution having density

$$f(x) = \frac{1}{\pi(1+x^2)}, \qquad -\infty < x < \infty.$$

Draw a graph of this score function.

9.1.9. Use a combinatorial argument to show that the median test defined by (9.1.4) has a null distribution given by

$$P_0[M=k] = \frac{\binom{n}{k}\binom{m}{[\![N/2]\!]-k}}{\binom{N}{[\![N/2]\!]}} \qquad \text{for } k = 0, \ldots, n.$$

9.1.10. Let $X_1, \ldots, X_m$ and $Y_1, \ldots, Y_n$ denote independent random samples from continuous distributions with respective c.d.f.'s $[F(x)]^{1+\Delta}$ and

$1-[1-F(x)]^{1+\Delta}$. Find the form of the locally most powerful rank test of

$$H_0: \Delta = 0 \text{ versus } H_1: \Delta > 0.$$

Express the test in terms of scores involving expected values of functions of uniform order statistics. [Hint: Proceed in a fashion similar to that in the proof of Theorem 9.1.20. You need not verify the conditions necessary to differentiate under the integral or for the resulting differentiated integral to be a continuous function of Δ.]

9.1.11. Let $X_1,\ldots,X_m$ and $Y_1,\ldots,Y_n$ be independent random samples from continuous distributions with c.d.f.'s $F(x)$ and $F(x-\Delta)$, respectively. Consider the test of H_0: $\Delta=0$ versus H_1: $\Delta>0$ that rejects H_0 for large values of the linear rank statistic $S_N=\Sigma_{i=1}^n a_N(R_i)$, where R_j is the combined sample rank of Y_j and $a_N(1)\leqslant\cdots\leqslant a_N(N)$ are any set of scores satisfying (9.1.2). Let $\hat{\Delta}(X_1,\ldots,X_m;Y_1,\ldots,Y_n)$ be the Hodges–Lehmann two-sample estimator of Δ based on S_N [see (7.1.21)]. Show that the distribution of $\hat{\Delta}$ is symmetric about Δ if $F(\cdot)$ corresponds to a distribution that is symmetric about some value η and

$$a_N(i)+a_N(N+1-i)=2\bar{a}_N, \qquad \text{for } i=1,\ldots,N.$$

[Hint: See Theorems 7.2.21 and 8.2.13.] Comment: Such a $\hat{\Delta}$ is also symmetrically distributed about Δ for any continuous distribution $F(\cdot)$ when $m=n$ [see Hodges and Lehmann (1963)].

9.1.12. For the setting of Exercise 9.1.11, let the scores $a_N(i)$ be the expected value F_0 scores of (9.1.8); that is,

$$a_N(i)=a^*_{F_0}(i)=E_{F_0}[V_{(i)}], \qquad i=1,\ldots,N,$$

where $V_{(1)}<\cdots<V_{(N)}$ are the order statistics for a random sample of size N from a distribution with c.d.f. $F_0(\cdot)$. Show that the Hodges–Lehmann estimator $\hat{\Delta}$ associated with $S_N=\Sigma_{i=1}^n a^*_{F_0}(R_i)$ has a distribution that is symmetric about Δ for every m and n, provided $F(\cdot)$ and $F_0(\cdot)$ correspond to continuous distributions that are symmetric about values η and η_0, respectively.

9.1.13. For the setting of Exercise 9.1.11, let the scores $a_N(i)$ be the quantile F_0 scores (9.1.7) for a c.d.f. $F_0(\cdot)$ of a continuous distribution that has support on an interval (α,β); thus,

$$a_N(i)=a_{F_0}(i)=F_0^{-1}\left(\frac{i}{N+1}\right), \qquad i=1,\ldots,N.$$

Show that the Hodges–Lehmann estimator $\hat{\Delta}$ associated with

$$S_N = \sum_{i=1}^{n} F_0^{-1}\left(\frac{R_i}{N+1}\right)$$

has a distribution that is symmetric about Δ for every m and n, provided $F(\cdot)$ and $F_0(\cdot)$ correspond to distributions that are symmetric about values η and η_0, respectively.

9.1.14. Find the forms of the quantile F_0 scores $a_{F_0}(i)$ of (9.1.7) when $F_0(\cdot)$ corresponds to the following distributions:

(i) uniform $(0,1)$
(ii) exponential $(0,\infty)$
(iii) Cauchy with density $f(t)=[\pi(1+t^2)]^{-1}$, $-\infty<t<\infty$.

9.1.15. Find the forms of the expected value F_0 scores $a^*_{F_0}(i)$ of (9.1.8) when $F_0(\cdot)$ corresponds to the following distributions:

(i) uniform $(0,1)$
(ii) exponential $(0,\infty)$.

9.2. Asymptotic Properties in the Location Problem

In this section we discuss the asymptotic properties of linear rank statistics for the two-sample location problem in which $X_1,\ldots,X_m$ and $Y_1,\ldots,Y_n$ are independent random samples from continuous populations with c.d.f.'s $F(x)$ and $F(x-\Delta)$, respectively. For testing H_0: $\Delta=0$ versus H_1: $\Delta>0$, we would reject H_0 for large values of a two-sample linear rank statistic,

$$S_N = \sum_{i=1}^{n} a_N(R_i), \tag{9.2.1}$$

where R_i is the rank of Y_i among all $m+n$ observations. We require the scores to satisfy

$$a_N(i) = b_N\phi\left(\frac{i}{N+1}\right) + d_N, \qquad i=1,\ldots,N, \tag{9.2.2}$$

where $b_N>0$ for every N and $\phi(\cdot)$ is a nondecreasing square integrable score function. The nondecreasing property of the $\phi(\cdot)$ function ensures that the scores $a_N(i)$ are nondecreasing in i. The asymptotic normality under H_0 of S_N (properly normed) follows from the theory developed in Chapter 8. In particular, the two-sample regression constants were shown to satisfy Noether's condition (8.3.2) in Example 8.3.3. Noting (8.3.7), we see that Hájek's theorem (8.4.9) is applicable to these two-sample linear rank statistics and implies that

$$\frac{\sum_{i=1}^{n} a_N(R_i) - n\bar{a}_N}{\sigma_{0:N}} \tag{9.2.3}$$

has a limiting standard normal distribution under H_0, where

$$\sigma_{0:N}^2 = \frac{mn}{N(N-1)} \sum_{i=1}^{N} (a_N(i) - \bar{a}_N)^2. \tag{9.2.4}$$

Remark 8.4.13 shows that when H_0 holds we could also use the asymptotic null standard deviation in expression (9.2.3); that is, we could substitute

$$\sigma_{a:N}^2 = \frac{mnb_N^2}{N} \int_0^1 \{\phi(u) - \bar{\phi}\}^2 du, \tag{9.2.5}$$

in place of the exact null variance $\sigma_{0:N}^2$ in (9.2.3). The b_N^2 enters into (9.2.5) because of the linear relationship assumed in (9.2.2) between $a_N(i)$ and $\phi(i/(N+1))$.

As a result, the asymptotic normality under H_0 of a properly normed two-sample linear rank statistic is ensured for the common choices of scores, including the Wilcoxon, median, quantile normal scores and the linear forms given in (9.1.9) through (9.1.12). In addition, Theorem 8.4.16 shows that a similar result holds when using any scores $a_N^*(1),\ldots,a_N^*(N)$ that are asymptotically equivalent to the scores $a_N(1),\ldots,a_N(N)$ satisfying (9.2.2) in the sense that

$$\lim_{N\to\infty} \frac{1}{Nb_N^2} \sum_{i=1}^{N} \{a_N(i) - a_N^*(i)\}^2 = 0. \tag{9.2.6}$$

[Here b_N^2 is used in the denominator to make this convergence independent of the scale factor used in (9.2.2) and hence correspond to Theorem 8.4.16.] Thus, for example, the expected value normal scores also yield a two-sample linear rank statistic that, when standardized, is asymptotically normal under H_0. The approximate normality under H_0 of (9.2.3) when N is large is often used to provide approximate critical regions for the associated test. Thus for the common sets of scores, we would test H_0: $\Delta=0$ versus H_1: $\Delta>0$ when N is large by rejecting H_0 iff

$$S_N > n\bar{a}_N + \sigma_{0:N} z_\alpha, \tag{9.2.7}$$

where z_α is the upper 100αth percentile of the standard normal distribution.

Numerous theorems can be used to show the asymptotic normality of a properly normed linear rank statistic under alternatives, as well as under H_0. The classic limit theorem of this type was proved by Chernoff and Savage (1958). Further refinements on the conditions of that theorem were developed by Govindarajulu, Le Cam, and Raghavachari (1966). Hájek (1968) proved the limiting normality when the square integrable score function is continuous, subject to mild regularity conditions on the underlying distribution. Pyke and Shorak (1968) and Dupac and Hájek (1969) prove similar results with weak restrictions on $\phi(\cdot)$ and different conditions on the underlying distributions.

One of the important asymptotic properties of linear rank statistics for the two-sample location problem is their asymptotic efficiencies against shift alternatives. We again consider the model in which $X_1,\ldots,X_m$ and $Y_1,\ldots,Y_n$ represent independent random samples from continuous populations with respective c.d.f.'s $F(x)$ and $F(x-\Delta)$, and we test H_0: $\Delta=0$ versus H_1: $\Delta>0$. In Section 5.2 we detailed the conditions for determining the efficacies needed to obtain asymptotic efficiencies. For each statistic S_N these conditions centered around the properties of two sequences of constants $\{\mu_{S_{N_i}}(\cdot)\}$ and $\{\sigma^2_{S_{N_i}}(\cdot)\}$ defined over the parameter space. In the current problem

$$S_N = \sum_{i=1}^{n} a_N(R_i), \tag{9.2.8}$$

where

$$a_N(i) = b_N\phi\left(\frac{i}{N+1}\right) + d_N, \qquad i = 1,\ldots,N,$$

for $b_N>0$ and $\phi(\cdot)$, a nondecreasing square integrable score function. Define

$$\mu_{S_N}(\Delta) = n\bar{a}_N + \frac{\Delta mn\,b_N}{N}\int_0^1 \phi(u)\phi(u,f)\,du \tag{9.2.9}$$

and

$$\sigma^2_{S_N}(\Delta) = \frac{mn\,b_N^2}{N}\int_0^1\left[\phi(u)-\bar{\phi}\right]^2 du. \tag{9.2.10}$$

where

$$\phi(u,f) = \frac{-f'(F^{-1}(u))}{f(F^{-1}(u))}. \tag{9.2.11}$$

Let us now examine the assumptions of Noether's Theorem 5.2.7. Assumptions A3 and A5 are obvious. Note that

$$\lim_{N\to\infty} \frac{\mu'_{S_N}(0)}{\sqrt{N\sigma^2_{S_N}(0)}} = \sqrt{\lambda(1-\lambda)}\ \frac{\int_0^1 \phi(u)\phi(u,f)\,du}{\sqrt{\int_0^1 [\phi(u)-\bar{\phi}]^2\,du}}, \tag{9.2.12}$$

where $\lambda = \lim_{N\to\infty} (m/N)$. Consequently, Assumptions A4 and A6 are met, provided $0<\lambda<1$ and

$$\int_0^1 \phi(u)\phi(u,f)\,du > 0. \tag{9.2.13}$$

Asymptotic normality under H_0, Assumption A2, follows from Theorem 8.4.9, as previously discussed. Asymptotic normality for Pitman alternatives $\Delta_N = k/\sqrt{N}$, where $k>0$, follows from several of the limit theorems for alternatives cited earlier. However, a version specifically directed toward alternatives collapsing to the null in the fashion of Pitman alternatives is given as Theorem VI.2.3 of Hájek and Šidák (1967). This theorem requires no additional restrictions on the square integrable score function $\phi(\cdot)$. Therefore, all the assumptions of Theorem 5.2.7 are met, and

$$\text{eff}(S) = \sqrt{\lambda(1-\lambda)}\ \frac{\int_0^1 \phi(u)\phi(u,f)\,du}{\sqrt{\int_0^1 [\phi(u)-\bar{\phi}]^2\,du}}. \tag{9.2.14}$$

Example 9.2.15. Consider the scores $a_L(i)$ defined in (9.1.9). They are associated with the square integrable score function

$$\begin{aligned} \phi_L(u) &= u - 1/4, \qquad u < 1/4 \\ &= 0, \qquad 1/4 \leqslant u \leqslant 3/4 \\ &= u - 3/4, \qquad u > 3/4, \end{aligned}$$

since $a_L(i)=(N+1)\phi_L(i/(N+1))$ for $i=1,\ldots,N$. Note that

$$\int_0^1 \left[\phi_L(u)-\bar{\phi}\right]^2 du = \int_0^1 \phi_L^2(u)\,du = 1/96,$$

and

$$\int_0^1 \phi_L(u)\phi(u,f)\,du$$

$$= -\int_0^{1/4}(u-1/4)\frac{f'(F^{-1}(u))}{f(F^{-1}(u))}\,du - \int_{3/4}^1 (u-3/4)\frac{f'(F^{-1}(u))}{f(F^{-1}(u))}\,du$$

$$= -\int_{\xi_0}^{\xi_{1/4}}(F(t)-1/4)f'(t)\,dt - \int_{\xi_{3/4}}^{\xi_1}(F(t)-3/4)f'(t)\,dt,$$

where ξ_p satisfies $F(\xi_p)=p$. For simplicity, we now make the assumption that the support of the continuous distribution $F(\cdot)$ is an interval $[a,b]$, with $a<b$, and that the density goes to zero at its endpoints; that is, $f(x)\to 0$ as $x\uparrow b$ and as $x\downarrow a$. Then, integration by parts yields

$$\int_0^1 \phi_L(u)\phi(u,f)\,du = \int_{\xi_0}^{\xi_{1/4}} f^2(t)\,dt + \int_{\xi_{3/4}}^{\xi_1} f^2(t)\,dt.$$

Thus, using the scores $a_L(i)$, we see that the efficacy of the test based on the resulting linear rank statistic is

$$\text{eff}(L) = \sqrt{96\,\lambda(1-\lambda)}\left[\int_{\xi_0}^{\xi_{1/4}} f^2(u)\,du + \int_{\xi_{3/4}}^{\xi_1} f^2(u)\,du\right].$$

Before we consider an important interpretation of the general efficacy expression in (9.2.14), we should make a remark about the efficacies of tests with scores that are asymptotically equivalent. Let S denote a linear rank statistic with scores $a_N(1),\ldots,a_N(N)$ satisfying (9.2.2). If S^* is a linear rank statistic with scores $a_N^*(1),\ldots,a_N^*(N)$ satisfying

$$\lim_{N\to\infty}\frac{1}{Nb_N^2}\sum_{i=1}^{N}\{a_N(i)-a_N^*(i)\}^2 = 0,$$

then $(S_N - n\bar{a}_N)$ and $(S_N^* - n\bar{a}_N^*)$, when identically normed, not only have the same null distribution but also have the same distribution for a sequence of Pitman alternatives. Thus these two tests will have the same efficacy expressions and hence the same asymptotic efficiencies. Details of this equivalence may be found in Sections V.1.4, V.1.6 and VI.2.3 of Hájek and Šidák (1967).

The following lemma leads to a useful interpretation of the efficacy expression (9.2.14) for a two-sample linear rank statistic, subject to certain restrictions on $F(\cdot)$.

Lemma 9.2.16. *Let $F(\cdot)$ and $f(\cdot)$ denote the c.d.f. and density, respectively, of a continuous distribution. Suppose that the derivative, $f'(\cdot)$, of the density exists at all but at most a countable number of values and satisfies*

$$\int_{-\infty}^{\infty} f'(x)dx = 0.$$

Then

$$\int_0^1 \phi(u,f)du = 0$$

and

$$\int_0^1 \phi^2(u,f)du = \int_{-\infty}^{\infty}\left[\frac{f'(x)}{f(x)}\right]^2 f(x)dx \equiv I(f),$$

provided $I(f)$ is finite. The quantity $I(f)$ is called **Fisher's Information for the distribution F(·).** [*Note: The condition that the integral of $f'(\cdot)$ is zero is not very restrictive. It is satisfied, for example, when the distribution has support on an interval and the density goes to zero at the endpoints of the support interval. More generally, the conclusions of this lemma are valid so long as $f(\cdot)$ is absolutely continuous and both $\int_{-\infty}^{\infty}|f'(x)|dx$ and $I(f)$ are finite. See Lemmas I.2.4a and I.2.4d of Hájek and Šidák* (1967).]

Proof: From the definition of $\phi(u,f)$, we see that

$$\int_0^1 \phi(u,f)du = \int_0^1 -\frac{f'(F^{-1}(u))}{f(F^{-1}(u))}du$$
$$= \int_{-\infty}^{\infty} \frac{-f'(t)}{f(t)} f(t)dt = 0.$$

Also,

$$\int_0^1 \phi^2(u,f)\,du = \int_0^1 \left[\frac{f'(F^{-1}(u))}{f(F^{-1}(u))}\right]^2 du$$
$$= \int_{-\infty}^{\infty} \left[\frac{f'(t)}{f(t)}\right]^2 f(t)\,dt \equiv I(F). \quad \blacksquare$$

Under the conditions of Lemma 9.2.16 we can thus express the efficacy of a linear rank statistic S as

$$\text{eff}(S) = \sqrt{\lambda(1-\lambda)I(F)}\left[\frac{\int_0^1 \phi(u)\phi(u,f)\,du}{\sqrt{I(F)\int_0^1 \{\phi(u)-\bar{\phi}\}^2 du}}\right]$$
$$= \sqrt{\lambda(1-\lambda)I(F)}\ \text{Corr}[\phi(U),\phi(U,f)], \qquad \textbf{(9.2.17)}$$

where U is a uniform $(0,1)$ random variable. By interpreting the efficacy as a constant times a correlation coefficient we can now make two useful observations. First, we see that condition (9.2.13) may be viewed as requiring $\phi(\cdot)$ to be positively correlated with the optimal score function. Second, to maximize the efficacy among score functions satisfying (9.2.13), we see that we should use a $\phi(\cdot)$ that is perfectly correlated with $\phi(u,f)$, that is, $\phi(u,f)$ itself. Thus the optimal score function $\phi(u,f)$ produces scores that are asymptotically best in the sense that they maximize the efficacy among all such choices of $\phi(\cdot)$.

An even stronger conclusion is possible. It can be shown that this efficacy value of $\sqrt{\lambda(1-\lambda)I(F)}$ is not only the largest value achievable among linear rank statistics but is actually the largest value possible for any α-level test. Therefore, a test based on a linear rank statistic with the optimal score function $\phi(u,f)$ is **asymptotically efficient** in that it produces the best possible ARE. This topic is treated rigorously in Theorem VII.1.3 of Hájek and Šidák (1967). What follows is a heuristic development of this optimality.

Let us consider an equivalent testing problem in which $X_1,\ldots,X_m$ and $Y_1,\ldots,Y_n$ are independent random samples from continuous populations with c.d.f.'s $F(x+(\Delta n/N))$ and $F(x-(\Delta m/N))$, respectively. It is still the case that the Y distribution is shifted Δ units to the right of the X distribution, but this formulation explicitly recognizes that information about Δ is supplied by both random samples. A test of H_0: $\Delta=0$ versus

H_1: $\Delta>0$ will be based on a test statistic $V_N = V_N(X_1,\ldots,X_m;\ Y_1,\ldots,Y_n)$ that in some sense measures Δ. Since the standardizing quantities $\{\mu_{V_N}(\cdot)\}$ and $\{\sigma^2_{V_N}(\cdot)\}$ used in the definition of efficacy are asymptotically equivalent (as $\Delta\to 0$) to the mean and variance of V_N, the square of the efficacy of the test base on V_N may be written as

$$[\text{eff}(V)]^2 = \lim_{N\to\infty} \frac{[\mu'_{V_N}(0)]^2}{N\sigma^2_{V_N}(0)}$$

$$= \lim_{N\to\infty} \frac{\left[\dfrac{d}{d\Delta} E_\Delta(V_N)\right]^2}{N\ \text{Var}_\Delta(V_N)}\Bigg|_{\Delta=0} \tag{9.2.18}$$

Now we seek to show that this quantity is bounded above by

$$\lambda(1-\lambda)I(F).$$

The general form of the Cramér–Rao inequality (Theorem A.3.4 in the Appendix) shows that

$$\frac{\left[\dfrac{d}{d\Delta} E_\Delta(V_N)\right]^2}{\text{Var}_\Delta(V_N)} \leqslant I^*_{(\mathbf{X},\mathbf{Y})}(\Delta), \tag{9.2.19}$$

where

$$I^*_{\mathbf{Z}}(\theta) = E\left[\left\{\frac{\partial}{\partial\theta}\ln[h(\mathbf{Z};\theta)]\right\}^2\right],$$

with $h(\cdot\,;\theta)$ being the joint density of $\mathbf{Z}$.

Lemma 9.2.20. *If Z_1 and Z_2 are independent, then*

$$I^*_{(Z_1,Z_2)}(\theta) = I^*_{Z_1}(\theta) + I^*_{Z_2}(\theta)$$

Proof: Exercise 9.2.9. ∎

By definition,

$$I^*_{X_1}(\Delta)\big|_{\Delta=0} = \left(\frac{n}{N}\right)^2 E\left[\left\{\frac{f'\left(X_1+\dfrac{\Delta n}{N}\right)}{f\left(X_1+\dfrac{\Delta n}{N}\right)}\right\}^2\right]_{\Delta=0}$$

$$= \left(\frac{n}{N}\right)^2 I(F).$$

Similarly,

$$I^*_{Y_1}(\Delta)\big|_{\Delta=0} = \left(\frac{m}{N}\right)^2 I(F).$$

Thus by Lemma 9.2.20 we have

$$I^*_{(\mathbf{X},\mathbf{Y})}(\Delta)\big|_{\Delta=0} = \frac{mn}{N} I(F).$$

Combining this result with (9.2.18) and (9.2.19) yields

$$[\text{eff}(V)]^2 \leqslant \lambda(1-\lambda) I(F),$$

showing that the test based on a linear rank statistic with the optimal score function achieves the greatest efficacy possible. As a result, we see the Mann–Whitney–Wilcoxon test is not only the locally most powerful *rank* test for detecting a shift in a logistic distribution (see Example 9.1.30), but it is also asymptotically efficient among *all* α-level tests for this problem. Similarly, in Example 9.1.29 we showed that the expected value normal scores yield the locally most powerful rank test for detecting a shift in a normal distribution. We now see that it is also asymptotically efficient for that problem. Naturally, any set of scores asymptotically associated with that same $\phi(\cdot)$ function in the sense of Theorem 8.4.16 will share that property. In particular, the quantile normal scores also produce a rank test that is asymptotically efficient for detecting a shift in a normal distribution.

Let us now compare the asymptotic efficiencies of several linear rank statistics for the two-sample location problem. Let W denote the Mann–Whitney–Wilcoxon test with efficacy described in (5.4.14). Using (9.2.14), we see that the normal scores (NS) two-sample test has efficacy

$$\text{eff(NS)} = \sqrt{\lambda(1-\lambda)} \int_0^1 \Phi^{-1}(u)\phi(u,f)\,du.$$

The light-tailed distribution test (L) using the scores defined in (9.1.9) has an efficacy expression that was derived in Example 9.2.15. Other competitors include a test designed for heavy-tailed distributions (H) with scores given in (9.1.10) and efficacy presented in Exercise 9.2.5, the median test (M) (see (9.1.4) and Exercise 9.2.3) and the test RS defined in (9.1.11) for right-skewed distributions (also see Exercise 9.2.2).

Evaluating these efficacy values, we obtain the asymptotic efficiencies in Table 9.2.21. In the computations the W test was used as the bench mark

TABLE 9.2.21

DISTRIBUTION	ARE(L, W)	ARE(NS, W)	ARE(H, W)	ARE(M, W)	ARE(RS, W)
Uniform	2.000	$+\infty$	0.500	0.333	0.800
Normal	0.927	1.047	0.870	0.667	0.800
Logistic	0.781	0.955	0.945	0.750	0.800
Double exponential	0.500	0.847	1.125	1.333	0.800
Cauchy	0.264	0.708	1.339	1.333	0.800
Exponential	2.000	$+\infty$	0.500	0.333	1.800

for all of the other tests. Note that the L test does well for the uniform and exponential distributions. Both the Mann-Whitney-Wilcoxon and the normal scores tests have good power characteristics. The results of their comparisons for uniform and exponential distributions are an anomaly caused by a distribution that has a support region that ends at a finite point and a $\phi(\cdot)$ function, such as $\Phi^{-1}(\cdot)$, that is unbounded at both ends. This causes the efficacy of the normal scores test to be infinite. The tests H and M are both effective against shifts in heavy-tailed symmetric distributions, with the H test being better than M for lighter-tailed distributions. The RS test is particularly effective for detecting shifts in the right-skewed exponential distribution.

Exercises 9.2.

9.2.1. Proceeding directly from (9.2.14), show that the efficacy of the Mann–Whitney–Wilcoxon test is

$$\text{eff}(W) = \sqrt{12\lambda(1-\lambda)} \int_{-\infty}^{\infty} f^2(x)\,dx,$$

as given earlier in (5.4.14). To simplify the derivation, assume that the support of the distribution $F(\cdot)$ is an interval $[a,b]$, where $a<b$, and that $f(x)\to 0$ as $x\uparrow b$ or $x\downarrow a$. (The same form results under more general assumptions.)

9.2.2. **(a)** Show that the scores $a_{RS}(i)$ defined in (9.1.11) are associated with the score function

$$\begin{aligned}\phi_{RS}(u) &= u - \tfrac{1}{2}, \qquad u < \tfrac{1}{2}\\ &= 0, \qquad u \geqslant \tfrac{1}{2}.\end{aligned}$$

(b) Using (9.2.14), show that the efficacy of the test based on a linear rank statistic with scores $a_{RS}(i)$ is

$$\text{eff}(RS) = \sqrt{\lambda(1-\lambda)\frac{192}{5}} \int_{-\infty}^{m^*} f^2(t)\,dt,$$

where m^* satisfies $F(m^*) = 1/2$. (To simplify the derivation you may make the same assumption about $F(\cdot)$ as in Exercise 9.2.1.)

9.2.3. **(a)** Show that the two-sample median test is asymptotically efficient for detecting a shift in a double exponential distribution.

(b) Use (9.2.14) to show that the efficacy of the median test for detecting a shift in a continuous distribution with median m^* and density $f(\cdot)$ is

$$\text{eff}(M) = 2\sqrt{\lambda(1-\lambda)}\, f(m^*).$$

[To simplify the derivation, assume $F(\cdot)$ satisfies the conditions given in Exercise 9.2.1 and that $f(x)$ is continuous at m^* satisfying $F(m^*) = 1/2$.]

9.2.4. Find the value of ARE(NS, T) when the underlying distributions are (a) uniform, (b) normal, (c) logistic and (d) double exponential. Here NS denotes the test based on a linear rank statistic with expected value normal scores and T denotes the two-sample Student t-test. [Comment: Chernoff and Savage (1958) showed that, subject to certain regularity conditions on the distributions, ARE$(NS, T) \geqslant 1$ with equality holding if and only if the underlying distributions are normal.]

9.2.5. **(a)** Show that the scores $a_H(i)$ defined in (9.1.10) are associated with the score function

$$\phi_H(u) = \begin{cases} -\dfrac{1}{4}, & 0 < u < \dfrac{1}{4} \\ u - \dfrac{1}{2}, & \dfrac{1}{4} \leqslant u \leqslant \dfrac{3}{4} \\ \dfrac{1}{4}, & \dfrac{3}{4} < u < 1. \end{cases}$$

(b) Show that the test based on the linear rank statistic with scores $a_H(\cdot)$ has an efficacy of the form

$$\text{eff}(H) = \sqrt{24\lambda(1-\lambda)}\left[\int_{\xi_{1/4}}^{\xi_{3/4}} f^2(x)\,dx\right],$$

where ξ_p satisfies $F(\xi_p)=p$. [To simplify the derivation, assume the distribution $F(\cdot)$ satisfies the conditions of Exercise 9.2.1.]

9.2.6. In Table 9.2.21 verify the values of ARE(L, W) in the two-sample location problem when the underlying distributions are **(a)** logistic and **(b)** uniform. Here W stands for the Mann–Whitney–Wilcoxon test and L represents the test based on the two-sample linear rank statistic that uses the scores $a_L(i)$ of (9.1.9).

9.2.7. Consider a two-sample linear rank statistic for the location problem that uses the quantile F_0 scores given in (9.1.7).

(a) To what score function do these scores correspond?

(b) Find an expression for the efficacy of the test based on a linear rank statistic with these scores. Assume $F(\cdot)$ and $F_0(\cdot)$ are c.d.f.'s of continuous distributions with interval support (not necessarily the same distribution or interval).

(c) Let $F_0(x)=1-e^{-x}$, for $x>0$, denote the c.d.f. of an exponential distribution. Find ARE(F_0, W) when the underlying distributions are exponential. Here F_0 is used to denote the test based on a linear rank statistic with quantile F_0 scores, and W represents the Mann–Whitney–Wilcoxon test.

9.2.8. Let W stand for the test based on the Mann–Whitney–Wilcoxon statistic, and let RS represent the test based on the two-sample linear rank statistic that uses the scores $a_{RS}(i)$ defined in (9.1.11). Use Exercise 9.2.2 and verify the values of ARE(RS, W) in Table 9.2.21 when the underlying distributions are **(a)** logistic and **(b)** exponential.

9.2.9. Prove Lemma 9.2.20.

9.2.10. Let W, M, and H be the tests based on the two-sample linear rank statistics with scores as defined in (9.1.3), (9.1.4), and (9.1.10), respectively. Use the results of Exercises 9.2.3 and 9.2.5 to verify the values of ARE(M, W) and ARE(H, W) in Table 9.2.21, when the underlying distributions are **(a)** logistic and **(b)** double exponential.

9.3. The Two-Sample Scale Problem

In some settings the objective is to test for a difference in scale between two continuous populations. Specifically, let $X_1,\ldots,X_m$ and $Y_1,\ldots,Y_n$ be independent random samples from continuous populations with c.d.f.'s $F(x-\theta_x)$ and $F((y-\theta_y)/\eta)$, respectively, where $\theta_x(\theta_y)$ denotes the median

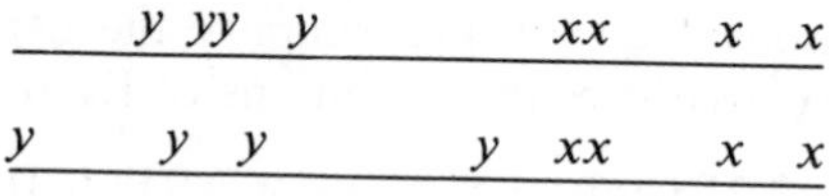

```
  y yy  y          xx    x  x
______________________________
y     y  y       y  xx    x  x
______________________________
```

FIGURE 9.3.1

of the distribution of $X_i(Y_j)$ and $\eta > 0$. We wish to test H_0: $\eta = 1$ versus H_1: $\eta > 1$; that is, we are testing to see if the Y_js tend to be more spread out than the X_is. (As before, the necessary modifications for alternatives $\eta < 1$ or $\eta \neq 1$ should be clear.) In general, one cannot accomplish the goal with a test based solely on ranks. As we see in Figure 9.3.1, the rank configuration may be the same for many different scale factors, provided that the two populations differ in location.

Hence we must focus our attention on the problem of testing H_0^*: $\theta_x = \theta_y$, $\eta = 1$ versus H_1^*: $\theta_x = \theta_y$, $\eta > 1$. By insisting that the populations have the same location parameter, the scale factor will have a definite effect on the probabilities of rank configurations. Under the alternative H_1^*, the Y_js will tend to be more spread out than the X_is and will therefore have greater attraction for the extreme (the largest *and* the smallest) ranks.

An intuitive rank test for scale was proposed by Mood (1954). **Mood's** statistic is the linear rank statistic

$$S_M = \sum_{j=1}^{n} \left\{ R_j - \frac{N+1}{2} \right\}^2, \tag{9.3.2}$$

where R_j is the rank of Y_j among all $m+n$ observations. Clearly, as the Y ranks tend to the extremes, the value of this test statistic increases. Thus to test H_0^* versus H_1^* we would reject for large values of S_M. The null distribution of S_M is easy to establish, since under H_0^* the populations are identical, and hence the rank vector $(Q_1, \ldots, Q_m, R_1, \ldots, R_n)$ is uniformly distributed over $\mathfrak{R}$, the set of permutations of the integers $1, \ldots, N$, where $N = m + n$.

More generally, we can test H_0^* versus H_1^* with a two-sample linear rank statistic

$$S_N = \sum_{j=1}^{n} a_N(R_j), \tag{9.3.3}$$

where $a_N(i)$ is nonincreasing (nondecreasing) for $i \leqslant (N+1)/2$ and nondecreasing (nonincreasing) for $i \geqslant (N+1)/2$. Mood's statistic is a special case of this. Other popular choices for the scores produce the **Ansari–Bradley**

(1960) statistic for which

$$a_{AB}(i) = \frac{N+1}{2} - \left| i - \frac{N+1}{2} \right|, \tag{9.3.4}$$

and the **Siegel–Tukey** (1960) statistic, for which

$$\begin{aligned} &a_{ST}(1) = 1, a_{ST}(N) = 2, a_{ST}(N-1) = 3, a_{ST}(2) = 4, \\ &a_{ST}(3) = 5, a_{ST}(N-2) = 6, a_{ST}(N-3) = 7, a_{ST}(4) = 8, \cdots, \end{aligned} \tag{9.3.5}$$

where the score values continue to increase in the designated pattern toward the middle ranks. Since extreme ranks are assigned small scores in both the Ansari–Bradley and Siegel–Tukey scoring schemes, the corresponding tests will reject H_0^* in favor of H_1^* for small values of the respective statistics. The method of scoring studied by Ansari–Bradley (1960) has the intuitive advantage that it assigns the same scores to corresponding extreme ranks, such as $N-1$ and 2. The Siegel–Tukey (1960) score assignments, however, are just a permutation of the integers $1,\ldots,N$. Hence, Theorem 8.2.11 shows that the H_0^* distribution of the Siegel–Tukey test statistic is the same as the null distribution of W in the Mann–Whitney–Wilcoxon test. The use of expected value normal scores in this problem was proposed by Capon (1961); specifically, he set

$$a_C(i) = E\left[Z_{(i)}^2 \right], \qquad i = 1,\ldots,N, \tag{9.3.6}$$

where $Z_{(1)} < \cdots < Z_{(N)}$ are the order statistics for a sample of size N from a standard normal distribution. Quantile normal scores of the form

$$a_K(i) = \left[\Phi^{-1}\left(\frac{i}{N+1} \right) \right]^2, \qquad i = 1,\ldots,N, \tag{9.3.7}$$

were proposed by Klotz (1962).

Tests for scale difference based on these linear rank statistics, S_N, all share the property of being nonparametric distribution-free. This follows because under H_0^* the rank vector $(Q_1,\ldots,Q_m,R_1,\ldots,R_n)$ is uniformly distributed over $\mathfrak{R}$ for any continuous underlying distribution $F(\cdot)$. Theorem 8.1.12 and Example 8.3.3 show that the null mean and variance of S_N are

$$\mu_{0:N} = n\bar{a}_N \text{ and } \sigma_{0:N}^2 = \frac{mn}{N(N-1)} \sum_{i=1}^{N} \left(a_N(i) - \bar{a}_N\right)^2$$

Now, suppose the scores satisfy

$$a_N(i) = b_N\phi\left(\frac{i}{N+1}\right) + d_N, \tag{9.3.8}$$

where $b_N > 0$ for every N (or $b_N < 0$ for every N) and $\phi(\cdot)$ is a square integrable score function that is nonincreasing for $0 < u \leqslant \frac{1}{2}$ and nondecreasing for $\frac{1}{2} \leqslant u < 1$. Then

$$(S_N - \mu_{0:N})/\sigma_{0:N}$$

has a limiting standard normal distribution under H_0^*. This fact is often used to create approximate (for large N) critical regions for the corresponding test. Under an alternative the limiting normality of a properly standardized version of S_N follows from the same theorems cited in the location problem (see Section 9.2).

To establish efficiency properties of linear rank statistics for the scale problem, we first note that if we add the same constant to every X_i and Y_j, the distribution of the ranks remains the same. Hence the distribution of the rank vector does not depend on the common median of the two distributions, namely, $\theta = \theta_x = \theta_y$. Therefore, in the following discussion of efficiency the common median is assumed (without loss of generality) to be $0 = \theta_x = \theta_y$, the median of $F(x)$. Let us re-express the parameter of interest in terms of Δ by $\eta = e^{\Delta}$. The conditions of Theorem 5.2.7 are then satisfied using the sequences of constants

$$\mu_{S_N}(\Delta) = n\bar{a}_N + \frac{\Delta mn\, b_N}{N}\int_0^1 \phi(u)\phi_{Sc}(u,f)\, du$$

and

$$\sigma_{S_N}^2(\Delta) = \frac{mn\, b_N^2}{N}\int_0^1 \left[\phi(u) - \bar{\phi}\right]^2 du,$$

where

$$\phi_{Sc}(u,f) = -1 - F^{-1}(u)\frac{f'(F^{-1}(u))}{f(F^{-1}(u))} \tag{9.3.9}$$

It follows in the same fashion as in Section 9.2 that the test based on the linear rank statistic S_N, with square integrable score function $\phi(u)$, will have efficacy

$$\text{eff}(S) = \sqrt{\lambda(1-\lambda)}\ \frac{\int_0^1 \phi(u)\phi_{Sc}(u,f)\,du}{\sqrt{\int_0^1 \left[\phi(u)-\bar{\phi}\right]^2 du}}, \tag{9.3.10}$$

provided $\lim_{N\to\infty}(m/N)=\lambda$, $0<\lambda<1$, and $\int_0^1\phi(u)\phi_{Sc}(u,f)\,du>0$.

Assumption A1 for finding asymptotic efficiencies with Noether's Theorem 5.2.7 is proved as Theorem VI.2.3 in Hájek and Šidák (1967). Theorem II.4.5 of that same reference shows that the scores

$$a_N^*(i) = E\left[\phi_{Sc}(U_{(i)},f)\right], \qquad i = 1,\dots,N,$$

where $U_{(1)}<\cdots<U_{(N)}$ are the order statistics for a random sample of size N from a uniform (0, 1) distribution, will yield a test statistic that is locally most powerful among all α-level rank tests of H_0^* versus H_1^*. Thus, $\phi_{Sc}(u,f)$, defined in (9.3.9), is termed the **optimal score function for scale,** since it plays the role analogous to that of $\phi(u,f)$ in the location problem. Also in analogy to the location problem, the test based on the linear rank statistic, S_N, with scores generated by the optimal score function, $\phi_{Sc}(u,f)$, is an asymptotically efficient test of H_0^* versus H_1^* for that particular $F(\cdot)$. Details are found in Section VII.1.3 of Hájek and Šidák (1967).

Let S_N denote a linear rank statistic for the scale problem, with scores $a_N(1),\dots,a_N(N)$ satisfying (9.3.8). Let S_N^* be a linear rank statistic with related scores $a_N^*(1),\dots,a_N^*(N)$ satisfying

$$\lim_{N\to\infty} \frac{1}{Nb_N^2} \sum_{i=1}^{N} \{a_N(i)-a_N^*(i)\}^2 = 0$$

Then, just as in the two-sample location problem, the two quantities $(S_N - n\bar{a}_N)$ and $(S_N^* - n\bar{a}_N^*)$, when identically normalized, have the same limiting normal distribution under H_0^* and also under a sequence of Pitman alternatives. Thus these two tests have the same efficacy expression. The Ansari–Bradley and the Siegel–Tukey statistics are asymptotically equivalent in this sense and hence they have the same efficacy. A similar statement applies to the Capon and Klotz tests.

Efficacies for the scale problem are computed much as they are in the location case. Let K, AB, and M denote the normal scores test due to Klotz, the Ansari–Bradley test, and Mood's test, respectively. Table 9.3.11 of asymptotic efficiencies was constructed by Klotz (1962).

TABLE 9.3.11

DISTRIBUTION	ARE(AB, K)	ARE(AB, M)
Uniform	0.000	0.600
Normal	0.608	0.800
Logistic	0.750	0.837
Double exponential	0.774	0.864
Cauchy	1.783	1.067
Exponential	0.000	0.806

For testing H_0: $\eta=1$ versus H_1: $\eta>1$ when one cannot assume that the two populations have equal locations, Sukhatme (1958) proposed a procedure that is equivalent to performing the Ansari–Bradley test on $X_1,\ldots,X_m$, $Y_1-\hat{\Delta},\ldots,Y_n-\hat{\Delta}$, with $\hat{\Delta}=\text{median}_{1\leqslant j\leqslant n}Y_j-\text{median}_{1\leqslant i\leqslant m}X_i$. That is, he proposed aligning the locations of the samples and then conducting an appropriate rank test. He studied the properties of this procedure, showing among other things that it is asymptotically distribution-free. A brief discussion of this paper and of many other contributions to the scale problem is given in a bibliography by Duran (1976).

Exercises 9.3.

9.3.1. Find an expression for the null hypothesis (H_0^*) mean and variance of the Ansari–Bradley statistic.

9.3.2. Suppose that $X_1,\ldots,X_m$ and $Y_1,\ldots,Y_n$ are independent random samples from continuous distributions with c.d.f.'s $F(x)$ and $F(x/\eta)$, respectively, where $F(0)=1/2$. Show that the optimal score function for testing H_0^*: $\eta=1$ versus H_1^*: $\eta>1$ does not depend on the scale of the underlying distribution $F(\cdot)$. That is, assume $F(x)=G(x/\eta_0)$ for all x and show that the optimal score function does not depend on η_0.

9.3.3. Determine the mean and variance of Mood's statistic under H_0^*.

9.3.4. Prove or disprove that the H_0^* distribution of Mood's statistic is symmetric about its null mean for every m and n.

9.3.5. Find the optimal score function for scale when the underlying distributions are normal with median zero. Graph this score function.

9.3.6. **(a)** Show that the Ansari–Bradley statistic is associated with the score function $\phi_{AB}(u)=|u-\frac{1}{2}|$, $0<u<1$.

(b) Use (9.3.10) to show that the efficacy of the Ansari–Bradley test is given by

$$\sqrt{48\lambda(1-\lambda)}\left[\int_{m^*}^{\infty} xf^2(x)\,dx - \int_{-\infty}^{m^*} xf^2(x)\,dx\right],$$

when $F(\cdot)$ has median m^*. To simplify the derivation, assume that the support of the distribution $F(\cdot)$ is an interval $[a,b]$, $a<b$, and that $xf(x)\to 0$ as $x\downarrow a$ and as $x\uparrow b$.

9.3.7. Find the optimal score function for scale when the underlying distributions are double exponential with median zero. Graph this score function.

9.3.8. **(a)** Show that Mood's statistic is associated with the score function

$$\phi_M(u) = \left(u-\tfrac{1}{2}\right)^2, \qquad 0<u<1.$$

(b) Use (9.3.10) to show that the efficacy of Mood's test is

$$\sqrt{720\lambda(1-\lambda)}\int_{-\infty}^{\infty}\left[F(x)-\frac{1}{2}\right]xf^2(x)\,dx,$$

when the median of $F(\cdot)$ is zero. To simplify the derivation, you may assume that $F(\cdot)$ has support on the interval $[a,b]$, $a<b$, and that $tf(t)\to 0$ as $t\downarrow a$ and as $t\uparrow b$.

9.3.9. Use the results of Exercises 9.3.6 and 9.3.8 to verify the $\text{ARE}(AB,M)$ values in Table 9.3.11 corresponding to the following distributions:

(a) Uniform $(-\frac{1}{2},\frac{1}{2})$.

(b) Standard normal.

(c) Double exponential with density $f(t)=\frac{1}{2}\exp(-|t|)$.

10 THE ONE-SAMPLE LOCATION PROBLEM

10.1. Linear Signed Rank Statistics

We now consider a location test in a one-sample setting. Let $X_1,\ldots,X_n$ denote a random sample from a continuous population with c.d.f. $F(x-\theta)$, where the associated density satisfies $f(x)=f(-x)$ for all x; hence, the X distribution is symmetric about the unknown location parameter θ. We wish to test

$$H_0\colon \theta = 0 \text{ versus } H_1\colon \theta > 0. \tag{10.1.1}$$

There is no loss of generality in dealing with the particular null value zero. If the null were $\theta=\theta_0$, we would perform a test of (10.1.1) on the random variables $X_1-\theta_0,\ldots,X_n-\theta_0$. We considered this same problem in Section 2.4. For each X_i we used its absolute value $|X_i|$ and whether or not it was positive, namely, $\Psi_i=\Psi(X_i)$, where $\Psi(t)=1,0$ as $t>,\leqslant 0$. The absolute rank vector, $\mathbf{R}^+=(R_1^+,\ldots,R_n^+)$, was constructed so that R_i^+ is the rank of $|X_i|$ among $|X_1|,\ldots,|X_n|$, and the signed ranks were defined to be $\Psi_1R_1^+,\ldots,\Psi_nR_n^+$.

One of the most widely used tests for the problem in (10.1.1) is the **Wilcoxon signed rank test**, based on the statistic

$$W^+ = \sum_{i=1}^{n} \Psi_i R_i^+. \tag{10.1.2}$$

In Section 2.4 we discussed many of the properties of this test. In particular, we noted that the null distribution of W^+ is an immediate consequence of the next result, which also plays a central role in this chapter.

Remark 10.1.3. Recall Theorem 2.4.4 which stated that under H_0 the $n+1$ random variables $\Psi_1,\ldots,\Psi_n,\mathbf{R}^+$ are mutually independent. In addition, each $\Psi_i=1,0$ with equal probability $1/2$, for $i=1,\ldots,n$, and $\mathbf{R}^+$ is uniformly distributed over $\mathfrak{R}$, the set of all permutations of the integers $1,\ldots,n$.

This important result shows that any statistic that is a function of $\boldsymbol{\Psi}$ and $\mathbf{R}^+$ alone is nonparametric distribution-free over the class of underlying distributions associated with H_0, namely, the class of distributions that are continuous and symmetric about zero. Moreover, the H_0 distributions of such statistics can be determined from the simple, discrete distributions of the independent random variables $\Psi_1,\ldots,\Psi_n$ and $\mathbf{R}^+$. In particular, we study the following general class of statistics.

Definition 10.1.4. Consider the set of **scores** $a(i)$, $i=1,2,\ldots,n$, satisfying

$$0 \leqslant a(1) \leqslant \cdots \leqslant a(n), \qquad a(n) > 0.$$

The **linear signed rank statistic** corresponding to these scores is

$$S^+ = \sum_{i=1}^{n} \Psi_i a(R_i^+).$$

Naturally, the Wilcoxon signed rank statistic is a member of this general class, corresponding to the Wilcoxon scores

$$a_{W^+}(i) = i, \qquad i = 1,\ldots,n.$$

Another popular member of this class uses the scores

$$a_B(i) = 1, \qquad i = 1,\ldots,n, \tag{10.1.5}$$

producing the **sign test** statistic

$$B = \sum_{i=1}^{n} \Psi_i$$

$$= [\text{number of positive } X_i\text{s}]. \tag{10.1.6}$$

Expected value normal scores for this one-sample location problem are of the form

$$a_{\text{NS}^*}(i) = E[|Z|_{(i)}], \qquad i = 1,\ldots,n, \tag{10.1.7}$$

where $|Z|_{(1)} < \cdots < |Z|_{(n)}$ denote the order statistics for the absolute values of a random sample of size n from a standard normal distribution. They were first proposed by Fraser (1957b) and have been tabled by Klotz (1963) and Govindarajulu and Eisenstat (1965). The **quantile normal scores** for this problem are

$$a_{\mathrm{NS}^+}(i) = \Phi^{-1}\left(\frac{1}{2} + \frac{1}{2}\left(\frac{i}{N+1}\right)\right), \qquad i = 1,\ldots,n, \qquad \textbf{(10.1.8)}$$

as first described by Van Eeden (1963).

Scores which emphasize extreme observations and hence are effective in detecting a shift in a light-tailed distribution are

$$\begin{aligned} a_{L^+}(i) &= 0 \qquad \text{if } i \leqslant [\![(n+1)/2]\!] \\ &= i - [\![(n+1)/2]\!], \qquad \text{if } i > [\![(n+1)/2]\!]. \end{aligned} \qquad \textbf{(10.1.9)}$$

These scores are analogues to the $a_L(\cdot)$ scores of (9.1.9) for the two-sample problem, and were proposed by Randles and Hogg (1973). The analogues to the heavy-tailed symmetric distribution scores of (9.1.10) are

$$\begin{aligned} a_{H^+}(i) &= i, \qquad i \leqslant (n+1)/2 \\ &= (n+1)/2, \qquad i > (n+1)/2, \end{aligned} \qquad \textbf{(10.1.10)}$$

which were developed by Policello (1974).

The following lemma yields an easy method of characterizing the null distribution of linear signed rank statistics.

Lemma 10.1.11. *Let $X_1,\ldots,X_n$ be i.i.d. continuous random variables symmetrically distributed about 0. If $\Psi_i = \Psi(X_i)$ and R_i^+ is the rank of $|X_i|$ among $|X_1|,\ldots,|X_n|$, then any linear signed rank statistic has the property that*

$$S^+ = \sum_{i=1}^{n} \Psi_i a(R_i^+) \stackrel{d}{=} \sum_{j=1}^{n} \Psi_j a(j).$$

Proof: For any arbitrary $\mathbf{r}^+ \in \mathfrak{R}$, the independence of $\Psi_1,\ldots,\Psi_n$ and $\mathbf{R}^+$ implies

$$\begin{aligned} \{S^+ | \mathbf{R}^+ = \mathbf{r}^+\} &= \left\{ \sum_{i=1}^{n} \Psi_i a(r_i^+) | \mathbf{R}^+ = \mathbf{r}^+ \right\} \\ &\stackrel{d}{=} \sum_{i=1}^{n} \Psi_i a(r_i^+). \end{aligned}$$

Define d_j to be the position of the integer j in the vector $(r_1^+, \dots, r_n^+)$, $j=1,\dots,n$. Thus

$$\sum_{i=1}^{n} \Psi_i a(r_i^+) = \sum_{j=1}^{n} \Psi_{d_j} a(j).$$

The fact that $\Psi_1,\dots,\Psi_n$ are i.i.d. implies $(\Psi_1,\dots,\Psi_n) \stackrel{d}{=} (\Psi_{d_1},\dots,\Psi_{d_n})$. (See Theorem 1.3.5.) It follows that

$$\{S^+ | \mathbf{R}^+ = \mathbf{r}^+\} \stackrel{d}{=} \sum_{j=1}^{n} \Psi_{d_j} a(j) \stackrel{d}{=} \sum_{j=1}^{n} \Psi_j a(j).$$

Since this holds for each $\mathbf{r}^+$ in $\mathfrak{R}$, the lemma is established. ■

Example 10.1.12. Since, under H_0, $\Psi_1,\dots,\Psi_n$ are i.i.d. Bernoulli random variables with $p=1/2$, the null distribution of

$$\sum_{i=1}^{n} \Psi_i a(i)$$

is easily determined. For example, let the scores be given by $a_{L^+}(i)$ in (10.1.9). Note that among these scores are the integers $1,2,\dots,n^*$, where $n^* = n - [\![(n+1)/2]\!]$. All the remaining scores are zero. Using Lemma 10.1.11, we see that the corresponding linear signed rank statistic S_L^+ satisfies

$$\begin{aligned} S_L^+ &\stackrel{d}{=} \sum_{i=1}^{n} \Psi_i a_{L^+}(i) \\ &\stackrel{d}{=} \Psi_1 + 2\Psi_2 + \cdots + n^*\Psi_{n^*}. \end{aligned}$$

Hence the null distribution of S_L^+ for sample size n is the same as that of the Wilcoxon signed rank statistic W^+ for sample size $n^* = n - [\![(n+1)/2]\!]$.

Lemma 10.1.11 is also useful in proving the following theorem.

Theorem 10.1.13. *Let $X_1,\dots,X_n$ be i.i.d. continuous random variables symmetrically distributed about zero. Then for any linear signed rank statistic S^+, its null mean and variance are*

$$E_0[S^+] = \left(\frac{1}{2}\right) \sum_{i=1}^{n} a(i) = \frac{n}{2}\bar{a}$$

and

$$\text{Var}_0[S^+] = \left(\frac{1}{4}\right)\sum_{i=1}^{n} a^2(i).$$

Proof: Exercise 10.1.2. ■

Example 10.1.14. Using $a_{W^+}(i)=i$, $i=1,\ldots,n$, we see that the null mean and variance of the Wilcoxon signed rank statistic are (as also shown in Section 2.4)

$$E_0[W^+] = \left(\frac{1}{2}\right)\sum_{i=1}^{n} i = \frac{n(n+1)}{4}$$

and

$$\text{Var}_0[W^+] = \left(\frac{1}{4}\right)\sum_{i=1}^{n} i^2 = \frac{n(n+1)(2n+1)}{24}.$$

Example 10.1.15. Using Examples 10.1.12 and 10.1.14 we see that the null mean and variance of the linear signed rank statistic with scores $a_{L^+}(\cdot)$, as defined in (10.1.9), are

$$E_0[S_L^+] = \frac{n^*(n^*+1)}{4}$$

and

$$\text{Var}_0[S_L^+] = \frac{n^*(n^*+1)(2n^*+1)}{24},$$

where $n^* = n - [\![(n+1)/2]\!]$.

We now establish the symmetry of the null distribution of every linear signed rank statistic.

Theorem 10.1.16. *Assume that $X_1,\ldots,X_n$ are i.i.d. according to some continuous distribution that is symmetric about zero. Then any linear signed rank statistic*

$$S^+ = \sum_{i=1}^{n} \Psi_i a(R_i^+)$$

is symmetrically distributed about its mean $(n/2)\bar{a}$.

Proof: Since $(\Psi_1,\ldots,\Psi_n) \stackrel{d}{=} (1-\Psi_1,\ldots,1-\Psi_n)$, it follows from Lemma 10.1.11 that

$$\begin{aligned} S^+ - \frac{n}{2}\bar{a} &\stackrel{d}{=} \sum_{i=1}^{n} \Psi_i a(i) - \frac{n}{2}\bar{a} \\ &\stackrel{d}{=} \sum_{i=1}^{n} (1-\Psi_i)a(i) - \frac{n}{2}\bar{a} \\ &= \frac{n}{2}\bar{a} - \sum_{i=1}^{n} \Psi_i a(i) \\ &\stackrel{d}{=} \frac{n}{2}\bar{a} - S^+. \quad \blacksquare \end{aligned}$$

The following lemma will be of use in the remainder of this chapter.

Lemma 10.1.17. *Let $F(\cdot)$ denote the c.d.f. of a continuous random variable with density satisfying $f(x)=f(-x)$ for all x. Then if $|X|_{(1)} < \cdots < |X|_{(n)}$ denote the order statistics of the absolute values for a random sample of size n from $F(\cdot)$, we have*

$$|X|_{(i)} \stackrel{d}{=} F^{-1}\left(\frac{1}{2}+\frac{1}{2}U_{(i)}\right),$$

where $U_{(1)} < \cdots < U_{(n)}$ are the order statistics for a random sample of size n from a uniform $(0,1)$ distribution.

Proof: Exercise 10.1.5. $\blacksquare$

This shows, for example, that the expected value normal scores of (10.1.7) may be written as

$$a_{\mathrm{NS}^*}(i) = E\left[\Phi^{-1}\left(\frac{1}{2}+\frac{1}{2}U_{(i)}\right)\right], \qquad i = 1,\ldots,n. \qquad \textbf{(10.1.18)}$$

The following theorem, evolved from initial work by Fraser (1957b), corresponds to Theorem 9.1.20 in the two-sample problem and provides a means of selecting a set of scores for a linear signed rank statistic.

Theorem 10.1.19. *Let $X_1,\ldots,X_n$ denote a random sample from a continuous population with c.d.f. $F(x-\theta)$, where the associated density $f(\cdot)$ satisfies $f(x)=f(-x)$ for all x and has a support set equal to the whole real line. Assume that $f'(x)$, the derivative of the density function $f(x)$, exists and is continuous at all $x \neq 0$. Setting $f'(0)=0$, we also assume that $-f'(x)/f(x)$ is a nondecreasing function of x. Finally, assume that for some $\delta > 0$ there exist* *functions* $g_1(\cdot)$ *and* $g_2(\cdot)$ *satisfying*

$$|f'(x-\theta)| \leqslant g_1(x)$$

and

$$f(x-\theta) \leqslant g_2(x)$$

for all $-\delta<\theta<\delta$ *and all* x *and, in addition, that*

$$E_F\left[\frac{g_1(W_1)\prod_{j=2}^{n} g_2(W_j)}{\prod_{i=1}^{n} f(W_i)}\right] < \infty,$$

where $W_1,\ldots,W_n$ *are i.i.d. nonnegative random variables with c.d.f.* $F(x)/(1-F(0))$. *Then the locally most powerful test based on* $\mathbf{\Psi}$ *and* $\mathbf{R}^+$ *for testing* H_0: $\theta=0$ *versus* H_1: $\theta>0$ *rejects for large values of*

$$S^+ = \sum_{i=1}^{n} \Psi_i a^*(R_i),$$

where

$$a^*(i) = E\left[-\frac{f'(|X|_{(i)})}{f(|X|_{(i)})}\right] \qquad \textbf{(10.1.20)}$$

and $|X|_{(1)}<\cdots<|X|_{(n)}$ *denote the order statistics for the absolute values of a random sample of size n from* $F(\cdot)$.

Proof: Exercise 10.1.11. ■

Note that Lemma 10.1.17 shows that the scores of this **locally most powerful signed rank test** for detecting a change in the location of $F(\cdot)$ may alternatively be written as

$$E\left[-\frac{f'\left(F^{-1}\left(\frac{1}{2}+\frac{1}{2}U_{(i)}\right)\right)}{f\left(F^{-1}\left(\frac{1}{2}+\frac{1}{2}U_{(i)}\right)\right)}\right], \qquad \textbf{(10.1.21)}$$

where $U_{(1)}<\cdots<U_{(n)}$ denote the order statistics for a random sample of

size n from a uniform $(0,1)$ distribution. Note also that, just as in the two-sample problem, the locally most powerful signed rank test for a change in location of the distribution $F(\cdot)$ does not depend on the scale parameter of that distribution. (See Exercise 10.1.4.)

Example 10.1.22. Suppose that $f(x)$ is the standard normal density function. From Example 9.1.29 we see that $[-f'(x)/f(x)]=x$, and hence that the expected value normal scores

$$a_{\mathrm{NS}^*}(i) = E\big[|Z|_{(i)}\big], \qquad i = 1,2,\ldots,n,$$

where $|Z|_{(1)}<\cdots<|Z|_{(n)}$ are the order statistics for the absolute values of a random sample of size n from a standard normal distribution, yield the locally most powerful test based on $\mathbf{\Psi}$ and $\mathbf{R}^+$ for detecting a change of location in a normal distribution.

Example 10.1.23. Consider a logistic distribution with c.d.f. $F(x)=[1+e^{-x}]^{-1}$. From Example 9.1.30 we see that $-f'(x)/f(x)=2F(x)-1$. Thus, according to (10.1.21), the scores

$$a^*(i) = E\big[U_{(i)}\big] = \frac{i}{n+1}, \qquad i = 1,\ldots,n,$$

provide the locally most powerful test based on $\mathbf{\Psi}$ and $\mathbf{R}^+$ for detecting a change of location in a logistic distribution. These scores clearly correspond to the Wilcoxon signed rank test.

In the remainder of this section we show that the null distribution properties of linear signed rank statistics hold under less restrictive conditions. Assume that $X_1,\ldots,X_n$ are independent, continuous random variables with X_i having a density function $f_i(x-\theta)$ satisfying $f_i(x)=f_i(-x)$, for all x values. (Thus the X_is are not necessarily identically distributed, but each has a continuous distribution that is symmetric about θ. These less restrictive assumptions might be appropriate, for example, when the n data values consist of measurements drawn from several populations on which the treatment should react similarly but not necessarily identically.) We wish to test

$$H_0\colon \theta = 0 \text{ versus } H_1\colon \theta > 0.$$

(Again, the value 0 is chosen merely for convenience. For H_0: $\theta=\theta_0$, we would base the test on the random variables $X_1-\theta_0,\ldots,X_n-\theta_0$.) The following lemma provides the key to relating this less restrictive case to the previous one.

Lemma 10.1.24. *If $X_1,\ldots,X_n$ are independent and each has some (not necessarily the same) continuous distribution that is symmetric about zero, then*

(i) $\Psi_1,\ldots,\Psi_n,\mathbf{R}^+$ *are* $n+1$ *mutually independent random variables and*
(ii) *each* $\Psi_i=1$ *or* 0 *with equal probability* $\frac{1}{2}$.

Proof: Using Lemma 2.4.2 we see that Ψ_i and $|X_i|$ are independent random variables. Since the X_is are mutually independent and since $\mathbf{R}^+$ depends only on $|X_1|,\ldots,|X_n|$, part (i) follows. Part (ii) is due to the fact that each X_i is symmetrically distributed about zero. ■

Comparing the results of Lemma 10.1.24 with Remark 10.1.3, we see that the only thing that is not the same in this less restrictive setting is that it no longer necessarily follows that $\mathbf{R}^+$ has a uniform distribution over $\mathcal{R}$, the set of all permutations of the integers $1,\ldots,n$. This, however, does not invalidate the following analogue to Lemma 10.1.11.

Lemma 10.1.25. *If $X_1,\ldots,X_n$ are independent and each X_i has some (not necessarily the same) distribution that is continuous and symmetric about zero, then for any linear signed rank statistic S^+ we have*

$$S^+ = \sum_{i=1}^{n} \Psi_i a(R_i^+) \stackrel{d}{=} \sum_{j=1}^{n} \Psi_j a(j).$$

Proof: The proof is identical with that of Lemma 10.1.11. ■

Note that the results of Theorems 10.1.13 and 10.1.16 followed from Lemma 10.1.11 and the fact that $\Psi_1,\ldots,\Psi_n$ are independent Bernoulli random variables with $p=1/2$. Hence their conclusions also remain valid under these less restrictive assumptions.

Exercises 10.1.

10.1.1. Consider the linear signed rank statistic S^+ with scores $a_{H^+}(i)$ given in (10.1.10).

(a) Derive expressions for the null mean and variance of S^+.

(b) Find the null distribution of S^+ when $n=3$.

10.1.2. Use Lemma 10.1.11 to prove Theorem 10.1.13.

10.1.3. With scores $a(i)=i^2$, for $i=1,\ldots,n$, form a linear signed rank statistic S^+.

(a) Derive expressions for the mean and variance of S^+ under H_0.

(b) Determine the H_0 distribution of S^+ for the case of $n=4$.

10.1.4. Let $f(t)$ be the density of a continuous random variable with support over the whole real line and satisfying $f(t)=f(-t)$ for all t. Let $X_1,\ldots,X_n$ be i.i.d. with density $(1/\delta)f((x-\theta)/\delta)$, where $\delta>0$ is unknown. Consider testing H_0: $\theta=0$ versus H_1: $\theta>0$, and assume that the conditions of Theorem 10.1.19 are satisfied. Show that the locally most powerful test of H_0 versus H_1 based on $\boldsymbol{\Psi}$ and $\mathbf{R}^+$ does not depend on $\delta>0$.

10.1.5. Prove Lemma 10.1.17.

10.1.6. Apply Theorem 10.1.19 to show that the sign test is the locally most powerful test based on $\boldsymbol{\Psi}$ and $\mathbf{R}^+$ for detecting a change in location for a double exponential distribution.

10.1.7. Let $S^+=\sum_{i=1}^n \Psi_i a(R_i^+)$ and $S'=\sum_{i=1}^n \Psi_i a'(R_i^+)$ be two linear signed rank statistics computed on the same data $X_1,\ldots,X_n$. Under H_0 show that

$$\operatorname{Cov}(S^+,S') = \left(\frac{1}{4}\right)\sum_{i=1}^{n} a(i)a'(i).$$

10.1.8. Use the result of Exercise 10.1.7 to find an expression for the correlation between the sign test statistic and the Wilcoxon signed rank test statistic under H_0. Evaluate this expression for the case $n=6$.

10.1.9. Let $X_1,\ldots,X_n$ be i.i.d. continuous random variables with density $f(x-\theta)$, where $f(x)=f(-x)$ for all x. For testing H_0: $\theta=0$ versus H_1: $\theta>0$ we could use any linear signed rank statistic S^+, as described in Definition 10.1.4. Show that such a test based on S^+ will have a monotone power function in θ and hence will be an unbiased test for this problem at all its natural significance levels.

10.1.10. Let $X_1,\ldots,X_n$ be a random sample from a continuous distribution with c.d.f. $F(x-\theta)$, where $F(\cdot)$ is the c.d.f. for a distribution that is symmetric about 0. Consider the test of H_0: $\theta=0$ versus H_1: $\theta>0$ that rejects H_0 for large values of a linear signed rank statistic $S^+=\sum_{i=1}^n \Psi_i a(R_i^+)$ as given in Definition 10.1.4. Let $\hat{\theta}(X_1,\ldots,X_n)$ be the Hodges–Lehmann one-sample estimator of θ based on S^+ [see (7.1.4)]. Show that the distribution of $\hat{\theta}$ is symmetric about θ.

10.1.11. Use the representation in Lemma 4.3.18 to prove Theorem 10.1.19. [Hint: The structure of the proof is similar to that of Theorem 9.1.20.]

10.2. Asymptotic Properties in the One-Sample Location Problem

We now consider the asymptotic behavior of linear signed rank statistics. We assume a condition on the scores similar to that used in the two previous chapters. Specifically, we assume

$$a_n(i) = b_n \phi^+\left(\frac{i}{n+1}\right), \tag{10.2.1}$$

where $b_n > 0$ for every n and $\phi^+(\cdot)$ is a nonnegative, nondecreasing score function that cannot depend on n and satisfies

$$0 < \int_0^1 [\phi^+(u)]^2 du < \infty. \tag{10.2.2}$$

The nonnegative and nondecreasing properties of $\phi^+(\cdot)$ are assumed here to ensure that the associated scores have those same properties.

The Wilcoxon scores are associated with

$$\phi_{W^+}^+(u) = u, \qquad 0 < u < 1,$$

since

$$a_{W^+}(i) = i = (n+1)\phi_{W^+}^+\left(\frac{i}{n+1}\right), \qquad i = 1, \ldots, n.$$

The quantile normal scores shown in (10.1.8) [and also the expected value normal scores in (10.1.7)] are associated with the score function

$$\phi_{\text{NS}}^+(u) = \Phi^{-1}\left(\frac{1}{2} + \frac{1}{2}u\right),$$

where $\Phi^{-1}(\cdot)$ denotes the inverse of the c.d.f. of a standard normal distribution. Using (10.1.21) and Theorem 10.1.19, we see that the **optimal one-sample score function** is of the form

$$\phi^+(u,f) = -\frac{f'\left(F^{-1}\left(\frac{1}{2} + \frac{1}{2}u\right)\right)}{f\left(F^{-1}\left(\frac{1}{2} + \frac{1}{2}u\right)\right)}, \qquad 0 < u < 1. \tag{10.2.3}$$

The score functions for the Wilcoxon and normal scores tests are special cases of this general form corresponding to the logistic and standard normal distributions, respectively. (See Examples 10.1.22 and 10.1.23.)

We now proceed to verify the asymptotic normality under H_0 of a large class of appropriately normed linear signed rank statistics. First we consider the following.

Lemma 10.2.4. *Let $a_n(i)$ be given by* (10.2.1), *for a nonnegative, nondecreasing $\phi^+(\cdot)$ satisfying* (10.2.2). *Then these scores satisfy the Noether-type condition that, as $n \to \infty$,*

$$\frac{\sum_{i=1}^{n} a_n^2(i)}{\max_{1 \leqslant j \leqslant n} a_n^2(j)} \to \infty.$$

Proof: Theorem 8.3.14 shows that

$$\lim_{n\to\infty} \frac{1}{nb_n^2} \sum_{i=1}^{n} a_n^2(i) = \int_0^1 [\phi^+(u)]^2\, du,$$

an integral that is positive and finite. Since $\phi^+(\cdot)$ is nondecreasing, we have

$$\left(\frac{1}{nb_n^2}\right) \max_{1\leqslant j\leqslant n} a_n^2(j) = \left(\frac{1}{n}\right)\left[\phi^+\left(\frac{n}{n+1}\right)\right]^2 \leqslant \frac{n+1}{n} \int_{\frac{n}{n+1}}^{1} [\phi^+(u)]^2\, du.$$

Thus

$$\lim_{n\to\infty} \frac{1}{nb_n^2} \max_{1\leqslant j\leqslant n} a_n^2(j) = 0. \quad \blacksquare$$

Verification of the asymptotic normality under H_0 for linear signed rank statistics is far easier than the argument for linear rank statistics given in Chapter 8.

Theorem 10.2.5. *Suppose $X_1, \ldots, X_n$ is a random sample from a continuous population with density satisfying $f(x) = f(-x)$ for all x. Set $\Psi_i = \Psi(X_i)$ and let R_i^+ be the rank of $|X_i|$ among $|X_1|, \ldots, |X_n|$, where $\Psi(t) = 1, 0$ as $t >, \leqslant 0$. Consider the linear signed rank statistic*

$$S_n^+ = \sum_{i=1}^{n} \Psi_i a_n(R_i^+),$$

where $a_n(\cdot)$ satisfies (10.2.1) *for some nonnegative and nondecreasing $\phi^+(\cdot)$*

for which (10.2.2) *holds. Then*

$$\left[S_n^+ - \frac{n}{2}\bar{a}_n\right] \Big/ \sqrt{\frac{1}{4}\sum_{i=1}^{n} a_n^2(i)}$$

has a limiting standard normal distribution.

Proof: Recall Lemma 10.1.11 which states that for each n

$$S_n^+ = \sum_{i=1}^{n} \Psi_i a_n(R_i^+) \stackrel{d}{=} S_n' = \sum_{i=1}^{n} \Psi_i a_n(i).$$

We now apply the Liapounov central limit theorem (A.3.10 in the Appendix) to S_n'. Note that

$$\sum_{i=1}^{n} \frac{E\left[|\Psi_i a_n(i) - \frac{1}{2}a_n(i)|^3\right]}{\left(\sum_{j=1}^{n} \operatorname{Var}(\Psi_j a_n(j))\right)^{3/2}} = \frac{\sum_{i=1}^{n} a_n^3(i)}{\left(\sum_{j=1}^{n} a_n^2(j)\right)^{3/2}}$$

$$\leqslant \frac{\sum_{i=1}^{n} a_n^2(i) \max_{1\leqslant k\leqslant n} |a_n(k)|}{\left(\sum_{j=1}^{n} a_n^2(j)\right)^{3/2}} = \left[\frac{\max_{1\leqslant k\leqslant n} a_n^2(k)}{\sum_{j=1}^{n} a_n^2(j)}\right]^{1/2} \to 0,$$

as $n\to\infty$, by Lemma 10.2.4. The asymptotic normality of the standardized S_n', and hence the standardized S_n^+, then follows. ■

Properly normed linear signed rank statistics are also asymptotically normal under alternative hypotheses, subject to certain regularity conditions. Theorems covering alternatives have been given by Govindarajulu (1960) and Puri and Sen (1969), among others.

Suppose S_n^+ is a linear signed rank statistic with scores $a_n(i)$ satisfying

$$a_n(i) = b_n\phi^+\left(\frac{i}{n+1}\right), \qquad \text{for } i = 1,\ldots,n,$$

where $b_n > 0$ for every n and $\phi^+(\cdot)$ is a nonnegative, nondecreasing score function satisfying (10.2.2). Let $a_n^*(i)$ be any set of scores satisfying

$$\lim_{n\to\infty} \frac{1}{nb_n^2}\sum_{i=1}^{n}\left[a_n^*(i) - a_n(i)\right]^2 = 0,$$

and define $S_n^* = \sum_{i=1}^n \Psi_i a_n^*(R_i^+)$. Then, just as in the two-sample problem, the standardized form of the statistics $(S_n^* - (n/2)\bar{a}_n^*)$ has the same limiting distribution as the corresponding standardized form of $(S_n^+ - (n/2)\bar{a}_n)$. This statement is valid under both null and Pitman translation alternative hypotheses. Hence the test based on S_n^+ and the one based on S_n^* have the same asymptotic efficiency properties. This applies, in particular, to linear signed rank statistics using the two types of normal scores, since they are asymptotically equivalent in this sense.

We now develop the asymptotic efficiency of these tests. Let S_n^+ denote a linear signed rank statistic with scores $a_n(i)$ given by (10.2.1) for some nonnegative and nondecreasing score function $\phi^+(\cdot)$ satisfying (10.2.2). Define

$$\mu_{S_n^+}(\theta) = \frac{n}{2}\bar{a}_n + \frac{\theta n b_n}{2}\int_0^1 \phi^+(u)\phi^+(u,f)\,du$$

and **(10.2.6)**

$$\sigma^2_{S_n^+}(\theta) = \frac{nb_n^2}{4}\int_0^1 [\phi^+(u)]^2\,du.$$

With these definitions, Assumptions A2–A6 of Theorem 5.2.7 are easily verified. Assumption A1, the asymptotic normality under sequences of Pitman alternatives, is more difficult, and we must employ Theorem VI.2.5 of Hájek and Šidák (1967). It follows that the efficacy of the test based on the linear signed rank statistic S_n^+ is

$$\operatorname{eff}(S^+) = \frac{\int_0^1 \phi^+(u)\phi^+(u,f)\,du}{\sqrt{\int_0^1 [\phi^+(u)]^2\,du}}, \qquad \textbf{(10.2.7)}$$

provided the numerator is positive.

Example 10.2.8. Consider the test based on the linear signed rank statistic with scores $a_{H^+}(i)$ given in (10.1.10). They satisfy

$$a_{H^+}(i) = (n+1)\phi^+_{H^+}\left(\frac{i}{n+1}\right), \qquad i = 1,\ldots,n,$$

where

$$\begin{aligned} \phi^+_{H^+}(u) &= u \qquad \text{for } 0 < u \leqslant \frac{1}{2} \\ &= \frac{1}{2} \qquad \text{for } \frac{1}{2} < u < 1. \end{aligned}$$

Using (10.2.7), we now derive the efficacy of this test. To simplify the derivation we assume that the underlying distribution, which is continuous and symmetric about zero, has support which forms an interval $[-d,d]$, for some $0<d\leqslant\infty$, and is such that its density $f(t)\to 0$ as $t\uparrow d$ and as $t\downarrow -d$. We note that

$$\int_0^1 [\phi_{H^+}^+(u)]^2\,du = \int_0^{1/2} u^2\,du + \int_{1/2}^1 \frac{1}{4}\,du = \frac{1}{6}.$$

The numerator of the efficacy expression in (10.2.7) is

$$\int_0^1 \phi^+(u)\phi^+(u,f)\,du = \int_0^{1/2} u\left\{-\frac{f'\left(F^{-1}\left(\frac{1}{2}+\frac{1}{2}u\right)\right)}{f\left(F^{-1}\left(\frac{1}{2}+\frac{1}{2}u\right)\right)}\right\}du$$

$$+\int_{1/2}^1 \frac{1}{2}\left\{-\frac{f'\left(F^{-1}\left(\frac{1}{2}+\frac{1}{2}u\right)\right)}{f\left(F^{-1}\left(\frac{1}{2}+\frac{1}{2}u\right)\right)}\right\}du,$$

which, using the change of variable $t=F^{-1}(\frac{1}{2}+\frac{1}{2}u)$, becomes

$$= -2\int_0^{\xi_{3/4}}(2F(t)-1)f'(t)\,dt - \int_{\xi_{3/4}}^{\xi_1} f'(t)\,dt$$

$$= 4\int_0^{\xi_{3/4}} f^2(t)\,dt,$$

where ξ_p satisfies $F(\xi_p)=p$ and the last equality follows from integration by parts. Using the fact that $f(t)=f(-t)$ for all t, we see that the efficacy of the test based on this linear signed rank statistic is

$$\mathrm{eff}(H^+) = 2\sqrt{6}\int_{\xi_{1/4}}^{\xi_{3/4}} f^2(t)\,dt. \qquad \textbf{(10.2.9)}$$

We note that the efficacy expression in (10.2.9) is just $[\lambda(1-\lambda)]^{-1/2}$ times the one for the heavy-tailed distribution scores in the two-sample problem (see Exercise 9.2.5). A similar relationship was previously obtained between the efficacy of the Wilcoxon signed rank test (5.4.5) and that of the two-sample Mann–Whitney–Wilcoxon test (5.4.14). We now explore the general relationship between the efficacy in (10.2.7) and the corresponding one for the two-sample location problem. Suppose the continuous distribution $F(\cdot)$ has a density satisfying $f(x)=f(-x)$ for all x.

Then,

$$\frac{-f'(x)}{f(x)}=\frac{f'(-x)}{f(-x)}$$

for all x, and hence for all $0<u<1$, we have

$$\phi(u,f)=\frac{-f'(F^{-1}(u))}{f(F^{-1}(u))}=\frac{f'(F^{-1}(1-u))}{f(F^{-1}(1-u))}=-\phi(1-u,f),$$

where $\phi(\cdot,f)$ is the optimal score function in the two-sample location problem. [See (9.1.25).] Thus the optimal one-sample score function satisfies

$$\phi^{+}(u,f)=\phi\left(\frac{1}{2}+\frac{1}{2}u,f\right)=-\phi\left(\frac{1}{2}-\frac{1}{2}u,f\right).$$

Let $\phi(u)$ denote a nondecreasing square integrable score function for the two-sample location problem such that $\phi(\cdot)$ is symmetric about $\frac{1}{2}$ in the sense that

$$\phi(u)-\bar{\phi}=\bar{\phi}-\phi(1-u)$$

for all u, where $\bar{\phi}=\int_0^1\phi(u)du$. Then,

$$\begin{aligned}\int_0^1(\phi(u)-\bar{\phi})^2du&=2\int_{1/2}^1(\phi(u)-\bar{\phi})^2du\\&=\int_0^1\left(\phi\left(\tfrac{1}{2}+\tfrac{1}{2}v\right)-\bar{\phi}\right)^2dv.\end{aligned}$$

Assuming that the conditions of Lemma 9.2.16 are satisfied (so that $\int_0^1\phi(u,f)du=0$), we also have

$$\begin{aligned}\int_0^1\phi(u)\phi(u,f)du&=\int_0^1(\phi(u)-\bar{\phi})\phi(u,f)du\\&=2\int_{1/2}^1(\phi(u)-\bar{\phi})\phi(u,f)du\\&=\int_0^1\left(\phi\left(\tfrac{1}{2}+\tfrac{1}{2}v\right)-\bar{\phi}\right)\phi\left(\tfrac{1}{2}+\tfrac{1}{2}v,f\right)dv.\end{aligned}$$

Therefore, if S_N denotes a linear rank statistic for the two-sample location problem with scores generated by the function $\phi(u)$ and if S_n^+ denotes a linear signed rank statistic for the one-sample location problem with scores generated by

$$\phi^+(u) = \phi\left(\tfrac{1}{2}+\tfrac{1}{2}u\right) - \bar{\phi}, \tag{10.2.10}$$

we see that the efficacies of the two tests are intimately related, since

$$\begin{aligned}\frac{\operatorname{eff}(S_N)}{\sqrt{\lambda(1-\lambda)}} &= \frac{\int_0^1 \phi(u)\phi(u,f)\,du}{\sqrt{\int_0^1 \left(\phi(u)-\bar{\phi}\right)^2 du}} \\ &= \frac{\int_0^1 \left[\phi\left(\frac{1}{2}+\frac{1}{2}v\right)-\bar{\phi}\right]\phi\left(\frac{1}{2}+\frac{1}{2}v,f\right)dv}{\sqrt{\int_0^1 \left[\phi\left(\frac{1}{2}+\frac{1}{2}v\right)-\bar{\phi}\right]^2 dv}} \\ &= \operatorname{eff}(S_n^+). \end{aligned} \tag{10.2.11}$$

The relationship displayed in (10.2.11) shows that slight modifications of some efficacy expressions developed in Section 9.2 will produce efficacy expressions for the corresponding tests in the one-sample location problem. For instance, as shown in Example 8.3.10 and in (9.1.3), the score function $\phi_W(u)=u$ in the two-sample location problem produces the Mann–Whitney–Wilcoxon test. From (10.2.10) we see that the one-sample analogue uses a score function

$$\phi^+(u) = \tfrac{1}{2}+\tfrac{1}{2}u-\tfrac{1}{2} = \tfrac{1}{2}u.$$

Since

$$\phi^+\left(\frac{i}{n+1}\right) = \frac{i}{2(n+1)},$$

we see that the one-sample linear signed rank test which uses these scores is equivalent to the Wilcoxon signed rank procedure. Therefore, using the expression for the efficacy of the Mann–Whitney–Wilcoxon test (see Exercise 9.2.1), we see that the efficacy of the Wilcoxon signed rank test is

$$\operatorname{eff}(W^+) = \sqrt{12}\int_{-\infty}^{\infty} f^2(x)\,dx, \tag{10.2.12}$$

as previously presented in (5.4.5).

Similar comparisons with associated linear rank statistics for the two-sample location problem show that

$$\text{eff}(NS^+) = -\int_0^1 \Phi^{-1}(u)\frac{f'(F^{-1}(u))}{f(F^{-1}(u))}\,du \tag{10.2.13}$$

is the efficacy of the one-sample location test based on a linear signed rank statistic with normal scores given by (10.1.7) or (10.1.8). The sign test described in (10.1.6) is the one-sample analogue of the median test given in (9.1.4). Thus, according to Exercise 9.2.3, we see that

$$\text{eff}(B) = 2f(m), \tag{10.2.14}$$

where m is the median of the distribution $F(\cdot)$. This agrees with the earlier derivation in Chapter 5 of the efficacy of the sign test (see(5.4.8)). Similarly, the light-tailed distribution scores of (10.1.9) are related to the two-sample scores in (9.1.9). Hence, by Example 9.2.15, we see that

$$\text{eff}(L^+) = \sqrt{96}\left[\int_{\xi_0}^{\xi_{1/4}} f^2(x)\,dx + \int_{\xi_{3/4}}^{\xi_1} f^2(x)\,dx\right], \tag{10.2.15}$$

where ξ_p satisfies $F(\xi_p)=p$.

The relationship in (10.2.11) between the efficacies of tests based on analogous one- and two-sample linear rank and signed rank statistics, respectively, yields some additional useful conclusions. If one-sample test scores are generated by $a_n^+(i)=\phi^+(i/(n+1),f)$, where

$$\phi^+(u,f) = -\frac{f'\left(F^{-1}\left(\frac{1}{2}+\frac{1}{2}u\right)\right)}{f\left(F^{-1}\left(\frac{1}{2}+\frac{1}{2}u\right)\right)} \tag{10.2.16}$$

is the optimal one-sample score function, then the resulting test is asymptotically efficient for detecting a change of location in the symmetric continuous population with c.d.f. $F(\cdot)$; that is, the optimal linear signed rank test achieves the best possible asymptotic efficiency among *all α-level tests* against a sequence of Pitman location alternatives. Thus, for example, both normal scores tests are asymptotically efficient for detecting a change of location in a normal distribution, and the Wilcoxon signed rank test is asymptotically efficient for detecting a change of location in a logistic distribution.

The relationship displayed in (10.2.11) between the efficacies of corresponding one- and two-sample rank tests for location problems shows that when the underlying distribution is symmetric, the efficiency comparisons

TABLE 10.2.17

DISTRIBUTION	ARE(L^+, W^+)	ARE(NS^+, W^+)	ARE(H^+, W^+)	ARE(B, W^+)
Uniform	2.000	$+\infty$	0.500	0.333
Normal	0.927	1.047	0.870	0.667
Logistic	0.781	0.955	0.945	0.750
Double exponential	0.500	0.847	1.125	1.333
Cauchy	0.264	0.708	1.339	1.333

will yield similar values. However, Table 10.2.17, the direct analogue of Table 9.2.21, is included for completeness. The symbol B stands for the sign test and W^+ for the Wilcoxon signed rank test. Likewise NS^+, L^+, and H^+ are signed rank tests using normal scores, light-tailed distribution scores [see (10.1.9)] and heavy-tailed distribution scores [see (10.1.10)].

Exercises 10.2.

10.2.1. Verify that Assumptions A2–A6 of Theorem 5.2.7 are satisfied for a linear signed rank statistic S_n^+ with scores satisfying (10.2.1) and (10.2.2) and with $\mu_{S_n^+}(\cdot)$ and $\sigma^2_{S_n^+}(\cdot)$ functions given in (10.2.6).

10.2.2. Using the efficacies in (10.2.12) and (10.2.14), verify that ARE(B, W^+)$=.333$ for the uniform $(-1,1)$ distribution, where B is the sign test and W^+ is the Wilcoxon signed rank test.

10.2.3. Using the efficacy expressions in (10.2.12) and (10.2.15), derive the numerical value ARE(L^+, W^+)$=.5$ for a double exponential distribution, where L^+ is the test associated with the linear signed rank statistic using the light-tailed distribution scores given in (10.1.9) and W^+ denotes the Wilcoxon signed rank test.

10.2.4. Find the following AREs for the one-sample location change problem:

(a) ARE(L^+, NS^+), when the distribution is standard normal,

(b) ARE(H^+, L^+), when the distribution is uniform $(-1,1)$,

(c) ARE(T^+, H^+), when the distribution is double exponential,

where L^+, NS^+, and H^+ are used here in the same fashion as in Table 10.2.17 and where T^+ denotes the one-sample Student t-test.

10.2.5. Consider a linear signed rank test corresponding to the score function

$$\begin{aligned}\phi^+(u) &= 0 \qquad \text{if } 0 < u \leqslant \delta \\ &= 1 \qquad \text{if } \delta < u < 1,\end{aligned}$$

where $0 < \delta < 1$ is a fixed constant.

(a) Describe the linear signed rank statistic corresponding to scores $a_n(i) = \phi^+(i/(n+1))$, $i = 1, \ldots, n$. What does it measure?

(b) Determine the null distribution of the test statistic in **(a)** for the case $n = 6$ and $\delta = 1/2$.

(c) Using (10.2.7), find and simplify the expression for the efficacy of this test. For ease in the derivation, assume the underlying distribution, which is continuous and symmetric about zero, also has support which forms an interval $[-d, d]$ and is such that its density $f(x) \to 0$ as $x \uparrow d$ and as $x \downarrow -d$ and $f(x)$ is continuous at $F^{-1}(\frac{1}{2} + \frac{1}{2}\delta)$.

(d) If B denotes the sign test and S^+ the test based on the statistic in **(a)** of this problem with $\delta = 1/2$, evaluate $\text{ARE}(S^+, B)$ when the underlying distribution is (i) logistic and (ii) double exponential.

10.2.6. Consider a linear signed rank statistic S^+ with scores $a_n(i) = i^2$, for $i = 1, \ldots, n$.

(a) With what score function $\phi^+(u)$ are these scores associated? Verify the relationship.

(b) Using (10.2.7) show that

$$\text{eff}(S^+) = 16\sqrt{5} \int_0^\infty \left[F(t) - \tfrac{1}{2}\right] f^2(t)\,dt.$$

To simplify the derivation, you may make the same assumptions about $F(\cdot)$ as in **(c)** of Exercise 10.2.5.

(c) Evaluate $\text{ARE}(S^+, W^+)$ when the underlying distribution is (i) uniform and (ii) double exponential. Here W^+ denotes the Wilcoxon signed rank statistic.

(d) What is the score function, $\phi(u)$, for the linear rank statistic in the two-sample change of location problem that corresponds to $\phi^+(u)$ in **(a)**? What is the efficacy of this two-sample test?

10.2.7. Consider the linear signed rank statistic with scores $a_n(i) = \phi^+(i/(n+1))$, $i = 1,2,\ldots,n$, where

$$\begin{aligned}\phi^+(u) &= c_1 u \qquad \text{if } 0 < u \leqslant \delta \\ &= c_2 u + (c_1 - c_2)\delta \qquad \text{if } \delta < u < 1,\end{aligned}$$

where $0 \leqslant \delta \leqslant 1$ is a fixed constant and c_1 and c_2 are nonnegative constants such that max $(c_1, c_2) > 0$.

(a) Using (10.2.7), find and simplify the expression for the efficacy of this test. For ease in the derivation, you may make the same assumptions about $F(\cdot)$ as in **(c)** of Exercise 10.2.5.

(b) Let S^+ denote the test based on this linear signed rank statistic with $c_1 = 1$, $c_2 = 1/2$ and $\delta = 1/2$. If W^+ denotes the Wilcoxon signed rank test, evaluate ARE(S^+, W^+) when the underlying distribution is (i) logistic and (ii) double exponential.

10.2.8. Suppose the underlying symmetric, continuous distribution has density

$$\begin{aligned}f(t) &= 1 \qquad \text{if } |t| < \tfrac{1}{4} \\ &= (16t^2)^{-1} \qquad \text{if } |t| \geqslant \tfrac{1}{4}.\end{aligned}$$

Find the form of the optimal score function for the one-sample change of location problem for this distribution. Graph the score function.

10.2.9. Let $\phi^+(u)$ be a nonnegative, nondecreasing score function satisfying (10.2.2). Consider a new score function

$$\phi_*^+(u) = k\phi^+(u) + k_*,$$

where $k > 0$ and $k_* \geqslant 0$ are fixed constants.

(a) Show that $\phi_*^+(u)$ is a nonnegative, nondecreasing score function satisfying (10.2.2).

(b) Find an expression for the efficacy of the test based on a linear signed rank statistic with scores generated by $\phi_*^+(u)$. When $k_* = 0$, compare this expression with the efficacy of the test based on a linear signed rank statistic with scores generated by $\phi^+(u)$. Is there a direct relationship between the two when $k_* > 0$?

10.2.10. Use the form of the optimal score function to show that the asymptotically efficient test for a change in location does not depend on the scale factor of the underlying continuous distribution which is symmetric about zero. That is, consider $G(x) = F(x/\eta)$ for $\eta > 0$, where $F(x) = 1 - F(-x)$ for all x.

11 ADDITIONAL METHODS FOR CONSTRUCTING DISTRIBUTION-FREE PROCEDURES

In Chapter 2 we introduced the concept of a distribution-free statistic and considered how to use such statistics to construct distribution-free tests of hypotheses. In particular, we studied three basic methods for obtaining a nonparametric distribution-free statistic, namely, counting, ranking, and a combination of these two techniques. In Chapter 6 we used these same ideas to develop nonparametric distribution-free confidence intervals and bounds.

These three methods form the foundation for the vast majority of nonparametric distribution-free statistical methods. However, there are other techniques for creating nonparametric distribution-free procedures, many of which are more involved (yet still very intuitive) adaptations of these three basic tools. In this chapter we briefly study a few of these additional methods for constructing nonparametric distribution-free procedures.

11.1. Conditional Nonparametric Distribution-Free Tests

In this section we describe a versatile method for creating hypothesis tests that have a nonparametric distribution-free property. The method is characterized by the fact that the construction of a critical region depends on certain information which is obtained from the observed data; that is, the critical region is a *conditional* one, created after the data have been observed. However, the associated test procedure has overall level α

because the critical region is constructed so that the *conditional* (given certain information in the data) probability of rejecting H_0 is α when H_0 is really true. For these reasons we associate the term **conditional tests** with these procedures.

We begin by motivating a general principle which often provides the method for constructing conditional tests. In developing distribution-free test procedures we have often used the property that, under the appropriate null hypothesis, every ordered arrangement of the single combined sample observations is equally likely; that is, the combined sample observations are exchangeable random variables when the null hypothesis is true. In fact, the distribution-freeness of any test based on the combined sample ranks of such observations is an immediate consequence of this exchangeability. However, replacing the actual observations by their combined sample ranks is not the only way to use the exchangeability property to construct distribution-free tests. We now use the actual sample values, not just the ranks, in the construction. Although we consider other examples later, we introduce this technique in the two-sample location setting.

Let $X_1,\ldots,X_m$ and $Y_1,\ldots,Y_n$ be independent random samples from continuous distributions with c.d.f.'s $F(x)$ and $F(x-\Delta)$, respectively, where $-\infty<\Delta<\infty$, and define

$$\begin{aligned} Z_i &= X_i, \qquad i = 1,\ldots,m \\ &= Y_{i-m}, \qquad i = m+1,\ldots,N, \end{aligned} \tag{11.1.1}$$

with $N=m+n$. Let $Z_{(1)} \leqslant \cdots \leqslant Z_{(N)}$ denote the combined sample order statistics and set

$$\mathbf{Z}_{(\cdot)} = (Z_{(1)},\ldots,Z_{(N)}).$$

We see from Theorem 2.3.3 that, under H_0: $\Delta=0$,

$$P_0((Z_1,\ldots,Z_N)=(Z_{(r_1^*)},\ldots,Z_{(r_N^*)})) = \frac{1}{N!}, \tag{11.1.2}$$

for every permutation $\mathbf{r}^*=(r_1^*,\ldots,r_N^*)$ of $(1,\ldots,N)$, where P_0 indicates that the probabilities are computed under H_0; that is, each of the $N!$ rankings is equally likely. However, Lemma 8.3.11 showed that the rank vector and the order statistics vector are stochastically independent under H_0. Hence,

$$P_0(Z_1=z_{(r_1^*)},\ldots,Z_N=z_{(r_N^*)}|\mathbf{Z}_{(\cdot)}=\mathbf{z}_{(\cdot)}) = \frac{1}{N!} \tag{11.1.3}$$

for each $\mathbf{r}^*$, a permutation of the integers $1,\ldots,N$. In other words, the *conditional* distribution of $(Z_1,\ldots,Z_N)$, given $\mathbf{Z}_{(\cdot)}=\mathbf{z}_{(\cdot)}$, is, under H_0, uniform over the $N!$ pairings of the elements in the vector $\mathbf{Z}=(Z_1,\ldots,Z_N)$ with those in the vector $\mathbf{z}_{(\cdot)}=(z_{(1)},\ldots,z_{(N)})$. The equally likely nature of these permutations of $\mathbf{z}_{(\cdot)}$ is often termed the **permutation principle**, and conditional tests that use this principle and condition on order statistics vectors like $\mathbf{z}_{(\cdot)}$ are called **permutation tests.**

We now describe how the permutation principle is used to construct a nonparametric distribution-free test of hypothesis. To decide whether to reject the null hypothesis H_0: $\Delta=0$ in view of the sample data $Z_1,\ldots,Z_N$, we look at where the actual observed association of the Z_is with the $z_{(j)}$s fits among the $N!$ possible associations. Let $S(X_1,\ldots,X_m; Y_1,\ldots,Y_n)$ be any statistic that measures the shift Δ, and let $s_{\mathbf{r}^*}$ denote the value of this statistic for the association $Z_1=z_{(r_1^*)},\ldots,Z_N=z_{(r_N^*)}$, with $\mathbf{r}^*=(r_1^*,\ldots,r_N^*)$; that is, $s_{\mathbf{r}^*}=S(Z_{(r_1^*)},\ldots,Z_{(r_m^*)}; Z_{(r_{m+1}^*)},\ldots,Z_{(r_{m+n}^*)})$. From (11.1.3) we see that

$$P_0(S(X_1,\ldots,X_m; Y_1,\ldots,Y_n)=s_{\mathbf{r}^*}|\mathbf{Z}_{(\cdot)}=\mathbf{z}_{(\cdot)}) = \frac{1}{N!},$$

for each of the $N!$ $s_{\mathbf{r}^*}$ values as $\mathbf{r}^*$ ranges over $\mathcal{R}$, the class of all permutations of the integers $1,\ldots,N$. If, in addition, the statistic S is unchanged after permuting the X values or the Y values separately (a natural property for measures of shift), then there are really at most $M=\binom{N}{m}$ distinct values for S, given $\mathbf{Z}_{(\cdot)}=\mathbf{z}_{(\cdot)}$, and each of these is still equally likely. That is, if $s_1^*,\ldots,s_M^*$ are the possible S values, corresponding to the $M=\binom{N}{m}$ choices of $Z_{(\cdot)}$ values to be assigned to the X_is, then

$$P_0(S(X_1,\ldots,X_m; Y_1,\ldots,Y_n)=s_i^*|\mathbf{Z}_{(\cdot)}=\mathbf{z}_{(\cdot)}) = \frac{1}{M}, \qquad \textbf{(11.1.4)}$$

for $i=1,\ldots,M$.

The construction of a conditional distribution-free test of H_0: $\Delta=0$ against, for example, the alternative H_1: $\Delta>0$ is then based on the equally likely property in (11.1.4) in the following manner. For an α-level test of H_0, with $\alpha=t/M$ for t an integer between 1 and M, we reject H_0 if and only if the observed value of S, say, s_{obs}^*, falls in a set C_α containing t elements chosen from the set $D=\{s_1^*,\ldots,s_M^*\}$ in a manner appropriate for the alternative H_1: $\Delta>0$ and the statistic S being used as a measure of Δ. The resulting procedure is *conditioned* on the observed value of the combined sample order statistics, namely, $\mathbf{Z}_{(\cdot)}=\mathbf{z}_{(\cdot)}$, and is an α-level test of H_0: $\Delta=0$ that is distribution-free over the class of all continuous distributions. We illustrate the details involved in conducting such a test by considering a specific example.

Example 11.1.5. Let $x_1=4.3$, $x_2=6.0$, $x_3=3.6$ and $y_1=7.4$, $y_2=5.5$, $y_3=6.2$ be the observed values of random samples of sizes $m=n=3$ from continuous distributions with c.d.f.'s $F(x)$ and $F(x-\Delta)$, respectively. The observed order statistics of the combined sample are then $z_{(1)}=3.6$, $z_{(2)}=4.3$, $z_{(3)}=5.5$, $z_{(4)}=6.0$, $z_{(5)}=6.2$ and $z_{(6)}=7.4$. To construct a conditional test that is a distribution-free permutation test of H_0: $\Delta=0$ against the alternative H_1: $\Delta>0$, we need to select a statistic $S(X_1,X_2,X_3;Y_1,Y_2,Y_3)$ that is a measure of Δ. For sake of our example, we take $S(X_1,X_2,X_3;Y_1,Y_2,Y_3)=\overline{Y}-\overline{X}$, with $\overline{X}=\frac{1}{3}\sum_{i=1}^{3}X_i$ and $\overline{Y}=\frac{1}{3}\sum_{j=1}^{3}Y_j$. Next we must compute the $M=\binom{6}{3}=20$ possible sample values of $\overline{Y}-\overline{X}$ corresponding to the ways in which we can assign three of the ordered values $z_{(1)},\ldots,z_{(6)}$ to be designated as x's. The resulting values of S, denoted by $s_1^*,\ldots,s_{20}^*$, are given in Table 11.1.6. The associated conditional null H_0 distribution of $\overline{Y}-\overline{X}$, given $Z_{(1)}=z_{(1)},\ldots,Z_{(6)}=z_{(6)}$, then assigns probability $\frac{1}{20}$ to each of the s_i^* values in Table 11.1.6. Since large values of $\overline{Y}-\overline{X}$ are indicative of the alternative H_1: $\Delta>0$, the critical regions for the

TABLE 11.1.6.
Permutation Values of $\bar{y}-\bar{x}$

VALUES ASSIGNED AS x_i'S	VALUES OF $\bar{y}-\bar{x}$
3.6, 4.3, 5.5	$s_1^*=$ 2.067
3.6, 4.3, 6.0	$s_2^*=$ 1.733
3.6, 4.3, 6.2	$s_3^*=$ 1.600
3.6, 4.3, 7.4	$s_4^*=$ .800
3.6, 5.5, 6.0	$s_5^*=$ .933
3.6, 5.5, 6.2	$s_6^*=$ .800
3.6, 5.5, 7.4	$s_7^*=$ .000
3.6, 6.0, 6.2	$s_8^*=$ .467
3.6, 6.0, 7.4	$s_9^*=$ −.333
3.6, 6.2, 7.4	$s_{10}^*=$ −.467
4.3, 5.5, 6.0	$s_{11}^*=$ .467
4.3, 5.5, 6.2	$s_{12}^*=$ .333
4.3, 5.5, 7.4	$s_{13}^*=$ −.467
4.3, 6.0, 6.2	$s_{14}^*=$ .000
4.3, 6.0, 7.4	$s_{15}^*=$ −.800
4.3, 6.2, 7.4	$s_{16}^*=$ −.933
5.5, 6.0, 6.2	$s_{17}^*=$ −.800
5.5, 6.0, 7.4	$s_{18}^*=$ −1.600
5.5, 6.2, 7.4	$s_{19}^*=$ −1.733
6.0, 6.2, 7.4	$s_{20}^*=$ −2.067

corresponding permutation test of H_0: $\Delta=0$ against H_1: $\Delta>0$ would contain an appropriate number of the largest s_i^* values. Thus, for example, an $\alpha=.10(=2/20)$ level critical region would be $C_{.10}=\{1.733, 2.067\}$, and the corresponding $\alpha=.05(=1/20)$ level critical region is $C_{.05}=\{2.067\}$. For our observed samples the value of $\bar{y}-\bar{x}$ is 1.733. Thus the smallest level at which this nonparametric distribution-free permutation test would reject H_0: $\Delta=0$ in favor of H_1: $\Delta>0$ is $\underline{\alpha}=.10$.

We note that the testing procedure (and hence the conclusion reached) in Example 11.1.5 depends on the measure of shift, namely, $S=\bar{Y}-\bar{X}$ that was employed. We could as easily have used the alternative measure

$$S_1=\left[\underset{1\leqslant j\leqslant n}{\text{median}}\, Y_j-\underset{1\leqslant i\leqslant m}{\text{median}}\, X_i\right]$$

or one of the Hodges–Lehmann two-sample shift estimators (see Section 7.1), such as $S_2=\text{median}\{Y_j-X_i\colon i=1,\ldots,m \text{ and } j=1,\ldots,n\}$. However, use of a different estimator for Δ could produce a different conditional rejection region. To help in the choice of the statistic for use in a permutation test, there is a theory of "most powerful permutation tests." See, for example, Chapter 5 of Lehmann (1959).

It is also important to emphasize that since it is a conditional procedure, the critical region for a permutation test cannot in general be constructed until *after* the data have been observed. Despite this inconvenience, permutation tests play an important role in nonparametric statistics. A primary reason for this is that the permutation principle, which we have discussed for the two-sample setting, can be applied to a wide variety of statistical problems. Thus we can often construct *conditional* tests that have a nonparametric distribution-free property in settings for which we would not be able to obtain this property with other test construction techniques. Such is the case, for example, in many of the basic problems of multivariate analysis. Counting and ranking techniques do not (at least with the current level of knowledge) provide convenient and effective nonparametric distribution-free methods for such problems when the sample sizes are small; yet conditional tests do. For a detailed survey of the use of conditional tests in dealing with multivariate data problems, see Puri and Sen (1971).

At this point we should also comment on the intimate relationship that often exists between permutation tests and rank tests. For example, in this two-sample setting a rank test may be viewed as a permutation test in which the order satistics of the combined sample (i.e., the $z_{(i)}$'s) are replaced by a standard set of numbers or scores, say, $a(i)$'s. By using these standard numbers, the required conditional distribution can be tabled once and for all, and it is no longer necessary to construct the null (H_0) distribution every time a new set of sample observations is obtained. In addition, when using these standard scores the involved conditional distribution is the same for each possible set of order statistics values. Thus the marginal distribution is the same as the conditional distribution, and the resulting test is not really conditional. We therefore see that the Mann-Whitney-Wilcoxon test (as well as the other tests based on linear rank statistics which were discussed in Chapter 9) may be considered as a permutation test in which the actual observation values are replaced by more convenient ones, namely the ranks (or scores).

As a second example of a conditional test, we consider a 2×2 contingency table problem. The resulting test procedure is referred to as **Fisher's exact test**, being named after R. A. Fisher, who was an early proponent of the permutation principle.

Example 11.1.7. Suppose that we have two populations, say, I and II, which consist of units belonging to one or the other of two mutually exclusive categories, say, C_1 and C_2. A random sample of N total observations are collected from the two populations and coded as to which population they are from and to which category they belong. Thus, letting

$$\begin{aligned} A &= \text{number of } C_1 \text{ observations from population I} \\ B &= \text{number of } C_2 \text{ observations from population I} \\ D &= \text{number of } C_1 \text{ observations from population II} \\ E &= \text{number of } C_2 \text{ observations from population II,} \end{aligned} \qquad \textbf{(11.1.8)}$$

we have the following data frequency table.

		Category		
		C_1	C_2	Total
Population	I	A	B	$A+B$
	II	D	E	$D+E$
	Total	$A+D$	$B+E$	N = total number of observations

Now, the question of interest with this data is whether the two populations are identical with respect to the C_1 and C_2 percentages. That is, we wish to test the null hypothesis H_0: [Populations I and II are identical] against an appropriate alternative. For sake of illustration, we consider here the one-sided alternative H_1: There is a greater percentage of C_1 category observations in population I than in population II. (The corresponding two-sided alternative can be handled by a similar approach.) To construct a conditional test of H_0 versus H_1 we consider the marginal totals $A+B$, $D+E$, $A+D$, and $B+E$ as all fixed (this is the conditional aspect of the procedure). Hence we have $A+D$ category C_1s and $B+E$ category C_2s which, under H_0, can be thought of as *randomly* separated into two groups of sizes $A+B$ and $D+E$. For a separation resulting in t population I units in category C_1, we see that the corresponding null hypothesis probability (conditioned on the fixed marginal totals) is given by the hypergeometric distribution, namely,

$$p(t) = P(t \text{ population I units in category } C_1)$$
$$= \frac{\binom{A+D}{t}\binom{B+E}{A+B-t}}{\binom{N}{A+B}}. \tag{11.1.9}$$

(Note that once we know there are t units from population I in category C_1, all the other entries in the table follow from the known fixed marginal totals.) So we see that it is relatively easy to compute the conditional probability of each possible outcome for the given marginal totals.

To complete the construction of our conditional test we need a statistic that is sensitive to whether H_0 or H_1 is true. One frequently applied measure is

$$S = AE - BD, \tag{11.1.10}$$

for which a large value is indicative of the alternative hypothesis H_1. To set up the corresponding conditional α-level critical region, we must first compute the value of the statistic $S(t) = t[(B+E)-(A+B-t)]-(A+B-t)(A+D-t)$ and the associated null (conditional) probability $p(t)$ from (11.1.9) for every *possible* (in view of the fixed marginal totals) value of t, the number of population I units belonging in category C_1. The associated α-level conditional test (known as Fisher's exact test) is, then, to:

$$\text{reject } H_0 \text{ in favor of } H_1 \text{ iff } S(A) \geqslant s_\alpha, \tag{11.1.11}$$

where $S(A)$ is the observed sample value for the statistic S, and s_α is the upper 100αth percentile point for the null conditional distribution of S, as enumerated using (11.1.9) and the formula for $S(t)$.

We consider one last example of a type of conditional test. In the one-sample location problem we assume $X_1,\ldots,X_n$ are i.i.d. continuous random variables with density $f(x-\theta)$ that is symmetric about the location parameter θ. We wish to test H_0: $\theta=0$ versus H_1: $\theta>0$. (There is no loss of generality in using the particular θ-value of 0, since the test can be based on the variables $X_1-\theta_0,\ldots,X_n-\theta_0$ if we seek to test H_0: $\theta=\theta_0$.) Lemma 2.4.2 shows that Ψ_i and $|X_i|$ are independent random variables under H_0, where each $\Psi_i(=\Psi(X_i)=1$ or 0 as $X_i>, \leqslant 0)$ is a Bernoulli random variable with $p=\frac{1}{2}$. This independence, under H_0, of the absolute value of an observation and its sign means that given $|X_i|=|x_i|$, the two observation values $-|x_i|$ and $+|x_i|$ are equally likely. Moreover, this fact applies to each of the n mutually independent observations. Thus we see that, given $|x_1|,|x_2|,\ldots,|x_n|$, the 2^n possible n vectors

$$(\pm|x_1|, \pm|x_2|,\ldots, \pm|x_n|) \tag{11.1.12}$$

are equally likely values for $(X_1,\ldots,X_n)$, where $\pm|x_i|$ implies that $|x_i|$ is assigned either a positive or negative sign.

To conduct a test based on this conditional distribution, we proceed much as we did in the two-sample setting. We first select a statistic $S(X_1,\ldots,X_n)$ that measures θ, and we compare the observed value of the statistic $S(x_1,\ldots,x_n)$ to the conditional distribution of $S(\cdot)$ derived from the 2^n equally likely possibilities represented by

$$S(\pm|x_1|,\ldots,\pm|x_n|).$$

If $S(x_1,\ldots,x_n)$ falls in an appropriately chosen critical region we reject H_0 in favor of H_1.

Example 11.1.13. Suppose we observe the values

$$x_1=4.2,\ x_2=-1.3,\ x_3=7.8,\ x_4=0.5,\ x_5=6.6.$$

The absolute value vector is

$$(4.2,\ 1.3,\ 7.8,\ 0.5,\ 6.6).$$

For testing H_0: $\theta=0$ versus H_1: $\theta>0$ we could use the statistic $\bar{X}$ and consider the 2^n possible values represented by

$$\bar{X}=\tfrac{1}{5}[\pm 4.2\pm 1.3\pm 7.8\pm 0.5\pm 6.6],$$

where $\pm|x|$ means either $+|x|$ or $-|x|$. The upper 25% of this conditional distribution of $\bar{X}$ is shown in Table 11.1.14. Since each of the $2^5=32$ assignments of signs is equally likely, we see that the smallest significance level at which the observed value of $\bar{x}=(4.2-1.3+7.8+0.5+6.6)/5=3.56$ would cause us to reject H_0 is $(3/32)=.094$. Thus, for example, if $\alpha=.10$, we would reject H_0, but if $\alpha=.05$, we would not.

One final comment is in order concerning conditional tests for the one-sample problem like the one we have just described. If we replace $|x_1|,\ldots,|x_n|$ with some nonnegative scores $a(1),\ldots,a(n)$ that are ordered among themselves in the same manner as $|x_1|,\ldots,|x_n|$, then the resulting test will have a critical region that can be tabulated once and for all. Hence such a test is no longer a conditional one. The Wilcoxon signed rank test and the other tests based on linear signed rank statistics (see Chapter 10) are tests of this type.

Remark 11.1.15. One additional important point should be stressed about the usefulness of conditional tests. They can be employed in a variety of settings, including those involving discrete (or mixtures of continuous and discrete) distributions. In particular, the result stated in (11.1.3) holds under any H_0 for which the random variables $Z_1,\ldots,Z_N$ are exchangeable; this includes any case for which they are i.i.d., no matter what type of distribution pertains. One important ramification of this fact demonstrates how to handle the occurrence of ties in rank tests. We couch the description in the two-sample location problem, but similar results hold for other problems as well. Suppose $X_1,\ldots,X_m$ and $Y_1,\ldots,Y_n$ are independent random samples from distributions with c.d.f.'s $F(x)$ and $F(x-\Delta)$,

TABLE 11.1.14.

SIGN ASSIGNED TO					
4.2	1.3	7.8	0.5	6.6	$\bar{X}$ VALUE
+	+	+	+	+	4.08
+	+	+	−	+	3.88
+	−	+	+	+	3.56
+	−	+	−	+	3.36
−	+	+	+	+	2.40
−	+	+	−	+	2.20
−	−	+	+	+	1.88
−	−	+	−	+	1.68

respectively. We wish to test H_0: $\Delta=0$ versus H_1: $\Delta>0$ using the Mann–Whitney–Wilcoxon test. If we observe

$$\begin{array}{lllll} x_1 = 10.4 & x_2 = 7.8 & x_3 = 8.4 & y_1 = 12.6 & y_2 = 8.4 \\ r_1^* = 4 & r_2^* = 1 & r_3^* = 2.5 & r_4^* = 5 & r_5^* = 2.5, \end{array}$$

we see that two observations are tied. The common way to deal with ties is to assign all tied observations the average of the ranks that would have been assigned to those observations had they been slightly different. (Thus in this case x_2 and y_2 both get rank $2.5=[2+3]/2$.) The sum of the ranks associated with the Y_js is then

$$W = 5 + 2.5 = 7.5.$$

Clearly, such a W-value cannot be compared to the usual null distribution of W, which assumes that W takes on only integer values. We must create a new null distribution. The construction of such an appropriate null distribution follows directly from the permutation principle in (11.1.3). Under H_0, each of the $N!=5!$ associations of the values in the combined order statistics

$$\mathbf{z}_{(\cdot)} = (7.8, 8.4, 8.4, 10.4, 12.6)$$

with the designations $(x_1, x_2, x_3, y_1, y_2)$ is equally likely. Therefore, conditioning on this vector of order statistics (i.e., given $\mathbf{Z}_{(\cdot)}=\mathbf{z}_{(\cdot)}$), the rank vector, $\mathbf{R}^*$, of $(X_1, X_2, X_3, Y_1, Y_2)$ assumes each of the $N!$ permutations of $(1, 2.5, 2.5, 4, 5)$ with probability $1/N!$. This fact yields an appropriate conditional null distribution for W. The point of this example is to show that this conditional test technique (often using the permutation principle) enables us to conduct nonparametric distribution-free tests in many problems where tied observations occur.

Exercises 11.1.

11.1.1. Using the setting and data of Example 11.1.5, construct the conditional (given $\mathbf{Z}_{(\cdot)} = \mathbf{z}_{(\cdot)}$) null H_0: $\Delta=0$ distribution of $S_1 = [\text{median}_{1 \leqslant j \leqslant n} Y_j - \text{median}_{1 \leqslant i \leqslant m} X_i]$. Use this to find the smallest significance level at which the associated (conditional) permutation test would reject H_0: $\Delta=0$ in favor of H_1: $\Delta>0$.

11.1.2. For the setting of Example 11.1.5, construct the conditional (given $\mathbf{Z}_{(\cdot)}=\mathbf{z}_{(\cdot)}$) null H_0: $\Delta=0$ distribution of $S_2=\text{median}\{Y_j - X_i$: $i=1,\ldots,m$ and $j=1,\ldots,n\}$. Use this to find the smallest significance level at which the associated (conditional) permutation test would reject H_0: $\Delta=0$ in favor of H_1: $\Delta>0$.

11.1.3. For the setting and data of Example 11.1.13, construct a sufficient portion of the conditional (given $|x_1|,\ldots,|x_n|$) null H_0: $\theta=0$ distribution of

$$S^* = \underset{1 \leqslant i \leqslant j \leqslant n}{\text{median}} \left(\frac{X_i + X_j}{2} \right),$$

so as to find the smallest significance level at which the conditional test would reject H_0: $\theta=0$ in favor of H_1: $\theta>0$.

11.1.4. Let $X_1,\ldots,X_m$ and $Y_1,\ldots,Y_n$ be independent random samples from a continuous distribution with c.d.f. $F(x)$, and let $\mathbf{Z}_{(\cdot)}=(Z_{(1)},\ldots,Z_{(N)})$ be the combined sample order statistics, with $N=m+n$. Find

$$E\left[\,\bar{Y}-\bar{X}|\mathbf{Z}_{(\cdot)}=\mathbf{z}_{(\cdot)}\right]$$

and

$$\text{Var}\left[\,\bar{Y}-\bar{X}|\mathbf{Z}_{(\cdot)}=\mathbf{z}_{(\cdot)}\right],$$

where $\bar{X}=(1/m)\Sigma_{i=1}^{m}X_i$, $\bar{Y}=(1/n)\Sigma_{j=1}^{n}Y_j$, and $\mathbf{z}_{(\cdot)}=(z_{(1)},\ldots,z_{(N)})$ is the observed value of $\mathbf{Z}_{(\cdot)}$.

[Hint: One way to approach this solution utilizes Theorem 8.1.12.]

11.1.5. A random sample of $N=20$ persons were interviewed and asked whether they were in favor of Title IX legislation giving more equal emphasis to men's and women's athletic programs. The null hypothesis of this experiment is that men's and women's attitudes toward Title IX are the same, whereas the alternative hypothesis is that women favor it more strongly than men. Of the 12 women interviewed, 8 favored Title IX and 4 did not. Among the 8 men interviewed only 2 favored Title IX. What is the smallest significance level at which we would reject H_0 in favor of H_1 for these data?

11.1.6. Show that when $m=A+B$ and $n=D+E$ are fixed and known, the null distribution of Fisher's exact test, as given in expression (11.1.9), comes from the permutation principle. [Hint: Assign all observations numerical values by considering all category C_1 responses as ones and all C_2 responses as twos.]

11.1.7. Let $(X_1, Y_1),\ldots,(X_n, Y_n)$ denote a random sample of n bivariate observations from a jointly continuous distribution with c.d.f. $F(x,y)$. Suppose that the null hypothesis is that X_i and Y_i are independent random variables for $i=1,\ldots,n$, while the alternative of interest is that they are positively correlated; that is, larger (smaller) values of X_i tend to be associated with larger (smaller) values of Y_i.

(a) Let $S(X_1, Y_1,\ldots,X_n, Y_n)$ denote some unspecified sample measure of association that tends to be large when there is indication of positive association and treats the n sample pairs $(X_1, Y_1),\ldots,(X_n, Y_n)$ symmetrically. Using the permutation principle, show how to construct a conditional nonparametric distribution-free test based on $S(\cdot)$.

(b) Let $n=4$ and suppose the observation vectors are $(x_1,y_1)=(2.8, 8.4)$, $(x_2,y_2)=(4.4, 12.0)$, $(x_3,y_3)=(1.2, 3.6)$ and $(x_4,y_4)=(3.5, 7.2)$. Use the Pearson product moment correlation coefficient,

$$S(X_1, Y_1,\ldots,X_n, Y_n) = \frac{\sum_{i=1}^{n}(X_i-\bar{X})(Y_i-\bar{Y})}{\left[\sum_{i=1}^{n}(X_i-\bar{X})^2\sum_{j=1}^{n}(Y_j-\bar{Y})^2\right]^{1/2}},$$

and find the smallest significance level at which the conditional test based on $S(\cdot)$ would reject H_0 in favor of H_1 for this data set.

11.2. Distribution-Free Rank-Like Tests

In the early development of nonparametric statistics virtually all distribution-free tests were based on rankings of observations in independent random samples from continuous distributions. Thus, under the appropriate null hypothesis, such a test deals with rankings of independent, identically distributed sample observations. The various procedures discussed in Chapters 8–10 and most of those to be considered in Chapter 12 are of this type. In this section we consider test procedures that involve ranking random variables which under the appropriate null hypothesis are exchangeable variables. (Hence, each possible ranking is equally likely under H_0.) However, the variables that are ranked are not the original observations but are, instead, functions of them. The term **rank-like** has been used to describe test procedures of this type. The term was coined by Moses (1963) in an article dealing with the two-sample scale problem. He considered random allocation of the samples observations into subgroups and proposed, among other suggestions, doing a Mann–Whitney–Wilcoxon test on the sample variances computed within these subgroups. (See Exercises 11.2.6 and 11.2.7.) Hollander (1970) used a similar random allocation idea to develop a distribution-free test for the parallelism of two straight line regression equations. Although both the Moses and Hollander procedures still use rankings of (under the respective null hypotheses) independent, identically distributed random variables, these variables are no longer the original sample observations. In a further development along these lines, Fligner and Killeen (1976) considered distribution-free rank-like procedures for the two-sample scale problem using statistics based on joint rankings of a set of exchangeable (under the appropriate null hypothesis), but not independent, functions of the original random samples. As an additional illustration that the independence is not essential to obtaining distribution-free tests, Fligner, Hogg and Killeen (1976) presented a general class of rank-like procedures for the two-sample setting.

In this section we consider a general approach to these rank-like procedures. Although our examples primarily deal with two-sample settings, we shall see that this idea of ranking exchangeable, but not independent, variables can be used in other settings as well (e.g., for k-sample or multivariate problems). We first establish the exchangeability of a broad class of statistics.

Let $\mathbf{x}_1,\ldots,\mathbf{x}_N$ denote N arbitrary p-tuples of real numbers.

Definition 11.2.1. If $g(\cdot)$ is a function (possibly vector-valued) of N p-tuples satisfying

$$g(\mathbf{x}_1,\ldots,\mathbf{x}_N) = g(\mathbf{x}_{d_1},\ldots,\mathbf{x}_{d_N}) \tag{11.2.2}$$

for every permutation $(d_1,\ldots,d_N)$ of $(1,\ldots,N)$ and every set of N p-tuples $(\mathbf{x}_1,\ldots,\mathbf{x}_N)$, then we say that $g(\cdot)$ is **symmetric in its N arguments**.

Using this symmetry property and equal in distribution arguments (see Section 1.3), we establish the following general theorem due to Smith and Wolfe (1977). [A similar result was given by Broffitt, Randles and Hogg (1976).]

Theorem 11.2.3. *Let $\mathbf{X}_i=(X_{i1},\ldots,X_{ip})$, $i=1,\ldots,N$, be a random sample from some p-variate continuous distribution. Let $g(\cdot)$ be any function of N p-vectors that is symmetric in its arguments. Let $h(\cdot,\cdot)$ be any real-valued function of a p-tuple and the function values of $g(\cdot)$ and define the random variables*

$$W_i = h(\mathbf{X}_i, g(\mathbf{X}_1,\ldots,\mathbf{X}_N)), \qquad i = 1,\ldots,N. \tag{11.2.4}$$

Then, $W_1,\ldots,W_N$ are exchangeable (see Definition 1.3.6) *random variables.*

Proof: Let $(d_1,\ldots,d_N)$ be any permutation of $(1,\ldots,N)$. Since $g(\cdot)$ is symmetric in its arguments, we have

$$g(\mathbf{X}_1,\ldots,\mathbf{X}_N) = g(\mathbf{X}_{d_1},\ldots,\mathbf{X}_{d_N}).$$

In addition, since $\mathbf{X}_1,\ldots,\mathbf{X}_N$ is a random sample, we have that $(\mathbf{X}_1,\ldots,\mathbf{X}_N) \stackrel{d}{=} (\mathbf{X}_{d_1},\ldots,\mathbf{X}_{d_N})$. These two facts combine to yield

$$\begin{aligned}(W_1,\ldots,W_N) &= (h(\mathbf{X}_1,g(\mathbf{X}_1,\ldots,\mathbf{X}_N)),\ldots,h(\mathbf{X}_N,g(\mathbf{X}_1,\ldots,\mathbf{X}_N)))\\ &\stackrel{d}{=} \left(h(\mathbf{X}_{d_1},g(\mathbf{X}_{d_1},\ldots,\mathbf{X}_{d_N})),\ldots,h(\mathbf{X}_{d_N},g(\mathbf{X}_{d_1},\ldots,\mathbf{X}_{d_N}))\right)\\ &= \left(h(\mathbf{X}_{d_1},g(\mathbf{X}_1,\ldots,\mathbf{X}_N)),\ldots,h(\mathbf{X}_{d_N},g(\mathbf{X}_1,\ldots,\mathbf{X}_N))\right)\\ &= (W_{d_1},\ldots,W_{d_N}),\end{aligned}$$

which establishes the exchangeability. ■

Theorem 11.2.3 leads immediately to the following important corollary that is useful in developing distribution-free rank-like statistics for many problems.

Corollary 11.2.5. Let $W_1,\ldots,W_N$ be as defined in Theorem 11.2.3 and let R_i^* denote the rank of W_i among $W_1,\ldots,W_N$. If $P[W_i = W_j]=0$ for every $i \neq j$, then

$$P[\mathbf{R}^* = \mathbf{r}] = \frac{1}{N!} \tag{11.2.6}$$

for every $\mathbf{r}$, a permutation of the integers $(1,\ldots,N)$. Thus any statistic that is a function of the sample observations $\mathbf{X}_1,\ldots,\mathbf{X}_N$ only through the ranks $R_1^*,\ldots,R_N^*$ is nonparametric distribution-free over the class of all p-variate continuous distributions.

Proof: These facts follow at once from Theorem 11.2.3. ■

Although the condition that $P(W_i = W_j)=0$ for every $i \neq j$ is not necessary to obtain nonparametric distribution-free rank-like test statistics, we shall see that it often allows us to conduct these tests without computing new null distribution tables. In view of the effort usually required to generate such tables, this can be an important plus for a nonparametric test procedure. The first rank-like procedure we consider was proposed by Fligner and Killeen (1976) for the two-sample scale problem.

Example 11.2.7. Let $X_1,\ldots,X_m$ and $Y_1,\ldots,Y_n$ be independent random samples from continuous populations with c.d.f.'s $F(x-\theta)$ and $F((x-\theta)/\eta)$, with $-\infty<\theta<\infty$ and $\eta>0$, where $F(0)=1/2$. We consider a nonparametric distribution-free rank-like procedure for testing H_0: $\eta=1$ against H_1: $\eta>1$. Let $\hat{\theta}=\hat{\theta}(X_1,\ldots,X_m,Y_1,\ldots,Y_n)=$ median$(X_1,\ldots,X_m,Y_1,\ldots,Y_n)$ and define

$$\begin{aligned} W_i &= |X_i-\hat{\theta}|, \qquad i=1,\ldots,m \\ &= |Y_{i-m}-\hat{\theta}|, \qquad i=m+1,\ldots,N, \end{aligned} \tag{11.2.8}$$

with $N=m+n$. Now, the function $\hat{\theta}(t_1,\ldots,t_N)$ is symmetric in its $N=(m+n)$ arguments. Hence, taking $g(\cdot)=\hat{\theta}(\cdot)$ and $h(s,t)=|s-t|$, we see from Theorem 11.2.3 that $W_1,\ldots,W_N$ are exchangeable random variables when H_0: $\eta=1$ is true (since then the X and Y distributions are identical). In view of Corollary 11.2.5, this implies that any test based on a statistic that is a function of $X_1,\ldots,X_m,Y_1,\ldots,Y_n$ only through the joint ranks of $W_1,\ldots,W_N$ will be nonparametric distribution-free over the class of continuous distributions, provided $P[W_i=W_j]=0$ for every $i\neq j$, a condition that will hold whenever $N=(m+n)$ is odd. Thus, for example, the statistic $V=\sum_{j=1}^{n}R_j$, where R_j is the rank of W_{m+j} among $W_1,\ldots,W_N$, is distribution-free when H_0: $\eta=1$ is true. (We note that V is simply the rank sum form of the Mann–Whitney–Wilcoxon statistic as applied to the Ws.) The corresponding α-level test of H_0: $\eta=1$ against H_1: $\eta>1$ is to

$$\text{reject } H_0 \text{ iff } V \geqslant c(\alpha,m,n),$$

where $c(\alpha,m,n)$ is the upper 100αth percentile for the null H_0 distribution of V. From (11.2.6) we see that when N is odd, the null distribution of V is identical with that of the usual Wilcoxon rank sum statistic, as discussed in Theorems 2.3.11 and 2.3.16. Therefore, $c(\alpha,m,n)$ is simply the upper 100αth percentile (previously denoted by $w(\alpha,m,n)$) for the null distribution of the Wilcoxon rank sum statistic.

When N is even, there is a tied rank among the W variables with probability one. Although the V test is still distribution-free [see Fligner and Killeen (1976)], the critical points for the usual Wilcoxon rank sum statistic can no longer be used for $c(\alpha,m,n)$. (See also Exercise 11.2.3.)

For a second example, we consider an application [proposed by Smith and Wolfe (1977)] of these rank-like procedures to a two-sample regression setting.

Example 11.2.9. Let $(X_{11}, Y_{11}), \ldots, (X_{1m}, Y_{1m})$ and $(X_{21}, Y_{21}), \ldots, (X_{2n}, Y_{2n})$ be independent random samples from continuous bivariate distributions satisfying the following conditions:

Condition 11.2.10. The distribution function of the conditional distribution of Y_{ij} given $X_{ij} = x_{ij}$, where x_{ij} is a constant, has the form

$$G_{ij}(y) = P(Y_{ij} \leqslant y | X_{ij} = x_{ij}) = G(y - \alpha - \beta_i x_{ij}), \tag{11.2.11}$$

for $i = 1$, $j = 1, \ldots, m$ and $i = 2$, $j = 1, \ldots, n$, where $G(\cdot)$ is the c.d.f. for a continuous distribution with zero median, and $-\infty < \alpha, \beta_1, \beta_2 < \infty$ are unknown parameters.

Condition 11.2.12. Each of the X_{ij}s, $i = 1$, $j = 1, \ldots, m$ and $i = 2$, $j = 1, \ldots, n$, has the same distribution.

Thus the median of the conditional distribution of Y_{ij}, given $X_{ij} = x_{ij}$, is simply $\alpha + \beta_i x_{ij}$, and we are in a two-sample straight line regression setting, with jointly random "independent" and "dependent" variables, and common, but unknown, intercept value α at $x_{ij} = 0$. We are interested in testing H_0: $\beta_1 = \beta_2 = \beta$, with β unknown, (i.e., the two regression lines have equal slopes) against the alternative H_1: $\beta_2 > \beta_1$. (The alternatives $\beta_2 < \beta_1$ or $\beta_2 \neq \beta_1$ would be handled similarly.) With our model, the null hypothesis H_0: $\beta_1 = \beta_2 = \beta$ is equivalent to H_0': $[F_1(x,y) = F_2(x,y)$, for all (x,y) pairs$]$, where $F_i(x,y)$, $i = 1,2$, is the joint c.d.f. for the population from which the ith sample is drawn; that is, under H_0: $\beta_1 = \beta_2 = \beta$, all of the bivariate observations in the two samples have a common joint distribution.

To develop distribution-free tests for H_0 we utilize Theorem 11.2.3. Set $N = m + n$ and let

$$\begin{aligned} \mathbf{Z}_s &= (U_s, V_s) = (X_{1s}, Y_{1s}), \qquad s = 1, \ldots, m \\ &= (U_s, V_s) = (X_{2,s-m}, Y_{2,s-m}), \qquad s = m+1, \ldots, N. \end{aligned} \tag{11.2.13}$$

(Thus $\mathbf{Z}_s, s = 1, \ldots, N$, represent the N bivariate observations obtained by combining the two samples.) Consider estimators $\hat{\alpha}$ and $\hat{\beta}$ of α and the common (under H_0) β, respectively, of the forms $\hat{\alpha} = g_1(\mathbf{Z}_1, \ldots, \mathbf{Z}_N)$ and $\hat{\beta} = g_2(\mathbf{Z}_1, \ldots, \mathbf{Z}_N)$, where both $g_1(\cdot)$ and $g_2(\cdot)$ are symmetric in their N arguments. Possible choices for $g_1(\cdot)$ and $g_2(\cdot)$ would be

$$\hat{\alpha} = \bar{V} - \hat{\beta}\bar{U} \text{ and } \hat{\beta} = \frac{\sum_{s=1}^{N} (U_s - \bar{U})(V_s - \bar{V})}{\sum_{s=1}^{N} (U_s - \bar{U})^2},$$

where $\bar{U}=\frac{1}{N}\Sigma_{i=1}^{N}U_i$ and $\bar{V}=\frac{1}{N}\Sigma_{j=1}^{N}V_j$. (For further elaboration on these and other possible choices see Exercises 11.2.4 and 11.2.5.) Define the N univariate statistics $W_1,\ldots,W_N$ by

$$W_s=\left[V_s-\hat{\beta}U_s-\hat{\alpha}\right]\text{sign}(U_s), \tag{11.2.14}$$

where

$$\begin{aligned}\text{sign}(t) &= 1, && \text{if } t>0,\\ &= 0, && \text{if } t=0,\\ &= -1, && \text{if } t<0.\end{aligned} \tag{11.2.15}$$

(Note that $V_s-\hat{\beta}U_s-\hat{\alpha}$ is simply the residual error resulting from predicting the value of V_s, say $\hat{V}_s$, with the estimated regression equation $\hat{V}_s=\hat{\beta}U_s-\hat{\alpha}$.)

Now, from Theorem 11.2.3 we see that $W_1,\ldots,W_N$ (as given in (11.2.14)) are exchangeable random variables when H_0: $\beta_1=\beta_2=\beta$ is true. Thus, using Corollary 11.2.5, any test based on a statistic that is a function of the original sample observations only through the joint ranks of $W_1,\ldots,W_N$ will be nonparametric distribution-free over the class of all bivariate continuous distributions, provided only that $P[W_i=W_j]=0$ for every $i\neq j$, a mild condition that depends on the choices of $\hat{\alpha}$ and $\hat{\beta}$. For example, if (as in Example 11.2.7) we set $T=\Sigma_{j=1}^{n}R_j$, where R_j is the rank of W_{m+j} among $W_1,\ldots,W_N$, then the test that

$$\text{rejects } H_0\text{: } \beta_1=\beta_2=\beta \text{ iff } T\geqslant d(\alpha,m,n), \tag{11.2.16}$$

where $d(\alpha,m,n)$ is the upper 100αth percentile for the null (H_0) distribution of T, is a nonparametric distribution-free α-level test of H_0: $\beta_1=\beta_2=\beta$ against H_1: $\beta_2>\beta_1$. Also, as in Example 11.2.7, whenever the condition $P(W_i=W_j)=0$ for every $i\neq j$ is satisfied, then $d(\alpha,m,n)$ is simply the upper 100αth percentile for the usual null distribution of the Wilcoxon rank sum location statistic.

Although the use of rank-like procedures is relatively new to nonparametric statistics, there have been several other applications considered in the literature. For example, Fligner and Killeen (1976) proposed an entire class of rank-like tests for the two-sample scale problem, while Smith and Wolfe (1977) utilized the general approach of Example 11.2.9 in constructing rank-like tests for a general k-sample straight-line regression setting. Broffitt, Randles, and Hogg (1976) used the idea of rank-like statistics to create nonparametric distribution-free procedures for partial discriminant analysis problems.

Exercises 11.2.

11.2.1. Let $X_1, \ldots, X_m$ and $Y_1, \ldots, Y_n$ be independent random samples from continuous distributions with c.d.f.'s $F(x)$ and $F(x-\Delta)$, respectively. Let R_i, $i=1,\ldots,n$, be the rank of Y_i among $X_1,\ldots,X_m, Y_1,\ldots,Y_n$, and let $S=\sum_{j=1}^{n} a(R_j)$ be an arbitrary two-sample linear rank statistic (9.1.1) for the location problem, with the scores $a(j)$ satisfying (9.1.2). Use Theorem 11.2.3 to argue that S is nonparametric distribution-free when H_0: $\Delta=0$ is true. [Hint: Consider appropriate choices for the functions $g(\cdot)$ and $h(\cdot)$.]

11.2.2. Let $V=\sum_{j=1}^{n} R_j$ be the test statistic discussed in Example 11.2.7, for N an odd integer. Obtain an expression for $E(V)$ when θ and η are arbitrary. To what does $E(V)$ simplify when H_0: $\eta=1$ is true? [Hint: Consider the Mann–Whitney form for the statistic V.]

11.2.3. Consider the statistic $V=\sum_{j=1}^{n} R_j$ of Example 11.2.7. Using the idea of average ranks to handle ties, find the null (H_0: $\eta=1$) distribution of V when $m=n=2$ (i.e., $N=4$ is even). Justify your computations. Compare this distribution with the corresponding null distribution of the Wilcoxon rank sum location statistic.

11.2.4. Consider the test statistic $T=\sum_{j=1}^{n} R_j$ in Example 11.2.9, and let

$$\hat{\beta} = \frac{\sum_{i=1}^{N} (U_i - \bar{U})(V_i - \bar{V})}{\sum_{i=1}^{N} (U_i - \bar{U})^2} \quad \text{and } \hat{\alpha} = \bar{V} - \hat{\beta}\bar{U}, \qquad \textbf{(11.2.17)}$$

where $\bar{U}=N^{-1}\sum_{i=1}^{N} U_i$ and $\bar{V}=N^{-1}\sum_{i=1}^{N} V_i$, be the usual combined samples least-squares estimators for α and the common (under the appropriate null H_0) β, respectively.

(a) Show that each of the estimators in (11.2.17) is symmetric in its use of the observations $\mathbf{Z}_1,\ldots,\mathbf{Z}_N$, where $\mathbf{Z}_s=(U_s, V_s)$, for $s=1,\ldots,N$.

(b) When will the statistic T in **(a)** have a null distribution that is the same as that of the Wilcoxon rank sum statistic for the two-sample location problem?

11.2.5. Consider the test statistic $T=\sum_{j=1}^{n} R_j$ in Example 11.2.9, and let

$$\hat{\beta} = \underset{1 \leq i < j \leq N}{\text{median}} \left\{ \frac{V_j - V_i}{U_j - U_i} \right\} \text{ and } \hat{\alpha} = \underset{1 \leq i \leq N}{\text{median}} \left\{ V_i - \hat{\beta} U_i \right\}. \quad \textbf{(11.2.18)}$$

[Estimators analogous to $\hat{\beta}$ and $\hat{\alpha}$ in (11.2.18) were proposed by Theil (1950b) and Adichie (1967) for the setting where the "independent" variables (i.e., the Xs) are fixed and not jointly random with the "dependent" Y variables.]

(a) Show that each of the estimators in (11.2.18) is symmetric in its use of the observations $\mathbf{Z}_1,\ldots,\mathbf{Z}_N$, where $\mathbf{Z}_s=(U_s,V_s)$, for $s=1,\ldots,N$.

(b) When will the statistic T have a null distribution that is the same as that of the Wilcoxon rank sum statistic for the two-sample location problem?

11.2.6. Let $X_1,\ldots,X_m$ and $Y_1,\ldots,Y_n$ be independent random samples from continuous distributions with c.d.f.'s $F(x-\theta_x)$ and $F((x-\theta_y)/\eta)$, with $-\infty<\theta_x,\theta_y<\infty$, $\eta>0$ and $F(0)=1/2$. For testing the hypothesis H_0: $\eta=1$ against H_1: $\eta>1$, Moses (1963) proposed the following procedure: Select a positive integer $k\geqslant 2$ and *randomly* divide the X and Y observations into m' and n' subgroups of size k, respectively. (Discard any extra observations.) For $i=1,\ldots,m'$, let $X_{i1},\ldots,X_{ik}$ denote the ith subgroup of k X observations, and for $j=1,\ldots,n'$, let $Y_{j1},\ldots,Y_{jk}$ denote the jth subgroup of k Y observations. Define

$$C_i=\sum_{s=1}^{k}\left(X_{is}-\bar{X}_i\right)^2,\qquad i=1,\ldots,m',\tag{11.2.19}$$

and

$$D_j=\sum_{t=1}^{k}\left(Y_{jt}-\bar{Y}_j\right)^2,\qquad j=1,\ldots,n',\tag{11.2.20}$$

with $\bar{X}_i=\frac{1}{k}\Sigma_{s=1}^{k}X_{is}$ and $\bar{Y}_j=\frac{1}{k}\Sigma_{t=1}^{k}Y_{jt}$. Let R_j, $j=1,\ldots,n'$, be the rank of D_j among $C_1,\ldots,C_{m'},D_1,\ldots,D_{n'}$ and set $T=\Sigma_{j=1}^{n'}R_j$. The α-level Moses test of H_0: $\eta=1$ against H_1: $\eta>1$ is then

$$\text{reject } H_0\colon \eta=1 \text{ iff } T\geqslant t(\alpha,m',n'),\tag{11.2.21}$$

where $t(\alpha,m',n')$ is the upper 100αth percentile for the null H_0 distribution of T.

(a) Argue that T is distribution-free for this setting when H_0: $\eta=1$ is true, regardless of the values of the unknown medians θ_x and θ_y.

(b) What is the null H_0: $\eta=1$ distribution of T? Is $t(\alpha,m',n')$ then an easy value to obtain?

(c) Compare the relative advantages and disadvantages of the following two-sample scale procedures: (i) the Moses test of this example, (ii) the Fligner–Killeen test considered in Example 11.2.7, and (iii) the Mood test defined in (and immediately following) equation (9.3.2).

11.2.7. Let T be the Moses test statistic discussed in Exercise 11.2.6.

(a) Find a general expression for $E(T)$. To what does this simplify when H_0: $\eta=1$ is true? [Hint: Consider the Mann–Whitney form for the statistic T.]

(b) For the setting of Exercise 11.2.6, let $F(x)$ be the c.d.f. for a standard normal distribution, and evaluate $E(T)$ in **(a)** for arbitrary values of θ_x, θ_y and η.

11.3. Distribution-Free Partially Sequential Tests

Thus far we have considered distribution-free tests only for settings where the number of sample observations is fixed; that is, for situations where we collect a fixed, preset number of sample observations to be used in a distribution-free test procedure. In this section we discuss a technique for constructing distribution-free two-sample tests for settings where one of the sample sizes is fixed while the second sample size is random, and hence not preset.

Let X and Y be independent, continuous random variables with c.d.f.'s $F(x)$ and $G(y)$, respectively, and consider the problem of testing H_0: $F \equiv G$ against some alternative of interest, say, H_1. (For example, H_1 could correspond to a difference in location between the X and Y distributions or perhaps a difference in scale with known locations.) In previous chapters (and earlier sections of this chapter) we have discussed several different methods for constructing distribution-free tests for H_0 versus H_1 based on independent random samples of sizes m and n from $F(x)$ and $G(y)$, respectively, where m and n are both preset integers. However, there are many settings in which the observations in one of these samples are easy and relatively inexpensive to collect, while the other sample observations are costly and/or difficult to collect. For example, the easily collected sample might be "standard" procedure observations, and the sample that is more difficult to collect could correspond to "new treatment" data. In such situations we would like to gather a sample (usually large) of these standard observations and then collect only enough difficult-to-obtain new treatment observations necessary in order to reach a decision regarding H_0 and H_1. To achieve these goals we consider collecting the new treatment observations in a sequential manner. Although there are many possible sequential sampling methods for obtaining these new treatment values, we consider here only one simple partially sequential scheme proposed by Wolfe (1977a). This sampling technique has an inverse binomial flavor and is truncated in the sense that it will never require more than a preset number of treatment observations.

Let $X_1,\ldots,X_m$ be a random sample from the continuous distribution with c.d.f. $F(x)$, where m is a fixed positive integer. For every m-tuple $(x_1,\ldots,x_m)$ of real numbers let $A(x_1,\ldots,x_m)$ be a subset of the reals, and define the indicator function $\varphi(\cdot)$ on the reals by

$$\varphi(y) = \begin{cases} 1, & \text{if } y \in A(x_1,\ldots,x_m), \\ 0, & \text{otherwise.} \end{cases} \tag{11.3.1}$$

Now, let Y be a random variable (independent of $X_1,\ldots,X_m$) from a second continuous distribution with c.d.f. $G(y)$. Applying $\varphi(y)$ to these random variables, we obtain the indicator random variable for the random set $A(X_1,\ldots,X_m)$, namely,

$$\varphi(Y) = \begin{cases} 1 & \text{if } Y \in A(X_1,\ldots,X_m) \\ 0 & \text{if } Y \notin A(X_1,\ldots,X_m). \end{cases} \tag{11.3.2}$$

The inverse binomial partially sequential sampling scheme is then to collect (independent) Y observations from $G(y)$ until a preset number, say, r, of these Ys are in the set $A(x_1,\ldots,x_m)$, where $(x_1,\ldots,x_m)$ is the observed value of the random vector $(X_1,\ldots,X_m)$. Define the statistic N_m (having random contributions from both the X and Y observations) to be

$$N_m = \{\text{number of } Y \text{ observations required to get } r\ Y\text{s in } A(X_1,\ldots,X_m)\}. \tag{11.3.3}$$

Thus if Y_i is the $G(y)$ observation at the ith stage of sampling, we see that N_m is the smallest integer such that $\sum_{i=1}^{N_m}\varphi(Y_i)=r$. The associated α-level test procedure is then to

$$\text{reject } H_0 \text{ iff } N_m \leqslant N_0(\alpha,r,m,A), \tag{11.3.4}$$

where $N_0(\alpha,r,m,A)$ is the lower 100αth percentile point for the null (H_0) distribution of N_m. [Note that this means that we need never collect more than $N_0(\alpha,r,m,A)$ Y observations.]

We comment that (11.3.4) really describes a large class of tests (including parametric as well as nonparametric procedures) varying as m, r, and

$A(x_1,\ldots,x_m)$ are changed. As a result, a great deal of freedom remains in selecting a particular test from this class. However, we can think of the choice of the set A as being closely tied to the alternatives to H_0 that are of interest in a given setting, while we can view r and m as being set as large as necessary to ensure a desired power level against a specific alternative. [For more discussion along these lines, see Wolfe (1977a, 1977b) and Orban and Wolfe (1978a, 1978b).] For our purposes we simply illustrate appropriate choices of the indicator set A for two particular nonparametric examples.

Example 11.3.5. Consider the two-sample location problem where $F(x)$ corresponds to any continuous distribution and the second population is simply shifted, with c.d.f. $G(x)=F(x-\Delta)$, $-\infty<\Delta<\infty$. For this setting the appropriate null hypothesis is H_0: $\Delta=0$, and we consider here the one-sided alternative H_1: $\Delta>0$. (The alternative $\Delta<0$ is handled by a similar approach and is the topic of Exercise 11.3.1.) With this location problem in mind, we would like to choose the indicator set to be sensitive to such differences. One such set $A(x_1,\ldots,x_m)$ is given by

$$A(x_1,\ldots,x_m) = \{y: y>\tilde{x}\}, \tag{11.3.6}$$

where $\tilde{x}=\text{median}\{x_1,\ldots,x_m\}$. The corresponding test statistic N_m (11.3.3) is then (for fixed r)

$$N_m = \{\text{number of } Ys \text{ necessary to get exactly } r\ Ys \text{ greater than } \tilde{X}\}, \tag{11.3.7}$$

with $\tilde{X}=\text{median}\{X_1,\ldots,X_m\}$. [We note that the procedure associated with N_m in (11.3.7) has an additional advantage in life-testing settings where both the X and Y observations are lifetimes of competing items placed on test. In such situations, the X items can be removed from the testing scheme once half of them have failed, since the median lifetime ($\tilde{X}$) would then be available.]

As a second example, we consider the two-sample scale problem when the locations are known.

Example 11.3.8. Let $F(x)=H(x-\mu_1)$, where μ_1 is known and $H(\cdot)$ corresponds to a continuous distribution that is symmetric about zero. In addition, let $G(x)=H((x-\mu_2)/\sigma)$, where μ_2 is known and $0<\sigma<\infty$ is an unknown scale parameter. Here the appropriate null hypothesis is H_0: $\sigma=1$, and we consider the one-sided alternative H_1: $\sigma>1$. (Similar approaches can be used for the alternatives $\sigma<1$ or $\sigma\neq1$.) We now require an indicator set that is designed for scale differences. One candidate would be

$$A(x_1,\ldots,x_m) = \{y: |y-\mu_2|>x^*\}, \qquad \textbf{(11.3.9)}$$

where $x^* = \text{median}\{|x_1-\mu_1|,\ldots,|x_m-\mu_1|\}$. For preset r the associated test statistic (11.3.3) is

$$N_m = \{\text{number of } Y\text{s necessary to obtain } r\ |Y-\mu_2|\text{'s greater than } X^*\}, \qquad \textbf{(11.3.10)}$$

with $X^* = \text{median}\{|X_1-\mu_1|,\ldots,|X_m-\mu_1|\}$.

To use such a partially sequential test procedure, we must be able to obtain the null (H_0) distribution of the test statistic N_m so that $N_0(\alpha,r,m,A)$ in (11.3.4) can be evaluated. We now derive a general expression for the distribution of N_m when the null hypothesis H_0: $F \equiv G$ is true.

Theorem 11.3.11. *Let N_m be as given in* (11.3.3) *for arbitrary* (*but fixed*) *integers r and m and indicator set $A(x_1,\ldots,x_m)$, and set $p_m(x_1,\ldots,x_m) = P_0(Y \in A(x_1,\ldots,x_m))$. If the X and Y observations are mutually independent and come from continuous distributions with c.d.f.'s $F(x)$ and $G(y)$, respectively, then the null H_0: $F\equiv G$ distribution of N_m can be expressed as*

$$P_0(N_m=n) = \int_{-\infty}^{\infty}\cdots\int_{-\infty}^{\infty}\binom{n-1}{r-1}\{p_m(x_1,\ldots,x_m)\}^r\{1-p_m(x_1,\ldots,x_m)\}^{n-r} \times \prod_{i=1}^{m} f(x_i)\,dx_i, \qquad n = r, r+1, \ldots$$

$$= 0, \qquad \text{elsewhere}, \qquad \textbf{(11.3.12)}$$

where $f(x) = dF(x)/dx$ and P_0 denotes that the probabilities correspond to the null distribution.

Proof: Conditional on $X_1 = x_1,\ldots,X_m = x_m$, the test statistic N_m simply counts the number of Y observations necessary to get r "successes," where success corresponds to Y falling in the set $A(x_1,\ldots,x_m)$. Hence, conditional on $X_1 = x_1,\ldots,X_m = x_m$, the variable N_m has a negative binomial distribution with parameters r and $p_m(x_1,\ldots,x_m) = P_0(Y\in A(x_1,\ldots,x_m))$. The expression in (11.3.12) then follows by multiplying this conditional negative binomial distribution by the joint distribution of $X_1,\ldots,X_m$ and then integrating out the Xs to obtain the unconditional distribution of N_m. ■

For most settings the X sample observations enter the expression for $p_m(X_1,\ldots,X_m)$ only through some statistic and not singly and separately. In such situations we need not use the entire joint distribution of $X_1,\ldots,X_m$ as specified in Theorem 11.3.11. We illustrate this fact with an example.

Example 11.3.13. Consider the statistic N_m in equation (11.3.7) of Example 11.3.5. If m is an odd integer, then the X sample median is simply the $((m+1)/2)$th ordered X observation, namely, $\tilde{X}=X_{((m+1)/2)}$. Moreover,

$$p_m(x_1,\ldots,x_m) = P(Y>\tilde{x}) = 1 - G(\tilde{x}).$$

Thus the expression for $P_0(N_m=n)$ in (11.3.12) is simply

$$E\left[\binom{n-1}{r-1}\{1-F(\tilde{X})\}^r\{F(\tilde{X})\}^{n-r}\right].$$

Consequently, instead of needing the entire joint distribution of $X_1,\ldots,X_m$, we can evaluate $P_0(N_m=n)$ by using the distribution of $\tilde{X}=X_{((m+1)/2)}$. From Theorem 1.2.4 we see that the marginal density of $\tilde{X}$ is

$$g(\tilde{x}) = \frac{m!}{\left[\left(\frac{m-1}{2}\right)!\right]^2}\left[F(\tilde{x})\right]^{\frac{m-1}{2}}\left[1-F(\tilde{x})\right]^{\frac{m-1}{2}}f(\tilde{x}), \qquad -\infty<\tilde{x}<\infty.$$

Combining these results we have that (for $n=r,r+1,\ldots$)

$$P_0(N_m=n) = \int_{-\infty}^{\infty}\left\{\binom{n-1}{r-1}\left[1-F(\tilde{x})\right]^r\left[F(\tilde{x})\right]^{n-r}\right\}$$
$$\times\left\{\frac{m!}{\left[\left(\frac{m-1}{2}\right)!\right]^2}\left[F(\tilde{x})\right]^{\frac{m-1}{2}}\left[1-F(\tilde{x})\right]^{\frac{m-1}{2}}f(\tilde{x})\right\}d\tilde{x},$$

which, on simplifying and using the change of variable $y=F(\tilde{x})$, yields the null distribution of N_m to be

$$P_0(N_m=n) = \frac{\binom{n-1}{r-1}m!\,\Gamma\left(\frac{m+2n-2r+1}{2}\right)\Gamma\left(\frac{2r+m+1}{2}\right)}{\left[\left(\frac{m-1}{2}\right)!\right]^2(m+n)!}, \qquad n=r,r+1,\ldots$$
$$= 0, \qquad \text{elsewhere.} \tag{11.3.14}$$

We note that this null distribution expression does not depend on the form of the common c.d.f. $F(x)$, so long as it is the c.d.f. for a continuous distribution. Thus the test (11.3.4) based on N_m (11.3.7) is nonparametric distribution-free over the class of all continuous distributions.

It can be shown [see, e.g., Wolfe (1977a)] that the limiting $(m \to \infty)$ distribution of a partially sequential test statistic N_m (11.3.3) is negative binomial with parameters r and $p(F,G)$ for arbitrary c.d.f.'s $F(\cdot)$ and $G(\cdot)$, provided $p_m(X_1,\ldots,X_m)$ converges $(m \to \infty)$ in probability to $p = p(F,G)$, $0 < p \leqslant 1$. Thus, for example, it follows (see Exercise 11.3.5) that the limiting distribution of N_m in (11.3.7) is negative binomial with parameters r and $p(F,G) = 1 - G(\xi_1)$, provided ξ_1 is the unique median of the X distribution (i.e., $F(\xi_1) = 1/2$), and $G(\xi_1) \neq 1$.

Such limiting distributions can be used to find approximate values for the cutoff point $N_0(\alpha, r, m, A)$ of the test in (11.3.4). In addition, they provide a means for selecting r so as to provide approximate asymptotic power guarantees against alternatives of interest. [For the median test of Examples 11.3.5 and 11.3.13, such power considerations are discussed in Orban and Wolfe (1978b).]

Another important property of a partially sequential test is the expected number of Y observations necessary to reach a conclusion regarding H_0 and H_1. Results along these lines can be found in Orban and Wolfe (1978a, 1978b).

The partially sequential technique discussed in this section employs a single indicator set, and the resulting test statistics have a strong counting flavor. However, more complicated partially sequential procedures using ranking statistics have been developed by Orban and Wolfe (1978c). (See also Exercise 11.3.7.)

Although this section has been devoted to test procedures in which only one of the samples is collected sequentially, there has been work on developing nonparametric distribution-free procedures where both samples are sequentially obtained. For the two-sample location setting such tests have been considered, for example, by Wilcoxon, Rhodes, and Bradley (1963) and Wilcoxon and Bradley (1964).

Exercises 11.3.

11.3.1. Consider the setting of Example 11.3.5. Discuss how you would change the indicator set $A(x_1,\ldots,x_m)$ to test H_0: $\Delta = 0$ against H_1: $\Delta < 0$. What is the null (H_0) distribution for the corresponding partially sequential test statistic N_m? [Hint: See Example 11.3.13.]

11.3.2. Find a closed form expression for the null H_0: $\sigma=1$ distribution of the partially sequential test statistic N_m (11.3.10) in Example 11.3.8. Is this procedure nonparametric distribution-free for testing H_0: $\sigma=1$? Justify your answer.

11.3.3. Consider the partially sequential test procedure given by (11.3.4), with r and m fixed and

$$A(x_1,\ldots,x_m) = \{y: y > x_{(j)}\} \tag{11.3.15}$$

for some $j\in\{1,\ldots,m\}$, where $x_{(1)}\leqslant\cdots\leqslant x_{(m)}$ are the ordered x's. (Note that the median procedure of Examples 11.3.5 and 11.3.13 is just a special case corresponding to m odd and $j=(m+1)/2$.) Find the exact null H_0: $F\equiv G$ distribution of the N_m test statistic corresponding to A in (11.3.15). [Hint: See Example 11.3.13.]

11.3.4. For the N_m test statistic of Exercise 11.3.3, show that

$$P_0(N_m=n+1) = \frac{n(n-r+j)}{(m+n+1)(n-r+1)} P_0(N_m=n),$$
$$\text{for any } n = r, r+1,\ldots,$$

where P_0 denotes that the probabilities are computed under the null H_0: $F\equiv G$ hypothesis. Discuss how this relationship can be used to construct null distribution tables for N_m.

11.3.5. Consider the setting of Example 11.3.13. Show that $p_m(X_1,\ldots,X_m)=1-G(\tilde{X})$ converges (as $m\to\infty$) in probability to $1-G(\xi_1)$, provided ξ_1 is the unique median of the X distribution. [Hint: See, for example, page 181 of Lehmann (1975).]

11.3.6. Consider the class (for $j\in\{1,2,\ldots,m\}$) of partially sequential test procedures presented in Exercise 11.3.3. Let

$$p_m^j(x_1,\ldots,x_m) = P(Y\in A(x_1,\ldots,x_m)),$$

where $A(x_1,\ldots,x_m)$ is given by (11.3.15). Discuss conditions (on F, G, and j) under which $p_m^j(X_1,\ldots,X_m)$ will converge ($m\to\infty$) in probability to some constant $p^j=p^j(F,G)$, $0<p^j\leqslant 1$. In view of the discussion following Example 11.3.13, what would these conditions (on F, G, and j) imply about the limiting ($m\to\infty$) distribution of the corresponding partially sequential test statistic N_m^j? [Hint: Under what conditions will a sample percentile converge in probability to the corresponding population percentile? See, for example, David (1970).]

11.3.7. Consider the two-sample location problem where $F(x)$ corresponds to any continuous distribution and the second population is shifted with c.d.f. $G(x)=F(x-\Delta)$, $-\infty<\Delta<\infty$. Let $X_1,\ldots,X_m$ be a random sample from $F(x)$, and let $Y_1, Y_2,\ldots$ be a sequence of independent random variables from $G(x)$ such that the Ys are also independent of $X_1,\ldots,X_m$. Set

$$N_m = \min_n \{n: S_{n,m} \geqslant c\},$$

where c is a preset constant and

$$S_{n,m} = \sum_{i=1}^{n} F_m(Y_i),$$

with $F_m(t)=\frac{1}{m}$[number of Xs less than or equal to t] being the empirical distribution function for the X sample. Consider the test of H_0: $\Delta=0$ against H_1: $\Delta>0$ given by

$$\text{reject } H_0 \text{ iff } N_m \leqslant N_0(\alpha,c,m), \qquad \textbf{(11.3.16)}$$

where $N_0(\alpha,c,m)$ is the 100αth percentile for the null H_0 distribution of N_m.

(a) The procedure in (11.3.16) is a partially sequential analogue to which fixed sample sizes two-sample location procedure?

(b) For fixed $N_0=N_0(\alpha,c,m)$, let U_{m,N_0} be the Mann–Whitney form of the Mann–Whitney–Wilcoxon statistic. Show how null H_0: $\Delta=0$ distribution tables of U_{m,N_0} (for varying N_0) can be used to determine the 100αth percentile $N_0(\alpha,c,m)$ needed in (11.3.16). [Hint: Consider the relationship between the events $\{N_m \leqslant N_0\}$ and $\{U_{m,N_0} \geqslant mc\}$.]

11.4. Distribution-Free Exceedance Statistics

Properties of many two-sample rank statistics (some linear and some nonlinear) can be easily obtained by studying the behavior of the empirical (sample) distribution functions for the two sets of observations. Some such test procedures are designed to detect *any* differences (including location or scale) whatsoever between the two underlying populations. We discuss a few of this type of tests in Section 11.5. A second important set of two-sample procedures which are based on the two empirical c.d.f.'s contains what are referred to as exceedance statistics, and these procedures are the topic of this section. Exceedance test statistics are functions of the combined sample rank vector and hence have a nonparametric distribution-free property. Their popularity stems in part from their simple structure which (i) makes them easy to compute and (ii) often yields a relatively simple null distribution.

Let $X_1, \ldots, X_m$ be a random sample from a distribution with c.d.f. $F(x)$, and let $F_m(x)$ be the empirical c.d.f. for the sample; that is, $F_m(t) = (1/m)[\text{number of } Xs \leqslant t]$, $-\infty < t < \infty$. The empirical distribution function plays a role in the sample similar to that played by the population c.d.f. $F(x)$ in the underlying distribution. For a fixed x-value, $mF_m(x)$ is a random variable having a binomial distribution.

Theorem 11.4.1. *Let $F_m(x)$ be the empirical distribution function for a random sample $X_1, \ldots, X_m$ from a distribution with c.d.f. $F(x)$. For a fixed x-value, say x_0, the random variable $mF_m(x_0)$ has a binomial$(m, F(x_0))$ distribution.*

Proof: For fixed x_0 the random variable $mF_m(x_0)$ is simply the number of X_is less than or equal to x_0, and the result follows immediately from the independence of the Xs. ■

The expected value and variance of $F_m(x_0)$, for fixed x_0, follow immediately from Theorem 11.4.1.

Corollary 11.4.2. Let $F_m(x)$ be the empirical distribution function for a random sample of size m from a distribution with c.d.f. $F(x)$. Then

$$E[F_m(x_0)] = F(x_0)$$

and

$$\text{Var}[F_m(x_0)] = \frac{F(x_0)[1 - F(x_0)]}{m},$$

for any fixed x_0.

Thus, for fixed x, the empirical distribution function $F_m(x)$ is an unbiased and consistent estimator of $F(x)$, the value of the population c.d.f. at x. For other properties of $F_m(x)$, where x is fixed, see Exercise 11.4.9 and Section 11.5.

In this section our primary interest in $F_m(x)$ is not for fixed x. Instead, we are concerned with distributional properties of random variables like $F_m(Y)$, where Y is also a random variable, independent of the sample

$X_1,\ldots,X_m$ used in constructing $F_m(x)$. Through developing such properties for continuous random variables we will be able to study a class of two-sample nonlinear rank statistics referred to as exceedance statistics.

Let $X_1,\ldots,X_m$ and $Y_1,\ldots,Y_n$ be independent random samples from the same continuous distribution with c.d.f. $F(x)$, and let $F_m(x)$ be the empirical distribution function for the X observations. We then have the following sample analogue to the probability integral transformation (Theorem 1.2.9).

Theorem 11.4.3. *Let $S_j=F_m(Y_j)$, $j=1,\ldots,n$. Then each S_j is uniformly distributed over the set of points $\{0,1/m,\ldots,(m-1)/m,1\}$.*

Proof: Let i and j be arbitrary integers from $\{1,\ldots,m-1\}$ and $\{1,\ldots,n\}$, respectively. Then

$$P\left(S_j=\tfrac{i}{m}\right) = P\left(X_{(i)}\leqslant Y_j<X_{(i+1)}\right)$$
$$=\int_{-\infty}^{\infty} P\left(X_{(i)}\leqslant Y_j<X_{(i+1)}\mid Y_j=t\right)f(t)\,dt,$$

where $f(x)$ is the density associated with $F(x)$ and $X_{(1)}\leqslant\cdots\leqslant X_{(m)}$ are the ordered Xs. However,

$$P\left(X_{(i)}\leqslant Y_j<X_{(i+1)}\mid Y_j=t\right) = \binom{m}{i}\left[F(t)\right]^i\left[1-F(t)\right]^{m-i},$$

since, given $Y_j=t$, this is nothing more than a binomial probability. Hence,

$$P\left(S_j=\tfrac{i}{m}\right) = \int_{-\infty}^{\infty}\binom{m}{i}\left[F(t)\right]^i\left[1-F(t)\right]^{m-i}f(t)\,dt.$$

Using the change of variable $w=F(t)$ in this integral, we obtain

$$P\left(S_j=\tfrac{i}{m}\right) = \int_0^1\binom{m}{i}w^i(1-w)^{m-i}\,dw$$
$$=\left(\frac{1}{m+1}\right)\int_0^1\frac{(m+1)!}{i!(m-i)!}w^i(1-w)^{m-i}\,dw,$$

and the result follows from the fact that the integrand in this last expression is the density for a Beta distribution with parameters $\alpha=i+1$ and $\beta=(m-i+1)$. The arguments for $i=0$ or m are similar and are left as Exercise 11.4.1. ■

Setting $S_{(j)}=F_m(Y_{(j)})$, $j=1,\ldots,n$, where $Y_{(1)}\leqslant\cdots\leqslant Y_{(n)}$ are the ordered Ys, we observe that $S_{(1)}\leqslant\cdots\leqslant S_{(n)}$ are the ordered S_js. These statistics play the role of sample analogues to the ordered $F(Y_{(j)})$'s discussed in Section 1.2. As we shall see, they also provide an important tool for the study of certain distribution-free two-sample rank statistics.

Theorem 11.4.4. *Let $X_1,\ldots,X_m$ and $Y_1,\ldots,Y_n$ denote independent random samples from the same continuous distribution, and set $S_{(j)}=F_m(Y_{(j)})$, $j=1,\ldots,n$. Then each $S_{(j)}$ has the distribution given by*

$$P\left(S_{(j)}=\frac{i}{m}\right)=\frac{\binom{m+n-i-j}{m-i}\binom{i+j-1}{i}}{\binom{m+n}{m}}, \qquad i=0,1,\ldots,m,$$

$$= 0, \qquad \text{elsewhere.}$$

[*This distribution is called the* **negative hypergeometric distribution,** *for which many of the basic properties are summarized in Guenther* (1975).]

Proof: Left as Exercise 11.4.2. ■

In view of Theorems 11.4.3 and 11.4.4, any statistic that is a function of S_j (or $S_{(j)}$) alone, for some $j\in\{1,\ldots,n\}$, will be distribution-free over $\mathcal{Q}=\{F\colon F \text{ is a c.d.f. for a continuous distribution}\}$. (Actually, any such statistic is a nonlinear two-sample rank statistic, from which the distribution-freeness also follows.) We illustrate these types of two-sample rank statistics, commonly called **exceedance statistics,** with two examples.

Example 11.4.5. Let $X_1,\ldots,X_m$ and $Y_1,\ldots,Y_n$ be independent random samples from continuous distributions with c.d.f.'s $F(x)$ and $G(y)$, respectively. For testing the null hypothesis H_0: $F\equiv G$ against the alternative that the medians of the two populations are different, Mathisen (1943) proposed procedures based on the statistic M=(number of Xs less than or equal to the sample median of the Ys). For the case of n odd, we can write $M=(\text{number of } X\text{s} \leqslant Y_{((n+1)/2)})=mF_m(Y_{((n+1)/2)})=mS_{((n+1)/2)}$, where $S_{((n+1)/2)}$ is as defined in Theorem 11.4.4. Thus $P(M=t)=P(S_{((n+1)/2)}=t/m)$, and when H_0: $F\equiv G$ is true we can use Theorem 11.4.4 to immediately establish the null distribution of M to be

$$P_0(M=t)=\frac{\left[\begin{matrix} m+\left(\frac{n-1}{2}\right)-t \\ m-t \end{matrix}\right]\left[\begin{matrix} \left(\frac{n-1}{2}\right)+t \\ t \end{matrix}\right]}{\binom{m+n}{m}}, \qquad t=0,1,\ldots,m,$$

$$= 0, \qquad \text{elsewhere.} \tag{11.4.6}$$

Example 11.4.7. Consider the same two-sample setting as in Example 11.4.5. Then for testing H_0: $F \equiv G$ against the alternative that the Y distribution is shifted to the right of the X distribution, Rosenbaum (1954) proposed basing a test on the statistic $T =$(number of Xs greater than the largest Y). Now, we can write $T = m(1 - F_m(Y_{(n)})) = m(1 - S_{(n)})$. Thus from Theorem 11.4.4 we have that the null (H_0) distribution of T is

$$P_0(T = t) = P_0\left(S_{(n)} = \frac{m-t}{m}\right) = \frac{\binom{m+n-t-1}{m-t}}{\binom{m+n}{m}}, \qquad t = 0, 1, \ldots, m,$$

$$= 0, \qquad \text{elsewhere.} \qquad \textbf{(11.4.8)}$$

In addition to providing the null distributions for these exceedance statistics, Theorems 11.4.3 and 11.4.4 can be used to easily obtain other null properties for a variety of two-sample rank statistics. For this purpose, we require the following two results.

Theorem 11.4.9. *Let* S_j, $j = 1, \ldots, n$, *be as defined in Theorem* 11.4.3. *Then*

$$E_0[S_j] = \frac{1}{2} \qquad \textbf{(11.4.10)}$$

and

$$\mathrm{Var}_0(S_j) = \frac{1}{12} + \frac{1}{6m}. \qquad \textbf{(11.4.11)}$$

Proof: Exercise 11.4.3. ■

Theorem 11.4.12. *Let* $S_{(j)}$, $j = 1, \ldots, n$, *be as defined in Theorem* 11.4.4. *Then*

$$E_0[S_{(j)}] = \frac{j}{n+1} \qquad \textbf{(11.4.13)}$$

and

$$\mathrm{Var}_0[S_{(j)}] = \frac{j(n-j+1)(m+n+1)}{m(n+1)^2(n+2)}. \qquad \textbf{(11.4.14)}$$

Proof: Exercise 11.4.4. ■

We demonstrate the use of these moment results in two examples.

Example 11.4.15. Let M be the Mathisen statistic in Example 11.4.5 for the case where n is odd. Using Theorem 11.4.12, we see that when H_0: $F \equiv G$ is true,

$$E_0(M) = mE_0(S_{((n+1)/2)}) = m\frac{\left(\frac{n+1}{2}\right)}{n+1} = \frac{m}{2},$$

and

$$\begin{aligned}\text{Var}_0(M) &= m^2\text{Var}_0(S_{((n+1)/2)}) \\ &= m^2\frac{\left(\frac{n+1}{2}\right)\left(n-\frac{n+1}{2}+1\right)(m+n+1)}{m(n+1)^2(n+2)} \\ &= \frac{m(m+n+1)}{4(n+2)}.\end{aligned}$$

Example 11.4.16. Let $U = U(X_1,\ldots,X_m; Y_1,\ldots,Y_n)$ be the Mann–Whitney form of the Mann–Whitney–Wilcoxon statistic. Then we see that U can be represented as

$$U = \sum_{j=1}^{n}\left[\text{number of } X\text{s} \leqslant Y_j\right] = m\sum_{j=1}^{n} S_{(j)} = m\sum_{j=1}^{n} S_j.$$

Hence when H_0: $F \equiv G$ is true, Theorem 11.4.9 (or Theorem 11.4.12) yields

$$E_0(U) = m\sum_{j=1}^{n} E_0(S_j) = \frac{mn}{2},$$

as we already know from previous work. To evaluate the null variance of U in this manner, we need to know the null covariances between S_i and S_j(or $S_{(i)}$ and $S_{(j)}$), for $i \neq j$, which is the topic of Exercises 11.4.6 and 11.4.10.

As a final example in this section, we consider applications of these sample c.d.f. techniques to the problem of obtaining a distribution-free prediction interval for a future sample median. Let $X_1,\ldots,X_m$ and $Y_1,\ldots,Y_n$ be independent random samples from a continuous distribution with c.d.f. $F(\cdot)$. Let $g(Y_1,\ldots,Y_n)$ be some function of the Y observations only.

Definition 11.4.17. Let $L_1(X_1,\ldots,X_m)$ and $L_2(X_1,\ldots,X_m)$ be statistics based only on the X observations. We say that $[L_1, L_2]$ is **a 100(1 − γ)**

percent prediction interval for $g(Y_1,\ldots,Y_n)$ if

$$P_F[L_1(X_1,\ldots,X_m) \leqslant g(Y_1,\ldots,Y_n) \leqslant L_2(X_1,\ldots,X_m)] = (1-\gamma). \tag{11.4.18}$$

If (11.4.18) holds for all $F \in \mathcal{Q} = \{\text{all c.d.f.'s of continuous distributions}\}$, we say that the prediction interval $[L_1, L_2]$ is **nonparametric distribution-free over $\mathcal{Q}$**.

Fligner and Wolfe (1976) discussed methods for obtaining such nonparametric distribution-free prediction intervals for several different $g(\cdot)$ functions. As an example, we consider the particular function $g(Y_1,\ldots,Y_n) = \text{median}\ (Y_1,\ldots,Y_n)$.

Example 11.4.19. Let $X_1,\ldots,X_m$ and $Y_1,\ldots,Y_n$ be independent random samples from a continuous distribution with c.d.f. $F(\cdot)$. Suppose n is an odd integer and set $g_1(Y_1,\ldots,Y_n) = \text{median}\ (Y_1,\ldots,Y_n) = Y_{((n+1)/2)}$, where $Y_{(1)} \leqslant \cdots \leqslant Y_{(n)}$ are the ordered Y observations. Letting $X_{(1)} \leqslant \cdots \leqslant X_{(m)}$ denote the X order statistics, we see that

$$\begin{aligned} P_F(X_{(r_1)} \leqslant Y_{((n+1)/2)} \leqslant X_{(r_2)}) &= P_F(X_{(r_1)} \leqslant Y_{((n+1)/2)} < X_{(r_2)}) \\ &= P_F\left(\frac{r_1}{m} \leqslant F_m(Y_{((n+1)/2)}) \leqslant \frac{r_2 - 1}{m}\right) \\ &= \sum_{i=r_1}^{r_2-1} \frac{\binom{m + \frac{n-1}{2} - i}{m-i}\binom{\frac{n-1}{2} + i}{i}}{\binom{m+n}{m}}, \end{aligned} \tag{11.4.20}$$

for any integers $1 \leqslant r_1 < r_2 \leqslant m$, where the last equality follows from (11.4.6). Thus for a given, appropriate γ, if we choose r_1 and r_2 so that

$$\sum_{i=r_1}^{r_2-1} \frac{\binom{m + \frac{n-1}{2} - i}{m-i}\binom{\frac{n-1}{2} + i}{i}}{\binom{m+n}{m}} = (1-\gamma),$$

then since (11.4.20) holds for every $F(\cdot)$ corresponding to a continuous distribution, we have that $[X_{(r_1)}, X_{(r_2)}]$ is a $100(1-\gamma)$ percent nonparametric distribution-free prediction interval for the Y sample median.

In general, there will be several pairs of integers (r_1, r_2) for which expression (11.4.20) is (at least roughly) equal to $1-\gamma$. Fligner and Wolfe (1979) showed that certain optimal properties are associated with the particular requirement that $r_2 = (m+1-r_1)$. (For the case where n is an even integer, similar arguments can be used to obtain a conservative prediction interval for the Y sample median. See Exercise 11.4.17.)

We note that the examples (with the exception of Example 11.4.16) in this section deal with statistics that are functions of only one $S_{(j)}$ or S_j. However, there are other two-sample exceedance statistics that are functions of two or more of the $S_{(j)}$s (or S_js). Since the $S_{(j)}$s (and, likewise, the S_js) are not mutually independent, to find the null H_0: $F \equiv G$ distributions (and related properties) for such statistics, we need the joint distribution of several of the $S_{(j)}$s (or S_js). This problem and resulting applications are considered for the case of two $S_{(j)}$s (or S_js) in Exercises 11.4.6–11.4.12. For more discussion about S_j and $S_{(j)}$ and tests based on them, see Fligner and Wolfe (1976). In addition, numerous examples of exceedance tests and a historical discussion of their development is given by Neave (1979).

Exercises 11.4.

11.4.1. Complete the proof of Theorem 11.4.3 by establishing the result for $i=0$ and $i=m$.

11.4.2. Prove Theorem 11.4.4.

11.4.3. Prove Theorem 11.4.9. Compare the values of $E[S_j]$ and $\text{Var}(S_j)$ with the corresponding population quantities $E[F(X)]$ and $\text{Var}(F(X))$, respectively, where X is a continuous random variable with c.d.f. $F(x)$. (See Theorem 1.2.9.)

11.4.4. Use the results of Corollary 11.4.2 to prove Theorem 11.4.12. [Hint: Consider conditional arguments given $Y_{(j)} = t$ and use properties for uniform order statistics, as obtained in Section 1.2.]

11.4.5. Let T be the Rosenbaum location statistic defined in Example 11.4.7. Find the null (H_0: $F \equiv G$) mean and variance of T.

11.4.6. Let S_j, $j=1,\ldots,n$, be as defined in Theorem 11.4.3, and let $k<l$ be arbitrary integers between 1 and n, inclusive. Find the joint distribution of S_k and S_l under the null hypothesis condition H_0: $F \equiv G$. From this, evaluate $\text{Cov}_0(S_k, S_l)$ and $\text{Corr}_0(S_k, S_l)$. [Hint: See Theorem 11.4.9.]

11.4.7. Let $S_{(j)}, j=1,\ldots,n$, be as defined in Theorem 11.4.4 and let $k<l$ be arbitrary integers between 1 and n, inclusive. Find the joint distribution of $S_{(k)}$ and $S_{(l)}$ under the null hypothesis condition $F\equiv G$.

11.4.8. Consider the setting of Exercise 11.4.7. Show that $S_{(l-k)} \stackrel{d}{=} S_{(l)} - S_{(k)}$. (Note that this corresponds to a sample analogue of the result in Exercise 1.2.3.) [Hint: Problem 12 on page 65 of Feller (1968) could be useful.]

11.4.9. Let $F_m(x)$ be the empirical distribution function for a random sample of size m from a distribution with c.d.f. $F(x)$. For $x<y$ arbitrary, show that

$$\operatorname{Cov}(F_m(x), F_m(y)) = \frac{F(x)[1-F(y)]}{m}.$$

Evaluate $\operatorname{Corr}(F_m(x), F_m(y))$.

11.4.10. Use the results of Exercises 1.2.2 and 11.4.9 to show that

$$\operatorname{Cov}_0(S_{(k)}, S_{(l)}) = \frac{k(n+1-l)(m+n+1)}{m(n+1)^2(n+2)}$$

and

$$\operatorname{Corr}_0(S_{(k)}, S_{(l)}) = \left[\frac{k(n+1-l)}{l(n+1-k)}\right]^{1/2}$$

under the null hypothesis condition $F\equiv G$, where $S_{(j)}$ is as defined in Theorem 11.4.4, and $k<l$ are arbitrary integers between 1 and n, inclusive. [Hint: The results in (1.2.14), Corollary 11.4.2, and Theorem 11.4.12, as well as the hint for Exercise 11.4.4, are useful.]

11.4.11. Compare the answers in Theorem 11.4.12 and Exercise 11.4.10 with the corresponding values for the variables $F(Y_{(1)}),\ldots,F(Y_{(n)})$, respectively, where $Y_{(1)} \leqslant \cdots \leqslant Y_{(n)}$ are the order statistics for a random sample $Y_1,\ldots,Y_n$ from a continuous distribution with c.d.f. $F(\cdot)$. What are the limits ($m\to\infty$) of the ratios of the corresponding values?

11.4.12. Let $X_1,\ldots,X_m$ and $Y_1,\ldots,Y_n$ be independent random samples from continuous distributions with c.d.f.'s $F(x)$ and $G(y)$, respectively. For testing the null hypothesis H_0: $F\equiv G$ against the alternative that the two populations differ in dispersion, Rosenbaum (1953) suggested using the statistic $V=[$(number of Xs greater than $Y_{(n)}$)+(number of Xs less than or equal to $Y_{(1)}$)], where $Y_{(1)} \leqslant \cdots \leqslant Y_{(n)}$ are the ordered Ys. Show that V is distribution-free over the class $\mathscr{Q}$ corresponding to H_0 by finding the form of the null (H_0) distribution of V. [Hint: See Exercise 11.4.8.]

11.4.13. Let U be the Mann-Whitney form of the Mann-Whitney-Wilcoxon statistic, as discussed in Example 11.4.16. Use the results of either Exercise 11.4.6 or Exercise 11.4.10 to obtain an expression for the null variance of U. (Also see Theorems 11.4.9 and 11.4.12.)

11.4.14. Let V be the Rosenbaum scale statistic of Exercise 11.4.12. Find the null (H_0: $F \equiv G$) mean and variance of V. [Hint: See Exercise 11.4.8.]

11.4.15. Let M be the Mathisen median statistic of Example 11.4.5, and let T be the Rosenbaum location statistic of Example 11.4.7. Find the null (H_0: $F \equiv G$) correlation between M and T. (See Exercise 11.4.10.) Comment on the value of this null correlation between two "location" statistics.

11.4.16. Consider the setting of Example 11.4.19, and let $[X_{(r_1)}, X_{(r_2)}]$ be a $100(1-\gamma)$ percent prediction interval for the Y sample median. We know (using the notation of Example 11.4.19) that this means

$$P_F\big(X_{(r_1)} \leqslant Y_{((n+1)/2)} \leqslant X_{(r_2)}\big) = 1 - \gamma$$

for every $F(\cdot)$ corresponding to a continuous distribution. Prove that this closed prediction interval $[X_{(r_1)}, X_{(r_2)}]$ is also conservative distribution-free for the entire class of underlying discrete distributions that have finite numbers of positive probability masses on any bounded interval of the real line; that is, show that

$$P_F\big(X_{(r_1)} \leqslant Y_{((n+1)/2)} \leqslant X_{(r_2)}\big) \geqslant 1 - \gamma$$

for every $F(\cdot)$ corresponding to such a discrete distribution. [Hint: See Example 6.1.16.]

11.4.17. Let $X_1, \ldots, X_m$ and $Y_1, \ldots, Y_n$ be independent random samples from a continuous distribution with c.d.f. $F(\cdot)$, and let $X_{(1)} \leqslant \cdots \leqslant X_{(m)}$ and $Y_{(1)} \leqslant \cdots \leqslant Y_{(n)}$ be the corresponding order statistics. Suppose n is an even integer, and, using the usual convention, set

$$g_2(Y_1, \ldots, Y_n) = \text{median}\,(Y_1, \ldots, Y_n) = \frac{1}{2}\big(Y_{(n/2)} + Y_{((n/2)+1)}\big).$$

(a) Let $1 \leqslant r_1 < r_2 \leqslant m$ and $1 \leqslant i < j \leqslant n$ be arbitrary integers. Show that

$$P(X_{(r_1)} \leqslant Y_{(i)} \leqslant Y_{(j)} \leqslant X_{(r_2)})$$

$$= \sum_{l=r_1}^{r_2-1} \sum_{k=l}^{r_2-1} \frac{\binom{l+i-1}{l}\binom{m+n-k-j}{m-k}\binom{j+k-l-i-1}{k-l}}{\binom{m+n}{m}}. \tag{11.4.21}$$

[Hint: Use the result of Exercise 11.4.7.]

(b) Let $1-\gamma$ be determined by

$$1-\gamma = \sum_{l=r_1}^{r_2-1} \sum_{k=l}^{r_2-1} \frac{\left[\begin{array}{c} l+\dfrac{n-2}{2} \\ l \end{array}\right]\left[\begin{array}{c} m+\dfrac{n-2}{2}-k \\ m-k \end{array}\right]}{\binom{m+n}{m}}.$$

Using this expression and the result in **(a)**, show that $[X_{(r_1)}, X_{(r_2)}]$ is a (possibly conservative) $100(1-\gamma)$ percent nonparametric distribution-free (over the set of all continuous distributions) prediction interval for the Y sample median $g_2(Y_1, \ldots, Y_n)$. That is,

$$P\left[X_{(r_1)} \leqslant \text{median}(Y_1, \ldots, Y_n) \leqslant X_{(r_2)}\right] \geqslant 1-\gamma$$

for any underlying continuous distribution. [We note that the conservative nature of this prediction interval diminishes as n increases.]

11.5. Distribution-Free Tests for General Alternatives

Throughout most of this text we have discussed test procedures designed with particular alternatives in mind. Thus, for example, in Chapter 9 we considered two-sample linear rank tests that were constructed to be sensitive to location or scale (with equal location) differences between the two underlying populations. When particular alternatives (such as a difference in location) are of interest, these tailor-made procedures perform very well. However, sometimes in practice we are concerned about detecting *any* possible deviations from the appropriate null hypothesis. To construct nonparametric distribution-free tests with the ability to detect such a general alternative, it is natural to turn to procedures based on sample distribution functions. Thus, for example, in a one-sample setting we would like our test statistic to be a measure (in some sense) of the difference between the empirical c.d.f. and the hypothesized theoretical c.d.f. For a general alternative two-sample problem, our test statistic should evaluate differences between the empirical c.d.f.'s for the two samples.

In this section we deal with several such test procedures for the one- and two-sample problems. Since all these tests are based directly on properties of sample cumulative distribution functions, we first state (without proof) a very important convergence property of a sample c.d.f.

Let $X_1,\ldots,X_n$ be a random sample from a distribution with c.d.f. $F(x)$, and let $F_n(x)$ be the empirical c.d.f. for the sample; that is, $F_n(t) = \frac{1}{n}[\text{number of } X\text{s} \leqslant t]$, $-\infty < t < \infty$.

Theorem 11.5.1. (Glivenko-Cantelli). *Let $F_n(x)$ be the empirical c.d.f. for a random sample of size n from a distribution with c.d.f. $F(x)$. Then*

$$P\left(\lim_{n\to\infty}\left\{\sup_{-\infty<x<\infty}|F_n(x)-F(x)|\right\}=0\right) = 1.$$

[*That is, with probability one, $F_n(x)$ converges to $F(x)$ uniformly in x.*]

Proof: See, for example, Loève (1963), pages 20–21. ■

We note that Theorem 11.5.1 is a much stronger result than those discussed in Section 11.4 for fixed x-values. However, for our purposes in this section, a somewhat weaker result (following immediately from Theorem 11.5.1) is adequate.

Corollary 11.5.2. Let $F_n(x)$ be the empirical c.d.f. for a random sample of size n from a distribution with c.d.f. $F(x)$. Then the random variable $\sup_{-\infty<x<\infty}|F_n(x)-F(x)|$ converges ($n\to\infty$) in probability to zero.

Proof: Follows at once from Theorem 11.5.1 and the fact that convergence with probability one implies convergence in probability. See, for example, Loève (1963), page 151. ■

We now illustrate, via several examples, techniques for basing general alternatives tests on sample c.d.f.'s. The first example deals with a one-sample setting referred to as a **goodness-of-fit problem**.

Example 11.5.3. Let $F_n(x)$ be the empirical c.d.f. for a random sample from a continuous distribution with c.d.f. $F(x)$. We are interested in testing H_0: $[F(x)=F_0(x)$ for all $x]$, against the general alternative H_1: $[F(x)\neq F_0(x)$ for at least one $x]$, where $F_0(x)$ is a completely specified c.d.f.

(Hence, the name "goodness-of-fit problem.") Define the statistic

$$D_n = \sup_{-\infty<x<\infty} |F_n(x) - F_0(x)|. \tag{11.5.4}$$

Then an α-level test of H_0 versus H_1, based on D_n and due to Kolmogorov (1933), consists of rejecting H_0: $[F(x)=F_0(x)$ for all $x]$ iff $D_n \geqslant d(n,\alpha)$, where $d(n,\alpha)$ is the upper 100αth percentile for the null H_0 distribution of D_n. Since the statistic D_n has the same null H_0: $F \equiv F_0$ distribution no matter which continuous population F_0 corresponds to (a fact which is left as Exercise 11.5.1), this statistic has a nonparametric distribution-free property. (Note that the term nonparametric distribution-free as used here is not the same as its use, for example, in referring to a two-sample rank test. In the present context the null hypothesis is a completely specified, known distribution. Moreover, that distribution is used to construct the test statistic. The nonparametric distribution-free property in this case comes from the fact that one null distribution for D_n applies to a very broad class of null hypothesis distributions F_0.) Using Corollary 11.5.2 it can be shown (Exercise 11.5.2) that D_n converges in probability to $\sup_{-\infty<x<\infty}|F(x)-F_0(x)|$, as $n\to\infty$. Then it follows from an argument similar to that in the proof of Theorem 4.2.18 that this α-level test based on D_n is consistent against any differences between the hypothesized X distribution, $F_0(\cdot)$, and the true X distribution, $F(\cdot)$.

The remaining portion of this section is devoted to general alternatives test procedures for the two-sample problem. Thus for the rest of this section we let $X_1,\ldots,X_m$ and $Y_1,\ldots,Y_n$ be independent random samples from continuous distributions with c.d.f.'s $F(x)$ and $G(y)$, respectively. We consider two examples of procedures designed to test H_0: $[F(x)=G(x)$ for all $x]$ against the general alternative H_1: $[F(x)\neq G(x)$ for at least one $x]$. Both of these procedures are based on examination of the differences between the values of the empirical distribution functions for the X and Y samples. Thus, letting $F_m(x)$ and $G_n(y)$ denote the empirical c.d.f.'s for the X and Y samples, respectively, we will be dealing with the differences $F_m(x)-G_n(x)$.

Example 11.5.5. In direct analogy to (11.5.4), set

$$D_{m,n} = \sup_{-\infty<x<\infty} |F_m(x) - G_n(x)|. \tag{11.5.6}$$

Smirnov (1939) proposed the α-level test of H_0: $[F(x)=G(x)$ for all $x]$ against the general alternative H_1 that rejects H_0 if and only if $D_{m,n} \geqslant d(m,n,\alpha)$, where $d(m,n,\alpha)$ is the upper 100αth percentile for the null distribution of $D_{m,n}$. This test is nonparametric distribution-free over the class of all continuous distributions (Exercise 11.5.4), and it follows from Exercise 11.5.5 and an argument similar to the proof of Theorem 4.2.18 that it is consistent against any differences between the X and Y distributions. For tables of the α-level cutoff points $d(m,n,\alpha)$, see, for example, Table A.23 in Hollander and Wolfe (1973).

In both Examples 11.5.3 and 11.5.5 the test statistics are based on the *greatest* deviations between the appropriate distribution functions. Cramér (1928) and von Mises (1931) suggested a goodness-of-fit competitor for the Kolmogorov procedure of Example 11.5.3 that directly uses the information from *all* (and not just the largest) deviations between the sample ($F_n(\cdot)$) and hypothesized ($F_0(\cdot)$) distribution functions. In the next example we consider a two-sample version, studied by Rosenblatt (1952), of this Cramér–von Mises scheme.

Example 11.5.7. Let $Z_{(1)} \leqslant \cdots \leqslant Z_{(N)}$ be the order statistics for the combined sample of $N=(m+n)$ X and Y observations, and define

$$C = \frac{mn}{(m+n)^2} \sum_{i=1}^{N} \left[F_m(Z_{(i)}) - G_n(Z_{(i)}) \right]^2. \qquad \textbf{(11.5.8)}$$

The corresponding Cramér–von Mises test, as considered by Rosenblatt, of H_0: $[F(x)=G(x)$ for all $x]$ against the general alternative H_1: $[F(x)\neq G(x)$ for at least one $x]$ is then given by

$$\text{reject } H_0 \text{ iff } C \geqslant c(\alpha,m,n), \qquad \textbf{(11.5.9)}$$

where $c(\alpha,m,n)$ is the upper 100αth percentile for the null H_0 distribution of C. As with the Smirnov test in Example 11.5.5, this Cramér–von Mises procedure is nonparametric distribution-free over the class of all continuous distributions (Exercise 11.5.8), and it can be shown [see, e.g., Rosenblatt (1952)] that it is consistent against *any* differences between the X and Y distributions.

These three examples provide only a brief introduction to the area of nonparametric statistics concerned with tests designed for general alternatives. There is a vast literature dealing only with properties of the statistics D_n (11.5.4), $D_{m,n}$ (11.5.6), and C (11.5.8) and the associated tests. In addition, many other general alternative two-sample procedures have been developed, and these ideas have been extended to other settings, including the case of k (>2) samples. For more study along these lines, we suggest the survey articles by Darling (1957) and Barton and Mallows (1965) and the advanced text by Hájek and Šidák (1967).

One final comment is in order. Although a general alternative test, such as those presented in Examples 11.5.5 and 11.5.7, is consistent against *any* alternative to the appropriate null hypothesis, its power may increase very slowly as the sample sizes become large. As a result, when we are really interested in a *specific* type of alternative, such as location differences in the two-sample setting, a test designed for that alternative will generally provide considerably better power for small or moderate sample sizes than one geared to detecting any differences. Thus, for example, when possible location differences are our prime concern in a two-sample setting, the Mann–Whitney–Wilcoxon test is preferred over either the Smirnov (Example 11.5.5) or the Cramér–von Mises-Rosenblatt (Example 11.5.7) procedures.

Exercises 11.5.

11.5.1. Argue that the Kolmogorov statistic D_n (11.5.4) in Example 11.5.3 is distribution-free under H_0: $F \equiv F_0$ (that is, D_n has the same null H_0: $F \equiv F_0$ distribution no matter which continuous distribution F_0 is used.) [Hint: Consider Theorem 1.2.9 and the relationship between D_n and the statistic $D_n^* = \sup_{-\infty < x < \infty} |F_n^*(x) - F^*(x)|$, where $F_n^*(x)$ is the empirical c.d.f. for the observations $F_0(X_1), \ldots, F_0(X_n)$ and $F^*(x)$ is the c.d.f. for the uniform distribution on (0, 1).]

11.5.2. In Example 11.5.3, show that D_n (11.5.4) converges in probability to $\sup_{-\infty < x < \infty} |F(x) - F_0(x)|$. [Hint: Consider the equalities $|F_n(x) - F_0(x)| = |F_n(x) - F(x) + F(x) - F_0(x)|$ and $|F(x) - F_0(x)| = |F(x) - F_n(x) + F_n(x) - F_0(x)|$.]

11.5.3. Consider the general two-sample setting of Example 11.5.5. Letting $Z_{(1)} \leqslant \cdots \leqslant Z_{(N)}$, with $N = m + n$, be the order statistics for the combined X and Y samples, set

$$\delta_i = \begin{cases} 1 & \text{if } Z_{(i)} \text{ is an } X \text{ observation,} \\ 0 & \text{if } Z_{(i)} \text{ is a } Y \text{ observation,} \end{cases}$$

for $i = 1, \ldots, N$. Define

$$s_i = \left[\frac{im}{N} - \sum_{j=1}^{i} \delta_j \right], \qquad i = 1, \ldots, N.$$

Show that

$$D_{m,n} = \frac{N}{mn} \max\{|s_1|, \ldots, |s_N|\}, \tag{11.5.10}$$

where $D_{m,n}$ (11.5.6) is the Smirnov statistic.

11.5.4. Let $D_{m,n}$ (11.5.6) be the Smirnov statistic considered in Example 11.5.5. Argue that $D_{m,n}$ is distribution-free under H_0: $[F(x) = G(x)$ for all $x]$. Find the null H_0 distribution of $D_{m,n}$ for $m = n = 3$. [Hint: See equation (11.5.10).]

11.5.5. Show that the two-sample Smirnov statistic $D_{m,n}$ (11.5.6) converges in probability to $\sup_{-\infty < x < \infty} |F(x) - G(x)|$, as m, $n \to \infty$. [Hint: Consider Corollary 11.5.2 and the equalities $|F_m(x) - G_n(x)| = |F_m(x) - F(x) + F(x) - G(x) + G(x) - G_n(x)|$ and $|F(x) - G(x)| = |F(x) - F_m(x) + F_m(x) - G_n(x) + G_n(x) - G(x)|$.]

11.5.6. Let $X_1, \ldots, X_m$ and $Y_1, \ldots, Y_n$ be independent random samples from continuous distributions with c.d.f.'s $F(x)$ and $G(y)$, respectively. Use the ideas of Example 11.5.5 to construct a Smirnov-type procedure for testing H_0: $F \equiv G$ against the one-sided general alternative H_1: $[F(x) \geqslant G(x)$ for all x with at least one strict inequality]. What is the consistency class for this test?

11.5.7. Consider the two-sample setting of Example 11.5.7. Show that the Cramér–von Mises–Rosenblatt statistic C (11.5.8) can be written as

$$C = \frac{\sum_{i=1}^{m} (R_i^* - i)^2}{n(m+n)} + \frac{\sum_{j=1}^{n} (S_j^* - j)^2}{m(m+n)} - \frac{4mn - 1}{6(m+n)}, \tag{11.5.11}$$

where $R_1^* < \cdots < R_m^*$ and $S_1^* < \cdots < S_n^*$ are the ordered combined sample ranks of $X_1, \ldots, X_m$ and $Y_1, \ldots, Y_n$, respectively.

11.5.8. Let C (11.5.8) be the Cramér–von Mises–Rosenblatt statistic discussed in Example 11.5.7. Argue that C is distribution-free under H_0: $[F(x) = G(x)$ for all $x]$. Find the null H_0 distribution of C for $m = n = 3$. [Hint: Consider equation (11.5.11).]

11.5.9. Let C (11.5.8) be the Cramér–von Mises–Rosenblatt statistic considered in Example 11.5.7. Find the null (H_0) expected value for C, namely, $E_0(C)$. [Hint: Consider Theorem 11.4.12.]

11.5.10. Let $F_m(x)$ and $G_n(y)$ be the empirical c.d.f.'s for independent random samples $X_1,\ldots,X_m$ and $Y_1,\ldots,Y_n$ from continuous distributions with c.d.f.'s $F(x)$ and $G(y)$, respectively. Set

$$T_i = [F_m(Y_{(i)}) - F_m(Y_{(i-1)})], \qquad i = 2,\ldots,n,$$

and define

$$T_1 = F_m(Y_{(1)}) \text{ and } T_{n+1} = [1 - F_m(Y_{(n)})],$$

where $Y_{(1)} \leqslant \cdots \leqslant Y_{(n)}$ are the ordered Ys.

(a) Show that

$$E_0[T_i] = \frac{1}{n+1}, \qquad i = 1,\ldots,n+1, \qquad \textbf{(11.5.12)}$$

where E_0 denotes that the expectation corresponds to H_0: $F \equiv G$ being true.

(b) Define

$$V = \sum_{i=1}^{n+1} \left[T_i - \frac{1}{n+1} \right]^2. \qquad \textbf{(11.5.13)}$$

Argue that V is distribution-free when H_0: $F \equiv G$ is true.

(c) Discuss how to use V to construct a distribution-free test of H_0: $F \equiv G$ against the general alternative H_1: [$F(x) \neq G(x)$ for at least one x]. [Note that Dixon (1940), Blum and Weiss (1957), and Blumenthal (1963) have considered two-sample general alternative tests based on a statistic very similar to V.]

11.5.11. Let V be the statistic given in (11.5.13). Find the null (H_0) expected value for V, namely, $E_0(V)$. [Hint: Consider Exercise 11.4.8 and Theorem 11.4.12.]

11.6. Adaptive Distribution-Free Tests

Persons involved in analyzing data often choose a statistical procedure after having examined the data. For example, it is not uncommon for a practitioner to transform or smooth data before applying a normal theory test of hypothesis. Such two-staged analyses are termed **adaptive** since the data determine the transformation used and then the same data (after transformation) are used in the testing procedure. We must recognize the dangers associated with such two-staged adaptive rules. Even though the final test is conducted at the desired level α, in the overall testing procedure (thinking of both stages) the true level of significance may be quite different from α. This is because in the second stage what we really should control is the conditional probability of a type I error, given that the data have determined the use of a particular transformation. This is usually quite difficult to do.

In this section we describe the use of rank statistics to construct adaptive testing procedures that are truly nonparametric distribution-free. That is, the two stages of the inference process are constructed in such a way that control of the overall α-level is maintained. For clarity of exposition we confine our attention to the two-sample location problem, with several other possible applications indicated in exercises. Suppose $X_1,\ldots,X_m$ and $Y_1,\ldots,Y_n$ are independent random samples from continuous populations with c.d.f.'s $F(x)$ and $F(x-\Delta)$, respectively. We wish to test

$$H_0\colon \Delta = 0 \text{ versus } H_1\colon \Delta > 0. \tag{11.6.1}$$

If we know the form of $F(\cdot)$, for example that it is a normal distribution, then we can use a parametric test. If, however, the form of $F(\cdot)$ is unknown we might wish to base the test on a linear rank statistic and reject H_0 if

$$S = \sum_{i=1}^{n} a(R_i) \geqslant c, \tag{11.6.2}$$

where c is the appropriate critical value, R_i is the rank of Y_i among the $N=(m+n)$ combined sample observations, and the scores satisfy $a(1) \leqslant \cdots \leqslant a(N)$ with $a(1)<a(N)$. (See Section 9.1.) This test would have the advantage of maintaining the desired α-level over the nonparametric class consisting of all continuous distributions. But how shall we choose the scores? We see from Table 9.2.21 that the performances of these rank tests vary considerably, depending on the nature of the underlying distribution. If we know $F(\cdot)$, we can use the score function that is asymptotically optimal for that case; however, we are not assuming that knowledge.

It would be desirable, therefore, to use the data itself to determine the nature of $F(\cdot)$, and on the basis of that information, we could choose an appropriate set of scores. We would then use that same data to perform the test. We must be sure that in the process we do not destroy our control over the α-level of the test. The following general result shows us how this can be done.

Lemma 11.6.3. *Let $\mathcal{Q}$ denote the class of distributions under consideration. Suppose that each of k testing procedures is distribution-free over $\mathcal{Q}$; that is, for the test based on S_j,*

$$P\left[S_j \geqslant c_j \mid H_0 \textit{ true}, F(\cdot) \textit{ is the distribution}\right] = \alpha$$

*for each $F(\cdot) \in \mathcal{Q}$ and for $j = 1, \ldots, k$. Let **V** be some statistic (or vector of statistics) that is independent of $(S_1, \ldots, S_k)$ under H_0 for each $F(\cdot) \in \mathcal{Q}$. Suppose we use **V** to decide which test (S_j) to conduct. Specifically, let $C_1, \ldots, C_k$ denote pairwise disjoint and mutually exhaustive collections of **V** values such that **V** in C_j corresponds to the decision to use the test based on S_j. The overall testing procedure is then defined by:*

*If **V** falls in set C_J,*
then reject H_0 in favor of H_1 iff $S_J \geqslant c_J$.

This two-staged, adaptive test is distribution-free under H_0 over the class $F(\cdot) \in \mathcal{Q}$; that is, it maintains the significance level α for every $F(\cdot) \in \mathcal{Q}$.

Proof: Note that

$$P\left[\text{reject } H_0 \mid H_0 \text{ true}, F(\cdot)\right] = P\left[\bigcup_{j=1}^{k} \{\mathbf{V} \in C_j \text{ and } S_j \geqslant c_j\} \mid H_0 \text{ true}, F(\cdot)\right]$$
$$= \sum_{j=1}^{k} P\left[\mathbf{V} \in C_j \text{ and } S_j \geqslant c_j \mid H_0 \text{ true}, F(\cdot)\right],$$

since the C_js are pairwise disjoint. Because **V** and S_j are independent under H_0, this expression becomes

$$\sum_{j=1}^{k} P\left[\mathbf{V} \in C_j \mid H_0 \text{ true}, F(\cdot)\right] P\left[S_j \geqslant c_j \mid H_0 \text{ true}, F(\cdot)\right]$$
$$= \alpha \sum_{j=1}^{k} P\left[\mathbf{V} \in C_j \mid H_0 \text{ true}, F(\cdot)\right]$$
$$= \alpha. \quad \blacksquare$$

Remark 11.6.4. In the previous lemma it is not necessary that the choice be made from merely a finite set of tests. The conclusion would be valid for a continuum as well.

Since each rank test described in (11.6.2) maintains its α-level over the broad class of continuous distributions, we see from Lemma 11.6.3 that to keep this nonparametric distribution-free property we should select the scores on the basis of statistics that are, under H_0, independent of the ranks $R_1,\dots,R_n$, for every $F(\cdot)$ in this class. The key to accomplishing this is found in Lemma 8.3.11. There it was shown that under H_0 the rank vector $\mathbf{R}^*$ is independent of $Z_{(1)} \leqslant \cdots \leqslant Z_{(N)}$, the order statistics of the combined sample of the $N=m+n$ observations $X_1,\dots,X_m, Y_1,\dots,Y_n$. By using the values of the combined sample order statistics to select the scores, the resulting adaptive rule will no longer be a rank test, since it will depend on more than just the vector of ranks. It will, however, have the same nonparametric distribution-free property as a rank test.

Since the performances of tests based on linear rank statistics depend on the tailweight and amount of asymmetry in the underlying distribution, we might wish to choose our scores by using sample information about those population characteristics. Thus, for example, we could measure asymmetry with the statistic

$$Q_3 = \frac{\overline{U}_{.05} - \overline{M}_{.50}}{\overline{M}_{.50} - \overline{L}_{.05}}$$

and tailweight with the statistic

$$Q_4 = \frac{\overline{U}_{.05} - \overline{L}_{.05}}{\overline{U}_{.50} - \overline{L}_{.50}},$$

where $\overline{U}_\gamma$ ($\overline{M}_\gamma$ and $\overline{L}_\gamma$) denotes the average of the γN largest (middle and smallest, respectively) combined order statistics $Z_{(1)} \leqslant \cdots \leqslant Z_{(N)}$. Fractional items are used when γN is not an integer. Thus, for example, if $N=30$, $\overline{U}_{.05}=[Z_{(30)}+.5Z_{(29)}]/1.5$. Figure 11.6.5 shows one possible scheme for selecting the scores. When a light-tailed, symmetric distribution is indicated ($Q_3 \leqslant 2$ and $1 \leqslant Q_4 < 2$), then the scores $a_L(\cdot)$ given in (9.1.9) are used and the associated test statistic is denoted by L. If the distribution is very heavy-tailed ($Q_4 > 7$), then the median test based on M (9.1.4) is appropriate. The test based on RS, corresponding to the scores given in (9.1.11), is used when the underlying distribution appears to be skewed to the right and not extremely heavy-tailed ($Q_3 > 2$ and $Q_4 \leqslant 7$). In other cases ($Q_3 \leqslant 2$ and $2 \leqslant Q_4 \leqslant 7$), the Mann–Whitney–Wilcoxon test is used. Its test statistic is denoted by W. This scheme does not include the possibility that the distribution is skewed to the left. If that is a possibility, an additional category ($Q_3 < \frac{1}{2}$ and $Q_4 \leqslant 7$) should be created for which the LS scores (9.1.12) are used.

This particular adaptive distribution-free test was proposed by Hogg, Fisher, and Randles (1975). Table 11.6.6 shows a Monte Carlo study taken

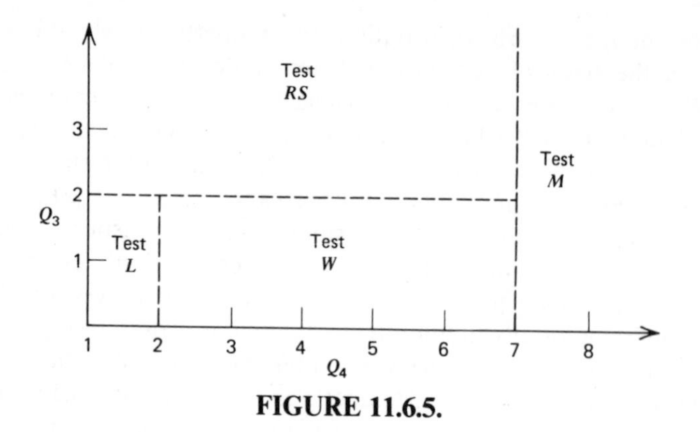

FIGURE 11.6.5.

from that paper. The symbol A denotes the adaptive test, T is the two-sample Student t-test, and the other tests included are the component tests of the adaptive rule. The underlying distributions are all members of the RST lambda family described in Exercise 4.1.6. They are indexed by their skewness (α_3) and kurtosis (α_4). Sample sizes for the two populations are $m=n=15$. The upper block ($\Delta=0$) in Table 11.6.6 shows the observed levels of significance for the tests. All procedures were run at a nominal $\alpha=.05$ level, with randomization being used so that the rank tests would achieve this nominal level. The lower block in Table 11.6.6 shows the powers for detecting a shift of magnitude $\Delta=.6\sigma$. Here σ denotes the standard deviation of the underlying distribution or, for the Cauchy-like distribution (having no moments) centered at zero, $\sigma=F^{-1}(.8413)$, where .8413 is the probability below $\mu+\sigma$ in a normal distribution. This Monte Carlo entailed 5000 replications. Further details are given in the paper.

We note that the power of the two-sample t-test (T) is reasonable for all distributions except the Cauchy-like case. However, the power of the adaptive rule (A) is excellent over a broad class of underlying distributions and in many instances greatly exceeds that of the t-test. The component tests which make up the adaptive one do the jobs for which they are designed. The light-tailed distribution test (L) is quite effective in detecting a shift in a light-tailed symmetric distribution. The test RS is very effective in detecting shifts in distributions that are skewed to the right, particularly if the distribution is rather heavy-tailed. The Mann–Whitney–Wilcoxon test has very good overall power properties and is particularly effective when the underlying distribution is medium-tailed and symmetric. In this study, the median test (M) is better than W only when the underlying distribution is so heavy-tailed that it has no moments. Finally, for these Monte Carlo results, the adaptive rule gives the best overall performance.

TABLE 11.6.6.
Empirical Power in Percent

	DISTRIBUTION													
α_3:	0	0	0	0	0	CAUCHY-LIKE	.5	1.5	.9	1.5	.8	2.0	3.16	3.88
α_4:	1.8	3.0	6.0	11.6	126		2.2	5.8	4.2	7.5	11.4	21.2	23.8	40.7
TEST						$\Delta = 0$								
T	5.2	5.2	5.3	5.2	5.0	3.1	5.1	5.1	5.2	4.9	5.1	5.1	5.1	5.0
A	4.7	5.2	5.2	5.1	5.0	5.1	4.8	4.9	4.9	5.0	5.0	5.0	4.9	4.9
W	5.1	5.2	5.2	5.2	5.1	5.1	5.1	5.1	5.2	5.2	5.1	5.1	5.1	5.1
L	4.9	5.0	4.9	4.9	4.9	4.9	4.9	4.8	5.1	5.0	4.9	5.0	5.0	4.9
RS	4.9	4.9	4.9	4.9	4.9	4.9	4.9	4.9	4.9	4.9	4.9	4.9	4.9	4.9
M	4.9	5.0	4.7	5.1	5.0	4.8	5.0	5.1	4.9	5.0	4.7	4.7	5.1	5.2
						$\Delta = .6\sigma$								
T	47	48	51	52	56	13	47	50	49	51	52	54	56	59
A	51	46	53	58	66	47	53	74	52	63	58	63	91	94
W	44	46	54	59	67	46	48	64	52	60	59	62	80	84
L	63	41	42	43	46	20	64	67	48	53	43	47	74	75
RS	37	39	46	51	58	40	53	78	55	66	54	63	93	95
M	25	36	45	52	60	54	28	42	39	46	52	54	61	65

Although it might not work well when a wider class of structures is considered, we see that this adaptive distribution-free procedure is extremely effective for the shift in location structure described in (11.6.1).

Basing an adaptive distribution-free test on linear rank statistics was suggested by Hájek (1962, 1970). In his original scheme, Hájek estimated the optimal score function for the two-sample location problem in a consistent fashion. However, the slow convergence of his estimator rendered the procedure impractical. Simpler schemes were subsequently developed by Randles and Hogg (1973) for both the one- and two-sample location problems. Beran (1974) suggested an alternative estimation scheme to replace the one used by Hájek. Policello and Hettmansperger (1976) proposed an adaptive rank test for the one-sample location problem that is not distribution-free but maintains its α-levels reasonably well. Jones (1976) considered a different adaptive rank test that is distribution-free for that same one-sample problem. Adaptive rules proposed for other statistical problems include those due to Hogg and Randles (1975) and Hogg (1976). A general discussion and bibliography of adaptive inference was given by Hogg (1974).

Exercises 11.6.

11.6.1. Theorem. *Let $X_1,\ldots,X_n$ be i.i.d. continuous random variables, each symmetrically distributed about zero. Let $\Psi_i=\Psi(X_i)=1,0$ as $X_i>, \leqslant 0$, and let R_i^+ denote the rank of $|X_i|$ among $|X_1|,\ldots,|X_n|$, for $i=1,\ldots,n$. The ordered absolute values $|X|_{(1)}\leqslant\cdots\leqslant|X|_{(n)}$ are then independent of $\mathbf{\Psi}=(\Psi_1,\ldots,\Psi_n)$, and $\mathbf{R}^+=(R_1^+,\ldots,R_n^+)$.*

Prove this theorem. [Hint: Note Lemmas 2.4.2 and 8.3.11.]

11.6.2. Let $X_1,\ldots,X_n$ denote a random sample from a continuous population with c.d.f. $F(x-\theta)$, where the underlying distribution is symmetric about the unknown location parameter θ. For testing H_0: $\theta=0$ versus H_1: $\theta>0$, discuss how to construct adaptive tests based on linear signed rank statistics of the form

$$S_n = \sum_{i=1}^{n} \Psi_i a(R_i^+),$$

where the scores satisfy $0\leqslant a(1)\leqslant\cdots\leqslant a(n)$ and $a(n)>0$. Use the results of Exercise 11.6.1 and state what information from the data may be used to select the scores while keeping the testing procedure nonparametric distribution-free over the indicated broad class of null hypothesis distributions. Specifically, suggest a statistic(s) that could be used to select the scores and present the rationale behind your selection.

11.6.3. Consider the two-sample problem in which $X_1,\ldots,X_m$ and $Y_1,\ldots,Y_n$ are independent random samples from continuous populations with c.d.f.'s $F(x)$ and $F(x/\eta)$, respectively, and where the density satisfies $f(x)=f(-x)$ for all x; that is, it is symmetric about zero. Propose a class of linear rank statistics based on the $|X_i|$s and $|Y_j|$s for testing H_0: $\eta=1$ versus H_1: $\eta>1$. What information from the sample may be used to select the scores while preserving the nonparametric distribution-free nature of the test? Justify your answer.

11.6.4. Let

$$X_i = \Delta t_i + E_i, \qquad i = 1,\ldots,n,$$

where $t_1<t_2<\cdots<t_n$ are known constants and $E_1,\ldots,E_n$ are i.i.d. continuous random variables, each symmetrically distributed about zero. Suppose that we consider testing H_0: $\Delta=0$ versus H_1: $\Delta>0$ with rank statistics of the form

$$S_n = \sum_{i=1}^{n} (t_i - \bar{t})(a(R_i) - \bar{a}),$$

where $\bar{t} = \frac{1}{n}\Sigma_{i=1}^n t_i$, $\bar{a} = \frac{1}{n}\Sigma_{i=1}^n a(i)$, R_i is the rank of X_i among $X_1, \ldots, X_n$, and the scores satisfy $a(1) \leqslant \cdots \leqslant a(n)$, $a(1) \neq a(n)$. What information from the data $X_1, \ldots, X_n$ can be used to choose the scores while retaining the nonparametric distribution-free property of these rank statistics? Justify your answer.

11.6.5. Let $(X_1, Y_1), \ldots, (X_n, Y_n)$ denote a random sample from a continuous bivariate population. Suppose that we wish to test H_0: [X_i and Y_i are independent] versus H_1: [X_i and Y_i are positively correlated] (larger [smaller] values of X_i tend to be associated with larger [smaller] values of Y_i). This test may be based on the rank statistic

$$S = \sum_{i=1}^{n} c(Q_i)a(R_i),$$

where Q_i denotes the rank of X_i among $X_1, \ldots, X_n$ and R_i denotes the rank of Y_i among $Y_1, \ldots, Y_n$. Here the scores are assumed to satisfy $c(1) \leqslant \cdots \leqslant c(n)$, $c(1) \neq c(n)$ and $a(1) \leqslant \cdots \leqslant a(n)$, $a(1) \neq a(n)$.

(a) Argue that the test statistic S is nonparametric distribution-free under H_0.

(b) What information from the data can be used to select the scores $c(1), \ldots, c(n)$ and $a(1), \ldots, a(n)$ without destroying the nonparametric distribution-free property of S under H_0? Justify your answer.

12 OTHER IMPORTANT PROBLEMS

In previous chapters we have discussed a variety of techniques for obtaining nonparametric distribution-free tests, and, subsequently, for deriving important properties of such procedures. Throughout the entire development we have emphasized the one- and two-sample settings, primarily because for such problems (i) the general theory is well-developed and (ii) many good, but simple, illustrative examples are available. However, the use of nonparametric distribution-free tests is not restricted to these two simple data types. In this chapter we briefly consider four additional settings where such procedures are available, namely, the one-way and two-way layouts and correlation and regression problems.

12.1. The One-Way Layout Problem

Let $X_{11},\ldots,X_{1n_1},\ldots,X_{k1},\ldots,X_{kn_k}$ be k independent random samples from continuous distributions with c.d.f.'s $F(x-\theta_1),\ldots,F(x-\theta_k)$, respectively, where θ_i denotes the median of the ith population. We consider here the problem of testing the null hypothesis H_0: $\theta_1=\cdots=\theta_k$; that is, the hypothesis that there are no differences between the k population medians. This is commonly referred to as the **one-way layout** problem. Now, the choice of a nonparametric distribution-free procedure will depend on the alternatives that are of primary interest. Although other alternatives have been studied (see, for example, Exercises 12.1.5 and 12.1.6), we consider here only the two alternatives that have received the most attention, namely, the general alternative

$$H_1\colon \theta_i \neq \theta_j \text{ for at least one } i \neq j \tag{12.1.1}$$

and the ordered alternative

$$H_2: \theta_1 \leqslant \cdots \leqslant \theta_k, \text{ with at least one strict inequality.} \qquad \textbf{(12.1.2)}$$

Before we proceed to examples dealing with nonparametric distribution-free procedures that are designed for one or the other of these alternatives, we need some brief preliminaries. Let R_{ij} be the rank of X_{ij} among $X_{11},\ldots,X_{1n_1},\ldots,X_{k1},\ldots,X_{kn_k}$; that is, R_{ij} is the rank of X_{ij} in the single combined sample of $N=n_1+\cdots+n_k$ observations. Now, if H_0: $\theta_1=\cdots=\theta_k$ is true, then all of the N sample observations arise from a single continuous distribution with c.d.f. $F(x-\theta)$, where θ represents the common population effect under H_0. Hence, letting $\mathbf{R}^*=(R_{11},\ldots,R_{1n_1},\ldots,R_{k1},\ldots,R_{kn_k})$, we see from Theorem 2.3.3 that $\mathbf{R}^*$ is uniformly distributed over $\mathfrak{R}=\{\mathbf{r}$: $\mathbf{r}$ is a permutation of the integers $1,2,\ldots,N\}$ when H_0 is true. It then follows via Corollary 2.3.6 that any statistic $V(\mathbf{R}^*)$ that is a function of $X_{11},\ldots,X_{kn_k}$ only through $\mathbf{R}^*$ will be nonparametric distribution-free over the class of continuous distributions when H_0 is true. Thus a test based on such a statistic will also be nonparametric distribution-free. We consider first a statistic $V(\mathbf{R}^*)$ that is appropriate for use in testing H_0 against the general alternative H_1 (12.1.1).

Example 12.1.3. Let

$$R_{i.}=\frac{1}{n_i}\sum_{j=1}^{n_i}R_{ij}$$

be the average of the ranks associated with the ith sample observations, for $i=1,\ldots,k$. When H_0: $\theta_1=\cdots=\theta_k$ is true, we see from Corollary 2.3.5 that

$$E_0[R_{i.}]=\frac{1}{n_i}\sum_{j=1}^{n_i}E_0[R_{ij}]=\frac{N+1}{2},$$

where E_0 indicates that this is the null expectation. It would thus be quite natural to base tests of H_0 against H_1 on statistics that measure differences between the expected $((N+1)/2)$ and observed average $(R_{i.})$ of the ith sample ranks. One such statistic (and associated test) was considered by Kruskal (1952) and Kruskal and Wallis (1952). They suggested using the statistical measure

$$V_1=\frac{12}{N(N+1)}\sum_{i=1}^{k}n_i\left(R_{i.}-\frac{N+1}{2}\right)^2 \qquad \textbf{(12.1.4)}$$

and the corresponding α-level test which

$$\text{rejects } H_0 \text{ in favor of } H_1 \text{ iff } V_1 \geqslant v_1(\alpha), \qquad \textbf{(12.1.5)}$$

where $v_1(\alpha)$ is the upper 100αth percentile for the null distribution of V_1.

To conduct the Kruskal–Wallis test (12.1.5) we need to be able to find the critical values $v_1(\alpha)$. This can be accomplished for given integers $k, n_1, \ldots, n_k$ by calculating the value of V_1 for each of the $N!$ permutations that are possible for the ranks $R_{11}, \ldots, R_{kn_k}$ (actually, the symmetry of this k-sample problem can be used to reduce the required calculations to $N!/n_1! \cdots n_k!$ permutations), and then tabulating the null distribution of V_1 resulting from the fact that each of these permutations is equally likely when H_0 is true. (See Example 2.3.13 for a detailed illustration of this process in a two-sample setting.) Such critical values can be found, for example, in Table A.7 of Hollander and Wolfe (1973) for $k=3$ and $n_i \leqslant 5$, $i=1,2,3$. Additional tables are given by Iman, Quade, and Alexander (1975). Kruskal and Wallis (1952) showed that the statistic V_1 has a limiting [minimum $(n_1, \ldots, n_k) \to \infty$, with $(n_j/N) \to \lambda_j$, $0 < \lambda_j < 1$, for $j = 1, \ldots, k$] distribution that is chi-square with $(k-1)$ degrees of freedom when H_0 is true. Hence the critical value $v_1(\alpha)$ can be approximated by the upper 100αth percentile for a chi-square distribution with $(k-1)$ degrees of freedom.

As an example of a test designed for the ordered alternative H_2 (12.1.2), we consider a procedure proposed independently by Terpstra (1952) and Jonckheere (1954).

Example 12.1.6. Let U_{uv}, $u < v = 2, 3, \ldots, k$, be the Mann-Whitney form of the Mann-Whitney-Wilcoxon statistic for the uth and vth samples; that is,

$$U_{uv} = \sum_{s=1}^{n_u} \sum_{t=1}^{n_v} \Psi(X_{vt} - X_{us}),$$

where $\Psi(t) = 1, 0$ as $t >, \leqslant 0$. The Jonckheere–Terpstra statistic is then

$$V_2 = \sum_{u<v}^{k}\sum U_{uv} \qquad \textbf{(12.1.7)}$$

and the associated α-level test rejects H_0 in favor of H_2 when V_2 is at least as large as $v_2(\alpha)$, the upper 100αth percentile for the null distribution of V_2. (Note that for $k=2$ the Jonckheere–Terpstra procedure is simply a one-sided Mann-Whitney-Wilcoxon test.)

The null distribution of V_2 can be obtained via computations similar to those mentioned following Example 12.1.3. Note, however, that for this Jonckheere–Terpstra statistic, Odeh (1971) was able to use the null hypothesis independence of certain Mann–Whitney sums to greatly reduce the required amount of work. (See Exercise 12.1.7.) The associated critical values $v_2(\alpha)$ are given, for example, in Table A.8 of Hollander and Wolfe (1973) for $k=3(1)6$ and a variety of n_j values, $j=1,\ldots,k$. It can be shown (see, for example, Exercise 12.1.7) that the limiting (minimum $(n_1,\ldots,n_k)\to\infty$, with $(n_j/N)\to\lambda_j$, $0<\lambda_j<1$, for $j=1,\ldots,k$) distribution of $(V_2-E_0(V_2))/\sqrt{\operatorname{Var}_0(V_2)}$ is standard normal when H_0 is true, where

$$E_0(V_2)=\frac{N^2-\sum_{j=1}^{k}n_j^2}{4} \tag{12.1.8}$$

and

$$\operatorname{Var}_0(V_2)=\frac{N^2(2N+3)-\sum_{j=1}^{k}n_j^2(2n_j+3)}{72} \tag{12.1.9}$$

are the null mean and variance, respectively, of V_2. (See Exercise 12.1.3.) Thus the critical value $v_2(\alpha)$ can be approximated by $E_0(V_2)+z_{(\alpha)}\sqrt{\operatorname{Var}_0(V_2)}$, where $z_{(\alpha)}$ is the upper 100αth percentile for the standard normal distribution.

Naturally, both of the statistics V_1 (12.1.4) and V_2 (12.1.7) can be generalized to allow for arbitrary scores, as considered in detail for the two-sample setting in Chapter 9. See, for example, Puri (1964) and Puri (1965) for such general scores versions of V_1 and V_2, respectively. For material regarding estimation of certain linear combinations (called contrasts) of the population medians $\theta_1,\ldots,\theta_k$, see, for example, Section 6.4 of Hollander and Wolfe (1973).

Exercises 12.1.

12.1.1. Let V_1 (12.1.4) be the Kruskal-Wallis statistic of Example 12.1.3.

(a) Show that V_1 can also be represented by

$$V_1=\left[\frac{12}{N(N+1)}\sum_{i=1}^{k}\frac{R_i^2}{n_i}\right]-3(N+1),$$

where $R_i = \Sigma_{j=1}^{n_i} R_{ij}$ = sum of ranks associated with the ith sample observations.

(b) Using the result in **(a)**, find an expression for the null H_0: $\theta_1 = \cdots = \theta_k$ mean of V_1.

[Hint: See Corollary 2.3.5.]

12.1.2. Obtain the exact null H_0: $\theta_1 = \cdots = \theta_k$ distribution of the Kruskal–Wallis statistic V_1 (12.1.4) for the case $k=3$, $n_1 = n_2 = n_3 = 2$. (Note that this requires only $6!/3!2!2!2! = 15$ different calculations of V_1. Since the sample sizes are equal, we can permute the roles of the k populations in this case, reducing the number of computations by a factor of $1/k!$.)

12.1.3. Let V_2 (12.1.7) be the Jonckheere–Terpstra statistic of Example 12.1.6. Verify the expressions for the null H_0: $\theta_1 = \cdots = \theta_k$ mean and variance of V_2, as given in (12.1.8) and (12.1.9), respectively. [Hint: For the null variance, first find the null covariance between the various Mann–Whitney statistics that make up V_2.]

12.1.4. Obtain the exact null H_0: $\theta_1 = \cdots = \theta_k$ distribution of the Jonckheere–Terpstra statistic V_2 (12.1.7) for the case $k=3$, $n_1 = n_2 = 2$ and $n_3 = 1$. (Note that this requires $5!/2!2!1! = 30$ different calculations of V_2.)

12.1.5. Let $X_{11}, \ldots, X_{1n_1}, \ldots, X_{k1}, \ldots, X_{kn_k}$ be k independent random samples from continuous distributions with c.d.f.'s $F(x-\theta_1), \ldots, F(x-\theta_k)$, respectively. For testing the null hypothesis H_0: $\theta_1 = \cdots = \theta_k$ against the alternative H_3: $\theta_1 \leqslant \cdots \leqslant \theta_{l-1} \leqslant \theta_l \geqslant \theta_{l+1} \geqslant \cdots \geqslant \theta_k$ (referred to as an "umbrella alternative"), where l is some *known* integer between 1 and k, inclusive, consider the procedure [studied by Mack and Wolfe (1978)]:

$$\text{reject } H_0 \text{ in favor of } H_3 \text{ iff } V_3 \geqslant v_3(\alpha),$$

where $v_3(\alpha)$ is the upper 100αth percentile point for the null H_0 distribution of

$$V_3 = \sum_{1 \leqslant u < v \leqslant l} \sum U_{uv} + \sum_{l \leqslant u < v \leqslant k} \sum U_{vu}, \qquad \textbf{(12.1.10)}$$

where U_{uv}, for $u < v = 2, \ldots, k$, are the Mann–Whitney statistics as defined in Example 12.1.6, and where $U_{vu} = n_u n_v - U_{uv}$.

(a) Argue that the test based on V_3 is nonparametric distribution-free when H_0: $\theta_1 = \cdots = \theta_k$ is true.

(b) Show that

$$V_3 = V_2(1,l) - V_2(l,k) + \sum\sum_{l \leqslant u < v \leqslant k} n_u n_v,$$

where $V_2(i,j)$, $i<j$, is the Jonckheere–Terpstra statistic V_2 (12.1.7) for testing the null hypothesis H_0: $\theta_i = \theta_{i+1} = \cdots = \theta_j$ against the ordered alternative H_2: $\theta_i \leqslant \theta_{i+1} \leqslant \cdots \leqslant \theta_j$, with at least one strict inequality. (Note that the usual Jonckheere–Terpstra statistic $V_2(1,k)$ is just a special case of V_3 corresponding to $l=k$.)

(c) Using the representation in **(b)** and the results of Exercise 12.1.3, find the null H_0: $\theta_1 = \cdots = \theta_k$ mean and variance of V_3. [Hint: For the null variance, first find the null covariance between U_{jl} and U_{lt} for $j<l<t$.]

12.1.6. Let $X_{11},\ldots,X_{1n_1},\ldots,X_{k1},\ldots,X_{kn_k}$ be k independent random samples from continuous distributions with c.d.f.'s $F(x-\theta_1),\ldots,F(x-\theta_k)$, respectively. Consider testing the null hypothesis H_0: $\theta_1 = \cdots = \theta_k$ against the alternative H_4: $\theta_j \geqslant \theta_1$ for $j=2,\ldots,k$, with at least one strict inequality. (The alternative H_4 corresponds to a one-sided $(k-1)$ treatments versus control setting, with population one playing the role of the control.) For this problem, Fligner and Wolfe (1978) proposed the procedure:

$$\text{reject } H_0 \text{ in favor of } H_4 \text{ iff } V_4 \geqslant v_4(\alpha),$$

where $v_4(\alpha)$ is the upper 100αth percentile for the null H_0 distribution of

$$V_4 = \sum_{j=2}^{k} U_{1j}, \tag{12.1.11}$$

where U_{1j}, for $j=2,\ldots,k$, are the Mann–Whitney statistics as defined in Example 12.1.6.

(a) Argue that the test based on V_4 is nonparametric distribution-free when H_0: $\theta_1 = \cdots = \theta_k$ is true.

(b) Show that the null H_0 distribution of V_4 is the same as that of a single Mann–Whitney statistic for sample sizes n_1 and $\sum_{j=2}^{k} n_j$.

(c) Use the result in **(b)** to obtain the null H_0 mean and variance of V_4.

12.1.7. Let U_{uv}, $u<v=2,3,\ldots,k$, be the Mann–Whitney statistics as defined in Example 12.1.6. Define

$$U_s^* = \sum_{i=1}^{s-1} U_{is}, \qquad s = 2,\ldots,k. \tag{12.1.12}$$

(Thus, U_s^* is like a Mann–Whitney statistic between the first $(s-1)$ samples combined and the sth sample.)

(a) Let V_2 be the Jonckheere–Terpstra statistic as defined in (12.1.7). Show that

$$V_2 = \sum_{s=2}^{k} U_s^*. \tag{12.1.13}$$

(b) It can be shown [see, e.g., Terpstra (1952)] that $U_2^*, \ldots, U_k^*$ are mutually stochastically independent when the null H_0: $\theta_1 = \cdots = \theta_k$ is true. Using this fact (without proof), verify the expressions for the null mean and variance of V_2, as given in (12.1.8) and (12.1.9), respectively.

(c) Explain how the mutual null independence of $U_2^*, \ldots, U_k^*$ and the appropriate Mann–Whitney null distribution tables can be used to obtain the corresponding null distribution of V_2. [Odeh (1971) used this scheme to create a fairly extensive set of null distribution tables for V_2.]

(d) Discuss how the mutual null independence of $U_2^*, \ldots, U_k^*$ could be used to establish the null asymptotic normality of the standardized Jonckheere–Terpstra statistic

$$\frac{V_2 - E_0(V_2)}{\sqrt{\text{Var}_0(V_2)}},$$

as presented in Example 12.1.6. (Note: An alternate method of proof for this limiting normality when $n_1 = n_2 = \cdots = n_k$ can be based on the relationship between the Terpstra–Jonckheere statistic V_2 and the k-sample U-statistic U_{11} discussed in Example 3.6.2.)

12.2. The Two-Way Layout Problem

Let X_{ij}, $i = 1, \ldots, k$ and $j = 1, \ldots, n$ be kn mutually independent random variables such that X_{ij} has a continuous distribution with c.d.f. $F(x - \tau_i - \beta_j)$. In this section we consider the problem of testing H_0: $\tau_1 = \cdots = \tau_k$; that is, the null hypothesis is that each X_{ij}, $i = 1, \ldots, k$, has the same distribution with c.d.f. $F(x - \tau - \beta_j)$, for $j = 1, \ldots, n$, where τ is the common null value of $\tau_1, \ldots, \tau_k$. This is referred to as a **two-way layout problem with no interaction**. If we think of the first and second subscripts on X_{ij} as designating treatment and block factor levels, respectively, then we are considering tests of the hypothesis that there are no differences among the k treatment levels, with our data consisting of *one* observation for every combination of treatment and block levels. (See Exercises 12.2.4 and 12.2.5 for settings with multiple observations for each such treatment-block pairing.)

As with the one-way layout problem considered in Section 12.1, the choice of a nonparametric distribution-free procedure for testing H_0 will depend on the alternatives that are important. The two most commonly studied alternatives are the general alternative

$$H_1: \tau_i \neq \tau_j \text{ for some } i \neq j \tag{12.2.1}$$

and the ordered alternative

$$H_2: \tau_1 \leqslant \cdots \leqslant \tau_k, \text{ with at least one strict inequality.} \tag{12.2.2}$$

Now, for each fixed $j \in \{1,\ldots,n\}$, let R_{ij} be the rank of X_{ij} among $X_{1j},\ldots,X_{kj}$; that is, R_{ij} is the rank of X_{ij} among the k observations belonging to the jth block. When H_0: $\tau_1 = \cdots = \tau_k$ is true, the variables $X_{1j},\ldots,X_{kj}$ constitute a random sample from the continuous distribution with c.d.f. $F(x-\tau-\beta_j)$. Thus, letting $\mathbf{R}_j^* = (R_{1j},\ldots,R_{kj})$, we see from Theorem 2.3.3 that $\mathbf{R}_j^*$ is uniformly distributed over $\mathcal{R} = \{\mathbf{r}$: $\mathbf{r}$ is a permutation of the integers $1,\ldots,k\}$ when H_0 is true. Since this is the case for each $j = 1,\ldots,n$, it follows from Corollary 2.3.6, and the fact that the observations in different blocks are mutually independent, that any statistic $S(\mathbf{R}_1^*,\ldots,\mathbf{R}_n^*)$ that is a function of the kn observations $X_{11},\ldots,X_{kn}$ only through the within-blocks ranks $\mathbf{R}_1^*,\ldots,\mathbf{R}_n^*$ will be nonparametric distribution-free over the class of continuous distributions when H_0 is true. Thus a test based on such a test statistic will also be nonparametric distribution-free. Our first example deals with the general alternative H_1 (12.2.1).

Example 12.2.3. Let $R_{i.} = \frac{1}{n}\sum_{j=1}^{n} R_{ij}$ be the average of the within-blocks ranks associated with the ith treatment observations, for $i = 1,\ldots,k$. When H_0: $\tau_1 = \cdots = \tau_k$ is true, we see from Corollary 2.3.5 that

$$E_0[R_{i.}] = \frac{1}{n}\sum_{j=1}^{n} E_0(R_{ij}) = \frac{1}{n}\sum_{j=1}^{n}\left(\frac{k+1}{2}\right) = \left(\frac{k+1}{2}\right),$$

where E_0 indicates that this is the null expectation. Thus statistics that measure deviations of the observed treatments within-blocks average ranks from their null expectation $((k+1)/2)$ would provide good bases for constructing tests of H_0 against H_1. One test of this type, proposed independently by Friedman (1937) and Kendall and Babington Smith (1939), is based on the statistic

$$S_1 = \frac{12n}{k(k+1)} \sum_{i=1}^{k} \left(R_{i.} - \frac{k+1}{2}\right)^2, \tag{12.2.4}$$

with the α-level procedure given by

$$\text{reject } H_0 \text{ in favor of } H_1 \text{ iff } S_1 \geqslant s_1(\alpha), \tag{12.2.5}$$

where $s_1(\alpha)$ is the upper 100αth percentile for the null distribution of S_1.

To obtain the critical values $s_1(\alpha)$ required in (12.2.5) for given integers k and n, we calculate S_1 for each of the $(k!)^n$ possible combinations of the n sets of permutations for the within-block ranks $(R_{1j},\ldots,R_{kj})$, $j=1,\ldots,n$, and then tabulate the null distribution of S_1 resulting from the fact that each of these combinations is equally likely when H_0 is true. (Symmetry considerations related to the statistic S_1 can be used to considerably reduce the actual number of calculations required for this tabulation.) Selected $s_1(\alpha)$ values can be found, for example, in Table A.15 of Hollander and Wolfe (1973). The asymptotic $(n\to\infty)$ null distribution of S_1 is chi-square with $(k-1)$ degrees of freedom, a fact which can be used to approximate the cutoff points $s_1(\alpha)$ for large n. [For a proof of this limiting null distribution, see, for example, pages 388–9 of Lehmann (1975).]

For the ordered alternative H_2 (12.2.2) we present a test due to Page (1963).

Example 12.2.6. Let $R_i = nR_{i.}$ be the sum of within-blocks ranks associated with the ith treatment, and set

$$S_2 = \sum_{i=1}^{k} iR_i. \tag{12.2.7}$$

Then the α-level Page test is defined by:

$$\text{reject } H_0 \text{ in favor of } H_1 \text{ iff } S_2 \geqslant s_2(\alpha), \tag{12.2.8}$$

where $s_2(\alpha)$ is the upper 100αth percentile for the null distribution of S_2. Tables of these $s_2(\alpha)$ critical values (obtained in the same manner as the $s_1(\alpha)$ critical values in Example 12.2.3) can be found, for example, in Table

A.16 of Hollander and Wolfe (1973). The limiting ($n \to \infty$) null distribution of $(S_2 - E_0(S_2))/\sqrt{\mathrm{Var}_0(S_2)}$ is standard normal (see Exercise 12.2.3), where

$$E_0(S_2) = \frac{nk(k+1)^2}{4} \tag{12.2.9}$$

and

$$\mathrm{Var}_0(S_2) = \frac{nk^2(k+1)^2(k-1)}{144} \tag{12.2.10}$$

are the null mean and variance, respectively, of S_2. (See Exercise 12.2.3.) Thus the critical value $s_2(\alpha)$ can be approximated by $E_0(S_2) + z_{(\alpha)}\sqrt{\mathrm{Var}_0(S_2)}$, where $z_{(\alpha)}$ is the upper 100αth percentile for the standard normal distribution.

Generalizations of the statistics S_1 (12.2.4) and S_2 (12.2.7) that allow for arbitrary scores have been studied, for example, by Sen (1968) and Puri and Sen (1968), respectively. Models that permit interaction between blocks and treatments have been treated by Lehmann (1963b) and Patel and Hoel (1973), among others. For material concerning estimation of contrasts in the treatment effects $\tau_1, \ldots, \tau_k$, see, for example, Sections 7.4 and 7.8 of Hollander and Wolfe (1973).

Exercises 12.2.

12.2.1. Let S_1 (12.2.4) be the Friedman statistic of Example 12.2.3.

(a) Show that

$$S_1 = \left[\frac{12}{nk(k+1)} \sum_{i=1}^{k} R_i^2 \right] - 3n(k+1),$$

where $R_i = \sum_{j=1}^{n} R_{ij} =$ sum of within-blocks ranks associated with the ith treatment observations.

(b) Using the representation in **(a)**, find an expression for the null H_0: $\tau_1 = \cdots = \tau_k$ mean of S_1. [Hint: Corollary 2.3.5 might be helpful.]

12.2.2. Tabulate the exact null H_0: $\tau_1 = \cdots = \tau_k$ distribution of the Page statistic S_2 (12.2.7) for the case $k=3$ and $n=2$.

12.2.3. **(a)** Show that the Page statistic S_2 (12.2.7) of Example 12.2.6 can be written as a sum of n independent random variables.

(b) Verify the expressions for the null H_0: $\tau_1 = \cdots = \tau_k$ mean and variance of S_2, as given in (12.2.9) and (12.2.10), respectively. [Hint: Use the representation in **(a)** and Corollary 2.3.5.]

(c) Use the representation in **(a)** to establish the limiting ($n \to \infty$) null normality of $(S_2 - E_0(S_2))/\sqrt{\text{Var}_0(S_2)}$, as stated in Example 12.2.6.

12.2.4. Let X_{ijl}, $i = 1, \ldots, k$, $j = 1, \ldots, n$ and $l = 1, \ldots, t_{ij} > 0$, be $\Sigma_{i=1}^{k} \Sigma_{j=1}^{n} t_{ij}$ mutually independent random variables such that X_{ijl} has a continuous distribution with c.d.f. $F(x - \tau_i - \beta_j)$. Thus we are considering a two-way layout problem with no interaction and t_{ij} observations (or replications) from the combination of the ith treatment and jth block levels. For testing H_0: $\tau_1 = \cdots = \tau_k$ against the ordered alternative H_2 (12.2.2) with this multiple replications data, Skillings and Wolfe (1977) considered the test that, at the α-level of significance,

$$\text{rejects } H_0 \text{ in favor of } H_2 \text{ iff } S_3 \geqslant s_3(\alpha), \qquad \textbf{(12.2.11)}$$

where $s_3(\alpha)$ is the upper 100αth percentile for the null distribution of the statistic

$$S_3 = \sum_{j=1}^{n} \left[\frac{V_{2j} - E_0(V_{2j})}{\sqrt{\text{Var}_0(V_{2j})}} \right], \qquad \textbf{(12.2.12)}$$

with V_{2j} being the Jonckheere statistic (12.1.7) for the observations in the jth block.

(a) Find the null H_0 mean and variance of S_3.

(b) Show that for n fixed the limiting ($\min(t_{11}, \ldots, t_{kn}) \to \infty$, with

$$\frac{t_{ij}}{\sum_{i=1}^{k} t_{ij}} \to \lambda_{ij},\ 0 < \lambda_{ij} < 1, \qquad \text{for } i = 1, \ldots, k \text{ and } j = 1, \ldots, n)$$

null distribution of $S_3/\sqrt{n}$ is standard normal. [Hint: See Example 12.1.6.]

12.2.5. For the multiple replications data setting of Exercise 12.2.4, consider the test of H_0: $\tau_1 = \cdots = \tau_k$ against the general alternative H_1 (12.2.1) given by:

$$\text{reject } H_0 \text{ in favor of } H_1 \text{ iff } S_4 \geqslant s_4(\alpha), \qquad \textbf{(12.2.13)}$$

where $s_4(\alpha)$ is the upper 100αth percentile for the null distribution of the statistic

$$S_4 = \sum_{j=1}^{n} V_{1j}, \tag{12.2.14}$$

with V_{1j} being the Kruskal–Wallis statistic (12.1.4) for the observations in the jth block.

(a) Derive the null mean of S_4. [Hint: The answer to Exercise 12.1.1 **(b)** is useful.]

(b) What is the limiting (n fixed and $\min(t_{11},\ldots,t_{kn})\to\infty$, with

$$\frac{t_{ij}}{\sum_{i=1}^{k} t_{ij}} \to \lambda_{ij},\ 0<\lambda_{ij}<1, \qquad \text{for } i=1,\ldots,k \text{ and } j=1,\ldots,n)$$

null distribution of S_4? Verify your answer. [Hint: See Example 12.1.3.]

12.3. The Independence (Correlation) Problem

In this section we discuss the common practical problem of determining if there is a dependence between a pair of random variables. Let $(X_1,Y_1),\ldots,(X_n,Y_n)$ be a random sample from a continuous bivariate distribution with c.d.f. $F(x,y)$. On the basis of these observations we wish to construct nonparametric distribution-free tests of H_0: $[F(x,y)=F_X(x)F_Y(y)$ for every (x,y) pair], where $F_X(x)$ and $F_Y(y)$ are the c.d.f.'s for the marginal distributions of $X_1,\ldots,X_m$ and $Y_1,\ldots,Y_n$, respectively. Thus the null hypothesis of interest is that the X and Y variables are independent, while appropriate alternatives correspond to one-sided positive or negative dependence or to two-sided general dependence.

We consider here two test procedures that are frequently used for this independence problem. The first of these, due to Spearman (1904), is a special case of a class of general scores procedures. Let Q_i, $i=1,\ldots,n$, be the rank of X_i among $X_1,\ldots,X_n$, and let R_i, $i=1,\ldots,n$, be the rank of Y_i among $Y_1,\ldots,Y_n$. In addition, let $a(1)\leqslant\cdots\leqslant a(n)$, with $a(1)\neq a(n)$, and $c(1)\leqslant\cdots\leqslant c(n)$, with $c(1)\neq c(n)$, be two sets of scores. Set

$$T_n = \frac{\sum_{i=1}^{n}\left(c(Q_i)-\bar{c}\right)\left(a(R_i)-\bar{a}\right)}{\left[\sum_{i=1}^{n}\left(c(Q_i)-\bar{c}\right)^2 \sum_{j=1}^{n}\left(a(R_j)-\bar{a}\right)^2\right]^{1/2}}, \tag{12.3.1}$$

where $\bar{a}=(1/n)\Sigma_{i=1}^{n}a(i)$ and $\bar{c}=(1/n)\Sigma_{j=1}^{n}c(j)$. Thus, T_n is the sample (Pearson product moment) correlation coefficient for the group of scored rank pairs $(c(Q_1),a(R_1)),\ldots,(c(Q_n),a(R_n))$. The associated α-level procedure for testing H_0 against the one-sided alternative H_1: [X and Y are positively dependent (related)] is:

$$\text{reject } H_0 \text{ in favor of } H_1 \text{ iff } T_n \geqslant t_n(\alpha), \quad \textbf{(12.3.2)}$$

where $t_n(\alpha)$ is the upper 100αth percentile for the null distribution of T_n. Actually, the procedure in (12.3.2) is equivalent (Exercise 12.3.1) to the one that

$$\text{rejects } H_0 \text{ in favor of } H_1 \text{ iff } S_n \geqslant s_n(\alpha), \quad \textbf{(12.3.3)}$$

where $s_n(\alpha)$ is the upper 100αth percentile for the null distribution of

$$S_n = \sum_{i=1}^{n} c(Q_i)a(R_i). \quad \textbf{(12.3.4)}$$

(Positively dependent means that large [small] values of X tend to be associated with large [small] values of Y.)

Now, when H_0 is true, the rank vectors $\mathbf{Q}=(Q_1,\ldots,Q_n)$ and $\mathbf{R}=(R_1,\ldots,R_n)$ are both uniformly distributed over $\mathfrak{R}$, the set of permutations of the integers $(1,\ldots,n)$, and moreover, $\mathbf{Q}$ and $\mathbf{R}$ are independent. Since this is true for any continuous marginal X and Y distributions, the test in (12.3.3) or, equivalently, in (12.3.2), is nonparametric distribution-free. In addition, these properties of $\mathbf{Q}$ and $\mathbf{R}$ imply (Exercise 12.3.2) that

$$S_n \stackrel{d}{=} \sum_{i=1}^{n} c(i)a(R_i) \quad \textbf{(12.3.5)}$$

when H_0 is true; that is, S_n has the same null distribution as does a linear rank statistic with scores $a(1),\ldots,a(n)$ and regression constants $c(1),\ldots,c(n)$. (See Section 8.1.) Consequently, all the null distribution theory developed in Chapter 8 is directly applicable to the statistic S_n (12.3.4). Thus, for example, sufficient conditions for S_n to have a symmetric null distribution are provided in Theorem 8.2.13, and the asymptotic normality of the properly standardized S_n follows from Theorem 8.4.9 (and related work in Section 8.4). Finally, the null mean and variance of S_n are as given in Theorem 8.1.12.

Example 12.3.6. The Spearman (1904) test statistic corresponds to taking $a(i)=c(i)=i$ (i.e., using Wilcoxon-type scores and regression constants), for which $S_{n,1}=\Sigma_{i=1}^{n}Q_iR_i \stackrel{d}{=} \Sigma_{i=1}^{n}iR_i$. The corresponding correlation coefficient $T_{n,1}$ (12.3.1) has form (Exercise 12.3.3)

$$T_{n,1} = 1 - \left[\frac{6}{n(n^2-1)} \sum_{i=1}^{n} (R_i - Q_i)^2\right], \tag{12.3.7}$$

and is known as the **Spearman sample correlation coefficient**. From the previously mentioned association with the results in Chapter 8, we have immediately that

(i) The null H_0 mean and variance for $S_{n,1}$ are, respectively,

$$E_0(S_{n,1}) = \frac{n(n+1)^2}{4} \text{ and } \mathrm{Var}_0(S_{n,1}) = \frac{n^2(n+1)^2(n-1)}{144}. \tag{12.3.8}$$

(ii) The null distribution of $S_{n,1}$ is symmetric about its mean

$$\frac{n(n+1)^2}{4}.$$

and

(iii) The limiting ($n \to \infty$) null distribution of

$$\frac{S_{n,1} - E_0(S_{n,1})}{\sqrt{\mathrm{Var}_0(S_{n,1})}} = \sqrt{n-1}\, T_{n,1} \qquad \text{is standard normal.} \tag{12.3.9}$$

A theory that leads to optimal (e.g., locally most powerful) ways to select the $a(\cdot)$ and $c(\cdot)$ scores can be found in Hájek and Šidák (1967). However, not every important rank statistic for testing the independence hypothesis H_0 fits into the general form given in (12.3.1). We consider one such statistic, studied by Kendall (1938), in the next example.

Example 12.3.10. For each pair (X_i, Y_i) and (X_j, Y_j), $i \neq j$, of observations from the random sample $(X_1, Y_1), \ldots, (X_n, Y_n)$, define

$$\xi(X_i, X_j; Y_i, Y_j) = \mathrm{sign}(Q_j - Q_i)\mathrm{sign}(R_j - R_i), \tag{12.3.11}$$

where the Qs and Rs are the separate X and Y ranks, respectively, as previously defined, and

$$\mathrm{sign}(x) = \begin{cases} 1, & \text{if } x > 0 \\ 0, & \text{if } x = 0 \\ -1, & \text{if } x < 0. \end{cases} \tag{12.3.12}$$

Then the α-level Kendall (1938) test of H_0 against the positive dependence alternative H_1 is given by

$$\text{reject } H_0 \text{ in favor of } H_1 \text{ iff } K \geqslant k(\alpha), \tag{12.3.13}$$

where $k(\alpha)$ is the upper 100αth percentile for the null distribution of the **Kendall statistic**

$$K = \sum_{i<j}^{n}\sum^{n} \xi(X_i, X_j; Y_i, Y_j). \tag{12.3.14}$$

The following are important null distribution properties of the statistic K:

(i) The null mean and variance for K are, respectively,

$$E_0(K) = 0 \text{ and } \operatorname{Var}_0(K) = \frac{n(n-1)(2n+5)}{18}. \tag{12.3.15}$$

(ii) The null distribution of K is symmetric about its mean 0.

and

(iii) The limiting ($n \to \infty$) null distribution of

$$\frac{K - E_0(K)}{\sqrt{\operatorname{Var}_0(K)}} = \frac{K}{\sqrt{\dfrac{n(n-1)(2n+5)}{18}}} \quad \text{is standard normal.} \tag{12.3.16}$$

The first two of these facts are left as Exercises 12.3.4 and 12.3.6, respectively, and (iii) follows from the fact that $K/\binom{n}{2}$ is a bivariate U-statistic (see Example 3.6.12). Hájek and Šidák (1967) show in Section II.3.1 that, up to a multiplicative constant, the Spearman statistic $S_{n,1}$ (see Example 12.3.6) is the projection of the Kendall statistic K into the family of linear rank statistics of the form S_n (12.3.4), and thus that $S_{n,1}$ and K are asymptotically equivalent.

To obtain the critical values necessary to carry out one of these rank tests for independence, we must find the null distribution of the associated test statistic. To accomplish this, we fix the vector $(Q_1, \ldots, Q_n) = (1, \ldots, n)$ and calculate the value of the test statistic for each of the $n!$ possible permutations of the Y ranks $(R_1, \ldots, R_n)$. The appropriate null distribution is then obtained by tabulating these computed values and making use of the fact that **R** is uniformly distributed over $\mathfrak{R}$ and independent of $\mathbf{Q}=(Q_1,\ldots,Q_n)$ when H_0 is true. Selected critical values $t_{n,1}(\alpha)$ and $k(\alpha)$ for the Spearman (based on $T_{n,1}$) and Kendall tests, respectively, can be found, for example, in Zar (1972) and Table A.21 of Hollander and Wolfe (1973), respectively.

Exercises 12.3.

12.3.1. Show that the α-level tests given in (12.3.2) and (12.3.3) are equivalent in the sense that one test rejects if and only if the other one also rejects.

12.3.2. Let $\mathbf{Q}=(Q_1,\ldots,Q_n)$ and $\mathbf{R}=(R_1,\ldots,R_n)$ be uniformly distributed over $\mathfrak{R}$, the set of permutations of the integers $(1,\ldots,n)$, and assume that **Q** and **R** are independent. Show that

$$\sum_{i=1}^{n} c(Q_i)a(R_i) \stackrel{d}{=} \sum_{i=1}^{n} c(i)a(R_i).$$

[Hint: Use arguments similar to those in Section 8.2.]

12.3.3. Establish the validity of the expression in (12.3.7).

12.3.4. Let $\hat{\tau}=2K/(n(n-1))$, where K is the Kendall statistic given in (12.3.14). The statistic $\hat{\tau}$ is referred to as the **Kendall sample correlation coefficient**.

(a) We say that the pair of bivariate observations $(X_i, Y_i),(X_j, Y_j)$ are **concordant** if $(Y_j - Y_i)(X_j - X_i) > 0$. Show that

$$\hat{\tau} = \frac{4C_n}{n(n-1)} - 1,$$

where C_n=(number of concordant pairs in the random sample $(X_1, Y_1),\ldots,(X_n, Y_n)$).

(b) Show that

$$\tau = E(\hat{\tau}) = 2P\{(X_2 - X_1)(Y_2 - Y_1) > 0\} - 1.$$

What is the value of τ when H_0: [X and Y are independent] is true? What are the minimum and maximum values of τ? Discuss why a large positive (negative) value of τ is indicative of positive (negative) dependence between the basic variables X and Y.

(c) Verify the expression for the null variance of K, as given in (12.3.15).

12.3.5. Tabulate the null distributions of both Spearman's statistic $T_{n,1}$ (12.3.7) and Kendall's statistic K (12.3.14) for the case $n=4$.

12.3.6. Show that the null distribution of Kendall's statistic K (12.3.14) is symmetric about 0. [Hint: Use Theorem 8.2.18 and a relationship similar to (12.3.5), but for the Kendall statistic K.]

12.3.7. Let (X_1, X_2, X_3) be a trivariate random variable having covariance matrix Σ with (i,j)th element $\sigma_{ij} = \rho_{ij}\sigma_i\sigma_j$.

(a) Define $Z = X_3 - X_2$. Assuming $\sigma_2 = \sigma_3$ and $\rho_{23} \neq 1$, show that $\rho_{1Z} = 0$ iff $\rho_{12} = \rho_{13}$, where ρ_{1Z} is the correlation coefficient between X_1 and Z.

(b) Let $(X_{11}, X_{21}, X_{31}), \ldots, (X_{1n}, X_{2n}, X_{3n})$ be a random sample from a continuous trivariate distribution. If $\mathrm{Var}(X_{21}) = \mathrm{Var}(X_{31})$ and $\mathrm{corr}(X_{21}, X_{31}) \neq 1$, discuss how Kendall's statistic could be used to test for whether X_{31} is more positively correlated with X_{11} than is X_{21}.

(c) If K^* represents the Kendall statistic to be used in **(b)**, what parameter τ^* does $2K^*/(n(n-1))$ estimate? Discuss why a large positive (negative) value of τ^* is indicative of X_{31} being more (less) positively related to X_{11} than is X_{21}.

12.4. The One-Sample Regression Problem

The primary interest in many statistical applications is to determine if there is a relationship between two variables. Of course, one way to handle such a problem is to use a test for correlation, such as those described in Section 12.3. However, it is often the case that we wish to know more than simply whether there *is* a relationship between the two variables. We might like to have some information about the *type* of relationship that exists; for example, we could be interested in whether the involved variables are linearly related. Such considerations are particularly important when we wish to use the value of one variable to predict the value of the other one. This general setting is referred to as the **one-sample regression problem** and in this section we consider some nonparametric distribution-free tests for the case of a straight line relationship between two variables.

Let $d_1 \leqslant \cdots \leqslant d_n$ be known constants, and let $Y_1, \ldots, Y_n$ be independent random variables such that Y_i has a continuous distribution with c.d.f. $F(y - \beta d_i - \gamma)$, where $F(0) = 1/2$. Thus, $\gamma + \beta d_i$ is a median of the Y_i distribution. (Note that this setting is equivalent to the model

$$Y_i = \beta d_i + \gamma + Z_i, \qquad i = 1, \ldots, n,$$

where $Z_1,\ldots,Z_n$ are i.i.d. continuous random variables with median 0 and c.d.f. $F(x)$.) We refer to this as the **one-sample straight line regression model** because there is a straight line relationship between the median of the Y_i population and the constant d_i. The parameters γ and β play the roles of the **intercept** and **slope**, respectively, of this straight line. In this section we concentrate on test procedures for the slope parameter β, while viewing γ as simply a nuisance parameter. In particular, we wish to test the null hypothesis H_0: $\beta=\beta_0$, where β_0 is some known value (often taken to be 0), against the one-sided alternative H_1: $\beta>\beta_0$. (Tests for the alternatives $\beta<\beta_0$ or $\beta\neq\beta_0$ would be constructed similarly.)

Let R_i be the rank of $Y_i-\beta_0 d_i$ among $Y_1-\beta_0 d_1,\ldots,Y_n-\beta_0 d_n$ and consider the sample Pearson product moment correlation coefficient for the pairs $(d_1,a(R_1)),\ldots,(d_n,a(R_n))$, namely,

$$W_n = \frac{\sum_{i=1}^{n}\left(d_i-\bar{d}\right)\left(a(R_i)-\bar{a}\right)}{\left[\sum_{i=1}^{n}\left(d_i-\bar{d}\right)^2\sum_{j=1}^{n}\left(a(j)-\bar{a}\right)^2\right]^{1/2}}, \tag{12.4.1}$$

where $a(1)\leqslant\cdots\leqslant a(n)$, with $a(1)\neq a(n)$, are scores, $\bar{a}=(1/n)\sum_{i=1}^{n}a(i)$ and $\bar{d}=(1/n)\sum_{i=1}^{n}d_i$. Because W_n is simply the sample correlation coefficient T_n (12.3.1) with $c(i)=d_i$, it follows from the arguments necessary to complete Exercise 12.3.1 that

$$W_n = a_{1n}S_n^* + a_{2n}, \tag{12.4.2}$$

where a_{1n} and a_{2n} are constants and

$$S_n^* = \sum_{i=1}^{n} d_i a(R_i). \tag{12.4.3}$$

The α-level test of H_0: $\beta=\beta_0$ against H_1: $\beta>\beta_0$ is then given by

$$\text{reject } H_0 \text{ in favor of } H_1 \text{ iff } S_n^* \geqslant s_n^*(\alpha), \tag{12.4.4}$$

where $s_n^*(\alpha)$ is the upper 100αth percentile for the null distribution of S_n^*.

Now, when H_0: $\beta=\beta_0$ is true, the variables $Y_1-\beta_0 d_1,\ldots,Y_n-\beta_0 d_n$ (which are being used to form the ranks $R_1,\ldots,R_n$) form a random sample from $F(x-\gamma)$. Therefore, the null hypothesis properties of the statistic S_n^* (12.4.3) follow immediately from the corresponding null hypothesis properties developed in Chapter 8 for a general linear rank statistic (Definition 8.1.1). Thus, that the test in (12.4.4) is nonparametric distribution-free follows from the fact that $\mathbf{R}=(R_1,\ldots,R_n)$ is uniformly distributed over $\mathcal{R}$, the collection of all $n!$ permutations of $(1,\ldots,n)$. In addition, sufficient conditions for S_n to have a symmetric null distribution are stated in Theorem 8.2.13, and the asymptotic normality of the properly standardized S_n^* (null mean and variance of S_n^* are given in Theorem 8.1.12) follows from Theorem 8.4.9. (Note that for this regression problem the constants $d_1,\ldots,d_n$ are determined during the experimentation leading to the observations $Y_1,\ldots,Y_n$, unlike the $c(i)$'s in the two-sample setting, for example, which are merely indicators for the two populations.)

Choosing optimal scores $a(i)$ to use in S_n^* (12.4.3) is accomplished in the same way as for the one- and two-sample problems (see Sections 9.1 and 10.1). In addition, the ARE results for two such rank tests (12.4.4) based on different score functions (but for the same constants $d_1,\ldots,d_n$) are identical with the ARE results of the two-sample location tests based on two linear rank statistics with corresponding scoring functions. Thus some of the material of Section 9.2 is applicable here as well. For details concerning these equivalences between the regression and two-sample location settings, see Hájek and Šidák (1967).

There is one major problem regarding the test procedures stated in (12.4.4). The test statistic S_n^* and consequently its null distribution depends on the particular values of the fixed d_i's. Thus a (potentially) different null distribution for S_n^* must be obtained whenever the d_i's are changed, and tables of the appropriate critical values $s_n^*(\alpha)$ cannot be prepared once and for all. One way to alleviate this problem is to score the d_i's just as we do the R_i's. That is, we replace $d_1 \leqslant \cdots \leqslant d_n$ with predetermined numbers $c(1) \leqslant \cdots \leqslant c(n)$. This use of "standard" scores in place of the d_i's enables us to prepare tables of the critical values of the associated statistics $\bar{S}_n = \sum_{i=1}^n c(i)a(R_i)$ that can be used for *any* set of d_i's. Tests of this form were developed by Hogg and Randles (1975). We consider here only a single illustration of their approach.

Example 12.4.5. For the regression statistic S_n^* (12.4.3), replace $d_1 \leqslant \cdots \leqslant d_n$ by the Wilcoxon scores $1 \leqslant \cdots \leqslant n$ to obtain the "standard scores" statistic $\bar{S}_{n,1} = \sum_{i=1}^n i a(R_i)$. If, in addition, we also take the Wilcoxon scores for $a(i)$, we have

$$\bar{S}_{n,1} = \sum_{i=1}^{n} iR_i. \tag{12.4.6}$$

Since $\bar{S}_{n,1}$ has the same null distribution as the Spearman correlation statistic considered in Example 12.3.6, the tables by Zar (1972) can be used to carry out an α-level test of H_0: $\beta = \beta_0$ based on $\bar{S}_{n,1}$ (12.4.6).

For additional references regarding nonparametric approaches to regression problems (including estimation of the intercept (γ) and slope (β) parameters, and procedures for more general regression models), see Chapter 9 of Hollander and Wolfe (1973).

Exercises 12.4.

12.4.1. Verify expression (12.4.2) by finding a_{1n} and a_{2n}.

12.4.2. Consider S_n^* (12.4.3). For $a(i)=i$ and $n=4$, find the null distribution of S_n^* if $d_1=-4$, $d_2=6$, $d_3=12$ and $d_4=27$.

12.4.3. Let $S_{n,1}^*$ be S_n^* (12.4.3) with the Wilcoxon scores $a(i)=i$. Suppose $d_1=\cdots=d_m=d$ and $d_{m+1}=\cdots=d_n=d^*$, with $d<d^*$, for some integer m strictly between 1 and n.

(a) Show that null distribution tables for the two-sample Mann–Whitney–Wilcoxon rank sum statistic can be used to obtain the necessary critical values $s_{n,1}^*(\alpha)$ for the test (12.4.4) of H_0: $\beta=\beta_0$ based on $S_{n,1}^*$. (Be sure to demonstrate that the use of these Mann–Whitney–Wilcoxon tables does *not* depend on the particular values of d and d^*.)

(b) Generalize the result in **(a)** to an arbitrary set of scores $a(i)$.

12.4.4. Let $d_1 \leqslant \cdots \leqslant d_n$ and $Y_1, \ldots, Y_n$ be as defined in this section for the straight-line regression problem. For fixed constant β_0, set

$$D_i = Y_i - \beta_0 d_i, \qquad i = 1, \ldots, n.$$

If no two d_i's are equal, discuss how we could use the Kendall statistic K (12.3.14), applied to the pairs $(d_1, D_1), \ldots, (d_n, D_n)$, to conduct a test of H_0: $\beta=\beta_0$ against H_1: $\beta>\beta_0$. Be sure to explain why this procedure would be reasonable, and to justify the fact that the usual null distribution tables for K (see the last paragraph in Section 12.3) could be used to carry out the test. (This procedure is the Kendall analogue to the test discussed in Example 12.4.5, and was proposed by Theil (1950a).)

APPENDIX

A.1. Distributions

The following list contains the major distribution types that are used in this text, along with additional information about them. Generally, these descriptions of the distributions do not include location (θ) or scale ($\eta > 0$) parameters. If such parameters are desired, the c.d.f. becomes $F((x-\theta)/\eta)$ and the density becomes $(1/\eta)f((x-\theta)/\eta)$, where $f(t)$ is as given in the listing. The mean would then be $\eta\mu + \theta$ and the variance $\eta^2\sigma^2$, where μ and σ^2 are the mean and variance, respectively, listed in the description that follows. The skewness (a measure of asymmetry) is defined to be the quantity $E[((X-\mu)/\sigma)^3]$, and the kurtosis (a measure of tailweight) is given by $E[((X-\mu)/\sigma)^4]$. These quantities are unaffected by location and scale changes. When it exists and can be written in closed form, we also include the optimal score function for the given distribution in the two-sample location shift problem.

A.1.1. *Beta Distribution*

Density: $f(t) = \dfrac{\Gamma(\alpha+\beta)}{\Gamma(\alpha)\Gamma(\beta)} t^{\alpha-1}(1-t)^{\beta-1}$, $0 < t < 1$, for $\alpha > 0$ and $\beta > 0$

Mean: $\dfrac{\alpha}{\alpha+\beta}$, Variance: $\dfrac{\alpha\beta}{(\alpha+\beta)^2(\alpha+\beta+1)}$,

Skewness: $\dfrac{2(\beta-\alpha)\sqrt{\alpha^{-1}+\beta^{-1}+(\alpha\beta)^{-1}}}{(\alpha+\beta+2)}$,

Kurtosis: $\dfrac{3(\alpha+\beta+1)\{2(\alpha+\beta)^2+\alpha\beta(\alpha+\beta-6)\}}{\alpha\beta(\alpha+\beta+2)(\alpha+\beta+3)}$

A.1.2. *Cauchy Distribution*

Density: $f(t)=[\pi(1+t^2)]^{-1}$, $-\infty<t<\infty$

It is symmetric about zero but has no moments.

A.1.3. *Double Exponential Distribution*

Density: $f(t)=\frac{1}{2}\exp[-|t|]$, $-\infty<t<\infty$

Mean: 0, Variance: 2, Skewness: 0, Kurtosis: 6

Two-Sample Score Function: $\phi(u,f)=\text{sign}(2u-1)$

A.1.4. *Exponential Distribution* (a scale parameter, β, is included in the description)

Density: $f(t)=\dfrac{1}{\beta}e^{-t/\beta}$, $0<t<\infty$, for $\beta>0$

Mean: β, Variance: β^2, Skewness: 2, Kurtosis: 9

A.1.5. *Gamma Distribution* (Exponential is a special case corresponding to $\alpha=1$; a scale parameter, β, is included in the description)

Density: $f(t)=\dfrac{t^{\alpha-1}}{\Gamma(\alpha)\beta^{\alpha}}e^{-t/\beta}$, $0<t<\infty$, for $\alpha>0$ and $\beta>0$

Mean: $\alpha\beta$, Variance: $\alpha\beta^2$, Skewness: $\dfrac{2}{\sqrt{\alpha}}$, Kurtosis: $\dfrac{3(\alpha+2)}{\alpha}$

A.1.6. *Logistic Distribution*

Density: $f(t)=\dfrac{e^{-t}}{(1+e^{-t})^2}$, $-\infty<t<\infty$

Mean: 0, Variance: $\pi^2/3$, Skewness: 0, Kurtosis: 4.2

Two-Sample Score Function: $\phi(u,f)=2u-1$

A.1.7. *Normal Distribution*

Density: $f(t)=\dfrac{1}{\sqrt{2\pi}}\exp[-t^2/2],\ -\infty<t<\infty$

Mean: 0, Variance: 1, Skewness: 0, Kurtosis: 3

Two-Sample Score Function: $\phi(u,f)=\Phi^{-1}(u)$, where $\Phi(t)$ is the c.d.f. for this standard normal distribution.

A.1.8. *RST (Ramberg–Schmeiser–Tukey) Lambda Distribution*

The density can only be written implicitly. Instead the distribution is defined in terms of its inverse c.d.f.

$$F^{-1}(u)=\lambda_1+\left[u^{\lambda_3}-(1-u)^{\lambda_4}\right]/\lambda_2, \qquad 0<u<1,$$

where λ_1 and λ_2 are location and scale parameters, respectively. The parameters λ_3 and λ_4 determine the distribution's shape. Restrictions on the four lambda parameters are described in Ramberg and Schmeiser (1974).

Mean: $\lambda_1+\dfrac{\lambda_4-\lambda_3}{\lambda_2(\lambda_3+1)(\lambda_4+1)}$

Variance:

$$\left[\frac{1}{2\lambda_3+1}+\frac{1}{2\lambda_4+1}-\frac{2\Gamma(\lambda_3+1)\Gamma(\lambda_4+1)}{\Gamma(\lambda_3+\lambda_4+2)}-\left\{\frac{\lambda_4-\lambda_3}{(\lambda_3+1)(\lambda_4+1)}\right\}^2\right]\Big/\lambda_2^2$$

The skewness and kurtosis have a closed form, but are complicated. See Ramberg and Schmeiser (1974) and Ramberg, Tadikamalla, Dudewicz, and Mykytka (1979).

A.1.9. *Uniform Distribution*

Density: $f(t)=1,\ 0<t<1$

Mean: $\frac{1}{2}$, Variance: $\frac{1}{12}$, Skewness: 0, Kurtosis: 1.8

A.1.10. *Weibull Distribution*

Density: $f(t)=\alpha t^{\alpha-1}\exp[-t^{\alpha}],\ t>0$, for $\alpha>0$

Mean: $\Gamma(1+\alpha^{-1})$, Variance: $\Gamma(1+2\alpha^{-1})-[\Gamma(1+\alpha^{-1})]^2$,

Skewness: $\{\Gamma(1+3\alpha^{-1})-3\Gamma(1+\alpha^{-1})\Gamma(1+2\alpha^{-1})+2[\Gamma(1+\alpha^{-1})]^3\}/(\text{Variance})^{3/2}$

Kurtosis: $\{\Gamma(1+4\alpha^{-1})-4\Gamma(1+\alpha^{-1})\Gamma(1+3\alpha^{-1})+6[\Gamma(1+\alpha^{-1})]^2\Gamma(1+2\alpha^{-1})-3[\Gamma(1+\alpha^{-1})]^3\}/(\text{Variance})^2$

A.2. Integrals

Let X denote a random variable with c.d.f. $F(x)$. If $g(\cdot)$ is a real-valued function, we write

$$E[g(X)] = \int g(x)\,dF(x),$$

provided this integral exists. This notation is common when dealing with such Lebesgue-Stieltjes integrals. A careful description of these integrals and their properties is found, for example, in Cramér (1946). For our purposes it suffices to simply mentally set

$$\int g(x)\,dF(x) = \int g(x)f(x)\,dx, \tag{A.2.1}$$

when X is a continuous random variable with density $f(x)$, and

$$\int g(x)\,dF(x) = \sum_x g(x)f(x), \tag{A.2.2}$$

when X is a discrete random variable with probability function $f(x)$.

When the underlying distributions are continuous we may also use, for example, integration by parts; that is,

$$F(b)G(b) - F(a)G(a) = \int_a^b F(x)\,dG(x) + \int_a^b G(x)\,dF(x).$$

The three theorems which follow provide general conditions under which we can make other manipulations with these integrals. They are applicable to situations in which $F(\cdot)$ is the c.d.f. of a continuous univariate or vector-valued random variable.

Theorem A.2.3. Continuity of an Integral. *Let $F(\cdot)$ be the c.d.f. of a continuous random variable. Suppose that $k(x,t)$ is a continuous function of t at t_0 for all but at most a countable number of x-values. If there exists a* *function* $K_*(x)$ *such that*

(i) *for some* $\delta > 0$,

$$|k(x,t)| \leqslant K_*(x) \text{ for all } x \text{ and for all } t \text{ such that } |t - t_0| < \delta,$$

and

(ii) $\int K_*(x)\,dF(x) < \infty$,

then

$$\lim_{t \to t_0} \int k(x,t)\,dF(x) = \int k(x,t_0)\,dF(x).$$

Proof: The result follows directly from the dominated convergence theorem (A.2.5). ∎

Theorem A.2.4. Differentiation of an Integral. *Let* $F(\cdot)$ *denote the c.d.f. of a continuous random variable. For the function* $k(x,t)$, *suppose that the following conditions are satisfied*:

(i) *for each fixed* t *in* (a,b), $\partial k(x,t)/\partial t$ *exists for all but at most a countable number of* x*-values*,

(ii) *for each fixed* x, $\partial k(x,t)/\partial t$ *exists and satisfies*

$$\left|\frac{\partial}{\partial t} k(x,t)\right| \leqslant K_*(x)$$

for all but at most a countable number of t*-values in* (a,b), *where* $K_*(\cdot)$ *is such that*

$$\int K_*(x)\,dF(x) < \infty,$$

and (*iii*) *for every* x *and* $a < t' < t'' < b$,

$$k(x,t'') - k(x,t') = \int_{t'}^{t''} \frac{\partial}{\partial t} k(x,t)\,dt,$$

(*i.e.*, $k(\cdot)$ *is an absolutely continuous function of* t). *Then*

$$\frac{d}{dt} \int k(x,t)\,dF(x) = \int \frac{\partial}{\partial t} k(x,t)\,dF(x)$$

for all t *in* (a,b).

Proof: If t and $t+h$ are in (a,b), conditions (ii) and (iii) imply

$$|k(x,t+h)-k(x,t)| = \left|\int_t^{t+h} \frac{\partial}{\partial s} k(x,s)\,ds\right| \leqslant |h| K_*(x)$$

for all x. Since

$$\frac{\partial}{\partial t}\int k(x,t)\,dF(x) = \lim_{h\to 0}\int\left[\frac{k(x,t+h)-k(x,t)}{h}\right]dF(x),$$

the result then follows from the dominated convergence theorem (A.2.5). ■

Theorem A.2.5. Dominated Convergence Theorem. *Let $F(\cdot)$ denote the c.d.f. of a continuous random variable, and let $\{g_\nu(x)\}$ be a sequence of functions satisfying*:

(i)
$$\lim_{\nu\to\infty} g_\nu(x) = g(x)$$

for all but at most a countable number of x-values, and

(ii)
$$|g_\nu(x)| \leqslant K_*(x)$$

for all x, where $K_*(\cdot)$ is such that

$$\int K_*(x)\,dF(x) < \infty.$$

Then

$$\lim_{\nu\to\infty}\int g_\nu(x)\,dF(x) = \int g(x)\,dF(x).$$

Proof: See, for example, Royden (1968), page 229. ■

A.3. Mathematical Statistics Results

Theorem A.3.1. Chebyshev's Inequality. *Let V denote a nonnegative random variable with finite expected value and let $\epsilon > 0$ be arbitrary. Then*

$$P[V \geqslant \epsilon] \leqslant \frac{E[V]}{\epsilon}.$$

Proof: See, for example, Hogg and Craig (1978), page 58. ■

Definition A.3.2. A function $k(u)$ is said to be convex if for each $u_1 \neq u_2$ and every $0 \leqslant \alpha \leqslant 1$,

$$k(\alpha u_1 + (1-\alpha)u_2) \leqslant \alpha k(u_1) + (1-\alpha)k(u_2).$$

Theorem A.3.3. Jensen's Inequality. *If $k(\cdot)$ is a convex function and X is a real-valued random variable such that $E[X]$ exists, then*

$$E[k(X)] \geqslant k(E[X]).$$

Proof: A general proof is given, for example, in Ferguson (1967), page 76. The only case used in this text is $k(u)=u^2$. For this special case the proof is simple since

$$\mathrm{Var}(X) = E[X^2] - \{E[X]\}^2 \geqslant 0. \quad \blacksquare$$

Theorem A.3.4. Cramér–Rao Inequality. *Let $\mathbf{Z}=(Z_1,\ldots,Z_N)$ denote a vector having a continuous distribution with joint density $h(\mathbf{z};\theta)$. Define the information in $\mathbf{Z}$ at θ to be*

$$I_{\mathbf{Z}}^*(\theta) = E\left[\left\{\frac{\partial}{\partial\theta}\ln[h(\mathbf{Z};\theta)]\right\}^2\right].$$

Then, if $V(Z_1,\ldots,Z_N)$ is any function of $Z_1,\ldots,Z_N$ with finite variance,

$$\frac{\left\{\dfrac{d}{d\theta}E_\theta(V)\right\}^2}{\mathrm{Var}_\theta(V)} \leqslant I_{\mathbf{Z}}^*(\theta).$$

Proof: See, for example, Lindgren (1976), page 261. ■

Theorem A.3.5. Helly–Bray Theorem. *Let $G(\cdot)$ and $\{G_n(\cdot)\}$, $n = 1,2,\cdots$ denote distribution functions such that $\lim_{n\to\infty} G_n(x) = G(x)$ at all x-values where $G(\cdot)$ is continuous. If $t(x)$ is continuous and bounded over the real line, then*

$$\lim_{n\to\infty}\int t(x)\,dG_n(x) = \int t(x)\,dG(x).$$

Proof: See, for example, Gnedenko (1962), page 267. ■

Theorem A.3.6. *Let $G(\cdot)$ denote the c.d.f. of a continuous distribution, and let $\{G_n(x)\}$ represent a sequence of c.d.f.'s satisfying $\lim_{n\to\infty} G_n(x) = G(x)$ for all x. Suppose $\{t_n(x)\}$, $n = 1, 2, \cdots$ is a sequence of functions such that $|t_n(x)| < M$ for all x and n, where M is a positive constant, and such that $\lim_{n\to\infty} t_n(x) = t(x)$ for all x, where $t(x)$ is some real-valued function. Then*

$$\lim_{n\to\infty} \int t_n(x)\,dG_n(x) = \int t(x)\,dG(x).$$

Proof: This result is a special case of Proposition 18 on page 232 of Royden (1968). ■

The following version of the central limit theorem includes the possibility that each of the i.i.d. random variables $X_{i:n}$ has a distribution that depends on n.

Theorem A.3.7. A Central Limit Theorem. *For $n = 1, 2, \cdots$ let $X_{1:n}, \ldots, X_{n:n}$ denote independent and identically distributed random variables, each with mean μ_n and finite variance σ_n^2. Assume also that $Y_{1:n} = (X_{1:n} - \mu_n)/\sigma_n$ has a limiting distribution which has mean zero and variance 1. Then*

$$\frac{\sum_{i=1}^{n} (X_{i:n} - \mu_n)}{\sqrt{n\sigma_n^2}}$$

has a limiting standard normal distribution.

Proof: We prove this theorem by showing that it satisfies the conditions of Theorem A.3.8. Let $G_n(x)$ denote the c.d.f. of $(X_{1:n} - \mu_n)/\sigma_n$ and suppose $G_n(x)$ converges to the c.d.f. $G(x)$ at all continuity points of $G(\cdot)$. To satisfy the conditions of Theorem A.3.8 we need to show that for every $\epsilon > 0$

$$\int_{|z| > \epsilon\sqrt{n}} z^2\,dG_n(z) \to 0$$

as $n \to \infty$. Since $G(\cdot)$ is the c.d.f. of a distribution with mean 0 and variance 1, for any δ satisfying $0 < \delta < 1$ there exists finite constants $c > 0$ and $d > 0$ such that

$$1 \geqslant \int_{-c}^{d} z^2\,dG(z) \geqslant 1 - \delta,$$

and such that $-c$ and d are both continuity points of $G(\cdot)$. Define

$$\begin{aligned} t(z) &= -c, & &\text{if } z \leqslant -c \\ &= z, & &\text{if } -c < z < d \\ &= d, & &\text{if } z \geqslant d. \end{aligned}$$

By Theorem A.3.5 we see that

$$\begin{aligned} \int t^2(z)\,dG_n(z) &= c^2G_n(-c) + d^2[1-G_n(d)] \\ &\quad + \int_{-c}^{d} z^2\,dG_n(z) \to \int t^2(z)\,dG(z) \\ &= c^2G(-c) + d^2[1-G(d)] + \int_{-c}^{d} z^2\,dG(z) \end{aligned}$$

as $n \to \infty$. Since $-c$ and d are continuity points of $G(\cdot)$, it follows that

$$\lim_{n\to\infty} \int_{-c}^{d} z^2\,dG_n(z) = \int_{-c}^{d} z^2\,dG(z) \geqslant 1-\delta.$$

Since $G_n(\cdot)$ is the c.d.f. of a distribution with mean zero and variance 1, it follows that

$$\lim_{n\to\infty} \int_{|z|>\epsilon\sqrt{n}} z^2\,dG_n(z) \leqslant \delta.$$

This $\delta > 0$ was arbitrary, so the proof is completed. ■

Theorem A.3.8. Lindeberg Central Limit Theorem. *For each n let $X_{1:n},\ldots,X_{n:n}$ be independent random variables with $X_{i:n}$ having c.d.f. $F_{i:n}(\cdot)$, mean $\mu_{i:n}$, and finite, positive variance $\sigma^2_{i:n}$, for $i=1,\ldots,n$. Then*

$$\frac{\sum_{i=1}^{n} (X_{i:n} - \mu_{i:n})}{\sigma_n}$$

has a limiting standard normal distribution, if for all $\epsilon > 0$,

$$\lim_{n\to\infty} \frac{1}{\sigma_n^2} \sum_{i=1}^{n} \int_{|x-\mu_{i:n}|>\epsilon\sigma_n} (x-\mu_{i:n})^2\,dF_{i:n}(x) = 0, \qquad \textbf{(A.3.9)}$$

where $\sigma_n^2 = \sum_{i=1}^{n} \sigma^2_{i:n}$.

Proof: See Gnedenko (1962), pages 321–2, where $\xi_{ni}=(X_{i:n}-\mu_{i:n})/\sigma_n$. ■

Theorem A.3.10. Liapounov Central Limit Theorem. *For each n let $X_{1:n},\ldots,X_{n:n}$ denote a sequence of n independent random variables where $X_{i:n}$ has mean $\mu_{i:n}$ and finite, positive variance $\sigma_{i:n}^2$, for $i=1,\ldots,n$. If there exists a $\delta>0$ such that*

$$\lim_{n\to\infty}\frac{\sum_{i=1}^{n}E\left[|X_{i:n}-\mu_{i:n}|^{2+\delta}\right]}{\sigma_n^{2+\delta}}=0, \qquad \textbf{(A.3.11)}$$

then

$$\frac{\sum_{i=1}^{n}(X_{i:n}-\mu_{i:n})}{\sigma_n}$$

has a limiting standard normal distribution, where $\sigma_n^2=\sum_{i=1}^{n}\sigma_{i:n}^2$.

Proof: The proof follows from Theorem A.3.8 and the fact that the Liapounov condition (A.3.11) implies the Lindeberg condition (A.3.9). See, for example, Gnedenko (1962), page 294. ■

Lemma A.3.12. *Suppose that $\{F_N(t)\}$ is a sequence of c.d.f.'s such that*

$$\operatorname*{limit}_{N\to\infty} F_N(t)=H(t)$$

for all t, where $H(\cdot)$ is the c.d.f. of a continuous random variable with support equal to some interval on the real line. Let $\{t_N\}$ be a sequence of real numbers. Then, for $0<\beta<1$,

$$\lim_{N\to\infty} F_N(t_N)=1-\beta$$

if and only if

$$\lim_{N\to\infty} t_N=h_\beta$$

where h_β is the unique value satisfying $H(h_\beta)=1-\beta$.

Proof: (a) Suppose $\lim_{N\to\infty}t_N=h_\beta$, and let $\epsilon>0$ be given. By the absolute continuity of $H(\cdot)$, there exists a $\delta>0$, such that

$$|H(x)-H(y)|<\epsilon/2 \text{ whenever } |x-y|<\delta.$$

Let N^* be such that $N > N^*$ implies $|t_N - h_\beta| < \delta/2$, $|F_N(h_\beta - \frac{\delta}{2}) - H(h_\beta - \frac{\delta}{2})| < \epsilon/2$ and $|F_N(h_\beta + \frac{\delta}{2}) - H(h_\beta + \frac{\delta}{2})| < \epsilon/2$. It follows, since $F_N(\cdot)$ and $H(\cdot)$ are c.d.f.'s, that for $N > N^*$ we have

$$F_N(t_N) \geqslant F_N\left(h_\beta - \tfrac{\delta}{2}\right) \geqslant H\left(h_\beta - \tfrac{\delta}{2}\right) - (\epsilon/2) \geqslant H(h_\beta) - \epsilon$$

and

$$F_N(t_N) \leqslant F_N\left(h_\beta + \tfrac{\delta}{2}\right) \leqslant H\left(h_\beta + \tfrac{\delta}{2}\right) + (\epsilon/2) \leqslant H(h_\beta) + \epsilon.$$

Thus

$$\lim_{N \to \infty} F_N(t_N) = H(h_\beta) = 1 - \beta.$$

(b) Assume $\lim_{N\to\infty} F_N(t_N) = 1 - \beta$, but $\lim_{N\to\infty} t_N \neq h_\beta$. Then there exists a $\delta > 0$ such that $|t_N - h_\beta| > \delta$ infinitely often. Choose $\epsilon > 0$ so that

$$\frac{3\epsilon}{2} \leqslant \min\left\{ H(h_\beta) - H(h_\beta - \delta), H(h_\beta + \delta) - H(h_\beta)\right\}.$$

Let N^* be such that $N > N^*$ implies

$$|F_N(h_\beta - \delta) - H(h_\beta - \delta)| < \epsilon/2 \text{ and } |F_N(h_\beta + \delta) - H(h_\beta + \delta)| < \epsilon/2.$$

If $t_N < h_\beta - \delta$ and $N > N^*$, then

$$F_N(t_N) \leqslant F_N(h_\beta - \delta) \leqslant H(h_\beta - \delta) + (\epsilon/2) \leqslant H(h_\beta) - \epsilon$$

and if $t_N > h_\beta + \delta$ and $N > N^*$, then

$$F_N(t_N) \geqslant F_N(h_\beta + \delta) \geqslant H(h_\beta + \delta) - (\epsilon/2) \geqslant H(h_\beta) + \epsilon.$$

Thus

$$|F_N(t_N) - (1 - \beta)| \geqslant \epsilon$$

occurs infinitely often, which violates the assumption, and the result follows. ■

Theorem A.3.13. Slutsky's Theorem. *Let $\{W_n\}$ be a sequence of random variables having a limiting distribution with c.d.f. $F(w)$. Let $\{X_n\}$ denote a sequence of random variables that converge in probability to the constant c. Then the first and second members of each pair listed below have* *the same limiting distribution*:

$$\begin{array}{llll} (a) & (W_n+X_n) & \text{and} & (W_n+c) \\ (b) & X_nW_n & \text{and} & cW_n \\ (c) & W_n/X_n & \text{and} & W_n/c, \quad \text{if } c\neq 0. \end{array}$$

Proof: We prove only part (c), the proofs of the other parts proceeding similarly. Without loss of generality assume $c>0$ and note that

$$G_n(t) = P[(W_n/c)\leqslant t] = P[W_n\leqslant ct]\to F(ct),$$

as $n\to\infty$ for every t such that ct is a continuity point of $F(\cdot)$. Now let t_0 denote an arbitrary nonnegative value such that ct_0 is a continuity point of $F(ct)$. For any given ϵ satisfying $0<\epsilon<c$, let ϵ_1 and ϵ_2 be chosen so that $0<\epsilon_1<\epsilon$ and $(c+\epsilon_1)t_0$ is a continuity point of $F(\cdot)$, and so that $0<\epsilon_2<\epsilon$ and $(c-\epsilon_2)t_0$ is a continuity point of $F(\cdot)$. Define

$$S_n = \{(X_n, W_n)|(W_n/X_n)\leqslant t_0\}$$

$$S_{1n} = \{(X_n, W_n)|(W_n/X_n)\leqslant t_0 \text{ and } -\epsilon_2<X_n-c<\epsilon_1\}$$

$$S_{2n} = \{(X_n, W_n)|(W_n/X_n)\leqslant t_0 \text{ and either } (X_n-c)\geqslant\epsilon_1 \text{ or } (X_n-c)\leqslant-\epsilon_2\}$$

and

$$S_n^* = \{(X_n, W_n)||X_n-c|\geqslant\epsilon^*\},$$

where $\epsilon^*=\min\{\epsilon_1,\epsilon_2\}$. Since S_{2n} is a subset of S_n^*, $P(S_{2n})\to 0$ as $n\to\infty$ by the fact that X_n converges to c in probability. Also,

$$P[W_n\leqslant(c-\epsilon_2)t_0, -\epsilon_2<X_n-c<\epsilon_1] \leqslant P(S_{1n}) \leqslant P[W_n\leqslant(c+\epsilon_1)t_0, -\epsilon_2<X_n-c<\epsilon_1].$$

Since $\{(X_n, W_n)|W_n\leqslant(c+\epsilon_1)t_0$ and either $(X_n-c)\geqslant\epsilon_1$ or $(X_n-c)\leqslant-\epsilon_2\}$ is also a subset of S_N^*, its probability goes to zero as well. Hence

$$P[W_n\leqslant(c+\epsilon_1)t_0, -\epsilon_2<X_n-c<\epsilon_1]\to F((c+\epsilon_1)t_0)$$

as $n\to\infty$. A similar argument for the lower bound then shows that

$$F((c-\epsilon)t_0) \leqslant F((c-\epsilon_2)t_0) \leqslant \lim_{n\to\infty} P(S_{1n}) = \lim_{n\to\infty} P(S_n) \leqslant F((c+\epsilon_1)t_0) \leqslant F((c+\epsilon)t_0).$$

Letting $\epsilon \to 0$ yields the desired result. A similar argument holds for $t_0 < 0$ and for $c < 0$. ■

Theorem A.3.14. Berry–Esséen Theorem. *For each n let $V_{1:n}, \ldots, V_{n:n}$ denote a random sample from a distribution with mean 0 and variance 1. Also, assume $\lim_{n\to\infty} n^{-1/2} E[|V_{1:n}|^3] = 0$. Then for all x and n*

$$|P(\sqrt{n}\,\overline{V}_n \leq x) - \Phi(x)| \leq A_0 \frac{E[|V_{1:n}|^3]}{\sqrt{n}},$$

where A_0 is some constant (independent of n), $\overline{V}_n = n^{-1}\sum_{i=1}^n V_{i:n}$ and $\Phi(x)$ is the c.d.f. for the standard normal distribution.

Proof: See Chung (1968), pages 206–11. ■

Theorem A.3.15. *Let $\{S_n\}$ and $\{T_n\}$ be independent sequences of random variables such that, as $n\to\infty$,*

(i) S_n has a limiting distribution that is normal with mean 0 and variance σ_1^2 (σ_1^2 does not depend on n)

and

(ii) T_n has a limiting distribution that is normal with mean 0 and variance σ_2^2 (σ_2^2 does not depend on n).

Then $S_n + T_n$ has a limiting distribution that is normal with mean 0 and variance $\sigma_1^2 + \sigma_2^2$.

Proof: Let $\phi_{S_n}(t)$ and $\phi_{T_n}(t)$ be the characteristic functions for the variables S_n and T_n, respectively. Then from (i) and (ii) we know [see, e.g., Loève (1963), page 191] that

$$\lim_{n\to\infty} \phi_{S_n}(t) = \exp(-t^2\sigma_1^2/2), \qquad -\infty < t < \infty$$

and

$$\lim_{n\to\infty} \phi_{T_n}(t) = \exp(-t^2\sigma_2^2/2), \qquad -\infty < t < \infty.$$

Set $V_n = S_n + T_n$ and let $\phi_{V_n}(t) = E[e^{itV_n}]$ be the characteristic function for V_n. Since S_n and T_n are independent, we have

$$\phi_{V_n}(t) = E[e^{itS_n}]E[e^{itT_n}] = \phi_{S_n}(t)\phi_{T_n}(t), \qquad -\infty < t < \infty,$$

which implies that

$$\lim_{n\to\infty} \phi_{V_n}(t) = \lim_{n\to\infty} \phi_{S_n}(t) \lim_{n\to\infty} \phi_{T_n}(t) = \exp(-t^2\sigma_1^2/2)\exp(-t^2\sigma_2^2/2)$$
$$= \exp(-t^2(\sigma_1^2+\sigma_2^2)/2), \qquad -\infty < t < \infty.$$

Since this limit is the characteristic function for a normal distribution with mean 0 and variance $\sigma_1^2+\sigma_2^2$, the proof is complete. ■

A.4. Mathematical Results

Lemma A.4.1. Fatou's Lemma. *Let $\{s_k(u)\}$ denote a sequence of nonnegative functions such that*

$$\lim_{k\to\infty} s_k(u) = s(u)$$

for all but at most a countable number of u-values in (0, 1). *Then*

$$\int_0^1 s(u)\,du \leqslant \liminf_{k\to\infty} \int_0^1 s_k(u)\,du.$$

Proof: See, for example, Royden (1968), page 226. ■

Lemma A.4.2. *Let a and b denote nonnegative constants. If $\{a_k\}$ and $\{b_k\}$ are sequences of nonnegative constants satisfying*

(i) $\limsup_{k\to\infty} \{a_k+b_k\} \leqslant (a+b)$

(ii) $\liminf_{k\to\infty} \{a_k\} \geqslant a$ *and* $\liminf_{k\to\infty} \{b_k\} \geqslant b$,

then

$$\lim_{k\to\infty} a_k = a \text{ and } \lim_{k\to\infty} b_k = b.$$

Proof: Assume $\limsup_{k\to\infty}\{a_k\}>a$. Then there exists an $\epsilon>0$ such that $a_k>a+\epsilon$ infinitely often. Also, $\liminf_{k\to\infty}\{b_k\}\geqslant b$ implies there exists a K such that $k>K$ yields

$$b_k > b - \frac{\epsilon}{2}.$$

Therefore, $a_k + b_k > a + b + (\epsilon/2)$ infinitely often. Hence, $\limsup_{k\to\infty}\{a_k + b_k\} > a + b$. This contradicts (i), and it follows that $\limsup_{k\to\infty}\{a_k\} \leqslant a$ and hence that

$$\lim_{k\to\infty} a_k = a.$$

A similar argument holds for $\{b_k\}$. ■

Lemma A.4.3. Hölder's Inequality. *If p and q are positive real numbers such that $(1/p)+(1/q)=1$, then*

$$\int |f(x)g(x)|\,dx \leqslant \left[\int |f(x)|^p\,dx\right]^{1/p} \cdot \left[\int |g(x)|^q\,dx\right]^{1/q},$$

provided that both integrals on the right are finite.

Proof: See, for example, Royden (1968), page 113. ■

Theorem A.4.4. (Hájek and Šidák (1967)). *Let $\{t_k(u)\}$ denote a sequence of square integrable functions defined on $(0,1)$ that converges to a square integrable function $t(u)$ for all but at most a countable number of u-values in $(0,1)$. In addition, assume that*

$$\limsup_{k\to\infty} \int_0^1 t_k^2(u)\,du \leqslant \int_0^1 t^2(u)\,du. \tag{A.4.5}$$

Then

(i) $$\lim_{k\to\infty} \int_0^1 t_k^2(u)\,du = \int_0^1 t^2(u)\,du$$

(ii) $$\lim_{k\to\infty} \int_0^1 t_k(u)t(u)\,du = \int_0^1 t^2(u)\,du$$

and (iii) $$\lim_{k\to\infty} \int_0^1 [t_k(u) - t(u)]^2\,du = 0.$$

Proof: (i) Fatou's Lemma A.4.1 implies that

$$\int_0^1 t^2(u)\,du \leqslant \liminf_{k\to\infty} \int_0^1 t_k^2(u)\,du,$$

which when combined with A.4.5 proves part (i).

(ii) Define the positive and negative parts of $t_k(u)t(u)$ as follows:

$$s_k^+(u) = t_k(u)t(u), \text{ if } t_k(u)t(u) \geqslant 0$$
$$= 0, \text{ otherwise},$$

and

$$s_k^-(u) = -t_k(u)t(u), \text{ if } t_k(u)t(u) \leqslant 0$$
$$= 0, \text{ otherwise},$$

so that

$$t_k(u)t(u) = s_k^+(u) - s_k^-(u).$$

From the convergence of $t_k(\cdot)$ to $t(\cdot)$ we see that both $\lim_{k\to\infty} s_k^+(u) = t^2(u)$ and $\lim_{k\to\infty} s_k^-(u) = 0$ hold at all but at most a countable number of u-values in $(0,1)$. Hölder's inequality A.4.3 shows that

$$\int_0^1 |t_k(u)t(u)|\,du \leqslant \left[\int_0^1 t_k^2(u)\,du \int_0^1 t^2(u)\,du\right]^{1/2},$$

and thus by part (i),

$$\limsup_{k\to\infty}\left\{\int_0^1 s_k^+(u)\,du + \int_0^1 s_k^-(u)\,du\right\} \leqslant \int_0^1 t^2(u)\,du.$$

Moreover, Fatou's lemma A.4.1 yields

$$\int_0^1 t^2(u)\,du \leqslant \liminf_{k\to\infty}\int_0^1 s_k^+(u)\,du \text{ and } 0 \leqslant \liminf_{k\to\infty}\int_0^1 s_k^-(u)\,du.$$

Hence Lemma A.4.2 shows that

$$\lim_{k\to\infty}\int_0^1 s_k^+(u)\,du = \int_0^1 t^2(u)\,du \text{ and } \lim_{k\to\infty}\int_0^1 s_k^-(u)\,du = 0,$$

which in turn implies that

$$\lim_{k\to\infty}\int_0^1 t_k(u)t(u)\,du = \lim_{k\to\infty}\int_0^1 \{s_k^+(u) - s_k^-(u)\}\,du = \int_0^1 t^2(u)\,du.$$

Part (iii) follows directly from parts (i) and (ii). ■

A.4.6. *Sums of Powers of Integers*

$$\sum_{k=1}^{n} k = \frac{n(n+1)}{2}$$

$$\sum_{k=1}^{n} k^2 = \frac{n(n+1)(2n+1)}{6}$$

$$\sum_{k=1}^{n} k^3 = \frac{n^2(n+1)^2}{4}$$

$$\sum_{k=1}^{n} k^4 = \frac{n}{30}(n+1)(2n+1)(3n^2+3n-1)$$

$$\sum_{k=1}^{n} k^5 = \frac{n^2}{12}(n+1)^2(2n^2+2n-1)$$

$$\sum_{k=1}^{n} k^6 = \frac{n}{42}(n+1)(2n+1)(3n^4+6n^3-3n+1)$$

BIBLIOGRAPHY

Adichie, J. N. (1967). Estimates of regression parameters based on rank tests. *Ann. Math. Statist.* **38**, 894–904.

Anderson, T. W. and Goodman, L. A. (1957). Statistical inference about Markov chains. *Ann. Math. Statist.* **28**, 89–109.

Andrews, D. F. (1974). A robust method for multiple linear regression. *Technometrics* **16**, 523–31.

Andrews, D. F., Bickel, P. J., Hampel, F. R., Huber, P. J., Rogers, W. H., and Tukey, J. W. (1972). *Robust Estimates of Location: Survey and Advances.* Princeton University Press, Princeton, New Jersey.

Andrews, F. C. (1954). Asymptotic behavior of some rank tests for analysis of variance. *Ann. Math. Statist.* **25**, 724–36.

Ansari, A. R. and Bradley, R. A. (1960). Rank-sum tests for dispersions. *Ann. Math. Statist.* **31**, 1174–89.

Antoniak, C. E. (1974). Mixtures of Dirichlet processes with applications to Bayesian nonparametric problems. *Ann. Statist.* **2**, 1152–74.

Archambault, W. A. T., Jr., Mack, G. A., and Wolfe, D. A. (1977). *K*-sample rank tests using pair-specific scoring functions. *Can. J. Statist.* **5**, 195–207.

Arnold, H. (1965). Small sample power for the one-sample Wilcoxon test for nonnormal shift alternatives. *Ann. Math. Statist.* **36**, 1767–78.

Bahadur, R. R. (1960a). Asymptotic efficiency of tests and estimators. *Sankhyā* **20**, 229–52.

Bahadur, R. R. (1960b). Stochastic comparison of tests. *Ann. Math. Statist.* **31**, 276–95.

Bahadur, R. R. (1967). Rates of convergence of estimates and test statistics. *Ann. Math. Statist.* **38**, 303–24.

Barry, J. (1968). General and comparative study of the psychokinetic effect on a fungus culture. *J. Parapsychology* **32**, 237–43.

Barton, D. E. and Mallows, C. L. (1965). Some aspects of the random sequence. *Ann. Math. Statist.* **36**, 236–60.

Beaton, A. E. and Tukey, J. W. (1974). The fitting of power series, meaning polynomials, illustrated on band-spectroscopic data. *Technometrics* **16**, 147–85.

Bell, C. B., Moser, J. M., and Thompson, R. (1966). Goodness criteria for two-sample distribution-free tests. *Ann. Math. Statist.* **37**, 133–42.

Beran, R. (1974). Asymptotically efficient adaptive rank estimates in location models. *Ann. Math. Statist.* **2**, 63–74.

Bick, R. L., Adams, T., and Schmalhorst, W. R. (1976). Bleeding times, platelet adhesion, and aspirin. *J. Clin. Path.* **65**, 69–72.

Bickel, P. J. (1967). Some contributions to the theory of order statistics. *Proc. Fifth Berkeley Symp.* **1**, 575–91.

Bickel, P. J. and Doksum, K. A. (1977). *Mathematical Statistics: Basic Ideas and Selected Topics.* Holden-Day, San Francisco.

Birnbaum, A. and Miké, V. (1970). Asymptotically robust estimators of location. *J. Amer. Statist. Assoc.* **65**, 1265–82.

Blum, J. R. and Weiss, L. (1957). Consistency of certain two-sample tests. *Ann. Math. Statist.* **28**, 242–6.

Blumenthal, S. (1963). The asymptotic normality of two test statistics associated with the two-sample problem. *Ann. Math. Statist.* **34**, 1513–23.

Boos, D. D. and Serfling, R. J. (1976). Development and comparison of *M*-estimators for location on the basis of the asymptotic variance functional. Tech. Rept. 108, Dept. of Statistics, Florida State University, Tallahassee, Florida.

Box, G. E. P. and Muller, M. E. (1958). A note on the generation of random normal deviates. *Ann. Math. Statist.* **29**, 610–1.

Broffitt, J. D., Randles, R. H., and Hogg, R. V. (1976). Distribution-free partial discriminant analysis. *J. Amer. Statist. Assoc.* **71**, 934–9.

Campbell, G. and Hollander, M. (1978). Rank order estimation with the Dirichlet prior. *Ann. Statist.* **6**, 142–53.

Capon, J. (1961). Asymptotic efficiency of certain locally most powerful rank tests. *Ann. Math. Statist.* **32**, 88–100.

Chernoff, H., Gastwirth, J. L., and Johns, M. V. (1967). Asymptotic distribution of linear combinations of functions of order statistics with applications to estimation. *Ann. Math. Statist.* **38**, 52–72.

Chernoff, H. and Savage, I. R. (1958). Asymptotic normality and efficiency of certain nonparametric test statistics. *Ann. Math. Statist.* **29**, 972–94.

Chung, K. L. (1963). *A Course in Probability Theory.* Harcourt, New York.

Cochran, W. G. (1952). The χ^2 test of goodness of fit. *Ann. Math. Statist.* **23**, 315–45.

Collins, J. R. (1976). Robust estimation of a location parameter in the presence of asymmetry. *Ann. Statist.* **4**, 68–85.

Conover, W. J. (1971). *Practical Nonparametric Statistics.* Wiley, New York.

Cramér, H. (1928). On the composition of elementary errors. *Skand. Aktuarietids* **11**, 13–74, 141–80.

Cramér, H. (1946). *Mathematical Methods of Statistics.* Princeton University Press, Princeton, New Jersey.

Crouse, C. F. (1964). Note on Mood's test. *Ann. Math. Statist.* **35**, 1825–6.

Darling, D. A. (1957). The Kolmogorov–Smirnov, Cramér–von Mises tests. *Ann. Math. Statist.* **28**, 823–38.

David, H. A. (1970). *Order Statistics.* Wiley, New York.

Dixon, W. J. (1940). A criterion for testing the hypothesis that two samples are

from the same population. *Ann. Math. Statist.* **11**, 199–204.
Doksum, K. A. (1972). Decision theory for some nonparametric models. *Proc. Sixth Berkeley Symp.* **1**, 331–41.
Dudewicz, E. J. (1976). *Introduction to Statistics and Probability*. Holt, New York.
Dupač, V. and Hájek, J. (1969). Asymptotic normality of simple linear rank statistics under alternatives II. *Ann. Math. Statist.* **40**, 1992–2017.
Duran, B. S. (1976). A survey of nonparametric tests for scale. *Comm. Statist.—Theor. Methods* **A5**, 1287–1312.
Feller, W. (1968). *An Introduction to Probability Theory and its Applications, Vol. I.* 3rd ed. Wiley, New York.
Ferguson, T. S. (1967). *Mathematical Statistics—A Decision Theoretic Approach.* Academic Press, New York.
Ferguson, T. S. (1973). A Bayesian analysis of some nonparametric problems. *Ann. Statist.* **1**, 209–30.
Ferguson, T. S. (1974). Prior distributions on the space of probability measures. *Ann. Statist.* **2**, 615–29.
Filippova, A. A. (1962). Mises' theorem on the asymptotic behavior of functionals of empirical distribution functions and its statistical applications. *Theor. Prob. App.* **7**, 24–57.
Fisher, R. A. (1935). *The Design of Experiments*. Oliver & Boyd, Edinburgh.
Fisher, R. A. and Yates, F. (1938). *Statistical Tables for Biological, Agricultural and Medical Research*, 1st ed. Oliver & Boyd, Edinburgh.
Fligner, M. A. and Hettmansperger, T. P. (1979). On the use of conditional asymptotic normality. To appear in *J. Roy. Statist. Assoc. Ser. B* **41**.
Fligner, M. A., Hogg, R. V., and Killeen, T. J. (1976). Some distribution-free rank-like statistics having the Mann-Whitney-Wilcoxon null distribution. *Comm. Statist.—Theor. Methods* **A5**, 573–6.
Fligner, M. A. and Killeen, T. J. (1976). Distribution-free two-sample tests for scale. *J. Amer. Statist. Assoc.* **71**, 210–3.
Fligner, M. A. and Wolfe, D. A. (1976). Some applications of sample analogues to the probability integral transformation and a coverage property. *Amer. Statistician* **30**(2), 78–85.
Fligner, M. A. and Wolfe, D. A. (1979). Nonparametric prediction intervals for a future sample median. To appear in *J. Amer. Statist. Assoc.* **74**.
Fligner, M. A. and Wolfe, D. A. (1978). Distribution-free treatments versus control tests. Tech. Rept., Department of Statistics, Ohio State University, Columbus, Ohio.
Flores, A. M. and Zohman, L. R. (1970). Energy cost of bedmaking to the cardiac patient and the nurse. *Amer. J. Nurs.* **70**, 1264–7.
Fraser, D. A. S. (1957a). *Nonparametric Methods in Statistics*. Wiley, New York.
Fraser, D. A. S. (1957b). Most powerful rank-type tests. *Ann. Math. Statist.* **28**, 1040–3.
Friedman, M. (1937). The use of ranks to avoid the assumption of normality implicit in the analysis of variance. *J. Amer. Statist. Assoc.* **217**, 929–32.
Gastwirth, J. L. (1965). Percentile modifications of two-sample rank tests. *J. Amer. Statist. Assoc.* **60**, 1127–41.

Gibbons, J. D. (1971). *Nonparametric Statistical Inference*. McGraw-Hill, New York.

Gibbons, J. D. (1976). *Nonparametric Methods in Quantitative Analysis*. Holt, New York.

Gnedenko, B. V. (1962). *The Theory of Probability* 2nd ed. Chelsea Publishing, New York.

Godwin, H. J. (1949). Some low moments of order statistics. *Ann. Math. Statist*. **20**, 279–85.

Goldstein, M. (1975). A note on some Bayesian nonparametric estimates. *Ann. Statist*. **2**, 615–29.

Govindarajulu, Z. (1960). Central limit theorems and asymptotic efficiency for one-sample nonparametric procedures. Tech. Rept. 11, Department of Statistics, University of Minnesota, Minneapolis, Minnesota.

Govindarajulu, Z. (1968). Distribution-free confidence bounds for $P(X<Y)$. *Ann. Inst. Statist. Math*. **20**, 229–38.

Govindarajulu, Z. and Eisenstat, S. (1965). Best estimates of location and scale parameters of a chi distribution using ordered observations. *Nippon Kagaku Gijutus*. **12**, 149–64.

Govindarajulu, Z., LeCam, L., and Raghavachari, M. (1966). Generalizations of theorems of Chernoff and Savage on the asymptotic normality of test statistics. *Proc. Fifth Berkeley Symp*. **1**, 609–38.

Gross, S. (1966). Nonparametric tests when nuisance parameters are present. Ph.D. thesis, University of California, Berkeley.

Guenther, W. C. (1975). The inverse hypergeometric—a useful model. *Statist. Neerl*. **29**, 129–44.

Gupta, M. K. (1967). Asymptotically nonparametric tests of symmetry. *Ann. Math. Statist*. **38**, 849–66.

Haga, T. (1960). A two-sample rank test on location. *Ann. Inst. Statist. Math*. **11**, 211–9.

Hájek, J. (1961). Some extensions of the Wald–Wolfowitz–Noether theorem. *Ann. Math. Statist*. **32**, 506–23.

Hájek, J. (1962). Asymptotically most powerful rank order tests. *Ann. Math. Statist*. **33**, 1124–47.

Hájek, J. (1968). Asymptotic normality of simple linear rank statistics under alternatives. *Ann. Math. Statist*. **39**, 325–46.

Hájek, J. (1969). *Nonparametric Statistics*. Holden-Day, San Francisco.

Hájek, J. (1970). Miscellaneous problems of rank tests. *Nonparametric Techniques in Statistical Inference*. (Ed. by M. L. Puri) Cambridge University Press.

Hájek, J. and Šidák, Z. (1967). *Theory of Rank Tests*. Academic Press, New York.

Hampel, F. R. (1968). Contributions to the theory of robust estimation. Ph.D. thesis, University of California, Berkeley.

Hampel, F. R. (1971). A general qualitative definition of robustness. *Ann. Math. Statist*. **42**, 1887–96.

Hampel, F. R. (1974). The influence curve and its role in robust estimation. *J. Amer. Statist. Assoc*. **69**, 383–93.

Harter, H. L. (1961). Expected values of normal order statistics. *Biometrika* **48**,

151–65.

Hayman, G. E. and Govindarajulu, Z. (1966). Exact power of the Mann-Whitney tests for exponential and rectangular alternatives. *Ann. Math. Statist.* **37**, 945–53.

Hettmansperger, T. P. and Utts, J. M. (1977). Robustness properties for simple class of rank estimates. Tech. Rept., Dept. of Statistics, Pennsylvania State University, State College, Pennsylvania.

Hodges, J. L., Jr. and Lehmann, E. L. (1956). The efficiency of some nonparametric competitors of the *t*-test. *Ann. Math. Statist.* **27**, 324–35.

Hodges, J. L., Jr. and Lehmann, E. L. (1963). Estimates of location based on rank tests. *Ann. Math. Statist.* **34**, 598–611.

Hodges, J. L., Jr. and Lehmann, E. L. (1970). Deficiency. *Ann. Math. Statist.* **41**, 783–801.

Hoeffding, W. (1948). A class of statistics with asymptotically normal distribution. *Ann. Math. Statist.* **19**, 293–325.

Hoeffding, W. (1951). Optimum nonparametric tests. *Proc. Second Berkeley Symp.* 83–92.

Hogg, R. V. (1960). Certain uncorrelated statistics. *J. Amer. Statist. Assoc.* **55**, 265–7.

Hogg, R. V. (1967). Some observations on robust estimation. *J. Amer. Statist. Assoc.* **62**, 1179–86.

Hogg, R. V. (1974). Adaptive robust procedures: a partial review and some suggestions for future applications and theory. *J. Amer. Statist. Assoc.* **69**, 909–23.

Hogg, R. V. (1976). A new dimension to nonparametric tests. *Comm. Statist.—Theor. Methods* **A5**, 1313–26.

Hogg, R. V. and Craig, A. T. (1978). *Introduction to Mathematical Statistics*. 4th ed. Macmillan, New York.

Hogg, R. V., Fisher, D. M., and Randles, R. H. (1975). A two-sample adaptive distribution-free test. *J. Amer. Statist. Assoc.* **70**, 656–61.

Hogg, R. V. and Randles, R. H. (1975). Adaptive distribution-free regression methods. *Technometrics* **17**, 399–408.

Hollander, M. (1967). Asymptotic efficiency of two nonparametric competitors of Wilcoxon's two sample test. *J. Amer. Statist. Assoc.* **62**, 939–49.

Hollander, M. (1968). Certain uncorrelated nonparametric test statistics. *J. Amer. Statist. Assoc.* **63**, 707–14.

Hollander, M. (1970). A distribution-free test for parallelism. *J. Amer. Statist. Assoc.* **65**, 387–94.

Hollander, M. and Proschan, F. (1972). Testing whether new is better than used. *Ann. Math. Statist.* **43**, 1136–46.

Hollander, M. and Wolfe, D. A. (1973). *Nonparametric Statistical Methods*. Wiley, New York.

Huber, P. J. (1964). Robust estimation of a location parameter. *Ann. Math. Statist.* **35**, 73–101.

Huber, P. J. (1972). Robust statistics: a review. *Ann. Math. Statist.* **43**, 1041–67.

Huber, P. J. (1973). Robust regression: asymptotics, conjectures and Monte Carlo. *Ann. Statist.* **1**, 799–821.

Huber, P. J. (1977). Robust covariances. In *Statistical Decision Theory and Related Topics II*, S. S. Gupta and D. S. Moore, Eds. Academic Press, New York.

Iman, R. L., Quade, D., and Alexander, D. A. (1975). Exact probability levels for the Kruskal–Wallis test. In *Selected Tables in Mathematical Statistics*, Vol. III, H. L. Harter and D. B. Owen, Eds., pp. 329–84. Markham, Chicago.

Jaeckel, L. A. (1971). Some flexible estimates of location. *Ann. Math. Statist.* **42**, 1540–52.

Jonckheere, A. R. (1954). A distribution-free k-sample test against ordered alternatives. *Biometrika* **41**, 133–45.

Jones, D. H. (1976). An efficient adaptive distribution-free test. Tech. Rept., Dept. of Statistics, Rutgers University, New Brunswick, New Jersey.

Kamat, A. R. (1956). A two-sample distribution-free test. *Biometrika* **43**, 377–87.

Kendall, M. G. (1938). A new measure of rank correlation. *Biometrika* **30**, 81–93.

Kendall, M. G. (1948). *Rank Correlation Methods*. 1st ed. Griffin, London.

Kendall, M. G. and Babington Smith, B. (1939). The problem of m rankings. *Ann. Math. Statist.* **10**, 275–87.

Klotz, J. (1962). Nonparametric tests for scale. *Ann. Math. Statist.* **33**, 498–512.

Klotz, J. (1963). Small sample power and efficiency for the one-sample Wilcoxon and normal scores tests. *Ann. Math. Statist.* **34**, 624–32.

Kolmogorov, A. N. (1933). Sulla determinazione empirica di una legge di distribuzione. *Giorn. Inst. Ital. Att.* **4**, 83–91.

Korwar, R. M. and Hollander, M. (1976). Empirical Bayes estimation of a distribution function. *Ann. Statist.* **4**, 581–8.

Kruskal, W. H. (1952). A nonparametric test for the several sample problem. *Ann. Math. Statist.* **23**, 525–40.

Kruskal, W. H. and Wallis, W. A. (1952). Use of ranks in one-criterion variance analysis. *J. Amer. Statist. Assoc.* **47**, 583–621.

Lamp, W. O. (1976). Statistical treatment of a study on the distribution of a stream insect by age. Masters thesis, Ohio State University, Columbus.

Lehmann, E. L. (1951). Consistency and unbiasedness of certain nonparametric tests. *Ann. Math. Statist.* **22**, 165–79.

Lehmann, E. L. (1953). The power of rank tests. *Ann. Math. Statist.* **24**, 23–43.

Lehmann, E. L. (1959). *Testing Statistical Hypotheses*. Wiley, New York.

Lehmann, E. L. (1963a). Robust estimation in analysis of variance. *Ann. Math. Statist.* **34**, 957–66.

Lehmann, E. L. (1963b). Asymptotically nonparametric inference: An alternative approach to linear models. *Ann. Math. Statist.* **34**, 1494–506.

Lehmann, E. L. (1963c). Nonparametric confidence intervals for a shift parameter. *Ann. Math. Statist.* **34**, 1507–12.

Lehmann, E. L. (1975). *Nonparametrics: Statistical Methods Based on Ranks*. Holden-Day, San Francisco.

Lindgren, B. W. (1976). *Statistical Theory*. 3rd ed. Macmillan, New York.

Loève, M. (1963). *Probability Theory*. 3rd ed. Van Nostrand, Princeton, New

Jersey.

Mack, G. A. and Wolfe, D. A. (1978). A distribution-free test for umbrella alternatives. I. Point of the umbrella known. Tech. Rept., Department of Statistics, Ohio State University, Columbus, Ohio.

Mann, H. B. and Whitney, D. R. (1947). On a test of whether one of two random variables is stochastically larger than the other. *Ann. Math. Statist.* **18**, 50–60.

Maronna, R. A. (1976). Robust M-estimators of location and scatter. *Ann. Statist.* **4**, 51–67.

Mathisen, H. C. (1943). A method of testing the hypothesis that two samples are from the same population. *Ann. Math. Statist.* **14**, 188–94.

Milton, R. C. (1970). *Rank Order Probabilities*. Wiley, New York.

Mood, A. M. (1950). *Introduction to the Theory of Statistics*. McGraw-Hill, New York.

Mood, A. M. (1954). On the asymptotic efficiency of certain nonparametric two-sample tests. *Ann. Math. Statist.* **25**, 514–22.

Mood, A. M., Graybill, F. A., and Boes, D. C. (1974). *Introduction to the Theory of Statistics*. 3rd ed. McGraw-Hill, New York.

Moses, L. E. (1963). Rank tests of dispersion. *Ann. Math. Statist.* **34**, 973–83.

Neave, H. R. (1979). A survey of some quick and simple statistical procedures based on numbers of extreme observations. *J. Qual. Techn.* **11**, 66–79.

Noether, G. E. (1949). On a theorem by Wald and Wolfowitz. *Ann. Math. Statist.* **20**, 455–8.

Noether, G. E. (1955). On a theorem of Pitman. *Ann. Math. Statist.* **26**, 64–8.

Noether, G. E. (1967). *Elements of Nonparametric Statistics*. Wiley, New York.

Noether, G. E. (1973). Some simple distribution-free confidence intervals for the center of a symmetric distribution. *J. Amer. Statist. Assoc.* **68**, 716–9.

Odeh, R. E. (1971). On Jonckheere's k-sample test against ordered alternatives. *Technometrics* **13**, 912–8.

Olshen, R. A. (1967). Sign and Wilcoxon tests for linearity. *Ann. Math. Statist.* **38**, 1759–69.

O'Meara, P. D. (1976). An investigation of linear rank statistics for the multiple linear regression model. Ph.D. thesis, University of Iowa, Iowa City.

Orban, J. and Wolfe, D. A. (1978a). Optimality criteria for the selection of a partially sequential indicator set. *Biometrika* **65**, 357–62.

Orban, J. and Wolfe, D. A. (1978b). Properties of a distribution-free partially sequential two-sample median test. Tech. Rept., Department of Statistics, Ohio State University, Columbus, Ohio.

Orban, J. and Wolfe, D. A. (1978c). Partially sequential two-sample linear placement statistics. Tech. Rept., Department of Statistics, Ohio State University, Columbus, Ohio.

Page, E. B. (1963). Ordered hypotheses for multiple treatments: a significance test for linear ranks. *J. Amer. Statist. Assoc.* **58**, 216–30.

Pitman, E. J. G. (1948). Notes on non-parametric statistical inference. Columbia University (duplicated).

Poland, A., Smith, D., Kuntzman, R., Jacobson, M., and Conney, A. H. (1970). Effect of intensive occupational exposure to DDT on phenylbutazone and

cortisol metabolism in human subjects. *Clin. Pharmacol. Ther.* **11**, 724–32.
Policello, G. E., II (1974). Adaptive robust procedures for the one-sample location problem. Ph.D. thesis, Pennsylvania State University, State College.
Policello, G. E., II and Hettmansperger, T. P. (1976). Adaptive robust procedures for the one-sample location problem. *J. Amer. Statist. Assoc.* **71**, 624–33.
Puri, M. L. (1964). Asymptotic efficiency of a class of c-sample tests. *Ann. Math. Statist.* **35**, 102–21.
Puri, M. L. (1965). Some distribution-free k-sample rank tests for homogeneity against ordered alternatives. *Comm. Pure Appl. Math.* **18**, 51–63.
Puri, M. L. and Sen, P. K. (1968). On Chernoff–Savage tests for ordered alternatives in randomized blocks. *Ann. Math. Statist.* **39**, 967–72.
Puri, M. L. and Sen, P. K. (1969). On the asymptotic normality of one sample rank order test statistics. *Teoria Veroyatnostey i ee Primenyia* **14**, 167–72.
Puri, M. L. and Sen, P. K. (1971). *Nonparametric Methods in Multivariate Analysis.* Wiley, New York.
Pyke, R. and Shorack, G. R. (1968). Weak convergence of a two-sample empirical process and a new approach to Chernoff-Savage theorems. *Ann. Math. Statist.* **39**, 755–71.
Raghavachari, M. (1965). The two-sample scale problem when locations are unknown. *Ann. Math. Statist.* **36**, 1236–42.
Ramberg, J. S. and Schmeiser, B. W. (1972). An approximate method for generating symmetric random variables. *Comm. of the A.C.M.* **15**, 987–90.
Ramberg, J. S. and Schmeiser, B. W. (1974). An approximate method for generating asymmetric random variables. *Comm. of the A.C.M.* **17**, 78–82.
Ramberg, J. S., Tadikamalla, P. R., Dudewicz, E. J., and Mykytka, E. F. (1979). A probability distribution and its uses in fitting data. To appear in *Technometrics* **21**.
Randles, R. H., Fligner, M. A., Policello, G. E., II, and Wolfe, D. A. (1979). An asymptotically distribution-free test for symmetry versus asymmetry. To appear in *J. Amer. Statist. Assoc.*
Randles, R. H. and Hogg, R. V. (1971). Certain uncorrelated and independent rank statistics. *J. Amer. Statist. Assoc.* **66**, 569–74.
Randles, R. H. and Hogg, R. V. (1973). Adaptive distribution-free tests. *Comm. in Statist.* **2**(4), 337–56.
Rosenbaum, S. (1953). Tables for a nonparametric test of dispersion. *Ann. Math. Statist.* **24**, 663–8.
Rosenbaum, S. (1954). Tables for a nonparametric test of location. *Ann. Math. Statist.* **25**, 146–50.
Rosenblatt, M. (1952). Limit theorems associated with variants of the von Mises statistic. *Ann. Math. Statist.* **23**, 617–23.
Royden, H. L. (1968). *Real Analysis.* 2nd ed. Macmillan, New York.
Sarhan, A. E. and Greenberg, B. G. (Eds.) (1962). *Contributions to Order Statistics.* Wiley, New York.
Scheffé, H. (1943). Statistical inference in the nonparametric case. *Ann. Math. Statist.* **14**, 305–32.
Sen, P. K. (1968). Asymptotically efficient tests by the method of n rankings. *J. R.*

Statist. Soc. B **30**, 312–7.

Siegel, S. (1956). *Nonparametric Statistics for the Behavioral Sciences*. McGraw-Hill, New York.

Siegel, S. and Tukey, J. W. (1960). A nonparametric sum of ranks procedure for relative spread in unpaired samples. *J. Amer. Statist. Assoc.* **55**, 429–45. Correction (1961) *J. Amer. Statist. Assoc.* **56**, 1005.

Skillings, J. H. and Wolfe, D. A. (1977). Testing for ordered alternatives by combining independent distribution-free block statistics. *Comm. Statist.—Theor. Methods A6*, 1453–63.

Skillings, J. H. and Wolfe, D. A. (1978). Distribution-free tests for ordered alternatives in a randomized block design. *J. Amer. Statist. Assoc.* **73**, 427–31.

Smirnov, N. V. (1939). On the estimation of the discrepancy between empirical curves of distribution for two independent samples. (Russian) *Bull. Moscow Univ.* **2**, 3–16.

Smith, T. M. and Wolfe, D. A. (1977). Distribution-free tests for equality of several regression equations. Tech. Rept., Department of Statistics, Ohio State University, Columbus, Ohio.

Spearman, C. (1904). The proof and measurement of association between two things. *Amer. J. Psychol.* **15**, 72–101.

Stigler, S. M. (1973). The asymptotic distribution of the trimmed mean. *Ann. Statist.* **1**, 472–7.

Stigler, S. M. (1977). Do robust estimators work with real data? *Ann. Statist.* **6**, 1055–98.

Sukhatme, B. V. (1957). On certain two-sample nonparametric tests for variances. *Ann. Math. Statist.* **28**, 188–94.

Sukhatme, B. V. (1958). Testing the hypothesis that two populations differ only in location. *Ann. Math. Statist.* **29**, 60–78.

Takeuchi, K. (1971). A uniformly asymptotically efficient estimator of a location parameter. *J. Amer. Statist. Assoc.* **66**, 292–301.

Terpstra, T. J. (1952). The asymptotic normality and consistency of Kendall's test against trend, when ties are present in one ranking. *Indag. Math.* **14**, 327–33.

Terry, M. E. (1952). Some rank order tests which are most powerful against specific parametric alternatives. *Ann. Math. Statist.* **23**, 346–66.

Theil, H. (1950a). A rank-invariant method of linear and polynomial regression analysis, I. *Proc. Kon. Ned. Akad. Wetensch. A* **53**, 386–92.

Theil, H. (1950b). A rank-invariant method of linear and polynomial regression analysis, III. *Proc. Kon. Ned. Akad. Wetensch. A* **53**, 1397–412.

Tukey, J. W. (1960). The practical relationship between the common transformations of percentages or counts and of amounts. Tech. Rept. 36, Statist. Res. Group, Princeton University, Princeton, New Jersey.

van der Waerden, B. L. (1952). Order tests for the two-sample problem and their power. *Indag. Math.* **14**, 453–8. Correction (1953) *Indag. Math.* **15**, 80.

van der Waerden, B. L. (1953a). Order tests for the two-sample problem, II. *Indag. Math.* **15**, 303–10.

van der Waerden, B. L. (1953b). Order tests for the two-sample problem, III. *Indag. Math.* **15**, 311–6.

van Eeden, C. (1963). The relation between Pitman's asymptotic relative efficiency of two tests and the correlation coefficient between their test statistics. *Ann. Math. Statist.* **34**, 1442–51.

van Eeden, C. (1970). Efficiency-robust estimation of location. *Ann. Math. Statist.* **41**, 172–81.

von Mises, R. (1931). Vorlesungen aus dem Gebiete der angewandten Mathematik. Vol. 1. *Wahrscheinlichkeitsrechnung und ihre Anwendung in der Statistik und Theoretischen Physik*, Franz Deutike, Leipzig and Vienna.

von Mises, R. (1947). On the asymptotic distributions of differentiable statistical functions. *Ann. Math. Statist.* **18**, 309–48.

Westenberg, J. (1948). Significance test for median and interquartile range in samples from continuous populations of any form. *Proc. Koninkl. Ned. Akad. Wetenshap.* **51**, 252–61.

Wilcoxon, F. (1945). Individual comparisons by ranking methods. *Biometrics* **1**, 80–3.

Wilcoxon, F. and Bradley, R. A. (1964). A note on the paper Two sequential two-sample grouped rank tests with applications to screening experiments. *Biometrics* **20**, 892–5.

Wilcoxon, F., Katti, S. K., and Wilcox, R. A. (1970). Critical values and probability levels for the Wilcoxon rank sum test and the Wilcoxon signed rank test. In: *Selected Tables in Mathematical Statistics*, Vol. I, H. L. Harter and D. B. Owen, Eds., pp. 171–259. Markham, Chicago.

Wilcoxon, F., Rhodes, L. J., and Bradley, R. A. (1963). Two sequential two-sample grouped rank tests with applications to screening experiments. *Biometrics* **19**, 58–84.

Wolfe, D. A. (1973). Some general results about uncorrelated statistics. *J. Amer. Statist. Assoc.* **68**, 1013–8.

Wolfe, D. A. (1974). A characterization of population weighted-symmetry and related results. *J. Amer. Statist. Assoc.* **69**, 819–22.

Wolfe, D. A. (1977a). On a class of partially sequential two-sample test procedures. *J. Amer. Statist. Assoc.* **72**, 202–5.

Wolfe, D. A. (1977b). Two-stage two-sample median test. *Technometrics* **19**, 495–501.

Zacks, S. (1971). *The Theory of Statistical Inference*. Wiley, New York.

Zar, J. H. (1972). Significance testing of the Spearman rank correlation coefficient. *J. Amer. Statist. Assoc.* **67**, 578–80.

Zelzano, P. R., Zelzano, N. A., and Kolb, S. (1972). "Walking" in the newborn. *Science* **176**, 314–5.

INDEX